Fields Institute Monographs

The Fields Institute for Research in Mathematical Sciences

Introduction to Orthogonal, Symplectic and Unitary Representations of Finite Groups

Carl R. Riehm

American Mathematical Society
Providence, Rhode Island

FIELDS

The Fields Institute
for Research in Mathematical Sciences
Toronto, Ontario

2000 *Mathematics Subject Classification.* Primary 20C15, 20C05, 11E81, 11E88, 16W22.

For additional information and updates on this book, visit
www.ams.org/bookpages/fim-28

Library of Congress Cataloging-in-Publication Data

Riehm, C. R. (Carl R.), 1935–
Introduction to orthogonal, symplectic and unitary representations of finite groups / Carl R. Riehm
p. cm. — (Fields Institute monographs ; 28)
Includes bibliographical references and index.
ISBN 978-0-8218-4271-3 (alk. paper)
1. Finite groups. 2. Characters of groups. 3. Representations of groups. I. Title.
QA176.R54 2011
512′.23—dc23

2011017501

∞ The paper used in this book is acid-free and falls within the guidelines
established to ensure permanence and durability.
This publication was prepared by the Fields Institute.
http://www.fields.utoronto.ca
Visit the AMS home page at http://www.ams.org/

10 9 8 7 6 5 4 3 2 1 16 15 14 13 12 11

Contents

Preface

Orthogonal, symplectic and unitary representations of finite groups combine two more traditional subjects of mathematics – linear representations of finite groups, and quadratic, skew symmetric and Hermitian forms – and thus inherit some of the characteristics of both. A little closer in approach to the latter, it could equally well be called the theory of quadratic, skew symmetric and Hermitian forms invariant under a (given) finite group.

This book is written as an introduction to the subject and not as an encyclopaedic reference text. The principal goal is an exposition of the known results on the equivalence theory, and related matters such as the Witt and Witt-Grothendieck groups, over the "classical" fields - algebraically closed, real closed, finite, local and global. I have included references to other results, especially those of a more "algebraic" nature.

I have chosen to ignore the integral theory entirely. This is unfortunate, especially from the point of view of surgery theory where the integral theory is essential ([71]) (and very well-developed – see [72]). But clearly either the size of this book or its level would have had to have been drastically changed to accommodate it.

Our exposition relies on the basic results in the linear representations of finite groups, in quadratic, symplectic and Hermitian forms and in involutions over simple algebras, which are outlined, mainly without proofs, in the first chapter. These subjects are blessed by many fine books on which we can rely for the background material – Curtis and Reiner's *Representation Theory of Finite Groups and Associative Algebras* [18] and their *Methods of Representation Theory with Applications to Finite Groups and Orders* [19], Reiner's *Maximal Orders* [55], Serre's *Linear Representations of Finite Groups* [68], Lam's *Introduction to Quadratic Forms over Fields* [39], O'Meara's *Introduction to Quadratic Forms* [54], Scharlau's *Quadratic and Hermitian Forms* [65], Knus's *Quadratic and Hermitian Forms over Rings* [35], and *The Book of Involutions* [36] of Knus, Merkurjev, Rost and Tignol.

It was A. Fröhlich who first gave a systematic organization of this subject, in a series of papers beginning in 1969. The paper *Orthogonal and symplectic representations of groups* [26] represents the culmination of his published work on orthogonal and symplectic representations. We have included most of the work from that paper, extending it to include unitary representations, and also providing new approaches, such as the use of the equivariant Brauer-Wall group in describing the principal invariants of orthogonal representations and their interplay with each other.

Chapter 1 is an outline, with few proofs, of the background material needed later. Most readers can skip it and refer to it later when necessary. The very brief

Chapter 2 introduces orthogonal, symplectic and unitary representations – which we refer to collectively as *isometry representations* – and gives the fundamental (and elementary) connection of these representations with Hermitian and skew Hermitian forms over the group algebra. Chapter 3 provides a firm foundation for the use of these latter forms in representation theory.

Chapter 4 begins the exposition of the theory proper, with a treatment of known results of the Witt-Grothendieck and Witt groups of isometry representations, followed by Chapter 5 in which the equivalence theory of isometry representations over finite, local and global fields is described, as well as a section on the isometry representations of the symmetric group. A final chapter 6 describes the equivariant Brauer-Wall group of equivariant central simple graded algebras and applies it to the description of the "classical invariants" of orthogonal representations and their interrelationships, through the medium of the equivariant Clifford algebra.

The main departure from the "usual" approach to the subject is the introduction, in Chapter 3, of a discriminant (or Gram) matrix for sesquilinear forms over simple algebras and its use in an explicit version of (Hermitian) Morita theory, which is critical in our treatment of this subject. In addition to the advantage of being completely elementary, it provides, in my opinion, more transparent proofs than does the more "invariant" and more general theory [28], and makes the detailed results of Chapter 5 possible.

I have tried to make this book as easy to read as possible by providing a list of notational conventions on p. ix and a glossary of symbols and a thorough index at the end.

There are many people to whom I owe a debt of gratitude. First of all, to O.T. O'Meara, A. Fröhlich and Ian Hambleton for their encouragement and for introducing me to and arousing my interest in this subject. On various occasions I have received help from Stefan Caenepeel, Richard Elman, Martin Epkenhans, Max Knus, Manfred Kolster, David Lewis and Winfried Scharlau. Matt Valeriote provided some support at a critical time. Emmanuel Lequeu checked large parts of the manuscript. But most of all, I am deeply indebted to my wife Elaine McKinnon Riehm for a lifetime of support and encouragement.

Notation

G is a finite group of order g except in Chapter 6 where it can be arbitrary. If X is a subset of a group, then $\langle X \rangle$ is the subgroup generated by X. If H is a subgroup of G, the index of H in G is denoted by $[G : H]$. The degree of a finite field extension L/K is denoted by $(L : K)$.

K is a field of characteristic not dividing $2g$, $\dot{K}$ its nonzero elements, and $^-$ an involution of K which is an automorphism of K which either has order 2 or is the identity. The fixed field $\{\alpha \in K : \bar{\alpha} = \alpha\}$ is denoted by K_0. If $(K : K_0) = 2$, $K = K_0(\sqrt{\delta})$ where $\delta \in \dot{K}_0 \setminus \dot{K}_0^2$, the nonsquares of K_0.

$\subset$ is inclusion, not necessarily strict. id_X or simply id is the identity map on the set X.

Rings (and algebras) will always be assumed to be nonzero and to have an identity element 1, denoted by 1_R if it is necessary to be more precise. A ring homomorphism $R \to R'$ is assumed to take 1_R to $1_{R'}$. All algebras over a field are assumed finite dimensional over that field. All modules will be assumed to be "unitary" in the sense that $1v = v$ for all v in the module, and will be assumed to be finitely generated, unless mentioned otherwise. The group of units of R is denoted by $R^\times$, although for a field K, the nonzero elements are denoted by $\dot{K}$.

$(R, ^-)$ stands for a ring R with involution $^-$, an antiautomorphism of order 1 or 2. A K-algebra A with $(K, ^-)$-involution $^-$ is an algebra satisfying $\overline{\alpha a} = \bar{\alpha}\bar{a}$ for all $\alpha \in K$ and $a \in A$. It is also called a $(K, ^-)$-involution algebra. It is a simple involution algebra if it has no two-sided ideals, other than 0 and A, which are stable under $^-$. It is a simple algebra with involution if it is a simple algebra (and has an involution).

The standard involution "conjugation" of the complex numbers $\mathbb{C}$, the algebraic closure $\mathbf{C}$ of a real closed field, or a quaternion algebra over an arbitrary field, is denoted by * : e.g. α^*. Conjugation * on a real closed field $\mathbf{R}$ – for example the real numbers $\mathbb{R}$ – is the identity $\mathrm{id}_{\mathbf{R}}$. If $(B, ^-)$ is a ring with involution, $^{\leftrightarrow}$ is the "interchange involution" on $B \oplus B$ – see Example 9, p. 26. Whenever confusion is unlikely, any other involution is denoted by $^-$: $\bar{\alpha}$.

If R is a ring, the $n \times m$ matrices over R are denoted by $R^{n\times m}$; $\mathrm{M}(n, R) = R^{n\times n}$ and I_n is the identity matrix. If $(R, ^-)$ is a ring with involution and $(r_{ij}) \in R^{n\times m}$, $(r_{ij})^- = (\bar{r}_{ij})$ and the "transpose involution" $(r_{ij}) \to (\bar{r}_{ji})$ from $R^{n\times m}$ to $R^{m\times n}$ is denoted by $a \to a^{t-}$, $a \in R^{n\times m}$. In the case of $A = \mathrm{M}(n, R)$, it is an involution, and the ring $(A, ^{t-})$ with involution is denoted by $\mathrm{M}(n, (R, ^-))$ and called the $n \times n$ matrix ring over $(R, ^-)$.

If $X, Y, \ldots$ are square matrices over a ring $(R, {}^-)$ with involution, $\langle X, Y, \ldots \rangle$ stands for the sesquilinear form

$$f : R^{1\times m} \times R^{1\times m} \to (R, {}^-), \quad f(u,v) = u \begin{pmatrix} X & 0 & 0 \\ 0 & Y & 0 \\ 0 & 0 & \ddots \end{pmatrix} v^{t-}$$

over $(R, {}^-)$ where m is the sum of the sizes of the matrices $X, Y, \ldots$. If the matrices are 1×1 matrices $(r_1), (r_2), \ldots$, the simpler notation $\langle r_1, r_2, \ldots \rangle$ is usually used.

If $\varphi : V \to W$ is a K-linear transformation, its matrix with respect to some basis is usually denoted by $\underline{\varphi}$. More generally if A is a simple algebra and $\varphi : V \to W$ is an A-module homomorphism, and if we identify A with a matrix algebra $\mathrm{M}(n, D)$ and V and W with matrix modules $D^{n\times m}$ and $D^{n\times k}$ respectively, then the matrix of φ (cf. Lem. 3.2.3, p. 119) is written $\underline{\varphi}$. And if $f : V \times V \to (A, {}^-)$ is a sesquilinear form, the discriminant ("Gram") matrix of f (cf. p. 119) is written $\underline{f}$.

The rational numbers are denoted by $\mathbb{Q}$ and the rational integers by $\mathbb{Z}$.

CHAPTER 1

Background material

1.1 Rings and modules

Good references for this section are [31], [11] and [10]. We assume known the rudiments of the theory of rings and modules, such as direct sums, factor rings and modules, exact sequences, algebras over a field, tensor products of modules over (possibly noncommutative) rings, tensor products of algebras, etc. Most statements will be made for left modules only, with the understanding that with the obvious changes, they apply to right modules as well.

We also assume the language and basic facts about categories and functors.

Let R be a ring. A left R-module M is sometimes denoted by ${}_RM$, a right R-module by M_R, and an (R, S)-bimodule by ${}_RM_S$.

A *simple*[1] R-module M is a nonzero module whose only submodules are 0 and M. It follows that a homomorphism between two simple modules must be either an isomorphism or 0, so the endomorphism ring $\mathrm{End}_R M$ of a simple module is a division ring ("Schur's Lemma ").

A Jordan-Hölder series for M is a series of submodules

$$\{0\} = M_0 \subset M_1 \subset M_2 \subset \cdots \subset M_n = M,$$

in which each factor M_i/M_{i-1} is a simple module. If M *has* a Jordan-Hölder series, M is said to be of *finite length*, and its *length*, $\mathrm{len}_R M = n$, is well-defined. If

$$0 \to N \to M \to Q \to 0$$

is an exact sequence of R-modules, M has finite length if and only if both N and Q do, and then

$$\mathrm{len}_R M = \mathrm{len}_R N + \mathrm{len}_R Q.$$

The ring of $n \times n$ matrices over R is denoted by $\mathrm{M}(n, R)$, and $\mathbf{GL}(n, R)$ is the "general linear group" of invertible matrices in $\mathrm{M}(n, R)$. The R-module – actually (R, R)-bimodule – of $n \times m$ matrices is $R^{n \times m}$; it is also an $(\mathrm{M}(n, R), \mathrm{M}(m, R))$-bimodule.

Let M be a right R-module and N a left R-module. A map $\varphi : M \times N \to P$ into an Abelian group P is *balanced* (*over* R) if it satisfies $\varphi(ur, v) = \varphi(u, rv)$ for all $u \in M, v \in N$, and $r \in R$. We sometimes also say that φ *admits* R. It is *biadditive* if it is additive in each variable.

The tensor product $M \otimes_R N$ is an Abelian group together with a balanced biadditive map can : $M \times N \to M \otimes_R N$ over R, with the following universal mapping property: *if* $\varphi : M \times N \to P$ *is a balanced biadditive map over* R *into an*

[1] sometimes called *irreducible*, as in [31].

Abelian group P*, there is a unique homomorphism* $\hat{\varphi} : M \otimes_R N \to P$ *of Abelian groups which makes the diagram*

$$\begin{array}{ccc} M \times N & \xrightarrow{\text{can}} & M \otimes_R N \\ & \searrow^{\varphi} & \downarrow{\hat{\varphi}} \\ & & P \end{array}$$

commutative. The image of (u, v) in $M \otimes_R N$ under the canonical map is denoted by $u \otimes_R v$ or simply $u \otimes v$.

If M is an (S, R)-bimodule, $M \otimes_R N$ is (in a natural way) a left S-module. Similarly it is a right T-module if N is an (R, T)-bimodule, and if P is a left T-module, there is a canonical isomorphism

$$(M \otimes_R N) \otimes_T P \cong M \otimes_R (N \otimes_T P).$$

And if M and N are both bimodules as above, then $M \otimes_R N$ is an (S, T)-bimodule.

There are also canonical isomorphisms

$$R \otimes_R M \cong M, \quad (\oplus_i M_i) \otimes_R N \cong \oplus_i (M_i \otimes_R N), \quad M \otimes_R (\oplus_i N_i) \cong \oplus_i (M \otimes_R N_i).$$

If M and N are (say) left R-modules, similar remarks apply to the Abelian group $\mathrm{Hom}_R(M, N)$. For example if M is an (R, S)-bimodule, then $\mathrm{Hom}_R(M, N)$ is a left S-module, while if N is an (R, T)-bimodule, $\mathrm{Hom}_R(M, N)$ is a right T-module, via

$$(s\varphi)(u) = \varphi(us) \text{ and } (\varphi t)(u) = \varphi(u)t.$$

We will generally write homomorphisms on the left, although in certain circumstances it is better to write them on the right (opposite to the scalars) and we will do so.

For finite direct sums, there are canonical isomorphisms

$$\begin{aligned} \mathrm{Hom}_R(\oplus_i M_i, N) &\cong \oplus_i \mathrm{Hom}_R(M_i, N) \\ \mathrm{Hom}_R(M, \oplus_i N_i) &\cong \oplus_i \mathrm{Hom}_R(M, N_i). \end{aligned} \tag{1.1}$$

Now write homomorphisms on the right, and consider the (internal) direct sum $M = M_1 \oplus \cdots \oplus M_m$ of left R-modules, and let

$$\iota_i = \iota_i^M : M_i \to M, \quad \pi_i = \pi_i^M : M \to M_i \tag{1.2}$$

be the associated injections and projections (written on the *right*). They satisfy

$$\iota_j \pi_i = 0 \text{ if } i \neq j, \quad \iota_i \pi_i = \mathrm{id}_{M_i}, \quad \text{and} \quad \sum_i \pi_i \iota_i = \mathrm{id}_M. \tag{1.3}$$

These conditions characterize the direct sum; that is to say if $M_1, \ldots, M_m$ are R-modules and homomorphisms (1.2) satisfying (1.3) are given, then M is the internal direct sum of its submodules $M_i \iota_i$ – or alternatively the map $\oplus_i \pi_i : M \to M_1 \oplus \cdots \oplus M_m$ is an isomorphism of M with the external direct sum.

If $N = N_1 \oplus \cdots \oplus N_n$ and $\varphi : M \to N$ is a homomorphism, define

$$\varphi_{ij} = \iota_i^M \varphi \pi_j^N : M_i \to N_j.$$

The matrix (φ_{ij}) describes φ completely since $\sum_{i,j} \pi_i \varphi_{ij} \iota_j = \varphi$. Conversely any matrix (φ_{ij}) of homomorphisms $\varphi_{ij} : M_i \to N_j$ corresponds uniquely to a homomorphism $\varphi : M \to N$ with the components φ_{ij}. Furthermore if $\psi : N \to P =$

$P_1 \oplus \cdots \oplus P_p$ is a second homomorphism, the matrix of $\varphi\psi$ is the matrix product $(\varphi_{ij})(\psi_{kl})$. Thus if M^m denotes the m-fold direct sum of M with itself and homomorphisms are written on the right,

Proposition 1.1.1 *(a)* $(\mathrm{End}_R M)^{m\times k}$ *is isomorphic to* $\mathrm{Hom}_R(M^m, M^k)$*, via "multiplication"*

$$(u_1, u_2, \ldots, u_m)(\varphi_{ij}) = \left(\sum_i u_i\varphi_{i1}, \ldots, \sum_i u_i\varphi_{ik}\right).$$

(b) The ring of endomorphisms $\mathrm{End}_R(M^m)$ *(operating on the* right *of* M^m*) is isomorphic to the matrix ring* $\mathrm{M}(m, \mathrm{End}_R M)$*. In particular if* M *is simple, then* $\mathrm{End}_R(M^m)$ *is isomorphic to the matrix ring* $\mathrm{M}(m, D)$ *where* D *is the division ring* $\mathrm{End}_R M$*.*

(c) The isomorphism

$$(\mathrm{End}_R M)^{m\times k} \to \mathrm{Hom}_R(M^m, M^k)$$

in (a) is an $(\mathrm{End}_R(M^m), \mathrm{End}_R(M^k))$*-bimodule isomorphism.*

(a) and (b) have already been proved. In (c), it is understood that $\mathrm{End}_R(M^m)$ operates on $(\mathrm{End}_R M)^{m\times k}$ through the isomorphism $\mathrm{End}_R(M^m) \to \mathrm{M}(m, \mathrm{End}_R M)$ of (b) and left multiplication, and that $\mathrm{End}_R(M^k)$ operates on $(\mathrm{End}_R M)^{m\times k}$ through the isomorphism $\mathrm{End}_R(M^k) \to \mathrm{M}(k, \mathrm{End}_R M)$ and right multiplication. (c) follows directly from this and the fact that, in general, $\mathrm{Hom}_R({}_R M_S, {}_R N_T)$ is an (S, T)-bimodule via $(s\varphi t)(u) = \varphi(us)t$. □

An R-*pairing* is an R-bilinear map

$$U \times W \to R, \quad (u, w) \to \langle u, w\rangle, \tag{1.4}$$

involving a left R-module U and a right R-module W. This means the map is biadditive and satisfies

$$\langle au, wb\rangle = a\langle u, w\rangle b \text{ for all } u \in U,\ w \in W \text{ and } a, b \in R.$$

The R-*dual* of a left R-module M is the right R-module $M^* = \mathrm{Hom}_R(M, R)$ where R is considered as a left R-module via multiplication. The module structure of M^* is determined by

$$(xr)(u) = (xu)r \quad (x \in M^*, r \in R, u \in M), \tag{1.5}$$

now writing homomorphisms on the left. We also use the pairing notation:

$$M \times M^* \to R, \quad (u, x) \to \langle u, x\rangle = x(u). \tag{1.6}$$

Then (1.5) becomes

$$\langle u, xr\rangle = \langle u, x\rangle r.$$

If M is an (R, S)-bimodule, then M^* is an (S, R)-bimodule via

$$\langle u, sxr\rangle = \langle us, x\rangle r.$$

Suppose that $M = N \oplus P$. If $N^o = \{x \in M^* |\ \langle N, x\rangle = 0\}$ is the annihilator of N and P^o that of P, then $M^* = P^o \oplus N^o$ (internal direct sum), $P^o \cong N^*$ and $N^o \cong P^*$ via the restriction maps $M^* \to N^*$, $x \to x|_N$, and $M^* \to P^*$, $x \to x|_P$. If we identify via these isomorphisms,

$$M^* = N^* \oplus P^*.$$

By induction if $M = M_1 \oplus M_2 \oplus \cdots \oplus M_n$, then

$$M^* = M_1^* \oplus M_2^* \oplus \cdots \oplus M_n^*, \tag{1.7}$$

where M_i^* is the annihilator of $\oplus_{j \neq i} M_j$, and is the dual of M_i when the pairing (1.6) is restricted to $M_i \times M_i^*$.

If $\varphi : M \to N$ is an R-homomorphism, then $\varphi^t : N^* \to M^*$ is the *transpose* of φ, the R-homomorphism determined by

$$\langle u, \varphi^t x \rangle = \langle \varphi u, x \rangle. \tag{1.8}$$

If $\psi : N \to P$ is another R-homomorphism, then

$$(\psi\varphi)^t = \varphi^t \psi^t.$$

It follows that φ^t is an isomorphism if φ is.

Definition 1.1.1 *An* involution *on the ring R is an automorphism $r \to \bar{r}$ of R as an additive group such that*
1. $\bar{\bar{r}} = r$,
2. $\overline{r_1 r_2} = \overline{r_2}\,\overline{r_1}$ for all $r, r_1, r_2 \in R$.

The pair $(R, ^-)$ is called a ring with involution, *or an* involution ring.

If R is a field K, then of course $(K, ^-)$ is called a field with involution, *or an involution field.*

If $(K, ^-)$ is a field with involution, a $(K, ^-)$-involution algebra, or an algebra with $(K, ^-)$-involution, is a pair $(A, ^-)$ where A is a K-algebra with an involution $^-$ which is compatible with that on K:
3. $\overline{\alpha a} = \bar{\alpha}\bar{a}$ for all $\alpha \in K$ and $a \in A$.

Briefly, an involution of a ring R is an antiautomorphism of R of period 1 or 2; when the period is 1 (i.e. the involution is the identity map id_R), then R must be commutative.

It is occasionally necessary to consider, a little more generally, an involution algebra $(A, ^-)$ over a commutative ring $(C, ^-)$ with involution, defined in the obvious way.

A *homomorphism* $\varphi : (R, ^-) \to (S, ^-)$ of rings with involution is a ring homomorphism $R \to S$ which commutes with the involutions: $\varphi(\bar{r}) = \overline{\varphi(r)}$ for all $r \in R$. And a homomorphism of involution algebras is one of rings with involution which commutes with the scalars. Thus an isomorphism of rings or algebras with involution is such a map which is an isomorphism of rings or algebras; we then also say that the involutions on R and S are *equivalent.*

A right R-module M can be "twisted" into a left R-module in the presence of an involution on R. Namely we define $ru = u\bar{r}$. It is often convenient to denote the twisted R-module M by $\bar{M}$ and its elements by $\bar{u}$. Thus $\bar{u} = u$ but $\bar{u}$ is considered to be a member of $\bar{M}$. This leads to the appealing equation $\overline{ur} = \bar{r}\bar{u}$, but we use it more often in the form $r\bar{u} = \overline{u\bar{r}}$. Similarly if M is an (R, S)-bimodule and R and S both have involutions, M can be twisted into an (S, R)-bimodule $\bar{M}$ via $s\bar{u}r = \overline{\bar{r}u\bar{s}}$.

Theorem 1.1.1 *Let V be a vector space over the field K spanned by the vectors $v_1, v_2, \ldots, v_n$ over K. Suppose further that there is a bilinear multiplication $V \times V \to V$ which is associative on the generators $v_1, v_2, \ldots, v_n$. Then V, with this multiplication, is an "algebra without 1" over K.* □

Theorem 1.1.2 *Suppose that A is an algebra over the field K, M is a right A-module and N a left A-module. If L/K is a field extension, there is a canonical isomorphism (of L-vector spaces)*

$$L \otimes_K (M \otimes_A N) \xrightarrow{\sim} (L \otimes_K M) \otimes_{L \otimes_K A} (L \otimes_K N),$$

given by

$$\lambda \otimes (u \otimes v) \to (\lambda \otimes u) \otimes (1 \otimes v).$$

It is understood that M is a right K-vector space through the action of A, and so also a left K-vector space since K is commutative: $\alpha u = u\alpha$ $(u \in M, \alpha \in K)$. Also $L \otimes_K M$, for example, is a right $L \otimes_K A$-module via the action $(\lambda \otimes u)(\mu \otimes a) = \lambda\mu \otimes ua$. And we assume that A is imbedded in $L \otimes_K A$ via $a \to 1 \otimes a$.

First of all, the map $M \times N \to (L \otimes_K M) \otimes_{L \otimes_K A} (L \otimes_K N)$, $(x, y) \to (1 \otimes M) \otimes (1 \otimes y)$, is balanced over A, and so we get a homomorphism $M \otimes_A N \to (L \otimes_K M) \otimes_{L \otimes_K A} (L \otimes_K N)$. Then the map $L \times (M \otimes_A N) \to (L \otimes_K M) \otimes_{L \otimes_K A} (L \otimes_K N)$ is balanced over K and so we get the map φ of the theorem.

On the other hand, the map $(\sum_i \lambda_i \otimes u_i, \sum_j \mu_j \otimes v_j) \to \sum_{i,j} (\lambda_i \mu_j) \otimes (u_i \otimes v_j)$ is a biadditive map $(L \otimes_K M) \times (L \otimes_K N) \to L \otimes_K (M \otimes_A N)$, balanced over $L \otimes_K A$; the only problem is to show that it is well-defined. Once we have done this, we will have a homomorphism $\psi : (L \otimes_K M) \otimes_{L \otimes_K A} (L \otimes_K N) \to L \otimes_K (M \otimes_A N)$ which is inverse to φ, and so both are isomorphisms. It is clear that they are L-isomorphisms, and so the theorem will have been proved.

Suppose then that $\lambda, \mu \in L$, $u \in M$, and $v \in N$. The map $(\lambda, u) \to \lambda\mu \otimes_K (u \otimes_A v)$ is biadditive and balanced over K, so gives a homomorphism $L \otimes_K M \to L \otimes_K (M \otimes_A N)$ of Abelian groups. Thus we have a map

$$L \times N \to \operatorname{Hom}(L \otimes_K M, L \otimes_K (M \otimes_A N)),$$

given by

$$(\mu, v) \to \left(\sum_i \lambda_i \otimes u_i \to \sum_i \lambda_i \mu \otimes (u_i \otimes v) \right).$$

This map is clearly biadditive and balanced over K, and so gives a homomorphism $\theta : L \otimes_K N \to \operatorname{Hom}(L \otimes_K M,\ L \otimes_K (M \otimes_A N))$ given by

$$(\theta(\sum_j \mu_j \otimes v_j))(\sum_i \lambda_i \otimes u_i) = \sum_{i,j} \lambda_i \mu_j \otimes (u_i \otimes v_j).$$

Now consider

$$(L \otimes_K M) \times (L \otimes_K N) \to L \otimes_K (M \otimes_A N)$$

given by $(\sum_i \lambda_i \otimes u_i, \sum_j \mu_j \otimes v_j) \to (\theta(\sum_j \mu_j \otimes v_j))(\sum_i \lambda_i \otimes u_i)$. It is easy to check that it is biadditive and balanced over $L \otimes_K A$, so the universal mapping property implies that ψ is well-defined. □

1.2 Simple and semisimple modules and algebras

1.2.1 Semisimple modules. References for this section are Chs. 3 and 4 in [31], Ch. 1 in [38], and for simple rings, section 1 of Ch. 4 in [39] and §§1-5 of Ch. 8 in [65].

Theorem 1.2.1 *The following are equivalent for an R-module M:*
(a) Every submodule of M is a direct summand of M.
(b) M is the direct sum of simple submodules.
(c) M is the sum of simple submodules. □

A module with these properties is called a *semisimple* module, or a *completely reducible* module. If S is a simple R-module, the sum of all submodules of an R-module M which are isomorphic to S is a (semisimple) submodule M_S of M, called the *isotypic component* or *homogeneous component* of type S of M.

Theorem 1.2.2 *Let M be a semisimple R-module.*
(a) M is the direct sum

$$M = \oplus_S M_S$$

of its homogeneous components, (where S runs over a set of representatives of the isomorphism classes of simple R-modules). In particular, every simple submodule of M_S is isomorphic to S.
(b) If $\varphi : M \to N$ is an R-homomorphism, $\varphi M_S \subset N_S$ for all S.
(c) M_S is a direct sum of simple modules

$$M_S = \oplus_{i \in I_S} S_i$$

each isomorphic to S. The cardinalities $\{|I_S|\}$ are well-defined and determine M up to isomorphism. The length $\mathrm{len}_R M_S$ of M_S is finite if and only if $|I_S|$ is finite, and then $\mathrm{len}_R M_S = |I_S|$.
(d) Every submodule of M is semisimple. □

Lemma 1.2.1 *If N and N′ are isomorphic submodules of a semisimple module M of finite length, there is a submodule Q of M such that $M = N \oplus Q = N' \oplus Q$.*

Among all submodules Q of M such that $N \cap Q = 0 = N' \cap Q$, choose one of maximal length, say Q. If $M \neq N \oplus Q$, len $M >$ len$(N \oplus Q) =$ len$(N' \oplus Q)$, so $M \neq N' \oplus Q$ as well. Since $N \oplus Q \cong N' \oplus Q$ and M has finite length, by Th. 1.2.2(c), there are isomorphic simple submodules S and S' of M such that $S \not\subset N \oplus Q$ and $S' \not\subset N' \oplus Q$. Let $S = Ru$, $S' = Ru'$ where the annihilators of u and u' are equal to the same maximal left ideal $\mathfrak{a}$ of R. If $u + u' = 0$, then $S = S'$ and $N \cap (Q \oplus S) = 0 = N' \cap (Q \oplus S)$, contradicting the maximality of Q. So $u + u' \neq 0$. Then $\mathfrak{a}$ is also the annihilator of $u + u'$, so $R(u + u') \cong R/\mathfrak{a}$ is isomorphic to S and S'. At least one of the modules $S, S', R(u+u')$ is not contained in either of $N \oplus Q$ and $N' \oplus Q$, again contradicting the maximality of Q. Therefore $M = N \oplus Q = N' \oplus Q$ as desired. □

1.2.2 Simple and semisimple rings.

Theorem 1.2.3 *The following are equivalent for a ring R:*
(a) All R-modules are semisimple.
(b) All finitely generated R-modules are semisimple.
(c) R is semisimple when considered as an R-module via left multiplication.
(d) All R-modules are projective.

(e) All short exact sequences of R-modules split. □

A *semisimple ring* R is a ring with these properties.

A finite direct product $R_1 \times \cdots \times R_n$ of semisimple rings is semisimple.

Let $M^* = \mathrm{Hom}_R(M, R)$ be the R-dual of R-module M. Let $\gamma_M : M \to M^{**}$ be the canonical mapping $u \to \langle u, \cdot \rangle$. It is an R-homomorphism.

Theorem 1.2.4 *Suppose that R is semisimple and that the R-module M has finite length.*

(a) $\mathrm{len}_R M^* = \mathrm{len}_R M$, in particular M^* is simple if and only if M is simple.

(b) γ_M is bijective, i.e. M is "reflexive".

(c) (i) If $M = N \oplus P$, then $M^* = N^o \oplus P^o$ and $N^o \cong P^*$ and $P^o \cong N^*$.

(ii) If $M^* = X \oplus Y$, then $M = X^o \oplus Y^o$ and $X^o \cong Y^*$ and $Y^o \cong X^*$.

If $M = M_1 \oplus M_2 \oplus \cdots \oplus M_t$, then by (1.7)

$$M^* = M_1^* \oplus M_2^* \oplus \cdots \oplus M_t^*, \tag{1.9}$$

so to prove (a) we can assume that M is simple. If $M = Ru$, the epimorphism $R \to M, r \to ru$, splits since M is projective, so there is a nonzero homomorphism $M \to R$. Thus $M^* \neq 0$.

Now let x and y be any nonzero elements of M^*. Then $x : M \to xM$ and $y : M \to yM$ are isomorphisms to the minimal nonzero left ideals xM and yM of R, and $yx^{-1} : xM \to yM$ is an isomorphism of R-modules. Since R is semisimple (and of finite length as an R-module since it is finitely generated), there are left ideals J_1 and J_2 such that $R = J_1 \oplus xM = J_2 \oplus yM$. Then $J_1 \cong J_2$ by Th. 1.2.2(c), so there is an automorphism of R as a left R-module which is yx^{-1} on xM. Such an automorphism is a right multiplication by a unit a of R, so for all $u \in M$

$$xu \to yu = (xu)a, \quad \text{i.e. } \langle u, y \rangle = \langle u, x \rangle a = \langle u, xa \rangle.$$

Thus $y = xa$. Since x and y are *arbitrary* nonzero elements of M^*, this implies that M^* is simple.

(b) Assume first that M is simple. Since $M^* \neq 0$ by (a), the canonical homomorphism $M \to M^{**}$ is $\neq 0$, hence must be an isomorphism since both modules are simple.

Now suppose $M = M_1 \oplus \cdots \oplus M_t$ is a decomposition of M into simple submodules. In the corresponding splitting (1.9) of M^*, M_i^* annihilates M_j if $j \neq i$. Thus the image of M_j under the canonical homomorphism γ_M annihilates all M_i^* with $i \neq j$ and so is contained in M_j^{**}, and since this map on M_j is simply γ_{M_j} itself and is $\neq 0$, it is an isomorphism and so therefore is γ_M.

(c) The first statement (i) holds without the assumption that R is semisimple – cf. p. 3. As for (ii) we can interpret M as the dual of M^* by (b), and so (ii) follows from (i). □

If X is any subset of a ring R, the *centralizer of X in R* is the subring

$$\mathrm{Z}_R(X) = \{r \in R : rx = xr \text{ for all } x \in X\}.$$

The *center* of R is the (commutative) subring $\mathrm{Z}(R) = \mathrm{Z}_R(R)$.

A ring R is *simple* if its only (2-sided) ideals are 0 and all of R.

R acts *faithfully* on a module M if $rM = 0$ implies that $r = 0$.

Theorem 1.2.5 *Let D be a division ring, and let R be the ring $\mathrm{M}(n, D)$ of $n \times n$ matrices over D. Then*

(a) *R is a simple and semisimple ring. If the center of D is the field K, then the center of R is $K1_R$.*
(b) *There is, up to isomorphism, only one simple R-module S, namely the column vector module $D^{n\times 1}$. R acts faithfully on S and the left R-module R is isomorphic to the direct sum S^n of n copies of S.*
(c) *$\mathrm{End}_R S \cong D$ (if R-endomorphisms are written on the right).* □

Theorem 1.2.6 Wedderburn's Theorem *for rings. If R is a semisimple simple ring, then $R \cong \mathrm{M}(n, D)$ for a uniquely determined positive integer n and division ring D (up to isomorphism), and so has, up to isomorphism, exactly one simple R-module $D^{n\times 1}$. If S is any simple R-module, $\mathrm{End}_R S \cong D$ (if endomorphisms are written on the right). As a left R-module, R is the direct sum of n simple R-modules.* □

We note also that a simple ring R is semisimple if and only if R is *Artinian*, that is to say, any chain of left ideals $I_1 \supset I_2 \supset I_3 \supset \cdots \supset I_k \supset \cdots$ is constant after a finite number of terms. A finite dimensional algebra over a field is certainly Artinian, and this is the case that is of most interest to us.

Theorem 1.2.7 *Let R be a semisimple ring.*

(a) *Then*

$$R = R_1 \oplus R_2 \oplus \cdots \oplus R_h$$

where the R_i are the minimal nonzero two-sided ideals of R.

(b) *The R_i are simple (and semisimple) rings, are uniquely determined and are called the "simple components of R".*
(c) *The R_i are also the homogeneous components of R as a left R-module, and if $R_i = R_{S_i}$ (S_i a simple R-module), then S_i is the unique (up to isomorphism) simple R_i-module.*
(d) *If M is an R-module, then*

$$M = R_1M \oplus R_2M \oplus \cdots \oplus R_hM,$$

and R_iM is the homogeneous component M_{S_i} of M. Moreover

$$M^* = \mathrm{Hom}_R(M, R) = (R_1M)^* \oplus (R_2M)^* \oplus \cdots \oplus (R_hM)^*$$

where $(R_iM)^* = \mathrm{Hom}_R(R_iM, R) = \mathrm{Hom}_{R_i}(R_iM, R_i)$. □

If $M = AM$ where A is one of the R_i or, more generally, where the ring A is a direct summand of R, we say that M is *A-isotypic*. In this case A is a direct sum of some of the R_i, and $R_jM = 0$ if $R_j \not\subset A$.

1.2.3 Simple and semisimple algebras. All algebras from now on will be assumed finite dimensional over the field of scalars, which is usually K.

An algebra A over K is called *central* if its center $\mathrm{Z}(A) = K$ (strictly speaking $K1_A$ but we generally identify $K1_A$ with K).

Now we consider *simple* K-algebras. Since they are assumed to be of finite dimension, they are semisimple.

Theorem 1.2.8 Wedderburn's Theorem *for algebras. If A is a simple K-algebra, then $A \cong \mathrm{M}(n, D)$ for a uniquely determined positive integer n and division K-algebra D (up to isomorphism). In particular A has only one simple module, up to isomorphism.*

The map $\alpha \to \alpha \mathrm{I}_n$ is an isomorphism $\mathrm{Z}(D) \to \mathrm{Z}(\mathrm{M}(n, D))$. In particular the center of A is a field and A is a central K-algebra if and only if D is. □

The tensor product $A \otimes_K B$ of two algebras is also an algebra, with multiplication determined by $(a \otimes b)(a' \otimes b') = aa' \otimes bb'$.

Theorem 1.2.9 *If A and B are simple K-algebras, their tensor product $A \otimes_K B$ is simple if at least one of A and B is central, and is central simple if (and only if) both are central simple. If $A \cong \mathrm{M}(m, K)$ and $B \cong \mathrm{M}(n, K)$, then $A \otimes_K B \cong \mathrm{M}(mn, K)$.* □

Proposition 1.2.1 *Let V be a finitely generated module over the simple algebra A. Then*

$$\dim_K \mathrm{Hom}_A(V, A) = \dim_K \mathrm{Hom}_K(V, K).$$

Since $\mathrm{Hom}_A(V, A)$ and $\mathrm{Hom}_K(V, K)$ are additive functors of V, we may assume that V is a simple A-module. As an A-module, A is the direct sum of copies of V, say n, so since $\mathrm{Hom}_A(A, A) = \mathrm{End}_A(A)$ consists of the right multiplications by elements of A,

$$\begin{aligned} \dim_K \mathrm{Hom}_A(V, A) &= \tfrac{1}{n} \dim_K \mathrm{Hom}_A(A, A) \\ &= \tfrac{1}{n} \dim_K A = \dim_K V = \dim_K \mathrm{Hom}_K(V, K). \end{aligned}$$ □

Theorem 1.2.10 *Suppose that A is a semisimple K-algebra and that $\mathrm{tr} : A \to K$ is a K-linear map with the properties*

1. *$\mathrm{tr}(ab) = \mathrm{tr}(ba)$ for all $a, b \in A$,*
2. *tr is nonzero on every simple component of A.*

If V is a finitely generated A-module, the map

$$\mathrm{tr}_* : \mathrm{Hom}_A(V, A) \to \mathrm{Hom}_K(V, K),$$

given by composition with tr, is an isomorphism of K-vector spaces.

By the additivity of Hom and the fact that A is semisimple, we can assume that V is simple. By Ths. 1.2.2, p. 6 and 1.2.7, the image of any A-homomorphism $V \to A$ lies in a simple component A_i of A. Since $A_j A_i$ and $A_j V$ are both 0 if A_j is some other simple component, $\mathrm{Hom}_A(V, A) = \mathrm{Hom}_{A_i}(V, A_i)$. Therefore we can also assume that A is a simple algebra.

By Prop. 1.2.1 it suffices to prove that tr_* is injective. Suppose that x is a nonzero functional in $\mathrm{Hom}_A(V, A)$. Then $x(V)$ is a minimal (nonzero) left ideal I of A. If $\mathrm{tr}\, I = 0$, then $\mathrm{tr}\, J = 0$ for all minimal left ideals J of A since by the proof of Theorem 1.2.4(a), $J = Ia$ for some $a \in A$ which implies that $\mathrm{tr} J = \mathrm{tr}(Ia) = \mathrm{tr}(aI) \subset \mathrm{tr}(I) = 0$. This contradicts the fact that $\mathrm{tr}(A) \neq 0$. Thus $\mathrm{tr}(I) \neq 0$, so $\mathrm{tr}(x(V)) \neq 0$. □

Suppose that we identify the two duals $\mathrm{Hom}_K V$ and $\mathrm{Hom}_A V$ via tr_* and denote them both by V^*. Then if we denote the corresponding pairings with V by $\langle \cdot, \cdot \rangle_K$ and $\langle \cdot, \cdot \rangle_A$,

$$\langle v, x \rangle_K = \mathrm{tr} \langle v, x \rangle_A \ \text{ for all } v \in V \text{ and } x \in V^*,$$

and if $\varphi \in \operatorname{Hom}_A(V, W)$,

$$\langle \varphi v, x\rangle = \operatorname{tr}\langle \varphi v, x\rangle = \operatorname{tr}\langle v, \varphi^t x\rangle = \langle v, \varphi^t x\rangle,$$

which implies

Corollary 1.2.1 *The restriction of the "K-transpose"*

$$\operatorname{Hom}_K(V, W) \xrightarrow{t} \operatorname{Hom}_K(W^*, V^*)$$

to the subgroup $\operatorname{Hom}_A(V, W)$ *of* $\operatorname{Hom}_K(V, W)$ *is the "A-transpose"*

$$\operatorname{Hom}_A(V, W) \xrightarrow{t} \operatorname{Hom}_A(W^*, V^*). \qquad \square$$

Proposition 1.2.2 *Suppose that V_1 and V_2 are left modules over the K-algebra A and W_1 and W_2 are left modules over the K-algebra B, all of them finitely generated. Write $\otimes_K = \otimes$ and assume that A and B-homomorphisms act on the right of the four modules.*
(a) There is a homomorphism

$$\operatorname{Hom}_A(V_1, V_2) \otimes \operatorname{Hom}_B(W_1, W_2) \to \operatorname{Hom}_{A\otimes B}(V_1 \otimes W_1, V_2 \otimes W_2) \tag{1.10}$$

determined by

$$\varphi \otimes \psi \to (v \otimes w \to v\varphi \otimes w\psi).$$

(b) If A and B are semisimple K-algebras, this homomorphism is an isomorphism.

(a) is straightforward.

(b) When $V_1 = V_2 = A$ and $W_1 = W_2 = B$, isomorphism is a consequence of the fact that $\operatorname{Hom}_A(A, A) \cong A$ – the right multiplications by elements of A – and the analogous isomorphisms for B and $A \otimes_K B$.

Now write $A = \oplus_i S_i$ and $B = \oplus_k T_k$ where S_i is a simple A-module for all i and T_k is a simple B-module for all k. Let $C = A \otimes_K B$. It is straightforward to check that

$$\begin{array}{ccc}
\operatorname{Hom}_A(\oplus_i S_i, \oplus_j S_j) \otimes_K \operatorname{Hom}_B(\oplus_k T_k, \oplus_l T_l) & \longrightarrow & \operatorname{Hom}_C((\oplus_i S_i) \otimes_K (\oplus_k T_k), (\oplus_j S_j) \otimes_K (\oplus_l T_l)) \\
\downarrow & & \downarrow \\
\oplus_{i,j,k,l}(\operatorname{Hom}_A(S_i, S_j) \otimes_K \operatorname{Hom}_B(T_k, T_l)) & \longrightarrow & \oplus_{i,j,k,l} \operatorname{Hom}_C(S_i \otimes_K T_k, S_j \otimes_K T_l)
\end{array}$$

is commutative, where the horizontal maps are particular cases of (1.10) and the vertical maps arise from the additivity of the functor Hom. The vertical maps are isomorphisms, and the upper horizontal map is too by the first case above, and so the lower horizontal map is also an isomorphism. This latter map is "defined component-wise",

$$\operatorname{Hom}_A(S_i, S_j) \otimes_K \operatorname{Hom}_B(T_k, T_l) \to \operatorname{Hom}_{A\otimes_K B}(S_i \otimes_K T_k, S_j \otimes_K T_l),$$

and so these component maps must also be isomorphisms. That is to say, the map (1.10) is an isomorphism when the modules are all simple, and it follows easily from this (and the additivity of Hom) that (1.10) holds in general. $\square$

An *inner automorphism* of a ring R is an automorphism of the form $a \to cac^{-1}$ for some $c \in R^\times$. We denote it by inn_c.

Theorem 1.2.11 The Skolem-Noether Theorem. *Let A be a central simple K-algebra and B a simple K-algebra. Any two algebra homomorphisms $\varphi, \psi : B \to A$ differ by an inner automorphism of A, i.e. there is an inner automorphism ι such that the diagram*

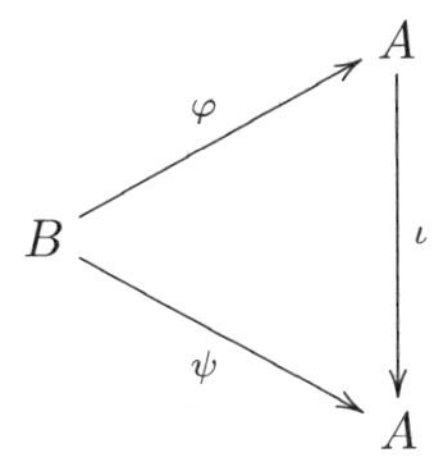

is commutative. In particular, any algebra automorphism of A is inner. □

Theorem 1.2.12 The Centralizer Theorem. *Let A be a central simple algebra over K and B a simple subalgebra of A. Let C be the centralizer $\mathrm{Z}_A(B)$ of B. Then:*

(a) C is simple, $\mathrm{Z}(C) = \mathrm{Z}(B)$, and $\dim_K A = \dim_K B \dim_K C$.

(b) If B is central simple, then so is C and $A \cong B \otimes_K C$.

(c) $\mathrm{Z}_A(\mathrm{Z}_A(B)) = B$. □

Now let L/K be a field extension. Then $A^L := L \otimes_K A$ is an L-algebra, the *algebra obtained from A by extension of scalars to L* and its center is $L \otimes_K \mathrm{Z}(A)$; thus $L \otimes_K A$ is central if A is. Moreover $L \otimes_K A$ is a central simple L-algebra if A is a central simple K-algebra.

We will need the following elementary result.

Lemma 1.1 *Let A be a K-algebra (finite dimensional, as usual), L/K an extension of fields, and $V \subset A$ a K-vector subspace of A. Then the canonical "inclusion homomorphism" $L \otimes_K \mathrm{Z}_A(V) \to L \otimes_K A = A^L$ has image $\mathrm{Z}_{A^L}(V^L)$.*

Let $v_1, \ldots, v_n$ be a K-basis of V and let $\varphi : A \to A^n$ be the K-linear map

$$\varphi(a) = (av_1 - v_1 a, av_2 - v_2 a \ldots, av_n - v_n a).$$

The kernel of φ is $\mathrm{Z}_A(V)$ and so we have an exact sequence

$$0 \to \mathrm{Z}_A(V) \to A \xrightarrow{\varphi} A^n$$

of K-vector spaces. Since $L \otimes_K \cdot$ is an exact functor from the category of K-vector spaces to the category of L-vector spaces,

$$0 \to L \otimes_K \mathrm{Z}_A(V) \to L \otimes_K A \xrightarrow{\mathrm{id}_L \otimes_K \varphi} L \otimes_K A^n$$

is also exact. The lemma will follow if we show that the kernel of $\mathrm{id}_L \otimes_K \varphi$ is $Z_{A^L}(V^L)$. Let

$$\sum_{i=1}^{m} \lambda_i \otimes a_i \in L \otimes_K A, \quad \lambda_i \in L \quad \text{for all } i.$$

Now

$$\begin{aligned}(\mathrm{id}_L \otimes_K \varphi)\left(\sum_i \lambda_i \otimes a_i\right) &= \sum_i \lambda_i \otimes \varphi(a_i)\\ &= \sum_i (\lambda_i \otimes (a_i v_1 - v_1 a_i, a_i v_2 - v_2 a_i, \dots, a_i v_n - v_n a_i))\\ &= \sum_i (\lambda_i \otimes (a_i v_1 - v_1 a_i, 0, \dots, 0))\\ &\quad + \sum_i (\lambda_i \otimes (0, a_i v_2 - v_2 a_i, \dots, 0)) + \dots\\ &\qquad + \sum_i (\lambda_i \otimes (0, 0, \dots, a_i v_n - v_n a_i)),\end{aligned}$$

which is $= 0$ if and only if

$$\sum_i \lambda_i \otimes (a_i v_j - v_j a_i) = 0 \quad \text{for all } j,$$

if and only if

$$\left(\sum_i \lambda_i \otimes a_i\right)(1 \otimes v_j) - (1 \otimes v_j)\left(\sum_i \lambda_i \otimes a_i\right) = 0 \quad \text{for all } j,$$

if and only if

$$\sum_i \lambda_i \otimes a_i \in \mathrm{Z}_{L \otimes_K A}(L \otimes_K V) = \mathrm{Z}_{A^L} V^L. \qquad \square$$

We will also need a slight extension of the notion of scalar extension – instead of a field extension L/K, we consider a commutative K-algebra C (of finite dimension, as usual). Then $A^C = C \otimes_K A$ is a C-algebra. If C is also separable, it is a direct sum of finite separable field extensions and A^C is a semisimple respectively separable K-algebra if A is.

The simple K-algebra A is said to be *split* if $A \cong \mathrm{M}(n, K)$ for some positive integer n, and is said to be *split by* L if it is central simple and $L \otimes_K A$ is split (as an L-algebra) (and so $\cong \mathrm{M}(n, L)$ for some n). If A is split by L, then it is also split by any extension of L. A central simple algebra always has a separable splitting field of finite degree over K, and so also a Galois splitting field of finite degree. If $A \cong \mathrm{M}(n, D)$ where D is a central division algebra over K, then any maximal subfield L of D is a splitting field of D and A, and $(D : K) = (L : K)^2$.

It follows that the dimension of a central simple algebra is a square integer $(nd)^2$, where $d^2 = (D : K)$; d is called the *(Schur) index* of A and nd is the *degree* of the algebra. The degree satisfies $A^L \cong \mathrm{M}(nd, L)$ for any splitting field L of A – in particular the degree of A^L as an L-algebra is the same as that of A. In fact if A is central simple and L/K is an arbitrary field extension, the degree of A^L is the same as that of A.

The degree or index of a simple algebra is defined to be its degree or index over its center. If A is a simple K-algebra with center L_1 which is a separable field extension of K, and L_2/K is an arbitrary algebraic extension, then by Th. 1.2.15 (p. 21)

$$L_2 \otimes_K A \cong L_2 \otimes_K L_1 \otimes_{L_1} A \cong (M_1 \otimes_{L_1} A) \oplus \cdots \oplus (M_m \otimes_{L_1} A)$$

for certain algebraic extensions $M_1, M_2, \ldots, M_m$ of K which contain L_1 and L_2. The degrees of the $M_i \otimes_{L_1} A$ are the same as the degree of A – thus the simple components of $L_2 \otimes_K A$ have the same degree as A.

Let A be a central simple K-algebra that it is split by the finite extension L/K. Consider A as a K-subalgebra of $L \otimes_K A$ by identifying $a = 1 \otimes a$. If $a \in A$, the trace $\operatorname{tr} a$ of a viewed as a matrix under an identification $L \otimes_K A = \mathrm{M}(nd, L)$, is independent of the identifications made and of the splitting field chosen. It is contained in K and is defined to be the *reduced trace* $\operatorname{trd} a = \operatorname{trd}_{A/K} a \in K$ of a. A similar fact holds for the determinant of $a \in A \subset \mathrm{M}(nd, L)$, which is by definition the *reduced norm* $\operatorname{nrd} a = \operatorname{nrd}_{A/K} a \in K$ of a. They have the properties

$$\begin{aligned} \operatorname{trd}(a+b) &= \operatorname{trd} a + \operatorname{trd} b & \operatorname{nrd}(ab) &= (\operatorname{nrd} a)(\operatorname{nrd} b) \\ \operatorname{trd}(\alpha a) &= \alpha \operatorname{trd} a & \operatorname{nrd}(\alpha a) &= \alpha^{nd} \operatorname{trd} a \\ \operatorname{trd}_{K_1 \otimes_K A/K_1} a &= \operatorname{trd}_{A/K} a & \operatorname{nrd}_{K_1 \otimes_K A/K_1} a &= \operatorname{nrd}_{A/K} a \\ \operatorname{trd}(ab) &= \operatorname{trd}(ba) & \operatorname{nrd}(ab) &= \operatorname{nrd}(ba) \end{aligned} \tag{1.11}$$

where $\alpha \in K$ and K_1/K is an arbitrary field extension of K.

The above relationships imply that $\operatorname{trd}(ab)$ is a symmetric bilinear form $A \times A \to K$, which can be shown to be nonsingular.

We shall also need the following lemma:

Lemma 1.2.2 *Let A and B be central simple algebras over K. Then if $a \in A$ and $b \in B$,*

$$\operatorname{trd}_{(A \otimes_K B)/K}(a \otimes b) = (\operatorname{trd}_{A/K} a)(\operatorname{trd}_{B/K} b).$$

If both A and B are split, say $A \cong \mathrm{M}(n, K)$ and $B \cong \mathrm{M}(m, K)$, then the formula holds because the "Kronecker product" $a \otimes_K b \in A \otimes_K B$ is an $nm \times nm$ matrix which can be realized as the block matrix $(a_{ij} B)$. In the general case, let L be a finite extension which splits both A and B – then it also splits $A \otimes_K B$ since

$$L \otimes_K (A \otimes_K B) \cong (L \otimes_K A) \otimes_L (L \otimes_K B),$$

and the formula follows at once. □

Except for the relation $\operatorname{nrd}(\alpha a) = \alpha^{nd} \operatorname{nrd} a$, all of the 8 properties in (1.11), as well as nonsingularity, are also enjoyed by a generalization of the reduced trace and norm to separable algebras. These are defined as follows. A *separable* K-algebra A is a semisimple K-algebra such that if

$$A = A_1 \oplus A_2 \oplus \cdots \oplus A_h$$

is its decomposition into simple algebras (cf. Th. 1.2.7), then the center K_i of each A_i is a separable field extension of K. Alternatively, it has the property that $L \otimes_K A$ is semisimple for every field extension L/K (Th. 7.18, p. 100, [55]), which implies that it is also separable for every such extension. The reduced trace and norm are then defined as

$$\operatorname{trd}_{A/K}(a) = \sum_i \operatorname{Tr}_{K_i/K} \operatorname{trd}_{A_i/K_i} a_i, \quad \operatorname{nrd}_{A/K}(a) = \prod_i \mathrm{N}_{K_i/K} \operatorname{nrd}_{A_i/K_i} a_i$$

where $a = \sum_i a_i$, $a_i \in A_i$, and $\operatorname{Tr}_{K_i/K}$ and $\mathrm{N}_{K_i/K}$ are the usual trace and norm in the field extension K_i/K. Of course each A_i is also a separable K-algebra, and we get

$$\operatorname{trd}_{A/K}(a) = \sum_i \operatorname{trd}_{A_i/K} a_i, \quad \operatorname{nrd}_{A/K}(a) = \prod_i \operatorname{nrd}_{A_i/K} a_i. \tag{1.12}$$

Suppose that $A = \mathrm{M}(n, D)$ is a separable K-algebra. If $a = (d_{ij}) \in \mathrm{M}(n, D)$ (with $d_{ij} \in D$ for all i, j), then it follows from the definition of the reduced trace and norm that

$$\operatorname{trd}_{\mathrm{A/K}}(a) = \sum_i \operatorname{trd}_{D/K}(d_{ii}) \tag{1.13}$$

and, if a is diagonal (or upper triangular),

$$\operatorname{nrd}_{\mathrm{A/K}}(a) = \prod_i \operatorname{nrd}_{D/K}(d_{ii}), \tag{1.14}$$

and that

$$\operatorname{nrd}_{A/K} A = \operatorname{nrd}_{D/K} D. \tag{1.15}$$

Exercise. Show that a commutative semisimple K-algebra is a direct sum of finite field extensions of K. □

If R is any ring – e.g. an algebra – R^o denotes the opposite ring, which $= R$ as an additive group, and whose multiplication is given by $a^o b^o = ba$, where a^o, for example, is a considered as an element of R^o. If R is a central simple algebra over K, then so is R^o.

Lemma 1.2.3 *Let A be a central simple algebra over K. Then* $\operatorname{trd}_{A^o/K} a^o = \operatorname{trd}_{A/K} a$ *and* $\operatorname{nrd}_{A^o/K} a^o = \operatorname{nrd}_{A/K} a$.

Let $o : A \to A^o$ be the antiisomorphism $a \to a^o$. It is clear that a splitting field L of A is also a splitting field of A^o. Consider the sequence of maps

$$L \otimes A \xrightarrow{\mathrm{id} \otimes o} L \otimes A^o \xrightarrow{\varphi} \mathrm{M}(nd, L) \xrightarrow{t} \mathrm{M}(nd, L)$$

where φ is an identifying isomorphism and t is transpose. We can use φ to compute $\operatorname{trd}_{A^o/K} a^o$ – namely it is $= \operatorname{tr}(\varphi(a^o))$ – and the composition of the three maps to compute $\operatorname{trd}_{A/K} a$. The image of a under the composition is simply $(\varphi(a^o))^t$, and so the statement about reduced traces follows from the fact that the matrix trace is preserved by the transpose. The proof for the reduced norm is entirely similar. □

Exercise. Suppose that $A = \mathrm{M}(n, R)$, and that

$$\varphi : A^o \to \mathrm{M}(n, R^o)$$

is the map defined by

$$\varphi\left((d_{ij})^o\right) = (d_{ij}^o)^t.$$

Show that φ is a ring isomorphism. □

There is another more elementary trace defined on a semisimple algebra – in fact defined on any finite dimensional algebra A over K – namely the "left regular algebra trace", or simply the "algebra trace". If $a \in A$, it is the trace of the K-linear transformation $x \to ax$ on A and is denoted by $\operatorname{tr}_{A/K} a$. The "left regular algebra norm" is defined similarly as $\mathrm{n}_{A/K} a = \det(x \to ax)$. The trace $\mathrm{Tr}_{L/K}$ and norm $\mathrm{N}_{L/K}$ in a finite extension L/K of fields are the regular trace and norm when L is viewed as a K-algebra.

If A is a simple separable algebra over K of degree nd, then

$$\mathrm{tr}_{A/K} a = nd\, \mathrm{trd}_{A/K} a \quad \text{for all } a \in A. \tag{1.16}$$

When A is central, see (9.17), p. 119, [55]. If A is *not* central, it follows from the fact that for *any* algebra A over L,

$$\mathrm{tr}_{A/K} = \mathrm{Tr}_{L/K} \circ \mathrm{tr}_{A/L}$$

for a subfield K of L for which L/K is finite and separable.

If G is a finite group, of order g, say, the *group algebra* KG of G is the set of all formal K-linear combinations of elements of G

$$\sum_{s \in G} \alpha_s s,$$

made into an algebra by extending the multiplication in G to all of KG using the distributive law:

$$\sum_s \alpha_s s \sum_t \beta_t t = \sum_{s,t} \alpha_s \beta_t st = \sum_r (\sum_{st=r} \alpha_s \beta_t) r.$$

KG is an (associative) algebra over K of dimension g, commutative if and only if G is.

It is clear that $\mathrm{tr}_{KG/K} s = 0$ if $s \in G$, $s \neq 1$, and $= g$ if $s = 1$, so

$$\mathrm{tr}_{KG/K} \left(\sum_{s \in G} \alpha_s s \right) = g \alpha_1.$$

Define the *canonical trace* $\mathrm{Tr}_{KG/K} = \mathrm{Tr} : KG \to K$ to be the K-linear functional such that $\mathrm{Tr}(s) = 1$ if $s = 1$, $\mathrm{Tr}(s) = 0$ otherwise. In other words

$$\mathrm{Tr}_{\mathrm{KG/K}}(\sum_{s \in G} \alpha_s s) = \alpha_1. \tag{1.17}$$

Thus

$$\mathrm{tr}_{KG/K} a = g(\mathrm{Tr}_{KG/K} a). \tag{1.18}$$

Suppose now that the characteristic of K does not divide g. We shall see later (Maschke's Theorem, Th. 1.7.1, p. 70) that KG is a separable K-algebra. Thus there is a decomposition

$$KG = A_1 \oplus A_2 \oplus \cdots \oplus A_h \tag{1.19}$$

of KG into simple algebras. There is also a simple relationship between $\mathrm{Tr}_{KG/K}$ and the reduced trace $\mathrm{trd}_{KG/K}$:

Theorem 1.2.13 *Assume that the characteristic of K does not divide g.*

(a) If $A \cong \mathrm{M}(n, D)$ is a simple component of KG, then for all $a \in A$,

$$\mathrm{Tr}_{KG/K}\, a = \tfrac{nd}{g} \mathrm{trd}_{A/K} a = \tfrac{\deg A}{g} \mathrm{trd}_{A/K} a. \tag{1.20}$$

(b) More generally, suppose that (1.19) is the decomposition of KG into simple algebras, where for each i, D_i is a division algebra of dimension d_i^2 such that $A_i \cong \mathrm{M}(n_i, D_i)$. If $a \in KG$, let $a = a_1 + \cdots + a_h$ be the decomposition of a into components $a_i \in A_i$. Then

$$\mathrm{Tr}_{KG/K}\, a = \tfrac{1}{g} \sum_i n_i d_i \mathrm{trd}_{A_i/K}\, a_i = \tfrac{1}{g} \sum_i (\deg A_i) \mathrm{trd}_{A_i/K}\, a_i.$$

Part (b) follows from (a) by the additivity of $\mathrm{Tr}_{KG/K}$. Consider (a). By (1.18) and (1.16),

$$\mathrm{Tr}_{KG/K}a = \tfrac{1}{g}\mathrm{tr}_{KG/K}a = \tfrac{1}{g}\mathrm{tr}_{A/K}a = \tfrac{1}{g}nd\,\mathrm{trd}_{A/K}. \qquad \square$$

Corollary 1.2.2 *If* $\mathrm{char}\,K \nmid g$, *then* $\mathrm{char}\,K \nmid nd$.

If a is a nonzero element of A, there is an $s \in G$ such that $\mathrm{Tr}_{KG/K}\,sa \neq 0$. Since $sa \in A$, the corollary follows from (a). □

One can also define a "reduced characteristic polynomial" $\mathrm{prd}_{A/K}a = \mathrm{prd}\,a$ for an element a in a central simple algebra A. If L is a splitting field of A, it is defined as the characteristic polynomial $\det(X\mathrm{I}_{nd} - a)$ where a is considered to be an element of $L \otimes_K A = \mathrm{M}(nd, L)$. It can be shown that it is independent of the splitting field chosen as well as the identification with $\mathrm{M}(nd, L)$, and that

$$\mathrm{prd}_{A/K}a = X^{nd} - (\mathrm{trd}_{A/K}a)X^{nd-1} + \cdots + (-1)^{nd}\mathrm{nrd}_{A/K}a \in K[X]. \qquad (1.21)$$

1.2.4 Quaternion algebras. As usual we assume here that $\mathrm{char}\,K \neq 2$. Quaternion algebras are particularly important in representation theory – see, for example, Th. 1.7.6, p. 78. They are 4-dimensional (associative) algebras A with a basis $\{1_A = 1, i, j, k\}$ and multiplication determined by two elements $\alpha, \beta \in \dot{K}$ and the rules

$$ij = k, \quad ji = -ij, \quad i^2 = \alpha\ (= \alpha 1_A), \quad j^2 = \beta.$$

It follows that i, j and k anticommute with each other, and that $k^2 = -\alpha\beta$. A is a central simple K-algebra and is denoted by $(\alpha, \beta)_K$ or simply (α, β). It is either a division algebra or is $\cong \mathrm{M}(2, K)$.

Conversely any 4-dimensional central simple algebra is a quaternion algebra.

The elements of the subspace $A_0 = Ki + Kj + Kk$ of A are called *pure quaternions*; the nonzero pure quaternions are characterized by the fact that they are not in K but their squares *are*. Thus the map * defined by $(a_0 + a_1)^* = a_0 - a_1$, where $a_0 \in K$ and $a_1 \in A_0$, is uniquely determined by the algebra structure of A, and is a (K, id)-involution[2] of A called *conjugation* (Example 6, p. 25).

The norm $\mathrm{N}_{A/K} : A \to K$ defined by $\mathrm{N}_{A/K}(a) = aa^*$ satisfies $\mathrm{N}_{A/K}(ab) = \mathrm{N}_{A/K}(a)\mathrm{N}_{A/K}(b)$. It is in fact the reduced norm $\mathrm{nrd}_{A/K}$. Similarly the map $a \to a + a^*$ is the reduced trace $\mathrm{trd}_{A/K}$, so $\mathrm{trd}_{A/K}(x_0 + x_1 i + x_2 j + x_3 k) = 2x_0$. The *norm form* of A is[3] the nonsingular quadratic form

$$\begin{aligned}\mathrm{nrd}_{A/K}(x) &= \mathrm{nrd}_{A/K}(x_0 + x_1 i + x_2 j + x_3 k)\\ &= x_0^2 - \alpha x_1^2 - \beta x_2^2 + \alpha\beta x_3^2 = \langle 1, -\alpha, -\beta, \alpha\beta\rangle.\end{aligned}$$

A is a division algebra if and only if the norm form is anisotropic; if it is *not* a division algebra, then the norm form is $\cong \langle 1, 1, -1, -1\rangle$ which is hyperbolic.

Quaternion algebras satisfy the following fundamental identities (where $(\ ,\ ) = (\ ,\ )_K$):

(a) $(\alpha, \beta) \cong (\gamma^2\alpha, \delta^2\beta), \quad (\alpha, \beta) \cong (\beta, \alpha)$,
(b) $(1, 1) \cong \mathrm{M}(2, K) \cong (\alpha, 1) \cong (\beta, -\beta) \cong (\gamma, 1 - \gamma)$ if $\gamma \neq 0$ or 1,
(c) $(\alpha, \alpha\beta) \cong (\alpha, -\beta), \quad (\alpha, \alpha) \cong (\alpha, -1)$.

---

[2]Involutions are defined in §1.1 and discussed in §1.3.
[3]See §1.5.1 for these notions about forms.

(d) $(\alpha, \beta) = 1$ if and only if $\alpha \in \mathrm{N}_{L/K}\dot{L}$.

Here $L = K(\sqrt{\beta})$ and, if $\beta \in \dot{K}^2$, $N_{L/K}(\gamma_1+\gamma_2\sqrt{\beta})$ is defined to be $\gamma_1^2-\gamma_2^2\beta$.

1.2.5 The Brauer group. By Wedderburn's theorem, any finite dimensional central simple K-algebra A is isomorphic to a matrix algebra $\mathrm{M}(n, D)$ where D is a central division algebra over K, *the division algebra belonging to* A. It is unique up to isomorphism. Two central simple algebras are said to be *similar* if their division algebras are isomorphic. The resulting equivalence classes are called *Brauer classes* and the collection of them is the *Brauer group* $\mathrm{Br}(K)$.

The product of two Brauer classes $[A]$ and $[B]$ in the Brauer group is the Brauer class $[A \otimes_K B]$ – the tensor product is central simple by Th. 1.2.9. The class $[K]$ of the trivial algebra K is the identity element. And the inverse of a Brauer class $[A]$ is the Brauer class $[A^o]$ of the opposite algebra A^o.

$\mathrm{Br}(K)$ is obviously Abelian, and it is also torsion: if the division algebra D in the Brauer class $[A]$ has dimension n^2, it can be shown that $[A]^n = [K]$. Thus a nonsplit quaternion algebra has order 2 in the Brauer group.

It is often more convenient to write $\mathrm{Br}(K)$ as an additive group.

Examples. 1. If K is algebraically closed, the only finite dimensional division algebra over K is K itself, so the Brauer group $\mathrm{Br}(K) = 0$.

2. If $K = \mathbb{R}$ (the real numbers), or if K is a real closed field $\mathbf{R}$ (cf. §1.4), the only nontrivial division algebra over K, up to isomorphism, is the quaternion algebra $\mathbb{H} = (-1, -1)_{\mathbb{R}}$ respectively $\mathbf{H} = (-1, -1)_{\mathbf{R}}$. Thus the Brauer group $\mathrm{Br}(K) = \mathbb{Z}/2$. We can think of this isomorphism as being an injective homomorphism $\mathrm{Inv}_K : \mathrm{Br}(K) \to \mathbb{Q}/\mathbb{Z}$ with image $\frac{1}{2}\mathbb{Z}/\mathbb{Z}$.

3. It is a theorem of Wedderburn that every finite division ring is commutative, so the Brauer group of a finite field is 0.

4. A field which is a finite extension of the field of p-adic numbers $\mathbb{Q}_p$ or of a power series field $k\langle x\rangle$ over a finite field k, is called a *local* field; it is *dyadic* if 2 is not a unit in the local ring of integers, otherwise *nondyadic.*

A local field K has a unique equivalence class of valuations, which is usually denoted by $\mathfrak{p}$ and referred to as a "prime". It induces the "$\mathfrak{p}$-adic topology" on K, under which the group of squares $\dot{K}^2$ is open (the "Local Square Theorem", p. 159, [54]).

There are several properties of local fields which are needed later. The first is (cf. p. 217, [65])

$$[\dot{K} : \dot{K}^2] = \begin{cases} 4 & \text{if } K \text{ is nondyadic,} \\ 2^{2+(K:\mathbb{Q}_2)} & \text{if } K \text{ is dyadic.} \end{cases} \tag{1.22}$$

If K is a local field, there is a canonical isomorphism

$$\mathrm{Inv}_K : \mathrm{Br}(K) \xrightarrow{\sim} \mathbb{Q}/\mathbb{Z}.$$

If α is a nonsquare in the local field K, then there is $\beta \in \dot{K}$ such that the quaternion algebra $(\alpha, \beta)_K$ is not split and so its image in $\mathrm{Br}(K)$ is $\neq 1$. See p. 203, [54].

We usually use the same symbol $(\alpha, \beta)_K$ – or even (α, β) – for the Brauer class of the quaternion algebra if it is unlikely to cause confusion.

Note that $\mathbb{Q}/\mathbb{Z}$ has a unique element of order 2, the Brauer class of a nonsplit quaternion algebra; thus there is, up to isomorphism, only one (central nonsplit) quaternion algebra over a local field.

Another important property of local fields is the existence of a (canonical) isomorphism $\dot{K}/\mathrm{N}_{L/K}\dot{L} \cong \mathrm{Gal}(L/K)$ if L/K is a finite Abelian extension (Th. 2.3, p. 94, [37]); in particular if L/K is a quadratic extension of local fields, then

$$[\dot{K} : \mathrm{N}_{L/K}\dot{L}] = 2. \tag{1.23}$$

5. A *global* field is either an algebraic number field (finite extension of the field $\mathbb{Q}$ of rational numbers) or a finite extension of a field $k(x)$ of rational functions over a finite field k. The completion of a global field with respect to a valuation is (isomorphic to) either a local field or, in the case of an algebraic number field, the field $\mathbb{R}$ of real numbers or the field $\mathbb{C}$ of complex numbers.

Suppose that K is a global field. To each equivalence class $\mathfrak{p}$ of valuations on K (which we usually refer to as a "prime" $\mathfrak{p}$), there corresponds a completion $K_\mathfrak{p}$. If $K_\mathfrak{p} \cong \mathbb{R}$ or $\mathbb{C}$, $\mathfrak{p}$ is called an Archimedean (or infinite) prime – more specifically a real or complex prime, while if $K_\mathfrak{p}$ is a local field, $\mathfrak{p}$ is nonArchimedean or finite.

We will need the *Approximation Theorem* (cf. p. 117, [53]):
Suppose $|\ |_1, |\ |_2, \dots, |\ |_n$ *are inequivalent valuations of the field* K *with completions* $K_1, K_2, \dots, K_n$ *respectively. If elements* $\alpha_1 \in K_1, \alpha_2 \in K_2, \dots, \alpha_n \in K_n$ *and a positive real number* ε *are given, there is an element* $\alpha \in K$ *such that* $|\alpha - \alpha_i|_i < \varepsilon$ *for* $i = 1, 2, \dots n$.

The *Global Square Theorem* (p. 182, [54]) states that a nonsquare in a global field K is a nonsquare at an infinite number of primes of K.

For each prime $\mathfrak{p}$ of a global field K, there is a homomorphism $\mathrm{Br}(K) \to \mathrm{Br}(K_\mathfrak{p})$, given by $[A] \to [A_\mathfrak{p}]$ where $A_\mathfrak{p} = K_\mathfrak{p} \otimes_K A$ – by Th. 32.1, p. 273, [55], $[A_\mathfrak{p}] = 0$ for all but a finite number of $\mathfrak{p}$. By 2. and 4. above, we get a sequence

$$0 \to \mathrm{Br}(K) \to \oplus_\mathfrak{p} \mathrm{Br}(K_\mathfrak{p}) \to \mathbb{Q}/\mathbb{Z} \to 0 \tag{1.24}$$

where the second-last map is the composite

$$\oplus_\mathfrak{p} \mathrm{Br}(K_\mathfrak{p}) \xrightarrow{\oplus_\mathfrak{p} \mathrm{Inv}_{K_\mathfrak{p}}} \oplus_\mathfrak{p} \mathbb{Q}/\mathbb{Z} \xrightarrow{\text{add}} \mathbb{Q}/\mathbb{Z}.$$

Here "add" means add the components of an element of $\oplus_\mathfrak{p}\mathbb{Q}/\mathbb{Z}$. This sequence is exact, and so implies in particular that for any central simple algebra A over K,

$$\sum_\mathfrak{p} \mathrm{Inv}_{K_\mathfrak{p}}[A_\mathfrak{p}] = 0,$$

the *Reciprocity Law* for K. The special case when A is a quaternion algebra $(\alpha, \beta)_K$ is often called the *Hilbert Reciprocity Law* and is usually written multiplicatively:

$$\prod_\mathfrak{p} (\alpha, \beta)_{K_\mathfrak{p}} = 1 \quad \text{for all } \alpha, \beta \in \dot{K}.$$

Equivalently, $(\alpha, \beta)_{K_\mathfrak{p}} \neq 1$ for an even number of $\mathfrak{p}$ (since in this notation, $(\alpha, \beta)_{K_\mathfrak{p}} = \pm 1$). One also sometimes writes $(\alpha, \beta)_{K_\mathfrak{p}} = (\alpha, \beta)_\mathfrak{p}$.

If S is a collection of an even number of primes of the global field K, and if $\beta \in \dot{K}$ is a nonsquare at each prime in S – that is to say, $\beta \notin \dot{K}_\mathfrak{p}^2$ for all $\mathfrak{p} \in S$ –

then there is an $\alpha \in \dot{K}$ such that $(\alpha,\beta)_\mathfrak{p} = -1$ for all $\mathfrak{p} \in S$ and is $= 1$ for all other primes. See p. 203, [54].

Lemma 1.2.4 *Let $\gamma \in \dot{K}$, an algebraic number field, and let (δ,γ) be a fixed quaternion Brauer class. Then the signs of δ at the real primes $\mathfrak{p}$ at which γ is a square (i.e. for which $\gamma >_\mathfrak{p} 0$) can be arbitrarily prescribed.*

If $\gamma \in \dot{K}^2$, the lemma is trivial, so suppose that $\gamma \notin \dot{K}^2$, and let S be the set of real primes at which γ is a square. Suppose $\varepsilon_\mathfrak{p} = \pm 1$ is given for each $\mathfrak{p} \in S$. Let $L = K(\sqrt{\gamma})$. For each $\mathfrak{p} \in S$, choose $\alpha_\mathfrak{p}, \beta_\mathfrak{p} \in K_\mathfrak{p}$ so that $(\alpha_\mathfrak{p}^2 - \beta_\mathfrak{p}^2\gamma)\varepsilon_\mathfrak{p} > 0$. By the approximation theorem for valuations (p. 18), there are $\alpha, \beta \in K$ such that $(\alpha^2 - \beta^2\gamma)\varepsilon_\mathfrak{p} > 0$ for all $\mathfrak{p} \in S$. Since $(\mathrm{N}_{L/K}\lambda, \gamma) = 1$ for all $\lambda \in \dot{L}$ (cf. (6), 57:9, p. 146, [54]), $(\mathrm{N}_{L/K}(\alpha + \beta\sqrt{\gamma})\delta, \gamma) = (\delta, \gamma)$. □

If the Brauer class of a division algebra D has order 2, then D is a quaternion algebra if K is a local or global field; in general the index of D is a power of 2 since the index of a division algebra D and the order of $[D]$ in the Brauer group have the same prime factors (p. 254, [55]). In fact a division algebra D of order 2 in the Brauer group is similar to a tensor product of quaternion algebras by "Merkurjev's Theorem" – cf. p. 81, [39].

We will also need the following lemma in §5.5.

Lemma 1.2.5 *If L/K is a quadratic extension of global fields, the group index $[\mathrm{N}_{L/K}\dot{L} : \dot{K}^2]$ is infinite.*

Let $L = K(\sqrt{\beta})$, and let S be the set of finite primes of K at which $\beta \notin K_\mathfrak{p}^2$. S is infinite by the Global Square Theorem. If $\mathfrak{p}$ is a prime of K, the nontrivial automorphism $\bar{\ }$ of L/K is continuous in the $\mathfrak{p}$-adic topology, so also the norm $\mathrm{N}_{L/K}(\lambda) = \lambda\bar{\lambda}$. Suppose $\mathfrak{p} \in S$. Then $\mathfrak{p}$ has a unique extension to L. The cosets in $(\mathrm{N}_{L_\mathfrak{p}/K_\mathfrak{p}}\dot{L}_\mathfrak{p})/\dot{K}_\mathfrak{p}^2$ are open in $\dot{K}_\mathfrak{p}$ (there are 2 of them if $\mathfrak{p}$ is nondyadic, $2^{1+(K_\mathfrak{p}:\mathbb{Q}_2)}$ if $\mathfrak{p}$ is dyadic). Let M be a positive integer, $\mathfrak{p}_1, \ldots, \mathfrak{p}_M$ distinct primes in S, and

$$(\mathrm{N}_{L_{\mathfrak{p}_i}/K_{\mathfrak{p}_i}}\lambda_i)\dot{K}_{\mathfrak{p}_i}^2 \in (\mathrm{N}_{L_{\mathfrak{p}_i}/K_{\mathfrak{p}_i}}\dot{L}_{\mathfrak{p}_i})/\dot{K}_{\mathfrak{p}_i}^2, \quad \text{for } i = 1, \ldots, M.$$

By the approximation theorem and the Local Square Theorem, there exists $\lambda_0 \in \dot{L}$ such that

$$(\mathrm{N}_{L_{\mathfrak{p}_i}/K_{\mathfrak{p}_i}}\lambda_0)\dot{K}_{\mathfrak{p}_i}^2 = (\mathrm{N}_{L_{\mathfrak{p}_i}/K_{\mathfrak{p}_i}}\lambda_i)\dot{K}_{\mathfrak{p}_i}^2 \quad \text{for } i = 1, \ldots, M. \tag{1.25}$$

Since $\mathrm{N}_{L_{\mathfrak{p}_i}/K_{\mathfrak{p}_i}}\lambda = \mathrm{N}_{L/K}\lambda$ for all $\lambda \in L$, the map $\dot{L} \to \prod_{i=1}^M (\mathrm{N}_{L_{\mathfrak{p}_i}/K_{\mathfrak{p}_i}}\dot{L}_{\mathfrak{p}_i})/\dot{K}_{\mathfrak{p}_i}^2$ given by

$$\lambda \to \left((\mathrm{N}_{L_{\mathfrak{p}_1}/K_{\mathfrak{p}_1}}\lambda)\dot{K}_{\mathfrak{p}_1}^2, \ldots, (\mathrm{N}_{L_{\mathfrak{p}_M}/K_{\mathfrak{p}_M}}\lambda)\dot{K}_{\mathfrak{p}_M}^2\right)$$

factors through $(\mathrm{N}_{L/K}\dot{L})/\dot{K}^2$ and so we have a map

$$(\mathrm{N}_{L/K}\dot{L})/\dot{K}^2 \to \prod_{i=1}^M (\mathrm{N}_{L_{\mathfrak{p}_i}/K_{\mathfrak{p}_i}}\dot{L}_{\mathfrak{p}_i})/\dot{K}_{\mathfrak{p}_i}^2$$

which is surjective by (1.25). Since M is arbitrary, the lemma follows. □

1.2.6 Compositums and tensor products of fields. A good reference is §8, pp. 99-102, [9].

Let L_1 and L_2 be extensions of K. A *compositum* or *composite extension field* of L_1 and L_2 is an extension field M/K along with K-homomorphisms $\varphi_i : L_i \to M$ such that M is generated (as a field) over K by the images $\varphi_1(L_1)$ and $\varphi_2(L_2)$. Thus a compositum is a triple

$$(M, \varphi_1, \varphi_2). \tag{1.26}$$

It is isomorphic to the compositum (N, ψ_1, ψ_2) of L_1 and L_2 if there is an isomorphism $\theta : M \to N$ such that both of the diagrams

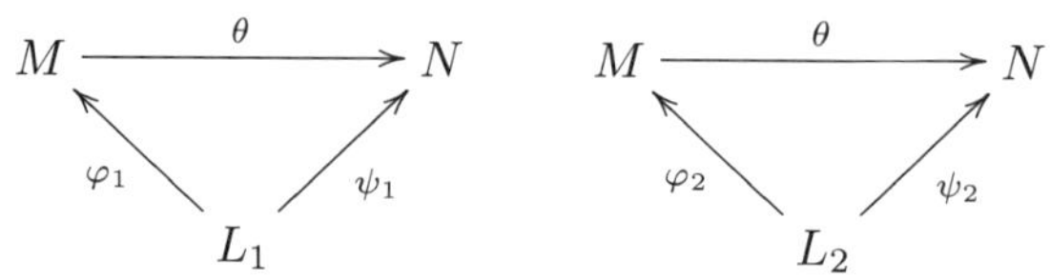

commute.

If $(M, \varphi_1, \varphi_2)$ is a compositum, there is a K-algebra homomorphism

$$(\varphi_1\varphi_2) : L_1 \otimes_K L_2 \to M$$

determined by $\varphi_1\varphi_2(\lambda_1 \otimes \lambda_2) = \varphi_1(\lambda_1)\varphi_2(\lambda_2)$. Let P be its kernel. It is a prime ideal of the commutative K-algebra $L_1 \otimes L_2$, and if $\pi_P : L_1 \otimes L_2 \to L_1 \otimes L_2/P$ is the canonical map, there is a unique homomorphism φ making the diagram

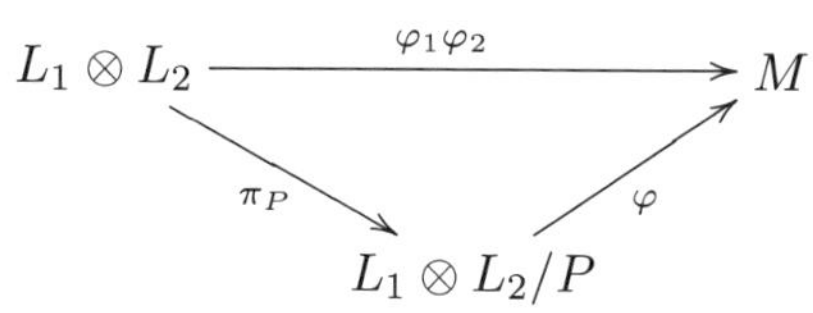

commutative. φ is an injective K-linear homomorphism of the integral domain $L_1 \otimes L_2/P$ into the field M, and has a unique K-linear extension $[\varphi] : [L_1 \otimes L_2/P] \to M$ to the field of fractions $[L_1 \otimes L_2/P]$. It makes the diagram

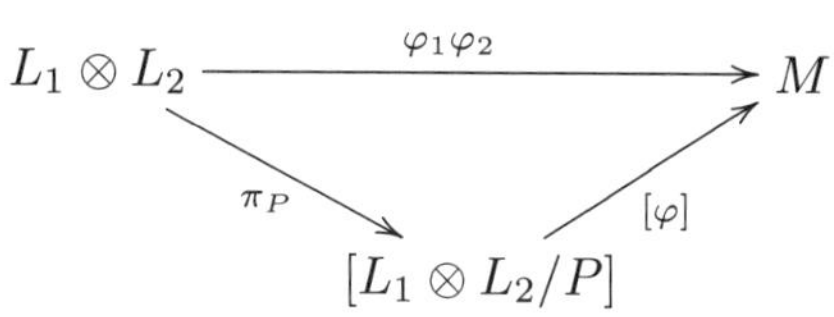

commutative, and is an isomorphism of fields. Next consider the diagrams

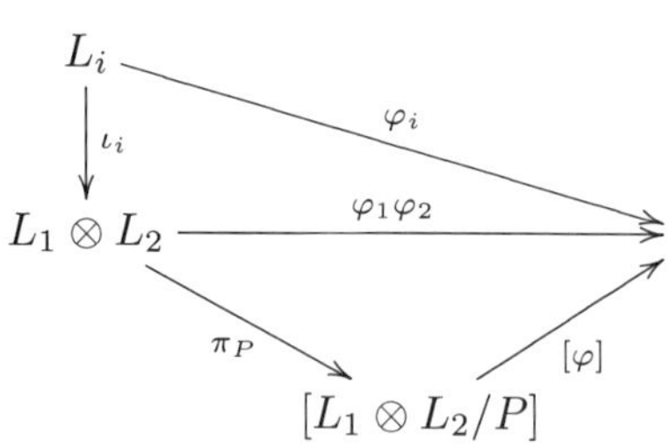

$(i = 1, 2)$

where ι_1 and ι_2 are the canonical maps $\lambda_1 \to \lambda_1 \otimes 1$ and $\lambda_2 \to 1 \otimes \lambda_2$. This implies that the compositum $(M, \varphi_1, \varphi_2)$ is isomorphic to the compositum $([L_1 \otimes L_2/P], \pi_P\iota_1, \pi_P\iota_2)$, so every compositum is isomorphic to one of the latter kind.

Suppose now that P and Q are prime ideals of $L_1 \otimes L_2$, and that $([L_1 \otimes L_2/P], \pi_P\iota_1, \pi_P\iota_2)$ and $([L_1 \otimes L_2/Q], \pi_Q\iota_1, \pi_Q\iota_2)$ are isomorphic compositums:

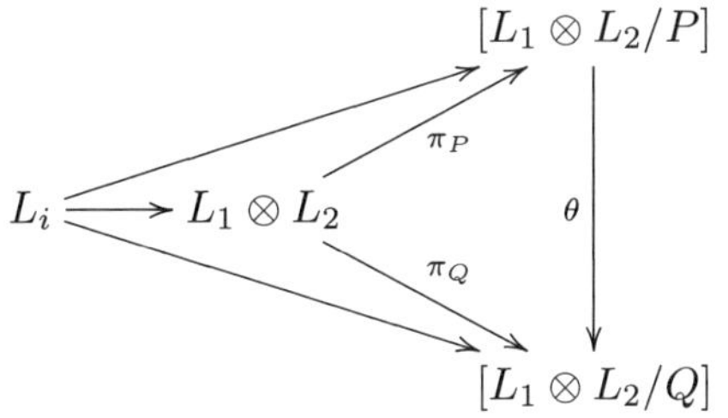

$(i = 1, 2)$

are commutative. Since θ is an isomorphism,

$$Q = \ker \pi_Q = \ker \pi_P = P.$$

Thus

Theorem 1.2.14 *As P runs over the prime ideals of $L_1 \otimes L_2$,*

$$([L_1 \otimes L_2/P], \pi_P\iota_1, \pi_P\iota_2)$$

runs over a system of representatives of the isomorphism classes of compositums of L_1 and L_2. □

Proposition 1.2.3 *Suppose that L_2 is algebraic over K, and that $\hat{L}_1$ is an algebraically closed extension of L_1. Then every compositum of L_1 and L_2 is isomorphic to one of the form $(M_\varphi, \mathrm{id}_{L_1}, \varphi)$ where φ is an isomorphism of L_2 into $\hat{L}_1$ and M_φ is the subfield of $\hat{L}_1$ generated over K by L_1 and $\varphi(L_2)$.*

If $(M, \varphi_1, \varphi_2)$ is a compositum, then M is an algebraic extension of $\varphi_1 L_1$, and so the map $(\varphi_1)^{-1} : \varphi_1 L_1 \to \hat{L}_1$ has an extension to a homomorphism $\theta : M \to \hat{L}_1$. Then θ is an isomorphism of $(M, \varphi_1, \varphi_2)$ with $(M_\varphi, \mathrm{id}_{L_1}, \varphi)$ where $\varphi = \theta\varphi_2$. □

Corollary 1.2.3 *If L_1 is algebraic over K and L_2 is a finite extension of K, the number of compositums of L_1 and L_2, distinct up to isomorphism, is $\leq (L_2 : K)$.*

This follows from the proposition since the number of distinct isomorphisms of L_2 into an algebraically closed extension of K is $\leq (L_2 : K)$. □

Theorem 1.2.15 *Let $(M_j, \varphi_{j1}, \varphi_{j2})$, $1 \leq j \leq n$, be a set of representatives of the isomorphism classes of compositums of L_1 and L_2. If L_2/K is finite and separable, the homomorphism*

$$\oplus_j(\varphi_{j1}\varphi_{j2})\colon\ L_1 \otimes_K L_2 \to M_1 \oplus \cdots \oplus M_n \tag{1.27}$$

is an isomorphism of K-algebras – in fact of L_1-algebras if each M_j is made into an L_1-algebra via $\varphi_{j1} : L_1 \to M_j$. Moreover, each M_j is a finite separable extension of $\varphi_{j1}L_1$.

Since L_2 is a finite dimensional separable K-algebra, $L_1 \otimes_K L_2$ is a semisimple L_1-algebra and so is the direct sum of simple L_1-algebras by Th. 1.2.7. It must in fact be the direct sum of extension fields $N_1, \ldots, N_k$ of L_1 since it is commutative. Thus its ideals are the partial direct sums of the N_i, and its prime ideals are those of the form $\oplus_{j\neq i}N_j$. Thus by Th. 1.2.14, $n = k$ and renumbering as necessary, we can assume that the epimorphism $\varphi_{i1}\varphi_{i2}\colon L_1 \otimes L_2 \to M_i$ has kernel $\oplus_{j\neq i}N_j$, and therefore takes N_i isomorphically onto M_i. The last statement is clear. □

Corollary 1.2.4 *Suppose L_1/K and L_2/K are both finite extensions, with relatively prime degrees over K. Then if at least one of L_i/K is separable, there is only one compositum L_1L_2, up to isomorphism, and $L_1 \otimes_K L_2 \cong L_1L_2$.*

If M is a compositum, then by the multiplicativity of field degrees, $(M : K)$ is divisible by both of $(L_1 : K)$ and $(L_2 : K)$ and so $(M : K) \geq (L_1 : K)(L_2 : K)$ by the relative primeness. But $L_1 \otimes_K L_2$ has K-dimension $(L_1 : K)(L_2 : K)$, and the corollary follows at once. □

Corollary 1.2.5 *Let $\mathbf{C} = \mathbf{R}(i)$ where $\mathbf{C}/\mathbf{R}$ is a quadratic field extension with $i^2 = -1$ and $\mathbf{C}$ has involution $(\alpha + \beta i)^* = \alpha - \beta i$. Then the involution algebra $(\mathbf{C}, \mathrm{id}) \otimes_{(\mathbf{R},\mathrm{id})} (\mathrm{M}(n, \mathbf{C}),\ ^{t*}\ )$ is isomorphic to the (hyperbolic) involution algebra $\mathrm{M}(n, \mathbf{C}) \oplus \mathrm{M}(n, \mathbf{C})$ with the interchange involution $\overleftrightarrow{(a, b)} = (b^t, a^t)$.*

See 4, p. 24, for the tensor product of involution algebras.

If $e_{ij} \in \mathrm{M}(n, \mathbf{C})$ is the usual matrix unit, then $(\alpha \otimes \beta e_{ij})^{\mathrm{id}\otimes(t*)} = \alpha \otimes \beta^* e_{ji}$. Thus

$$(\mathbf{C}, \mathrm{id}) \otimes_{(\mathbf{R},\mathrm{id})} (\mathrm{M}(n, \mathbf{C}),\ ^{t*}\ ) \cong (\mathrm{M}(n, \mathbf{C} \otimes_{\mathbf{R}} \mathbf{C}),\ ^{t-}\ )$$

where the involution $^-$ on $\mathbf{C} \otimes_{\mathbf{R}} \mathbf{C}$ is

$$\overline{\alpha \otimes \beta} = \alpha \otimes \beta^*.$$

By the theorem and Prop. 1.2.3, $\mathbf{C} \otimes_{\mathbf{R}} \mathbf{C} \cong \mathbf{C} \oplus \mathbf{C}$ via the map $\alpha \otimes \beta \to (\alpha\beta, \alpha\beta^*)$ (since the image of any $\mathbf{R}$-linear homomorphism of $\mathbf{C}$ into an algebraic closure of $\mathbf{C}$ is $\mathbf{C}$ itself). This is a homomorphism of involution algebras if the involution on $\mathbf{C} \oplus \mathbf{C}$ is the interchange involution $\overleftrightarrow{(\alpha, \beta)} = (\beta, \alpha)$, and the corollary follows. □

Corollary 1.2.6 *Assume that L_1 and L_2 are finite Galois extensions of K, with Galois groups G_1 and G_2. Let $G = G_1 \times G_2$ act as automorphisms of the K-algebra $L_1 \otimes_K L_2$ via $(\sigma_1, \sigma_2)(\lambda_1 \otimes \lambda_2) = \sigma_1\lambda_1 \otimes \sigma_2\lambda_2$, and suppose G acts on $M_1 \oplus M_2 \oplus \cdots \oplus M_n$ via the isomorphism $\oplus_j(\varphi_{j1}\varphi_{j2})$ in (1.27). If $H_j \subset G$ is the stabilizer of the direct summand M_j, the restriction map $H_j \to \mathrm{Gal}(M_j/K)$ is onto.*

Let $\sigma \in \mathrm{Gal}(M_j/K)$, and define for $i = 1, 2$

$$\sigma_i = \varphi_{ji}^{-1} \left(\sigma|_{\varphi_{ji}L_i}\right) \varphi_{ji} \in \mathrm{Gal}(L_i/K).$$

The diagrams

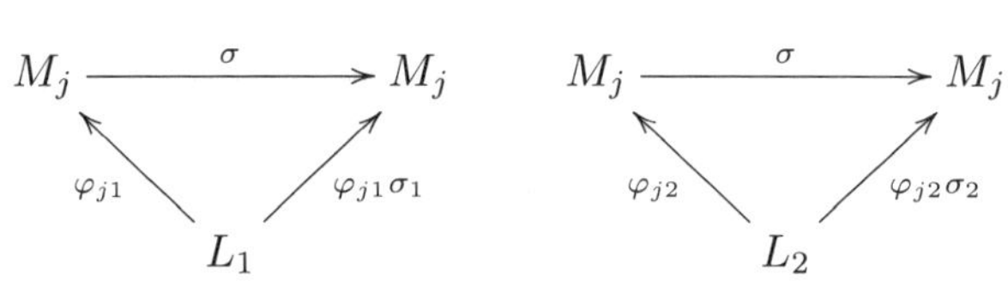

both commute and therefore the compositums $(M_j, \varphi_{j1}, \varphi_{j2})$ and $(M_j, \varphi_{j1}\sigma_1, \varphi_{j2}\sigma_2)$ are isomorphic. This means that the action of (σ_1, σ_2) on the direct sum stabilizes M_j, i.e. $(\sigma_1, \sigma_2) \in H_j$. The commutativity of the triangles also shows that the restrictions of σ to $\varphi_{j1}L_1$ and $\varphi_{j2}L_2$ are the automorphisms induced by (σ_1, σ_2), and so σ itself must be the automorphism on M_j induced by (σ_1, σ_2) since M_j is a compositum of L_1 and L_2. □

We mention here another related fact – see Prop. 8.3, p. 164, [53]:

Theorem 1.2.16 *Let L/K be a finite separable extension. Suppose that $\mathfrak{p}$ is a prime (equivalence class of valuations, Archimedean or nonArchimedean) of K and that $\mathfrak{P}_1, \dots, \mathfrak{P}_t$ are the primes of L which lie above $\mathfrak{p}$. Then*

$$K_\mathfrak{p} \otimes_K L \cong L_{\mathfrak{P}_1} \oplus \cdots \oplus L_{\mathfrak{P}_t} = \oplus_{\mathfrak{P}|\mathfrak{p}} L_\mathfrak{P}$$

as $K_\mathfrak{p}$-algebras, where $K_\mathfrak{p}$ and $L_{\mathfrak{P}_j}$ denote completions of these fields. If $\iota_i : K_\mathfrak{p} \to L_{\mathfrak{P}_i}$ is the canonical map, the isomorphism is given by

$$\alpha \otimes \lambda \to ((\iota_1\alpha)\lambda, \dots, (\iota_t\alpha)\lambda). \qquad \square$$

There is an important special case of this which arises later on – that of a quadratic extension L/K. In this case $t = 2$ or 1, so $K_\mathfrak{p} \otimes_K L = L_{\mathfrak{P}_1} \oplus L_{\mathfrak{P}_2}$ or $L_\mathfrak{P}$, depending on whether $\mathfrak{p}$ splits in L or not. In the first case $L_{\mathfrak{P}_1} = K_\mathfrak{p} = L_{\mathfrak{P}_2}$ and in the second, $L_\mathfrak{P}/K_\mathfrak{p}$ is a quadratic extension. Moreover it follows from the Global Square Theorem (p. 18) that the second case holds for an infinite number of $\mathfrak{p}$ when K is a global field.

Exercise. Suppose L/K is a finite Galois extension, with Galois group $G = \{\sigma_1 = 1, \sigma_2, \dots, \sigma_n\}$, and let L act on the first factor of $L \otimes_K L$ and G on the second. Suppose $\sigma \in G$ acts on $L \oplus L \oplus \cdots \oplus L$ via the permutation which takes the ith component to the jth where $\sigma_j\sigma = \sigma_i$. Show that the additive homomorphism

$$\varphi : L \otimes_K L \to L \oplus L \oplus \cdots \oplus L, \quad \varphi(\lambda \otimes \mu) = (\lambda\sigma_1(\mu), \lambda\sigma_2(\mu), \dots, \lambda\sigma_n(\mu)),$$

is an isomorphism of L-algebras commuting with the action of G, and that

$$\begin{array}{ccc} L \otimes_K L & \xrightarrow{\ \varphi\ } & L \oplus L \oplus \cdots \oplus L \\ {\scriptstyle \mathrm{id}\otimes\mathrm{Tr}_{L/K}}\downarrow & & \downarrow{\scriptstyle \mathrm{add}} \\ L \otimes_K K & \xrightarrow{\ \mathrm{can}\ } & L \end{array}$$

is a commutative diagram of L-vector spaces. $\square$

1.3 Involutions on algebras

A good treatment of this subject can be found in Ch. 8, [65], or in [36], especially §§1–8.

The center of a K-algebra $(A, ^-)$ with involution is stable under the involution, and $(A, ^-)$ (or the involution $^-$ itself) is said to be *of the first kind* if $^-$ is the identity on the center of A, otherwise *of the second kind.*

The *symmetric elements* respectively *skew symmetric elements* of $(A, ^-)$ are

$$(A, ^-)_+ = \{a \in A : \bar{a} = a\}, \quad (A, ^-)_- = \{a \in A : \bar{a} = -a\}.$$

They are sometimes denoted simply by A_+ and A_- if the involution is clear from the context, and are the ± 1-eigenspaces of the involution viewed as a K_0-linear transformation of A. Thus

$$A = (A, ^-)_+ \oplus (A, ^-)_-,$$

and it is clear that the involution is determined by A_+ and A_-.

An isomorphism $\varphi : (A, ^-) \cong (B, ^\sim)$ of $(K, ^-)$-involution algebras is a K-algebra isomorphism $\varphi : A \to B$ which commutes with the involutions, i.e. $\overline{\varphi(a)} = \widetilde{\varphi a}$ for all $a \in A$.

Examples. 1. The "transpose bar" $(\alpha_{ij}) \to (\bar{\alpha}_{ij})^t$ on $\mathrm{M}(n, K)$ is an involution. We usually denote it by ${}^{t-}$: $(\alpha_{ij})^{t-} = (\bar{\alpha}_{ij})^t = (\bar{\alpha}_{ji})$.

2. More generally if $(B, {}^-)$ is a $(K, {}^-)$-involution algebra, then "transpose bar"

$$(b_{ij})^{t-} = (\bar{b}_{ij})^t = (\bar{b}_{ji})$$

is a $(K, {}^-)$-involution on $\mathrm{M}(n, B)$; we write the resulting involution algebra as $\mathrm{M}(n, (B, {}^-))$ and call it the $n \times n$ *matrix (involution) algebra over* $(B, {}^-)$.

Exercise. Let $(R, {}^-)$ and $(S, {}^\sim)$ be rings with involution, and $\varphi : (R, {}^-) \to \mathrm{M}(m, (S, {}^\sim))$ a homomorphism. Define the map $\varphi_n : \mathrm{M}(n, R) \to \mathrm{M}(mn, S)$ in the obvious way, by replacing each entry r_{kl} of $(r_{ij}) \in \mathrm{M}(n, R)$ by its image $\varphi(r_{kl})$. Show that φ_n is a homomorphism $\mathrm{M}(n, (R, {}^-)) \to \mathrm{M}(mn, (S, {}^\sim))$ of rings with involution. □

3. Let b be an invertible element of a $(K, {}^-)$-involution algebra $(A, {}^-)$ satisfying $\bar{b}b^{-1} \in \mathrm{Z}(A)^\times$. Then

$$\widetilde{a} = b\bar{a}b^{-1}$$

defines another involution on A. If $\varepsilon = \bar{b}b^{-1}$, then $b = \overline{\varepsilon b} = \bar{\varepsilon}\varepsilon b$, so $\varepsilon\bar{\varepsilon} = 1$.

Suppose A is simple. Then $L = \mathrm{Z}(A)$ is a field – in fact a finite extension of K since we are assuming all algebras have finite dimension over the field of scalars. If ${}^-$ is of the first kind, $\varepsilon = \pm 1$. If ${}^-$ is of the second kind, we use Hilbert's Theorem 90 ((p. 281, [53]) to find $\eta \in L$ such that $\eta\bar{\eta}^{-1} = \varepsilon$; then if we replace b by ηb, the involution ${}^\sim$ does not change but $\bar{b} = b$; thus we can assume that $\varepsilon = 1$ in this case.

4. If $(A, {}^-)$ and $(B, {}^-)$ are $(K, {}^-)$-involution algebras, we define their sum to be

$$(A, {}^-) \oplus (B, {}^-) = (A \oplus B, {}^- \oplus {}^-),$$

and their tensor product to be

$$(A, {}^-) \otimes_{(K,{}^-)} (B, {}^-) = (A \otimes_K B, {}^- \otimes {}^-).$$

Both are involution algebras over $(K, {}^-)$ – the involution on the tensor product is given by

$$\overline{a \otimes b} = \bar{a} \otimes \bar{b}.$$

In particular if $(K, {}^-)$ is an "involution subfield" of $(L, {}^-)$ in the obvious sense, and if $(A, {}^-)$ is a $(K, {}^-)$-involution algebra, then $(L, {}^-) \otimes_{(K,{}^-)} (A, {}^-)$ is an $(L, {}^-)$-involution algebra, referred to as the involution algebra obtained from $(A, {}^-)$ by extending the scalars and sometimes denoted by $(A, {}^-)^{(L,{}^-)}$. If it is clear from the context what the involutions on K and L are, we sometimes write

$$(L, {}^-) \otimes_{(K,{}^-)} (A, {}^-) = L \otimes_K (A, {}^-).$$

Slightly more generally if $\varphi : (K, {}^-) \to (L, {}^-)$ is a homomorphism of fields with involution, one can also extend the scalars in $(A, {}^-)$ to $(L, {}^-)$ by making K act on L via $\lambda.\alpha = \lambda\varphi(\alpha)$ $(\alpha \in K, \lambda \in L)$. □

Exercises. 1. Let $(A, {}^-)$ and $(B, {}^\sim)$ be $(K, {}^-)$-involution algebras, and let $\varphi : B \otimes_K \mathrm{M}(n, A) \to \mathrm{M}(n, B \otimes_K A)$ be the canonical isomorphism of K-algebras. Show that φ is an isomorphism

$$(B, {}^\sim) \otimes_{(K,{}^-)} \mathrm{M}(n, (A, {}^-)) \to \mathrm{M}\Big(n, (B, {}^\sim) \otimes_{(K,{}^-)} (A, {}^-)\Big)$$

of involution algebras.

2. Let $(A, {}^-)$ be a $(K, {}^-)$-involution algebra, $(B, {}^-)$ both a $(K, {}^-)$ involution algebra and an $(L, {}^-)$-involution algebra where the action of the fields K and L commute, and $(C, {}^-)$ an $(L, {}^-)$-involution algebra. Show that the canonical isomorphisms $K \otimes_K A \to A$ and $(A \otimes_K B) \otimes_L C \to A \otimes_K (B \otimes_L C)$ are also isomorphisms

$$(K, {}^-) \otimes_{(K,{}^-)} (A, {}^-) \to (A, {}^-)$$
$$((A, {}^-) \otimes_{(K,{}^-)} (B, {}^-)) \otimes_{(L,{}^-)} (C, {}^-) \to (A, {}^-) \otimes_{(K,{}^-)} ((B, {}^-) \otimes_{(L,{}^-)} (C, {}^-)).$$

3. Let $K = K_0(\sqrt{\delta})$, with the nontrivial involution ${}^-$, and let $(L, {}^-)$ be another involution field extension of K_0, with $L_0 = (L, {}^-)_+$. Show that

$$(K, {}^-) \otimes_{(K_0,\mathrm{id})} (L_0, \mathrm{id}) \cong \begin{cases} ((L_0, \mathrm{id}) \oplus (L_0, \mathrm{id}), \leftrightarrow) & \text{if } \sqrt{\delta} \in L_0, \\ (L_0(\sqrt{\delta}), {}^-) = (L, {}^-) & \text{if } \sqrt{\delta} \in L \setminus L_0. \end{cases}$$

Also show that

$$(K, {}^-) \otimes_{(K_0,\mathrm{id})} (L, \mathrm{id}) = (L(\sqrt{\delta}), {}^{-\,-})$$

if $\sqrt{\delta} \notin L$, if ${}^{-\,-}$ is the involution $\alpha + \beta\sqrt{\delta} \to \bar{\alpha} + \bar{\beta}(-\sqrt{\delta})$ where $\alpha, \beta \in L$. Suggestion: use Th. 1.2.15, p. 21. □

5. If G is a finite group, the group algebra KG (cf. p. 15) has a canonical $(K, {}^-)$-involution, which we also denote by ${}^-$, given by

$$\overline{\sum_{s \in G} \alpha_s s} = \sum_{s \in G} \bar{\alpha}_s s^{-1}.$$

It is a $(K, {}^-)$-involution and we refer to it as the "standard $(K, {}^-)$-involution" on KG. This is the most important involution algebra in this book.

6. Conjugation * on a quaternion algebra $A = (\alpha, \beta)_K$ is an involution of the first kind whose spaces of symmetric and skew symmetric elements are $A_+ = K1$ and $A_- = Ki + Kj + Kk = A_0$, the "pure quaternions".

If K has an involution ${}^- \neq$ the identity, with fixed field K_0, then the quaternion algebra $(\alpha_0, \beta_0)_K$, $\alpha_0, \beta_0 \in K_0$ is canonically isomorphic – if we consider i, j, k as given in both $(\alpha_0, \beta_0)_K$ and $(\alpha_0, \beta_0)_{K_0}$ – to $K \otimes (\alpha_0, \beta_0)_{K_0}$ and so has a canonical involution of the second kind ${}^- \otimes {}^*$. Conversely (cf. Th. 8.11.2(ii), p. 314, [65]), if a quaternion algebra A with center K has an involution ${}^\sim$ of the second kind, then it has a K_0-subalgebra A_0 which is a quaternion algebra such that

$$(A, {}^\sim) \cong (K, {}^-) \otimes_{(K_0,\mathrm{id})} (A_0, {}^*).$$

Exercise. Determine $((\alpha_0, \beta_0)_K, {}^- \otimes {}^*)_+$ and $((\alpha_0, \beta_0)_K, {}^- \otimes {}^*)_-$. □

7. If $(K, {}^-)$ is field with involution and A is any K-algebra, define

$$\mathrm{H}(A) = A \oplus \overline{A^o},$$

where $\overline{A^o}$ is the opposite algebra A^o with the action of K twisted by its involution (cf. p. 4). Denote by a^o the element a of A considered as an element of $\overline{A^o}$ – thus

$$\alpha a^o = (\bar{\alpha} a)^o \text{ and } a_1^o a_2^o = (a_2 a_1)^o.$$

$\mathrm{H}(A)$ has a canonical involution given by

$$\overline{(a_1, a_2^o)} = (a_2, a_1^o),$$

which is easily checked to be a $(K, {}^-)$-involution. It is called the *hyperbolic* $(K, {}^-)$*-involution algebra based on* A.

Note that $\mathrm{H}(A) = A \oplus \bar{A}$ where here $\bar{A}$ is the image of A under the involution ${}^-$ of $\mathrm{H}(A)$. Moreover it is clear that if $(B, {}^-)$ is *any* $(K, {}^-)$-involution algebra which has a two-sided ideal A such that $B = A \oplus \bar{A}$, then $(B, {}^-) \cong \mathrm{H}(A)$ and so "is" hyperbolic.

When A is a simple K-algebra, $\mathrm{H}(A)$ is a *simple involution algebra*, in the sense that the only two-sided ideals which are stable under the involution are 0 and $\mathrm{H}(A)$. By contrast a *simple algebra with involution* will mean that the algebra itself is simple – and is of course a simple involution algebra.

Conversely a simple involution algebra $(A, {}^-)$ which is a semisimple algebra, is either simple as an algebra or a hyperbolic algebra based on a simple algebra – see p. 36.

8. Let $(B, {}^-)$ be a $(K, {}^-)$-involution algebra such that $B = A \oplus A'$ where the involution interchanges the two summands: $\bar{A} = A'$ and $\overline{A'} = A$. The map ${}^- : A' \to A$ is an isomorphism $A' \cong \overline{A^o}$ of K-algebras and $\mathrm{id}_A \oplus {}^- : (B, {}^-) \to (\mathrm{H}(A), {}^-)$ is an isomorphism of $(K, {}^-)$-involution algebras, so in particular $(B, {}^-)$ is hyperbolic.

This kind of involution algebra often occurs as a direct summand of $(KG, {}^-)$ and so plays an important role in this book.

A useful corollary of this example is that the tensor product

$$(C, {}^-) \otimes_{(K,{}^-)} (\mathrm{H}(A), {}^-)$$

is hyperbolic for any involution algebra $(C, {}^-)$.

9. Suppose that $(B, {}^-)$ is a $(K, {}^-)$-involution algebra. Then $B \oplus B$ has a $(K, {}^-)$-involution

$$\overleftrightarrow{(b_1, b_2)} = (\bar{b}_2, \bar{b}_1)$$

called the "interchange"[4]. The map $\varphi : B \to \overline{B^o}$, $\varphi(b) = (\bar{b})^o$, is an isomorphism of K-algebras, and

$$\mathrm{id} \oplus \varphi : (B \oplus B, \leftrightarrow) \to (B \oplus \overline{B^o}, {}^-) = (\mathrm{H}(B), {}^-)$$

is an isomorphism of $(K, {}^-)$-involution algebras. Thus in this case also, $(B \oplus B, \leftrightarrow)$ is hyperbolic. (A more precise notation would be

$$((B, {}^-) \oplus (B, {}^-), \leftrightarrow),$$

but this is usually not necessary).

An important special case is $(B, {}^-) = (K, {}^-)$.

Exercise. (a) Show that

$$(K \oplus K, \leftrightarrow) \otimes_{(K,{}^-)} (B, {}^-) \cong (B \oplus B, \leftrightarrow).$$

(b) More generally, show that if $(A, {}^-)$ is a $(K, {}^-)$-involution algebra,

$$(A \oplus A, \leftrightarrow) \otimes_{(K,{}^-)} (B, {}^-) \cong \big((A \otimes_{(K,{}^-)} B) \oplus (A \otimes_{(K,{}^-)} B), \leftrightarrow\big). \qquad \square$$

[4] The interchange involution (as it is also called in [43]) is called the "exchange involution" in [36].

Lemma 1.3.1 *Let $(L, {}^-)/(K, {}^-)$ be a (not necessarily finite) extension of involution fields, and assume $(K, {}^-) \neq (K, \mathrm{id})$. If A is a finite dimensional K-algebra, the canonical map*

$$L_0 \otimes_{K_0} A \to L \otimes_K A, \quad \lambda_0 \otimes_{K_0} a \to \lambda_0 \otimes_K a \tag{1.28}$$

is an isomorphism of L_0-algebras. Furthermore if $(A, {}^-)$ is a $(K, {}^-)$-involution algebra, (1.28) is an isomorphism

$$(L_0, \mathrm{id}) \otimes_{(K_0, \mathrm{id})} (A, {}^-) \cong (L, {}^-) \otimes_{(K, {}^-)} (A, {}^-)$$

of involution algebras.

Since $L_0 \otimes_{K_0} A$ and $L \otimes_K A$ have the same finite L_0-dimension, in order to prove that (1.28) is an isomorphism of L_0-vector spaces, it suffices to show that it is onto. Let $K = K_0(\sqrt{\delta})$. Then $L = L_0(\sqrt{\delta})$, and $(\alpha_0 + \beta_0\sqrt{\delta}) \otimes a$ $(\alpha_0, \beta_0 \in L_0)$ is the image of $\alpha_0 \otimes a + \beta_0 \otimes \sqrt{\delta}a$. That it is an isomorphism of involution algebras is easily checked. □

1.3.1 Involutions on simple algebras. Suppose now that A is a central simple K-algebra. Then there is a converse to Example 3 above, namely

10. Let ${}^-$ and ${}^\sim$ be $(K, {}^-)$-involutions on A. Their composition $\tilde{\bar{\ }}$ is a K-automorphism of A, hence by the Skolem-Noether Theorem (Th. 1.2.11) it is an inner automorphism, say $a \to bab^{-1}$. This implies that $\widetilde{a} = \bar{b}^{-1}\bar{a}\bar{b}$, and that $b^{-1}\bar{b} \in \dot{K}$ since ${}^\sim$ is an involution. As we saw in Example 3, this implies that $\bar{b} = \varepsilon b$ where $\varepsilon = \pm 1$ if ${}^-$ is of the first kind, and we can assume that $\varepsilon = 1$ if ${}^-$ is of the second kind. □

Suppose that A is central simple over K of dimension n^2 and has a $(K, {}^-)$-involution. Then there are precisely 3 possibilities for the dimensions of A_+ and A_- (cf. Th. 8.7.5, p. 303, [65]):

(a) If the involution is of the second kind, then

$$\dim_{K_0} A_+ = \dim_{K_0} A_- = n^2,$$

and the involution and the involution algebra $(A, {}^-))$ are said to be *unitary* or *of unitary type.* (If $(B, {}^-)$ is a simple algebra with involution, we also say that the hyperbolic simple involution algebra $(B \oplus B, {}^\leftrightarrow)$ is unitary).

(b) If the involution is of the first kind and

$$\dim_K A_+ = \tfrac{1}{2}n(n+1), \quad \dim_K A_- = \tfrac{1}{2}n(n-1),$$

the involution and the involution algebra $(A, {}^-)$ are said to be *orthogonal* or *of orthogonal type.*

(c) If the involution is of the first kind and

$$\dim_K A_+ = \tfrac{1}{2}n(n-1), \quad \dim_K A_- = \tfrac{1}{2}n(n+1),$$

the involution and the involution algebra $(A, {}^-)$ are said to be *symplectic* or *of symplectic type.*

If the involution is of the second kind, say $K = K_0(\sqrt{\delta})$, then multiplication by $\sqrt{\delta}$ interchanges the K_0-spaces A_+ and A_-, and so (a) follows at once from $A = A_+ \oplus A_-$.

In the case of an involution of the *first* kind, this is shown as follows: If $A = \mathrm{M}(n, K)$, by Examples 1 and 10 above the involution is of the form $\bar{a} = ba^tb^{-1}$

for some $b \in \mathbf{GL}(n, K)$ satisfying $b^t = \varepsilon b$ with $\varepsilon = \pm 1$. The condition $\bar{a} = a$ is equivalent to $(ab)^t = \varepsilon ab$, and it follows easily that its eigenspaces A_+ and A_- have the dimensions in (b) if $\varepsilon = 1$, i.e. if b is symmetric, and have the dimensions in (c) if $\varepsilon = -1$, i.e. if b is skew symmetric. (This shows of course how the terminology in (b) and (c) arises). The bilinear form[5] $\hat{b}(u, v) = u^t bv$ on $K^{n\times 1}$ is nonsingular and ε-symmetric, and it is easy to see that $^-$ is the *adjoint* with respect to $\hat{b}$, namely

$$\hat{b}(au, v) = \hat{b}(u, \bar{a}v) \quad \text{for all } a \in A,$$

(cf. (1.34), p. 46). If A is *not* split, the K-dimensions of the eigenspaces A_+ and A_- will of course be the same as the L-dimensions of $(A^L)_+$ and $(A^L)_-$ for any splitting extension L/K, namely those given in (b) and (c).

As examples, the identity on K and the transpose on $\mathrm{M}(n, K)$ are orthogonal involutions.

If A is simple but not necessarily central over K, $(A, ^-)$ is called *orthogonal, symplectic* or *unitary* if it is such over its center. The simple involution algebra $\mathrm{H}(B)$ based on the simple K-algebra B will also be called unitary. Thus if $(A, ^-)$ is a simple involution algebra, it is unitary if and only if $^-$ is not the identity on its center.

Theorem 1.3.1 *Let $(D, ^-)$ be a (possibly split) quaternion algebra over K with an involution of the first kind. The involution is symplectic if and only if it is conjugation. The conjugate transpose on $\mathrm{M}(n, D)$ is also symplectic.*

For a symplectic involution, D_+ and D_- have dimensions 1 and 3 respectively. Thus it is clear that conjugation on D is symplectic.

Conversely suppose $^-$ is any symplectic involution on the central quaternion K-algebra D. Since $^-$ is the identity on $K1_D$, $D_+ = K1_D$. On the other hand the 3-dimensional subspace $Ki + Kj + Kk$ of pure quaternions is stable under $^-$ since it is $= \{a \in D : a \notin K, a^2 \in K\}$. Thus it must $= D_-$ and so the involution is conjugation.

The last statement is a special case of the following lemma:

Lemma 1.3.2 *Let $(A, ^-)$ be a central simple algebra with involution over $(K, ^-)$. Then the type of $\mathrm{M}(r, (A, ^-))$ is the same as that of $(A, ^-)$.*

Clearly the "kind" is the same, so $(A, ^-)$ unitary implies that $\mathrm{M}(r, (A, ^-))$ is unitary. Suppose that $(A, ^-)$ is orthogonal and $\dim_K A = n^2$. Then the dimension of the "diagonal elements" fixed under $^{t-}$ + the dimension of the "off-diagonal elements" fixed by it is

$$r\tfrac{1}{2}n(n+1) + \tfrac{1}{2}r(r-1)n^2 = \tfrac{1}{2}rn(rn+1),$$

which implies that $\mathrm{M}(r, (A, ^-))$ is orthogonal since its dimension is $(rn)^2$. The proof for the symplectic case is similar. □

Exercises. 1. Let A be a central simple algebra over K with a $(K, ^-)$-involution. Say that the involution (or $(A, ^-)$) *has type* 1 if the involution is orthogonal, *type* -1 if it is symplectic, and *type* 0 if it is unitary. Show that the type of $(A^o, ^-)$ is the same as that of $(A, ^-)$, and if $(B, ^-)$ is another central simple algebra with a $(K, ^-)$-involution, that the type of $(A, ^-) \otimes_{(K,^-)} (B, ^-)$ is (type of $(A, ^-)$)(type of $(B, ^-)$).

[5] The basic notions of bilinear and sesquilinear forms are given in §1.5.1.

2. Show that the index of a simple algebra with an involution of the first kind is a power of 2. (Use the fact ([55], p. 254) that the exponent of A (its order in $\mathrm{Br}(\mathrm{Z}(A))$) and its index have the same prime factors).

3. Let $(\mathrm{M}(2,K),\,^-)$ be a symplectic involution algebra. Suppose that D is a quaternion algebra over K and that $\varphi : D \to \mathrm{M}(2,K)$ is an isomorphism of K-algebras. Show that φ is also an isomorphism $(D,\,^*\,) \to (\mathrm{M}(2,K),\,^-)$ of involution algebras. □

4. If $(A,\,^-)$ is symplectic, show that its degree is even.

Lemma 1.3.3 *Let $(A,\,^-)$ be a $(K,\,^-)$-involution algebra with A simple, and $(M,\,^-)/(K,\,^-)$ an extension of fields with involution. Suppose that*

$$(A^M,\,^-) = (M,\,^-) \otimes_{(K,^-)} (A,\,^-)$$

and that

$$A^M = B \oplus \bar{B} \oplus \cdots \tag{1.29}$$

where B (and $\bar{B}$) are simple algebras. Let V be an A-module (finitely generated, as usual). Then

$$\operatorname{len}_{A^M} BV^M = \operatorname{len}_{A^M} \bar{B}V^M. \tag{1.30}$$

We may assume that V is simple. Suppose that $B \cong \mathrm{M}(n,D)$, so $\operatorname{len}_{A^M} B = n$. Since $\bar{B} \cong B^o$ and $B^o \cong \mathrm{M}(n,D^o)$, $\operatorname{len}_{A^M} \bar{B}$ is also $= n$. Thus if we view (1.29) as the decomposition of the left A^M-module A^M into isotypic components, the isotypic components B and $\bar{B}$ have the same length as A^M-modules. On the other hand $V^m \cong A$ (as an A-module) for some m, so $(V^M)^m \cong A^M = B \oplus \bar{B} \oplus \cdots$, and therefore its isotypic components of type S_B (a simple B-module) and of type $S_{\bar{B}}$ have the same length, and therefore the same must be true of V^M, i.e. (1.30) holds. □

Theorem 1.3.2 *Let $(A,\,^-)$ be a separable simple $(K,\,^-)$-involution algebra and $(M,\,^-)/(K,\,^-)$ an extension of involution fields of the same kind. Then the simple (involution) factors of $(M,\,^-) \otimes_{(K,^-)} (A,\,^-)$ are separable M-algebras and have the same type as $(A,\,^-)$.*

Suppose first that $(A,\,^-)$ is hyperbolic, say $A = B \oplus \bar{B}$. Then B is a simple separable K-algebra. So $M \otimes_K B$ is semisimple and therefore is a direct sum of simple M-algebras. If L is the center of B, the center $M \otimes_K L$ of $M \otimes_K B$ is a direct sum of finite separable extensions of M since L/K is finite and separable (Th. 1.2.15, p. 21). and it follows that $(M,\,^-) \otimes_{(K,^-)} (A,\,^-)$ is a direct sum of simple separable hyperbolic M-algebras.

Therefore we may assume that A is simple (and separable). Then $M \otimes_K A$ is semisimple and so a direct sum of simple $(M,\,^-)$-involution algebras which are also separable as M-algebras by the foregoing argument. If L is the center of A, the center of $M \otimes_K A$ is

$$M \otimes_K L = L_1 \oplus \cdots \oplus L_k$$

where the L_i are finite separable extensions of M (*ibid*), and are the centers of the simple components of $M \otimes_K A$:

$$M \otimes_K A \cong (M \otimes_K L) \otimes_L A \cong (L_1 \otimes_L A) \oplus \cdots \oplus (L_k \otimes_L A).$$

If $1 \le i \le k$, the involution $^- \otimes ^-$ of $M \otimes_K A$ either stabilizes $L_i \otimes_L A$ or interchanges it with $L_j \otimes_L A$ for some $j \neq i$. Thus $(M,\,^-) \otimes_{(K,^-)} (A,\,^-)$ is a direct sum of separable simple involution algebras, and we must show that they are all of the same type as A.

If $(A, {}^{-})$ is unitary, the involution is nontrivial on its center L. Then either the involution interchanges $L_i \otimes_L A$ with $L_j \otimes_L A$, $j \neq i$, or $L_i \otimes_L A$ is stable and the involution is nontrivial on L_i since

$$L \to M \otimes_K L = L_1 \oplus \cdots \oplus L_k \xrightarrow{\text{proj.}} L_i$$

is a homomorphism of involution algebras. This proves the theorem in the unitary case (since hyperbolic algebras are considered to be unitary).

Suppose then that $(A, {}^{-})$ is orthogonal or symplectic. Since $(M, {}^{-})$ has the same kind as $(K, {}^{-})$ – namely the first kind – the involution on $M \otimes_K L$ is the identity, so also on each L_i. Thus $A_i = L_i \otimes_L A$ is stable under the involution and it follows immediately that

$$\dim_{L_i}(A_i)_+ = \dim_L A_+ \quad \text{and} \quad \dim_{L_i}(A_i)_- = \dim_L A_- \quad \text{for all } i,$$

whence the theorem. □

Theorem 1.3.3 *Let A and B be central simple algebras over K which are similar. Then A admits a $(K, {}^{-})$-involution if and only if B admits a $(K, {}^{-})$-involution. In particular if A has a $(K, {}^{-})$-involution, then so does the division algebra D such that $A \cong \mathrm{M}(n, D)$.*

See Corollary 8.3, p. 306, [65]. □

Theorem 1.3.4 *Let A be a central simple K-algebra with a $(K, {}^{-})$-involution, and B a simple subalgebra of A with a $(K, {}^{-})$-involution ${}^{\sim}$. Then ${}^{\sim}$ can be extended to an involution on A.*

This is Th. 10.1, p. 311 in [65]. Th. 4.14, p. 51, [36] is a somewhat sharper form of this theorem, giving the possible types of the extension. □

Theorem 1.3.5 *Let A be a simple separable K-algebra with an involution ${}^{-}$ of the first kind. If K is a global field and $K_\mathfrak{p}$ is the completion of K at a prime $\mathfrak{p}$, then $(K_\mathfrak{p}, \mathrm{id}) \otimes_{(K,\mathrm{id})} (A, {}^{-})$ is a direct sum of involution algebras of the same type (orthogonal or symplectic) as $(A, {}^{-})$ – namely if $\mathfrak{P}_1, \ldots, \mathfrak{P}_t$ are the primes of the center L of A lying above $\mathfrak{p}$, and for each i, ${}^{-}$ denotes the extension of the involution on A to an involution of the first kind on $A_{L_{\mathfrak{P}_i}}$, then*

$$(A_{K_\mathfrak{p}}, {}^{-}) \cong (A_{L_{\mathfrak{P}_1}}, {}^{-}) \oplus \cdots \oplus (A_{L_{\mathfrak{P}_t}}, {}^{-}).$$

By Th. 1.2.16, there is a canonical isomorphism $K_\mathfrak{p} \otimes_K L \cong L_{\mathfrak{P}_1} \oplus \cdots \oplus L_{\mathfrak{P}_t}$. Thus

$$K_\mathfrak{p} \otimes_K A \cong (K_\mathfrak{p} \otimes_K L) \otimes_L A \cong (L_{\mathfrak{P}_1} \otimes_L A) \oplus \cdots \oplus (L_{\mathfrak{P}_t} \otimes_L A).$$

Since the involution on $K_\mathfrak{p} \otimes_K L$ is the identity, the involution on the right side stabilizes all of the summands, and so the theorem follows by dimensions – for example, $\dim_{L_{\mathfrak{P}_i}}(L_{\mathfrak{P}_i} \otimes_L A)_+ = \dim_L A_+$ – or by Ex. 1 on p. 29. □

Theorem 1.3.6 *Let A be a central simple algebra with two $(K, {}^{-})$-involutions ${}^{-}$ and ${}^{\sim}$.*

(a) Then $\tilde{a} = b^{-1}\bar{a}b$ for some $b \in A$ such that $\bar{b} = \pm b$; if ${}^{-}$ is of the second kind, then b can be chosen so that $\bar{b} = b$.

(b) If $(A, {}^{-})$ is orthogonal, then the involution ${}^{\sim}$ is orthogonal if $\bar{b} = b$, symplectic if $\bar{b} = -b$.

(c) *If* $(A, ^-)$ *is symplectic, then the involution* $^\sim$ *is symplectic if* $\bar{b} = b$, *orthogonal if* $\bar{b} = -b$.

Part (a) is Example 10 above.

Suppose $\bar{b} = \varepsilon_0 b$ where $\varepsilon_0 = \pm 1$. Since $\widetilde{a} = \varepsilon a$ if and only if $\overline{ba} = \varepsilon_0 \varepsilon ba$, i.e. $a \in b^{-1} A^{(-)}_{\varepsilon_0\varepsilon}$, $\dim_K A^{(\sim)}_{\varepsilon} = \dim_K A^{(-)}_{\varepsilon_0\varepsilon}$ (where $A^{(-)}_+$, for example, is the space of elements of A which are symmetric with respect to $^-$). Thus the dimensions of the eigenspaces A_+ and A_- of $^\sim$ are the same as those of the corresponding eigenspaces of $^-$ if $\varepsilon_0 = 1$, and are interchanged if $\varepsilon_0 = -1$. □

Exercise. Let $(A, ^-)$ be a simple algebra with an involution of the first kind, and let $(B, ^-) = \mathrm{M}(n, (A, ^-))$. Suppose that $^\sim$ is a second involution of the first kind on B, and that $b_0 \in B^\times$ satisfies $\widetilde{b} = b_0 \bar{b} b_0^{-1}$ for all $b \in B$. Show that $\bar{b}_0 = \varepsilon_0 b_0$ where $\varepsilon_0 = \pm 1$, and that

$$\mathrm{type}(B, ^\sim) = \varepsilon_0 \mathrm{type}(A, ^-). \qquad \square$$

Corollary 1.3.1 (cf. Prop. 7.1, p. 81, [36]) *Let A be a central simple K-algebra of even degree with an orthogonal involution* $^-$. *If $a, b \in A_-$ are invertible elements of A, then* $\mathrm{nrd}\, a \equiv \mathrm{nrd}\, b \bmod \dot{K}^2$.

The involution $\widetilde{x} = a\bar{x}a^{-1}$ is symplectic by the theorem, and $\widetilde{ab} = ab$. Therefore it suffices to show that $\mathrm{nrd}\, c \in K^2$ for all $c \in A^{(\sim)}_+$. We do this by showing that the reduced characteristic polynomial $\mathrm{prd}_A c$ is a square in $K[X]$. The result then follows from (1.21), since the degree nd is even.

Let L be a Galois splitting field of A, $L \otimes_K A \cong \mathrm{M}(nd, L)$. It suffices to show that $\mathrm{prd}_A c$ is a square in $L[X]$, since the monic square root of $\mathrm{prd}_A c$ is invariant under the Galois group of L/K since $L[X]$ is a unique factorization domain, and so must be in $K[X]$. Therefore we can assume that A is split, $A = \mathrm{M}(n, K)$.

Since transpose is an orthogonal involution, there is a skew symmetric matrix r (in the usual sense, i.e. with respect to the transpose) so that $\widetilde{x} = rx^t r^{-1}$ for all $x \in A$. Let A_- denote the subspace of skew symmetric matrices, of dimension $\frac{1}{2}n(n-1)$. Then rA_- consists of matrices which are symmetric with respect to $^\sim$, and since it also has dimension $\frac{1}{2}n(n-1)$ and $^\sim$ is symplectic, it must be the *entire* space $A^{(\sim)}_+$ of matrices which are symmetric with respect to $^\sim$. Therefore there is $e \in A_-$ such that $c = re$, and so

$$\mathrm{prd}_A c = \det(X\mathrm{I}_n - c) = (\det r)\det(Xr^{-1} - e).$$

Since r and $Xr^{-1} - e$ are skew-symmetric, $\det r$ and $\det(Xr^{-1} - e)$ are squares, in $\mathrm{M}(n, K)$ and $\mathrm{M}(n, K[X])$ respectively, of their Pfaffians (cf. p. 45). □

This corollary makes it possible to define the determinant $\det_K(^-)$ and discriminant $\mathrm{disc}_K(^-) = (-1)^{nd/2}\det_K(^-)$ of an orthogonal involution $^-$ on a separable simple K-algebra of *even* degree nd:

$$\det_K(^-) = (\mathrm{nrd}_{A/K} a)\dot{K}^2 \in \dot{K}/\dot{K}^2, \quad \text{where } a \in A_- \cap A^\times$$

(cf. p. 81, [36]). If A is central simple, the determinant and discriminant are well-defined by the corollary, and if it is separable simple, say with center L, then they are well defined since $\mathrm{nrd}_{A/K} = \mathrm{N}_{L/K}\mathrm{nrd}_{A/L}$.

This definition will be used in Ch. 5 in the case of $A = \mathrm{M}(n, D)$ where D is either the field L or a quaternion algebra over K. In the latter case $\bar{a} = h_0^{-1} a^{t*} h_0$ where $h_0^{t*} = -h_0$ by Th. 1.3.6(c), and so

$$\det_K(^-) =_2 \mathrm{nrd}_{A/K} h_0$$

since $\bar{h}_0 = -h_0$ also. Suppose that $A = \mathrm{M}(n, K)$. Note that $\det_K(^t) =_2 1$ since a skew symmetric matrix has square determinant (cf. p. 45). Write $\bar{a} = h_0^{-1} a^t h_0$ where $h_0^t = h_0$ (Th. 1.3.6(b)). Then it is straightforward to check that

$$A_-^{(-)} = h_0^{-1} A_-^{(t)},$$

whence

$$\det_K(^-) =_2 \mathrm{nrd}_{A/K} h_0^{-1} =_2 \mathrm{nrd}_{A/K} h_0.$$

This formula also holds when $A = \mathrm{M}(n, L)$ where L/K is a finite separable extension.

Two further definitions for orthogonal involutions used in Ch. 5 are of the Hasse and Hasse-Witt invariants, albeit in more special circumstances, namely when A is split of even degree and $\mathrm{disc}(^-) = 1$ for the Hasse invariant, and when A is split of odd degree in the case of the Hasse-Witt invariant. Thus when $A \cong \mathrm{M}(n, K)$, $\mathrm{disc}(^-) = 1$ and n is even, we define

$$\mathrm{s}(^-) = \mathrm{s}(\langle h_0 \rangle). \tag{1.31}$$

(See p. 53 for the definition of the Hasse invariant $\mathrm{s}(b)$ of a symmetric bilinear form.) This is well-defined since, by Prop. 1.5.1(a) (p. 53), if $\alpha \in \dot{K}$

$$\begin{aligned} \mathrm{s}(\langle \alpha h_0 \rangle) &= \mathrm{s}(\langle h_0 \rangle)(\alpha, -1)_K^{n(n-1)/2} (\alpha, \det h_0)_K^{n-1} \\ &= \mathrm{s}(\langle h_0 \rangle)(\alpha, \mathrm{disc}\, h_0)_K = \mathrm{s}(\langle h_0 \rangle). \end{aligned}$$

If K is a global field and $\mathfrak{p}$ a prime of K, $\mathrm{s}_\mathfrak{p}(^-)$ stands for the Hasse invariant $\mathrm{s}(^-)$ of the involution of $\mathrm{M}(n, K_\mathfrak{p})$.

If $A \cong \mathrm{M}(n, K)$, and n is odd, we define the Hasse-Witt invariant

$$\mathrm{w}(^-) = \mathrm{w}(\langle h_0 \rangle).$$

This is well-defined by Lem. 1.5.2, p. 55. In fact it is also well-defined when n is even and $\mathrm{disc}(^-) = 1$.

If it is necessary to make the field K explicit, we will of course write $\mathrm{s}_K(^-)$ and $\mathrm{w}_K(^-)$.

One can similarly define the determinant of a unitary involution $^-$ on a simple separable K-algebra $A \cong \mathrm{M}(n, D)$ when n itself is even: identify A with $\mathrm{M}(n, D)$ and write $\bar{a} = h_0^{-1} a^{t-} h_0$, where $^-$ also denotes an $(L, ^-)$-involution on D, L is the center of A (and D), and $h_0^{t-} = h_0$ (possible by Hilbert's Theorem 90 – cf. p. 24). By Lem. 1.3.7 (p. 34), $\mathrm{nrd}_{A/K} h_0 \in \dot{K}_0$. Define

$$\det_K(^-) = (\mathrm{nrd}_{A/K} h_0) \mathrm{N}_{K/K_0} \dot{K} \in \dot{K}_0 / \mathrm{N}_{K/K_0} \dot{K}.$$

Lemma 1.3.4 *Let $(K, ^-)$ be of the second kind. If $(A, ^-)$ is a separable simple K-algebra $\cong \mathrm{M}(n, D)$ with n even and $(K, ^-)$-involution $^-$, then $\det_K(^-)$ is well-defined.*

It suffices to prove the lemma when A is central simple since $\mathrm{nrd}_{A/K} = \mathrm{N}_{L/K} \circ \mathrm{nrd}_{A/L}$ and, by Lem. 1.3.5 below, $\mathrm{N}_{L/K} L_0 \subset \mathrm{K}_0$ where L_0 and K_0 are, as usual, the fixed elements of L and K under the involution.

Fix the identification $A = \mathrm{M}(n, D)$ for the moment. Any other choice of h_0 is of the form αh_0 where $\alpha \in \dot{K}$. But $\alpha h_0 = (\alpha h_0)^{t-} = \bar{\alpha} h_0^{t-} = \bar{\alpha} h_0$ and so $\alpha \in \dot{K}_0$. Let $\hat{K}$ be a splitting field of D, so $\hat{K} \otimes_K \mathrm{M}(n, D) \cong \mathrm{M}(nd, \hat{K})$ where $\dim_K D = d^2$. Under this isomorphism, $\alpha \mathrm{I}_n \in \mathrm{M}(n, D)$ corresponds to $\alpha \mathrm{I}_{nd}$, and so $\mathrm{nrd}_{A/K} \alpha \mathrm{I}_n = \alpha^{nd}, \in \mathrm{N}_{K/K_0} \dot{K}$ since $K_0^2 \subset \mathrm{N}_{K/K_0} K$ and nd is even. Thus $\det_K(-)$ is independent of the choice of h_0.

Suppose $(A, ^-) \to (\mathrm{M}(n, D), {}^{t-}\mathrm{inn}_{g_0^{-1}})$, $g_0^{t-} = g_0$, is another identification. It gives rise to an isomorphism

$$\mathrm{inn}_c : (\mathrm{M}(n, D), {}^{t-}\mathrm{inn}_{g_0^{-1}}) \to (\mathrm{M}(n, D), {}^{t-}\mathrm{inn}_{h_0^{-1}})$$

of involution algebras, which implies that

$$c g_0^{-1} a^{t-} g_0 c^{-1} = h_0^{-1} (c a c^{-1})^{t-} h_0 = h_0^{-1} (c^{-1})^{t-} a^{t-} c^{t-} h_0 \ \text{ for all } a \in \mathrm{M}(n, D),$$

and this in turn implies that, for some $\alpha \in \dot{K}$,

$$\alpha g_0 c^{-1} = c^{t-} h_0, \text{ i.e. } \alpha g_0 = c^{t-} h_0 c.$$

But $(c^{t-} h_0 c)^{t-} = c^{t-} h_0 c$ so $\alpha \in \dot{K}_0$. Also $\mathrm{nrd}_{A/K}(c^{t-}) = \overline{\mathrm{nrd}_{A/K} c}$ by Lem. 1.3.7, and so

$$\mathrm{nrd}_{A/K} g_0 \mathrm{N}_{K/K_0} \dot{K} = \mathrm{nrd}_{A/K} h_0 \mathrm{N}_{K/K_0} \dot{K}.$$

Thus $\det(^-)$ does not depend on the identification $A = \mathrm{M}(n, D)$.

It remains to show that it is also independent of the choice of the involution $^-$ on D. Suppose $^\sim$ is another $(K, ^-)$-involution on D, say $\widetilde{d} = c\bar{d}c^{-1}$ for all $d \in D$. Again we can assume that $\bar{c} = c$ by Hilbert's Theorem 90, so $\mathrm{nrd}_{D/K} c \in \dot{K}_0$. If $a = (d_{ij}) \in \mathrm{M}(n, D)$,

$$\begin{aligned} \bar{a} = h_0^{-1} (d_{ij})^{t-} h_0 &= h_0^{-1} (\bar{d}_{ji}) h_0 \\ &= h_0^{-1} (c^{-1} \widetilde{d}_{ji} c) h_0 = (c\mathrm{I}_n h_0)^{-1} (d_{ij})^{t\sim} (c\mathrm{I}_n h_0). \end{aligned}$$

Again using the splitting field $\hat{K}$ to calculate reduced norms, we see that

$$\mathrm{nrd}_{A/K}(c\mathrm{I}_n) = (\mathrm{nrd}_{D/K} c)^n$$

and so

$$\mathrm{nrd}_{A/K}(c\mathrm{I}_n h_0) \mathrm{N}_{K/K_0} \dot{K} = (\mathrm{nrd}_{D/K} c)^n (\mathrm{nrd}_{A/K} h_0) \mathrm{N}_{K/K_0} \dot{K} = (\mathrm{nrd}_{A/K} h_0) \mathrm{N}_{K/K_0} \dot{K}$$

since n is even and $\mathrm{nrd}_{D/K} c \in \dot{K}_0$. □

Lemma 1.3.5 *If L/K is a finite separable extension and $^-$ is an involution on L such that $\bar{K} = K$, then*

$$\mathrm{Tr}_{L/K} \bar{\lambda} = \overline{\mathrm{Tr}_{L/K} \lambda} \quad \textit{and} \quad \mathrm{N}_{L/K} \bar{\lambda} = \overline{\mathrm{N}_{L/K} \lambda}.$$

This is a simple exercise, using the definition of $\mathrm{Tr}_{L/K}$ and $\mathrm{N}_{L/K}$ in terms of the trace and determinant of multiplication $\alpha \to \lambda\alpha$ on L – and the fact that if $\omega_1, \ldots, \omega_l$ is a basis of L/K, then so is $\bar{\omega}_1, \ldots, \omega_l$. □

Lemma 1.3.6 *Suppose A is a central simple K-algebra with $(K,\bar{\ })$-involutions $\sim$ and $\bar{\ }$. Then $(A, \sim) \cong (A, \bar{\ })$ if and only if there is a unit $b \in A^\times$ such that $\widetilde{a} = (b\bar{b})\bar{a}(b\bar{b})^{-1}$ for all $a \in A$.*

Suppose the involution algebras are isomorphic: by the Skolem-Noether Theorem, there is $b \in A^\times$ such that

$$b^{-1}\widetilde{a}b = \overline{b^{-1}ab} = \bar{b}\bar{a}\bar{b}^{-1} \quad \text{for all } a \in A.$$

This proves the necessity and the sufficiency is also clear from this expression. □

Theorem 1.3.7 *Let A be a central simple K-algebra.*

(a) ("Albert's Theorem") A admits an involution of the first kind if and only if it is isomorphic to its opposite algebra A^o, i.e. the Brauer class of A has order 1 or 2.

(b) [6] If the involution on K is $\neq$ the identity, A admits an involution of the second kind if and only if the "corestriction" $\operatorname{Cor} A$ over the fixed field K_0 splits.

(c) If A admits a $(K,\bar{\ })$-involution, then every algebra similar to A (in particular the division algebra in the Brauer class of A) admits a $(K,\bar{\ })$-involution.

See p. 306, 8.8.3, 8.8.4, and p. 309, 8.9.5, in [65], or Th. 3.1, p. 31, [36]. □

If K is a local field, a (noncommutative) division algebra D over K does not admit an involution of the second kind (Theorem 2.2 (ii), p. 353, [65]). Furthermore the only noncommutative central division algebra over K with an involution of the *first* kind is the quaternions. In fact any such algebra is isomorphic to its opposite algebra and so of order 2 in the Brauer group; and since the latter is $\cong \mathbb{Q}/\mathbb{Z}$ (cf. p. 17), there is only one element of order 2.

Lemma 1.3.7 (2.2 and 2.16, [36]) *Let $(A,\bar{\ })$ be a simple separable K-algebra with a $(K,\bar{\ })$-involution. Then*

$$\overline{\operatorname{trd}_{A/K}a} = \operatorname{trd}_{A/K}\bar{a}, \quad \overline{\operatorname{nrd}_{A/K}a} = \operatorname{nrd}_{A/K}\bar{a}.$$

Suppose first that K is the center of A. Let M be a splitting field of A which is Galois over the subfield K_0 of K of elements fixed by $\bar{\ }$, and $\varphi : M \otimes_K A \to \mathrm{M}(n, M)$ an M-isomorphism. Choose $\sigma \in \mathrm{Gal}(M/K_0)$ which restricts to $\bar{\ }$ on K. Then

$$M \times A \to M \otimes_K A, \quad (\lambda, a) \to \sigma(\lambda) \otimes \bar{a},$$

is biadditive and balanced over K, so we get a homomorphism

$$\Sigma : M \otimes_K A \to M \otimes_K A, \quad \Sigma(\lambda \otimes a) = \sigma(\lambda) \otimes \bar{a},$$

of Abelian groups. It is in fact a σ-linear anti-automorphism of M-algebras. Now consider

$$\begin{array}{ccccc} M \otimes A & \xrightarrow{\Sigma} & M \otimes A & \xrightarrow{i} & M \otimes A^o \\ \downarrow{\scriptstyle\varphi} & & & & \vdots{\scriptstyle\psi} \\ \mathrm{M}(n, M) & & \xrightarrow{\hat{\sigma}} & & \mathrm{M}(n, M) \end{array}$$

where $\hat{\sigma}$ is the map which applies σ to every entry of the matrix, and $i(\lambda \otimes a) = \lambda \otimes a^o$. Since the four undotted maps are bijective, there is a unique map ψ making the diagram commutative; it is an isomorphism of M-algebras. Thus $\psi(1 \otimes a^o) =$

[6] This part is not used in this book, so we will not define "corestriction".

$\hat{\sigma}\varphi\Sigma^{-1}i^{-1}(1\otimes a^o) = \hat{\sigma}\varphi(1\otimes\bar{a})$, and so since the matrix trace $\mathrm{tr}(\varphi(1\otimes\bar{a})) = \mathrm{trd}_{A/K}\bar{a}$ lies in K,

$$\begin{aligned}\overline{\mathrm{trd}_{A/K}\bar{a}} = \overline{\mathrm{tr}(\varphi(1\otimes\bar{a}))} &= \sigma(\mathrm{tr}(\varphi(1\otimes\bar{a}))) \\ &= \mathrm{tr}(\hat{\sigma}\varphi(1\otimes\bar{a})) = \mathrm{tr}(\psi(1\otimes a^o)) \\ &= \mathrm{trd}_{A^o/K}a^o = \mathrm{trd}_{A/K}a.\end{aligned}$$

The last step follows from Lem. 1.2.3, p. 14.

If A is not central, one applies Lem. 1.3.5 to the central case just proved. □

The property $\overline{\mathrm{trd}_{A/K}a} = \mathrm{trd}_{A/K}\bar{a}$ is important in the subject of isometry representations, and we make the following definition:

Definition 1.3.1 *If* $(A, ^-)$ *is a* $(K, ^-)$*-involution algebra, an involution trace*

$$\mathrm{t} : (A, ^-) \to (K, ^-)$$

is a K*-linear map* $\mathrm{t} : A \to K$ *which commutes with the involutions:*

$$\mathrm{t}(\bar{a}) = \overline{\mathrm{t}(a)} \quad \textit{for all } a \in A,$$

is symmetric – i.e. $\mathrm{t}(ab) = \mathrm{t}(ba)$ *for all* $a, b \in A$ *– and is nonsingular in the sense that the (symmetric or Hermitian) form* $\mathrm{t}(a, b) = \mathrm{t}(a\bar{b})$ *on* A *is nonsingular.*

The form $\mathrm{t}(a, b)$ is called the *trace form* of the algebra A/K – or more precisely of the involution trace t. (The basic properties of symmetric and Hermitian forms are dealt with in §1.5).

Examples 1. To show that $\mathrm{t} = \mathrm{trd}_{A/K}$ is an involution trace when A is a separable algebra, we begin with the simple case. It remains to check that t is nonsingular since symmetry follows by (1.11), p. 13. It suffices to show that it is nondegenerate ($\mathrm{t}(A, a) = 0 \Rightarrow a = 0$) since this means that the right adjoint $\mathrm{t}_r : A \to \overline{A^*}$ is injective and therefore bijective. The kernel

$$\{a \in A : \mathrm{t}(Aa) = 0\}$$

is a two-sided ideal $\neq A$ and so must be 0, hence $\mathrm{trd}_{A/K}$ is indeed nonsingular.

Now suppose that $(A, ^-)$ is a separable algebra such that $A = B \oplus \bar{B}$ where B is a simple K-algebra. By Example 8, p. 26, $(A, ^-)$ is hyperbolic, and by the foregoing argument, the right adjoint t_r is injective on both B and $\bar{B}$, and since it takes them to $(\bar{B})^*$ and B^* respectively, it is injective on A itself. Thus $\mathrm{t} = \mathrm{trd}_{A/K}$ is an involution trace in this case too.

It follows now that the reduced trace is also an involution trace when A is a separable K-algebra. For in this case $(A, ^-)$ is a direct sum of separable simple algebras with $(K, ^-)$-involutions and separable hyperbolic algebras – see the next section.

2. If $(L, ^-)/(K, ^-)$ is a finite separable extension of fields with involution, then $\mathrm{Tr}_{L/K}$ is an involution trace – this is a special case of Example 1.

3. Let $A = KG$ with the standard $(K, ^-)$-involution. Then the canonical trace

$$\mathrm{Tr}_{KG/K}(\sum_s \alpha_s s) = \alpha_1$$

is an involution trace. It is clear that it commutes with the involutions and symmetry follows from

$$\mathrm{Tr}_{KG/K}\left((\sum_s \alpha_s s)(\sum_s \beta_s s)\right) = \sum_s \alpha_s \beta_{s^{-1}}.$$

Nonsingularity results from the fact that G is an orthonormal basis of KG with respect to $\mathrm{Tr}(a\bar{b})$: if $s, t \in G$, $\mathrm{Tr}(s\bar{t}) = 1$ if $s = t$ and $= 0$ if $s \neq t$. It also follows from Example 1 and Th. 1.2.13 (p. 15).

4. Suppose that $\mathrm{t} : (A, ^-) \to (K, ^-)$ is an involution trace, and that the involution algebra $(B, ^-)$ is a direct summand of $(A, ^-)$. Then $\mathrm{t}|_B : (B, ^-) \to (K, ^-)$ is an involution trace. Only the last property of involution trace needs verification; and this is clear since if $(A, ^-) = (B, ^-) \oplus (C, ^-)$, then as reflexive spaces over $(K, ^-)$, $A = B \perp C$ with respect to the nonsingular form $\mathrm{t}(a, a')$. □

1.3.2 Involutions on semisimple algebras. Now suppose that A is a semisimple K-algebra with $(K, ^-)$-involution $^-$. Let

$$A = A_1 \oplus A_2 \oplus \cdots \oplus A_h$$

be the decomposition of A into its simple (algebra) components (cf. Th. 1.2.7). Since $\bar{A}_i$ is also a minimal nonzero 2-sided ideal of A, hence one of the A_j, the involution permutes these simple algebras, and so A is the direct sum of two kinds of algebras – those A_i which are stable under the involution, and algebras of the form $A_i \oplus \bar{A}_i$. Thus these components of A, in either case, are simple $(K, ^-)$-involution algebras (cf. p. 26). An algebra of the form $A_i \oplus \bar{A}_i$ is (isomorphic to) the hyperbolic (involution) algebra $\mathrm{H}(A_i)$ by Ex. 8, p. 26.

In particular a simple involution algebra $(A, ^-)$ which is semisimple as an algebra, is either a simple algebra or a hyperbolic algebra $\mathrm{H}(B)$ on a simple algebra B.

Therefore after a possible renumbering of the A_i, we get a splitting

$$(A, ^-) = (A_1, ^-) \oplus \cdots \oplus (A_k, ^-) \oplus \mathrm{H}(A_{k+1}) \oplus \cdots \oplus \mathrm{H}(A_l)$$

into simple involution algebras. These direct summands are called the *simple (involution) components* of $(A, ^-)$. This will generally be written as (with another change in notation)

$$(A, ^-) = (A_1, ^-) \oplus \cdots \oplus (A_k, ^-) \oplus (A_{k+1}, ^-) \oplus \cdots \oplus (A_l, ^-) \tag{1.32}$$

where now $A_1, \ldots, A_k$ are still simple algebras and $(A_{k+1}, ^-), \ldots, (A_l, ^-)$ are simple involution algebras which are hyperbolic.

1.4 Formally real fields

A *formally real field* is a field in which -1 is *not* a sum of squares. In this case we say that the *level* of K, $s(K)$, is ∞. If K is *not* formally real, the level is the least integer n such that -1 is a sum of n squares in K.

If K is a finite field with q elements, then

$$s(K) = \begin{cases} 1 & q \equiv 1 \bmod 4 \text{ or is a power of } 2, \\ 2 & q \equiv 3 \bmod 4, \end{cases}$$

(cf. p.71, [65]). Thus a formally real field has characteristic 0.

Let K be a local field. By Examples 2.4, p. 380, [39], if p is the characteristic of its residue class field,

$$s(K) = \begin{cases} 1 & p \equiv 1 \bmod 4 \text{ or } \sqrt{-1} \in K, \\ 2 & p \equiv 3 \bmod 4 \text{ and } \sqrt{-1} \notin K, \\ 4 & p = 2, \ (K : \mathbb{Q}_2) \text{ odd}, \\ 2 & p = 2, \ (K : \mathbb{Q}_2) \text{ even and } \sqrt{-1} \notin K, \end{cases}$$

It is a theorem of A. Pfister (p. 379, *ibid*) that the level of a field is a power of 2 or ∞.

An *ordering* of the field K is a subset $P \subset \dot{K}$ such that
(a) $\alpha, \beta \in P$ implies $\alpha + \beta \in P$.
(b) $\alpha, \beta \in P$ implies $\alpha\beta \in P$.
(c) $P \cup -P = \dot{K}$

Note that $P \cap (-P) = \emptyset$ since if α were in the intersection, then α and $-\alpha \in P$, so $0 = \alpha - \alpha \in P$.

An *ordered field* is a field K with a given ordering P – more formally a pair (K, P). The elements of P are called positive and those of $-P$ negative. An ordering in the usual sense of the word is defined by saying that $\alpha > \beta$ if $\alpha - \beta \in P$. It has the familiar properties:

$\alpha, \beta > 0$ implies that $\alpha + \beta$ and $\alpha\beta$ are > 0.
If $\alpha, \beta \in K$, then either $\alpha < \beta, \alpha = \beta$, or $\alpha > \beta$.
$\alpha > \beta > \gamma$ implies that $\alpha > \gamma$.
$\alpha > \beta,\ \gamma > 0$ implies that $\alpha\gamma > \beta\gamma$.
$\alpha \neq 0$ implies that $\alpha^2 > 0$.

If it is not clear from the context which ordering P is being used to define $>$, we write $\alpha >_P \beta$ or say that $\alpha > \beta$ at P.

Details on orderings and formally real fields can be found in [65], especially §4, §6 and §10 of Ch. 2, and Ch. 3, and also in Ch. VIII, [39]. We list here some of their basic properties:

1. Suppose K is ordered. Then any sum of squares $\sum \alpha_i^2 + 1^2$ is $\in P$, hence is nonzero, so -1 is not a sum of squares. Thus an ordered field is formally real. The converse is also true:

Theorem 1.4.1 Artin-Schreier. *A formally real field has at least one ordering. The set of all sums of squares of nonzero elements of a formally real field K is the intersection $\cap P$ of all orderings of K.* □

2. Let P be an ordering on K, and let $b : V \times V \to K$ be a nonsingular symmetric bilinear form. Choose an orthogonal basis $v_1, v_2, \dots, v_n$ for V over K. Let $r^+(b) =$ the number of $b(v_j, v_j)$ which are > 0 at P, and $r^-(b) =$ the number which are < 0. Thus $r^+(b) + r^-(b) = n$. Define the *signature* of b (at P) to be $\text{sgn}(b) = \text{sgn}_P(b) = r^+(b) - r^-(b)$.

Theorem 1.4.2 Inertia Theorem of Jacobi and Sylvester. $\text{sgn}_P(b)$ *is independent of the basis used to define it.* □

The positive index $r^+(b)$ and the negative index $r^-(b)$ are also independent of the choice of basis. b is called *positive definite* if $r^+(b) = n$, i.e. if $r^-(b) = 0$ or if $\text{sgn}_P(b) = n$, and *negative definite* if $r^-(b) = n$ or $\text{sgn}_P(b) = -n$.

3. The rationals $\mathbb{Q}$ and reals $\mathbb{R}$ have only one ordering, the usual ones.

4. The field K is called *Euclidean* if the set of nonzero squares $\dot{K}^2$ is an ordering of K – and therefore the *only* ordering of K. In particular then any sum of squares is a square. When K is Euclidean, two nonsingular symmetric bilinear forms are equivalent if and only if their ranks and their signatures are equal (cf. 4.8, p. 42, [65]). And the signature provides an isomorphism

$$\mathrm{sgn}_{\dot{K}^2} : \mathrm{W}_1(K) \to \mathbb{Z}$$

of the Witt ring.

5. If K has an ordering P, it can be extended to the field $K(x)$ of rational functions. For example we can define $f(x)/g(x)$ to be positive if $f(x)g(x)$ has positive leading coefficient. This implies then that the ordering on K can be extended to *any* purely transcendental extension. □

A formally real field K is *real closed* if no proper algebraic extension of K is formally real. This is the same as saying that there is no extension of an ordering on K to any proper algebraic extension. In this case the set of nonzero squares $\dot{K}^2$ is an ordering in K, so in particular a real closed field is Euclidean (3.1.14, p. 111, [65]) and remark 4. above applies.

An extension $\mathbf{R}$ of a formally real field K is a *real closure* of K if it is real closed and algebraic over K. It follows from Zorn's Lemma that, if P is an ordering of K, there is a real closure $\mathbf{R}$ in any algebraically closed extension whose ordering $\dot{\mathbf{R}}^2$ is an extension of P – i.e. $P = \dot{\mathbf{R}}^2 \cap K$; it is called a real closure of K at P.

If $\mathbf{R}$ is real closed, $\mathbf{R}(i)$, where $i^2 = -1$, is an algebraic closure of $\mathbf{R}$. If K_P is a real closure of K at P, $K_P(i)$ is called an *algebraic closure of K at P*.

Theorem 1.4.3 (Artin-Schreier) *If $\mathbf{R}_1$ and $\mathbf{R}_2$ are real closures for the same ordering of the field K (so $K \cap \dot{\mathbf{R}}_1^2 = K \cap \dot{\mathbf{R}}_2^2$), there is a unique K-isomorphism $\mathbf{R}_1 \cong \mathbf{R}_2$.* □

Real closures $\mathbf{R}_1$ and $\mathbf{R}_2$ for *different* orderings P_1 and P_2 of K cannot be K-isomorphic, since such an isomorphism would carry the ordering $\dot{\mathbf{R}}_1^2$ onto the ordering $\dot{\mathbf{R}}_2^2$, and hence the orderings $\dot{\mathbf{R}}_1^2 \cap K = P_1$ and $\dot{\mathbf{R}}_2^2 \cap K = P_2$ would be equal. Thus there is a canonical bijection between the orderings of K and the K-isomorphism classes of real closures of K.

Theorem 1.4.4 *(p. 113, ibid) The following are equivalent:*
(a) K is real closed.
(b) K is Euclidean and every polynomial of odd degree in $K[X]$ has a zero in K.
(c) K is not algebraically closed but $K(\sqrt{-1})$ is. □

Theorem 1.4.5 *If $\mathbf{C}$ is an algebraically closed field of characteristic $\neq 2$ which has an automorphism of order 2, then the fixed field of the automorphism is a real closed field $\mathbf{R}$, and $\mathbf{C} = \mathbf{R}(i)$ where $i^2 = -1$.*

Since $(\mathbf{C} : \mathbf{R}) = 2$, $\mathbf{R}$ contains nonsquares; let δ be one of them. Then $\mathbf{C} = \mathbf{R}(\sqrt{\delta})$. Let $\alpha + \beta\sqrt{\delta}$ $(\alpha, \beta \in \mathbf{R})$ be a fourth root of δ in $\mathbf{C}$. Then $\alpha \neq 0 \neq \beta$ and

$$(\alpha + \beta\sqrt{\delta})^4 = (\alpha^4 + 6\alpha^2\beta^2\delta + \beta^4\delta^2) + 4\alpha\beta(\alpha^2 + \beta^2\delta)\sqrt{\delta} \in \mathbf{R}.$$

This implies that $\alpha^2+\beta^2\delta=0$, so $\delta=-\frac{\alpha^2}{\beta^2}\in-\dot{\mathbf{R}}^2$. Thus $\mathbf{R}$ is the union of its squares and their negatives, and $-1\notin\dot{\mathbf{R}}^2$.

We next show that $P=\dot{\mathbf{R}}^2$ is an ordering on $\mathbf{R}$. First of all $P\cap -P=\emptyset$ since otherwise we would have an equality $\beta^2=-\gamma^2$ with $\beta,\gamma\in\dot{\mathbf{R}}$, which is impossible since -1 is not a square in $\mathbf{R}$. Of course $PP\subset P$, so it remains to show that $P+P\subset P$, in other words the sum of 2 nonzero squares $\alpha^2+\beta^2$ is a square. Consider the polynomial

$$\begin{aligned}&(x+\sqrt{\alpha+\beta i})(x-\sqrt{\alpha+\beta i})(x+\sqrt{\alpha-\beta i})(x-\sqrt{\alpha-\beta i})\\ =\;&(x^2-(\alpha+\beta i))(x^2-(\alpha-\beta i))\\ =\;&x^4-2\alpha x^2+(\alpha^2+\beta^2)\in\mathbf{R}[x].\end{aligned}$$

Since neither $\alpha+\beta i$ nor $\alpha-\beta i$ is in $\mathbf{R}$ and since the maximum degree of an irreducible polynomial in $\mathbf{R}[x]$ is 2, the irreducible factors of this polynomial over $\mathbf{R}$ must be of degree 2. And so at least one of the polynomials

$$\begin{aligned}(x+\sqrt{\alpha+\beta i})(x+\sqrt{\alpha-\beta i})&=x^2+(\sqrt{\alpha+\beta i}+\sqrt{\alpha-\beta i})x+\sqrt{\alpha^2+\beta^2},\\ (x+\sqrt{\alpha+\beta i})(x-\sqrt{\alpha-\beta i})&=x^2+(\sqrt{\alpha+\beta i}-\sqrt{\alpha-\beta i})x-\sqrt{\alpha^2+\beta^2}\end{aligned}$$

is an irreducible polynomial over $\mathbf{R}$. It follows that $\sqrt{\alpha^2+\beta^2}\in\mathbf{R}$, i.e. that $\alpha^2+\beta^2$ is a square in $\mathbf{R}$.

Therefore $\mathbf{R}$ is formally real and so is real closed since its only algebraic extension is $\mathbf{C}=\mathbf{R}(i)$. □

There is in fact a much stronger theorem, also due to E. Artin and O. Schreier – see 2.21, p. 250, [39]:

Theorem 1.4.6 *If $\mathbf{C}$ is algebraically closed and K is a proper subfield of finite codimension $(\mathbf{C}:K)$, then* char $\mathbf{C}=0$, *K is real closed, and $\mathbf{C}=K(\sqrt{-1})$ (so $(\mathbf{C}:K)=2$).* □

Lemma 1.4.1 *Suppose that $\mathbf{R}$ is real closed, and let $\mathbf{C}=\mathbf{R}(\sqrt{-1})$, its algebraic closure. Let ζ be a primitive n^{th} root of unity in $\mathbf{C}$. Then if * is the nonidentity automorphism of $\mathbf{C}/\mathbf{R}$, $\zeta^*=\zeta^{-1}$.*

We may suppose that $n>2$. Let $\eta\in\mathbf{C}$, $\eta^2=\zeta$. Now $\eta\eta^*=\mathrm{N}_{\mathbf{C}/\mathbf{R}}\eta\in\mathbf{R}$ and $(\eta\eta^*)^2$ is an nth root of unity; if $(\eta\eta^*)^2\neq 1$, then

$$((\eta\eta^*)^2)^{n-1}+((\eta\eta^*)^2)^{n-2}+\cdots+(\eta\eta^*)^2=-1$$

which is impossible. Therefore $\zeta\zeta^*=(\eta\eta^*)^2$ must be 1. □

If $L=K(\theta)$ is an algebraic extension of K and the minimal polynomial of θ over K is $p(X)$, then the number of extensions of an ordering P on K to L is the number of zeros of p in a real closure K_P, and is also equal to the number of K-imbeddings of L into K_P (cf. 3.2.7 and 3.2.8, pp. 115-116, [65]). Thus the number of extensions of a given ordering is $\leq(L:K)$; if the number is actually $=(L:K)$ for the ordering P, L/K is called a *totally real* extension at P, and if it is $=(L:K)$ for *every* ordering of K (and K is formally real), L/K is called a *totally real* extension. If P is an ordering of K for which no extension to L/K is possible, then L/K is called an *imaginary* extension at P, and if this is true for *all* P (in other words, L is not formally real but K is), then L/K is called a *totally imaginary* extension.

Theorem 1.4.7 *Let P be an ordering of K, $\mathbf{R}$ a real closure of K at P, and $\mathbf{C} = \mathbf{R}(i)$ its algebraic closure. If $L = K(\theta)$ is a finite extension, then*

$$\mathbf{R} \otimes_K L = \overbrace{\mathbf{R} \oplus \cdots \oplus \mathbf{R}}^{r} \oplus \overbrace{\mathbf{C} \oplus \cdots \oplus \mathbf{C}}^{s}$$

where r is the number of conjugates of θ in $\mathbf{R}$ (and therefore $2s$ is the number in $\mathbf{C} \setminus \mathbf{R}$). In particular P has $(L : K)$ extensions to L (i.e. L/K is totally real at P) if

$$\mathbf{R} \otimes_K L = \mathbf{R} \oplus \cdots \oplus \mathbf{R},$$

and has no extensions (i.e. L/K is imaginary at P) if

$$\mathbf{R} \otimes_K L = \mathbf{C} \oplus \cdots \oplus \mathbf{C}.$$

Apply Prop. 1.2.3 (p. 21) with $L_2 = L$, $L_1 = \mathbf{R}$ and $\hat{L}_1 = \mathbf{C}$: the compositums $M_1, \ldots, M_t$ of L and $\mathbf{R}$ are of the form $\varphi L \cdot \mathbf{R}$ (the subfield of $\mathbf{C}$ generated by φL and $\mathbf{R}$) where φ runs over the $t = r + 2s$ imbeddings $L \to \mathbf{C}$ – which correspond to the conjugates of θ in $\mathbf{C}$. By Th. 1.2.15 (p. 21), the imbeddings of $\mathbf{R}$ and L into their composites give rise to an isomorphism

$$\mathbf{R} \otimes_K L \xrightarrow{\sim} M_1 \oplus \cdots \oplus M_t,$$

and it is clear that $M_i = \mathbf{R}$ or $\mathbf{C}$, and $= \mathbf{R}$ if and only the corresponding conjugate of θ is in $\mathbf{R}$.

Alternatively one can apply the functor $\mathbf{R}\otimes_K$ to the exact sequence

$$0 \to (p(X)) \to K[X] \to L \to 0$$

where $(p(X))$ is the principal ideal generated by the minimal polynomial $p(X) \in K[X]$ of θ over K, giving the exact sequence

$$0 \to (p(X)) \to \mathbf{R}[X] \to \mathbf{R} \otimes_K L \to 0$$

where $(p(X))$ is now the principal ideal in $\mathbf{R}[X]$. One now factors $p(X)$ into irreducibles

$$p(X) = l_1(X) \ldots l_{r_1}(X) q_1(X) \ldots q_{r_2}(X)$$

in $\mathbf{R}[X]$, where the l_i are linear and the q_i quadratic, and applies the Chinese Remainder Theorem to obtain

$$\begin{aligned} \mathbf{R} \otimes_K L &\cong \mathbf{R}[X]/(l_1 \ldots l_{r_1} q_1 \ldots q_{r_2}) \\ &\cong \big(\oplus_i \mathbf{R}[X]/(l_i)\big) \oplus \big(\oplus_i \mathbf{R}[X]/(q_i)\big). \end{aligned}$$ □

Corollary 1.4.1 *If (K_0, P) is an ordered field and $\delta \in \dot{K}_0 \setminus \dot{K}_0^2$, then P can be extended to the quadratic extension $K = K_0(\sqrt{\delta})$ if and only if $\delta >_P 0$.*

Since the ordering of a real closure $\mathbf{R}$ at P is $\dot{\mathbf{R}}^2$ and $P = \dot{\mathbf{R}}^2 \cap K_0$, $\delta >_P 0$ if and only if δ is a square in $\mathbf{R}$ if and only if $\mathbf{R}[X]/(X^2 - \delta) = \mathbf{R} \oplus \mathbf{R}$. So the corollary follows from the second part of Th. 1.4.7. □

It turns out that the case when an ordering *cannot* be extended is more important later on (cf. §4.10) in the situation of a field $(K, {}^-)$ with involution (with subfield $K_0 = (K, {}^-)_+$ of symmetric elements). If P is an ordering of K_0, we will say that ${}^-$ is a *definite involution at P* if $a\bar{a} >_P 0$ for all $a \in \dot{K}$, or that K/K_0 is a *definite extension at P*; this is equivalent to saying that either $K = K_0$ (and P is an ordering of K), or that $K = K_0(\sqrt{\delta})$ with $\delta < 0$ at P. The involution ${}^-$

is *definite* – or K/K_0 is a definite extension – if there *is* an ordering P of K_0 at which K/K_0 is definite, and it is *totally definite* if K_0 is formally real and K/K_0 is definite at each ordering of K_0.

If $^-$ or K/K_0 is not definite at P and $K \neq K_0$, we call it *indefinite at* P. If $K \neq K_0$ and K/K_0 is indefinite at each ordering P, or if $K = K_0$ and K_0 is not formally real, K/K_0 is simply called *indefinite*. In particular, K/K_0 is indefinite if K/K_0 is a totally real quadratic extension.

If K/K_0 is a definite quadratic extension at P and $h : V \times V \to (K, ^-)$ is an Hermitian form, the positive and negative indices $r^+(h)$ and $r^-(h)$ are again defined using an orthogonal basis of (V, h), exactly as in the case of a symmetric bilinear form, and the signature $\operatorname{sgn}(h) = \operatorname{sgn}_P(h)$ is defined to be $r^+(h) - r^-(h)$. That they are well-defined follows from the symmetric bilinear case since if $\alpha \in \dot{K}_0$ and $\langle\alpha\rangle$ is the corresponding Hermitian form,

$$r^+(\langle\alpha\rangle) = \tfrac{1}{2} r^+(\operatorname{Tr}_{K/K_0}(\langle\alpha\rangle))$$

since $\operatorname{Tr}_{K/K_0}(\langle\alpha\rangle)$ is the symmetric form $\langle 2\alpha, -2\delta\alpha\rangle$ over K_0.

There is also a signature defined when h is an Hermitian form over the Hamiltonian quaternions $(\mathbf{H}, {}^*)$ over a real closed field $\mathbf{R}$. In this case $h \cong \langle\alpha_1, \ldots, \alpha_n\rangle$ with all $\alpha_i \in \mathbf{R}$, and $\operatorname{sgn}(h)$ is again defined as the number of positive α_i minus the number of negative α_i.

Another important property of formally real fields is that *if* $(K, ^-)$ *is definite and* G *is a finite group, then all simple components of* KG *are stable under the standard* $(K, ^-)$*-involution of* KG. This follows from the fact that G is an orthonormal basis of the Hermitian space $(KG, \operatorname{Tr}_{KG/K}(u\bar{v}))$; thus it is a positive definite space and so $\operatorname{Tr}_{KG/K}(u\bar{u}) > 0$ if $u \neq 0$ – in particular if u is in a simple component A of KG, then $\bar{u} \in A$ also since $u\bar{u} \neq 0$.

The following lemma will be useful later on:

Lemma 1.4.2 *Let* $(K, ^-)$ *be a field with involution* $^- \neq \mathrm{id}$, *and* $K = K_0(\sqrt{\delta})$.

(a) *If* $\mathbf{C}$ *is an algebraic closure of* K, *then* $\mathbf{C} \otimes_{K_0} K \cong \mathbf{C} \oplus \mathbf{C}$ *via the map* $(\alpha \otimes \lambda) \to (\alpha\lambda, \alpha\bar{\lambda})$, *and the involution induced on* $\mathbf{C} \oplus \mathbf{C}$ *by* $\mathrm{id} \otimes {}^-$ *on* $\mathbf{C} \otimes_{K_0} K$ *is the interchange* $\overleftrightarrow{(\beta_1, \beta_2)} = (\beta_2, \beta_1)$.

(b) *Let* $\mathbf{R}$ *be a real closed field containing* K_0 *and* P *the restriction to* K_0 *of the ordering of* $\mathbf{R}$. *Let* $\mathbf{C}$ *be an algebraic closure* $\mathbf{R}(i)$ *containing* K. *Then*

$$(\mathbf{R} \otimes_{K_0} K, \mathrm{id}_{\mathbf{R}} \otimes {}^-) \cong \begin{cases} (\mathbf{C}, {}^*) & \text{if } K/K_0 \text{ is definite at } P, \\ (\mathbf{R} \oplus \mathbf{R}, \leftrightarrow) & \text{if } K/K_0 \text{ is indefinite at } P. \end{cases}$$

The isomorphism in (a) follows from Prop. 1.2.3 and Th. 1.2.15, p. 21. And it is easy to see that it is an isomorphism

$$(\mathbf{C} \otimes_{K_0} K, \mathrm{id} \otimes {}^-) \cong (\mathbf{C} \oplus \mathbf{C}, \leftrightarrow),$$

so (a) is proved.

If K/K_0 is definite at P, that is to say $\delta < 0$ at P where $K = K_0(\sqrt{\delta})$, $\mathbf{R}(\sqrt{\delta}) = \mathbf{C}$ and (b) follows easily in this case. If K/K_0 is indefinite at P, then $\delta > 0$ at P, $K \subset \mathbf{R}$ and

$$\mathbf{R} \otimes_{K_0} K \xrightarrow{\sim} \mathbf{R} \oplus \mathbf{R} \quad \text{via} \quad \rho \otimes \alpha \to (\rho\alpha, \rho\bar{\alpha}).$$

It is clear this is the isomorphism of involution algebras in (b). □

Proposition 1.4.1 *Suppose that K is a formally real algebraic number field (of finite degree over $\mathbb{Q}$) and that $\sigma_i : K \to \mathbb{R}$, $1 \leq i \leq r$, are the distinct imbeddings of K into the real numbers. Then the orderings on K are $P_1, \ldots, P_r$ where $P_i = \sigma_i^{-1}\dot{\mathbb{R}}^2$ and they are distinct.*

Let $\mathbf{R}$ be the algebraic closure of $\mathbb{Q}$ in $\mathbb{R}$. Suppose we show that it is a real closure of $\mathbb{Q}$ (for its unique ordering). Since $\sigma_i K \subset \mathbf{R}$, the σ_i are the distinct imbeddings $K \to \mathbf{R}$, so by pp. 115-116, [65], the distinct orderings of K are the $\sigma_i^{-1}\dot{\mathbf{R}}^2$; but $\sigma_i^{-1}\dot{\mathbf{R}}^2 = \sigma_i^{-1}\dot{\mathbb{R}}^2$ since $\mathbf{R} \cap \dot{\mathbb{R}}^2 = \dot{\mathbf{R}}^2$.

To show that $\mathbf{R}$ is a real closure of $\mathbb{Q}$, it suffices to show that $\mathbf{R}(i)$ is algebraically closed (Th. 1.4.4, p. 38).

Suppose $p(X) \in \mathbb{Q}[X]$ is monic of degree > 1. Factor it over $\mathbb{R}$:

$$p(X) = l_1 \ldots l_r q_1 \ldots q_s$$

where the l_i are monic linear polynomials and the q_i are monic irreducible quadratic polynomials. Since the roots of p are algebraic (over $\mathbb{Q}$), $p_i, q_j \in \mathbf{R}[X]$ for all i, j. Suppose $q_i = X^2 + \beta X + \gamma$. Its roots are $\frac{1}{2}(-\beta \pm \sqrt{\beta^2 - 4\gamma})$. Since $\beta^2 - 4\gamma < 0$, $\sqrt{\beta^2 - 4\gamma} = (\sqrt{4\gamma - \beta^2})i \in \mathbf{R}i$ since $\sqrt{4\gamma - \beta^2}$ is algebraic over $\mathbb{Q}$, and so $\mathbf{R}(i)$ is algebraically closed since it is an algebraic closure of $\mathbb{Q}$. □

Thus the orderings on a number field K are in canonical bijective correspondence with the real primes of K – which are the equivalence classes of the valuations $\alpha \to |\sigma_i \alpha|$ of K (where $|\cdot|$ is the absolute value on $\mathbb{R}$).

Corollary 1.4.2 *Suppose the ordering P_1 of the algebraic number field K_1 (of finite degree over $\mathbb{Q}$) can be extended to an ordering of the finite Galois extension K/K_1. Then the extensions of P_1 to K are the conjugates $\{\sigma P : \sigma \in \mathrm{Gal}(K/K_1)\}$ of any one of them.*

Let $\tau_1 : K_1 \to \mathbb{R}$ be an imbedding such that $P_1 = \tau_1^{-1}\dot{\mathbb{R}}^2$, let $\mathfrak{p}_1$ be the corresponding real prime – the equivalence class of the valuation $\alpha_1 \to |\tau_1(\alpha_1)|$ of K_1 – and let $\mathfrak{P}$ be a real prime of K lying over $\mathfrak{p}$. It is the equivalence class of a valuation $\alpha \to |\tau(\alpha)|$ of K given by an imbedding $\tau : K \to \mathbb{R}$ extending τ_1. By [53], Prop. 9.1, p. 167, the distinct primes of K lying over $\mathfrak{p}$ are the conjugates $\mathfrak{P}^\sigma$, the equivalence classes of the valuations $\alpha \to |\tau\sigma\alpha|$, $\sigma \in \mathrm{Gal}(K/K_1)$. Thus the extensions of P_1 to K are the conjugates

$$(\tau\sigma)^{-1}\dot{\mathbb{R}}^2 = \sigma^{-1}(\tau^{-1}\dot{\mathbb{R}}^2)$$

of the ordering $P = \tau^{-1}\dot{\mathbb{R}}^2$. □

Theorem 1.4.8 Frobenius' Theorem *If $\mathbf{R}$ is real closed, then up to isomorphism, the only finite dimensional division algebras over $\mathbf{R}$ are $\mathbf{R}, \mathbf{C} = \mathbf{R}(i)$, and the Hamiltonion quaternions $\mathbf{H} \cong (-1, -1)_{\mathbf{R}}$.*

The only finite extension fields of $\mathbf{R}$ are $\mathbf{R}$ itself and $\mathbf{C}$. The center of any division algebra finite dimensional over $\mathbf{R}$ must be one of these fields, and since there are no nontrivial finite dimensional division algebras over an algebraically closed field, any noncommutative division algebra over $\mathbf{R}$ must have center $\mathbf{R}$. Since the only proper finite extension field of $\mathbf{R}$ has degree 2, the maximal subfields of a proper division algebra have degree 2, so the division algebra has dimension 4 over $\mathbf{R}$ (cf. p. 12), hence is a quaternion algebra $(\alpha, \beta)_{\mathbf{R}}$ over $\mathbf{R}$ (cf. p. 16). This quaternion algebra splits if α or β is positive since the norm form is then isotropic, so both must be negative. Since positive elements of $\mathbf{R}$ are squares, the quaternion algebra is $\cong (-1,-1)_{\mathbf{R}}$, the Hamiltonian quaternion algebra $\mathbf{H}$. □

1.5 Bilinear and sesquilinear forms

This section consists of an outline of the algebraic theory of sesquilinear, bilinear, Hermitian, skew symmetric and symmetric forms, with few proofs. References are Chs. 1 and 2 in [39] and Chs. 1,2 and 7 in [65]. Later in Ch. 3 we shall give proofs for most of these matters in the more general setting of forms over a separable K-algebra with involution – such forms are necessary to deal with questions about orthogonal, symplectic and unitary representations over K.

1.5.1 Definitions and basic properties. Let D be a finite dimensional division algebra over the field K, with a $(K,\bar{\ })$-involution $\bar{\ }$. (Keep in mind our blanket assumption that $\operatorname{char} K \neq 2$). Let V be a finite dimensional (left) vector space over D, say of dimension n, and let $V^* = \operatorname{Hom}_D(V, D)$ be its dual space. The canonical pairing $V \times V^* \to D$ is denoted by $\langle v, x \rangle$.

A *sesquilinear form* on V is a map

$$f:\ V \times V \to D$$

which is biadditive, linear in the first variable and semilinear with respect to $\bar{\ }$ in the second. Thus f satisfies

$$\begin{aligned} f(u+u',v) &= f(u,v)+f(u',v),\\ f(u,v+v') &= f(u,v)+f(u,v'),\\ f(du,d'v) &= df(u,v)\bar{d}'. \end{aligned}$$

We usually write

$$f : V \times V \to (D, \bar{\ })$$

to indicate the dependence of f on the involution.

In this section, we write all mappings on the left including homomorphisms

$$\varphi : V \to W$$

from one *left* D-space to another – this accounts for the unorthodox relationship

$$\operatorname{mat}(\psi\varphi) = (\operatorname{mat}\varphi)(\operatorname{mat}\psi), \quad \text{i.e.} \quad \underline{\psi\varphi} = \underline{\varphi}\,\underline{\psi},$$

if $\psi : W \to U$ is another such homomorphism and $\operatorname{mat}\varphi = \underline{\varphi} = (\underline{\varphi}_{ij}) \in D^{n\times m}$, for example, is defined by $\varphi v_i = \sum_j \underline{\varphi}_{ij} w_j$ with respect to the bases $\{v_1,\ldots,v_n\}$ and $\{w_1,\ldots,w_m\}$ of V and W respectively. This also means that if we use these bases to identify V and W with $D^{1\times n}$ and $D^{1\times m}$ respectively, then

$$\varphi v = v\underline{\varphi}.$$

If the involution $\bar{\ }$ is the identity, then D is a field since $\overline{dd'} = \bar{d}'\bar{d}$ is $dd' = d'd$. In this case f is usually called a *bilinear form.*

Let $\varepsilon = \pm 1$. A sesquilinear form f which satisfies the symmetry condition

$$f(u,v) = \varepsilon\overline{f(v,u)} \quad \text{for all } u,v \in V,$$

will be called a *reflexive* form (over D or $(D,\bar{\ })$), or more precisely an ε*-reflexive* form. If the involution is known to be the identity, it is called an ε*-symmetric* bilinear form, or a *symmetric* bilinear form if $\varepsilon = 1$ and a *skew symmetric* bilinear form if $\varepsilon = -1$. If the involution is *not* the identity, it is called an *ε-Hermitian* form, or an *Hermitian* form if $\varepsilon = 1$ and a *skew Hermitian* form if $\varepsilon = -1$.

The map

$$f_r : V \to V^* \text{ given by } f_r(v) = f(\cdot, v)$$

is the *(right) adjoint* of the sesquilinear form f. Thus

$$\langle u, f_r(v)\rangle = f(u, v).$$

V^* is naturally a right vector space over D, and f_r satisfies

$$f_r(dv) = f_r(v)\bar{d} \quad \text{for all } d \in D \text{ and } v \in V.$$

Such a map, from a left D-space to a right D-space, is called a *$(D, {}^-)$-homomorphism*. Alternatively we can twist V^* into a *left* vector space $\overline{V^*}$ over D by defining a left D-action on V^*:

$$dx = x\bar{d}.$$

We often write $\bar{x}$ for x when we are considering it to be in $\overline{V^*}$; then this equation becomes

$$d\bar{x} = \overline{x\bar{d}} \quad \text{or} \quad \overline{xd} = \bar{d}\bar{x}.$$

Then the adjoint is a *D-linear* map

$$f_r : V \to \overline{V^*}.$$

The sesquilinear form f is *nonsingular* if its (right) adjoint is an isomorphism, and in this case (V, f) is called a *sesquilinear space* – or a *bilinear space* if ${}^-$ is known to be the identity (and D a field). We also say, loosely, that V is nonsingular if f is so. If f is ε-reflexive and nonsingular, the pair (V, f) is called a *reflexive space*, or an *ε-reflexive space*. Symmetric, skew symmetric, Hermitian and skew Hermitian spaces are similarly defined.

There is similarly a *left* adjoint

$$f_l : V \to \overline{V^*} \quad \text{given by } f_l(v) = \overline{f(v, \cdot)},$$

that is to say,

$$\langle u, f_l(v)\rangle = \overline{f(v, u)}.$$

It is also a D-linear transformation.

The *discriminant matrix*[7] of (V, f) or f with respect to a basis $v_1, v_2, \ldots, v_n$ of V is the $n \times n$ matrix

$$\text{mat}\, f = \underline{f} = (f(v_i, v_j)) \in D^{n\times n}.$$

It is easy to see that f is symmetric or Hermitian if and only if $\underline{f}^{t-} = \underline{f}$, i.e. the matrix $\underline{f}$ is respectively symmetric or Hermitian, and is skew symmetric or skew Hermitian if and only if $\underline{f}^{t-} = -\underline{f}$, i.e. $\underline{f}$ is respectively skew symmetric or skew Hermitian. If we identify V with $D^{1\times n}$ using the basis, then

$$f(u, v) = u\underline{f}v^{t-}. \tag{1.33}$$

If $u_1, u_2, \ldots, u_n$ is another basis, related to the first by

$$u_i = \sum_k d_{ik} v_k,$$

[7] Also often called the *Gram matrix* of f.

then

$$f(u_i, u_j) = \sum_{k,\,l} d_{ik} f(v_k, v_l) \overline{d_{jl}}.$$

Thus the discriminant matrix of f with respect to the new basis is

$$\Delta \underline{f} \Delta^{t-}$$

where Δ is the matrix (d_{ij}) of the basis change. Assume now that D is a *separable* division algebra. Since $\operatorname{nrd}_{\mathrm{M}(n,D)/K}\Delta^{t-} = \overline{\operatorname{nrd}_{\mathrm{M}(n,D)/K}\Delta}$ by Lem. 1.3.7, p. 34, the *determinant* $\det f$ of f,

$$\det f = \operatorname{nrd}_{\mathrm{M}(n,D)/K}(\underline{f}),$$

is well defined up to a nonzero "norm" $\alpha\bar{\alpha}$ $(\alpha \in \operatorname{nrd}_{\mathrm{M}(n,D)/K}\mathbf{GL}(n,D) = \operatorname{nrd}_{D/K}\dot{D}$ – cf. (1.15), p. 14), and so can be considered to lie in $K/\mathrm{N}\dot{K} = \{0\} \cup \dot{K}/\mathrm{N}\dot{K}$ where $\mathrm{N}\dot{K}$ is the group of nonzero "norms" $\alpha\bar{\alpha}$ of K:

$$\mathrm{N}\dot{K} = \begin{cases} \dot{K}^2 & \text{if } ^- = \text{id on } K, \\ \mathrm{N}_{K/K_0}\dot{K} & \text{otherwise.} \end{cases}$$

Note that $\det f \in K_0/\mathrm{N}_{K/K_0}\dot{K}$ if f is Hermitian.

If f is a nonsingular skew symmetric form, one can show that $\det f \in \dot{K}^2$, and so det is not a useful invariant in this case. On the other hand for each $n \geq 1$, there is a polynomial $\mathrm{Pf}_n \in \mathbb{Z}[X_{ij}; i,j = 1,\ldots,n, i \neq j]$ of degree n, called the *Pfaffian*, such that

$$\det(a_{ij}) = (\mathrm{Pf}_n(a_{ij}))^2$$

for any skew symmetric matrix $\in \mathrm{M}(2n, K)$. See Th. 3.27, p. 141, [2].

If $\det f = \alpha\mathrm{N}\dot{K}$, we often write $\det f = \alpha$. And if $\beta \in \alpha\mathrm{N}\dot{K}$, we also write $\det f = \beta =_{\mathrm{N}} \alpha$, or if $\mathrm{N}\dot{K} = \dot{K}^2$, $\det f = \beta =_2 \alpha$.

When $D = K$, we also define the *discriminant* of a symmetric or Hermitian form $h : V \times V \to (K, ^-)$ to be $\operatorname{disc} h = (-1)^{n(n-1)/2} \det h$ where $n = \dim V$, and that of an even-dimensional skew Hermitian form to be $\operatorname{disc} h = \operatorname{disc} \sqrt{\delta} h \in \dot{K}_0/\mathrm{N}_{K/K_0}\dot{K}$; of course $\sqrt{\delta}h$ is an Hermitian form, and it depends on the choice of δ and $\sqrt{\delta}$ but $\operatorname{disc}\sqrt{\delta}h$ does not. The discriminant of an odd-dimensional skew Hermitian form is not defined.

Suppose that the sesquilinear form f is nonsingular, so $f(\cdot, v_1), \ldots, f(\cdot, v_n)$ is a basis of V^*. Let $v_1^*, \ldots, v_n^*$ be the dual basis of V^*. Then

$$f_r(v_i) = f(\cdot, v_i) = \sum_{j=1}^{n} v_j^* f(v_j, v_i) = \sum_{j=1}^{n} \overline{f(v_j, v_i)}\, v_j^*,$$

so the matrix of f_r with respect to the bases $\{v_i\}$ and $\{v_i^*\}$ is $(\operatorname{mat} f)^{t-}$. It follows that the determinant of f is $\neq 0$ if and only if f is nonsingular. Similarly

$$f_l(v_i) = \overline{f(v_i, \cdot)} = \sum_{j=1}^{n} f(v_i, v_j) v_j^*,$$

so the matrix of f_l is the discriminant matrix and f is nonsingular if and only if f_l is an isomorphism – in particular f_r is an isomorphism if and only if f_l is.

If U is a subspace of V and the form f is restricted to $U \times U$, we shall generally say it has been *restricted to* U. Such a subspace is called nonsingular if the restriction of f to it is nonsingular. Let

$$U^{\perp} = \{v \in V : f(U, v) = 0\},$$

the subspace of V *orthogonal to* U, which we also refer to as the *orthogonal complement* of U (in V). (Note that it is not necessarily the same as $\{v \in V : f(v, U) = 0\}$ unless f is reflexive.) Consider the sequence

$$0 \to U^{\perp} \xrightarrow{\text{incl}} V \xrightarrow{f_r} \overline{V^*} \xrightarrow{\text{restr}} \overline{U^*} \to 0$$

where the second map is inclusion and the fourth is restriction, $x \to x|_U$. The restriction is onto and if f is nonsingular, the kernel of the composite map $V \xrightarrow{f_r} \overline{V^*} \xrightarrow{\text{restr}} \overline{U^*}$ is $U^{\perp}$ and so

$$\dim U + \dim U^{\perp} = \dim \overline{U^*} + \dim U^{\perp} = \dim V.$$

If U also is nonsingular, $V = U \oplus U^{\perp}$ since then $U \cap U^{\perp} = 0$. In general if $f(U, W) = 0$ and $V = U \oplus W$, we shall write $V = U \perp W$ and say that V is the *orthogonal direct sum of* U and W. In this case U and W are both nonsingular if and only if V is. When this is so, $W = U^{\perp}$ and

$$\operatorname{mat} f = \begin{pmatrix} \operatorname{mat} g & 0 \\ * & \operatorname{mat} h \end{pmatrix}$$

where g is the restriction of f to U and h that to W. If we also have $f(W, U) = 0$, then of course

$$\operatorname{mat} f = \begin{pmatrix} \operatorname{mat} g & 0 \\ 0 & \operatorname{mat} h \end{pmatrix}.$$

This is in particular the case when $V = U \perp W$ and f is reflexive – which is almost always the case in the remainder of this book.

There is also an *external* orthogonal direct sum of sesquilinear spaces $(V_1, h_1) \perp (V_2, h_2) =$ (by definition) $(V_1 \perp V_2, h_1 \perp h_2)$ where $V_1 \perp V_2$ is the external direct sum $V_1 \oplus V_2$ and $(h_1 \perp h_2)(u_1 + u_2, v_1 + v_2) = h_1(u_1, v_1) + h_2(u_2, v_2)$. Here we have identified V_1 and V_2 with subspaces of $V_1 \oplus V_2$, so that $(v_1, v_2) = v_1 + v_2$. The external orthogonal direct sum is often denoted simply by $V_1 \perp V_2$ or $h_1 \perp h_2$. Obviously h_1 and h_2 do not have to be nonsingular in order for their external orthogonal direct sum to exist.

Let f be a nonsingular sesquilinear form on V. If $\varphi \in \operatorname{End}_D V$, its *adjoint* $\overline{\varphi} \in \operatorname{End}_D V$ is the transformation uniquely determined by

$$f(\varphi u, v) = f(u, \overline{\varphi} v) \quad \text{for all } u, v \in V. \tag{1.34}$$

The map $\varphi \to \overline{\varphi}$ is a $(K, {}^-)$-antiautomorphism of $\operatorname{End}_D V$, and is a $(K, {}^-)$-involution if f is reflexive. Using (1.33) and (1.34), one easily shows that the matrix $\underline{\overline{\varphi}}$ of the adjoint $\overline{\varphi}$ is given by

$$\begin{aligned} \underline{\overline{\varphi}} &= \underline{f}^{t-} \underline{\varphi}^{t-} (\underline{f}^{t-})^{-1} \\ &= \underline{f}\, \underline{\varphi}^{t-} \underline{f}^{-1} \ \text{ if } f \text{ is reflexive.} \end{aligned} \tag{1.35}$$

Exercises 1. Conversely, show that if $(D, {}^-)$ is a division algebra with involution and ${}^-$ is an involution on $\mathrm{M}(n, D)$ given by $\bar{a} = ba^{t-}b^{-1}$ for some $b \in \mathbf{GL}(n, D)$, it is the matrix version of the adjoint equation (1.35) with respect to a nonsingular reflexive form f on a vector space of dimension n over D.

2. If $(A, {}^-)$ is a central simple algebra with involution, show that it is isomorphic to an involution algebra of the form $(\mathrm{M}(n, D), {}^\sim)$ where D is a division algebra and ${}^\sim$ is the adjoint involution of some reflexive form. $\square$

Proposition 1.5.1 *If $f : V \times V \to (K, {}^-)$ is a nonsingular ε-reflexive form, the adjoint involution on $\mathrm{End}_K V$ is orthogonal if f is symmetric, symplectic if f is skew symmetric, and unitary if ${}^- \neq \mathrm{id}_K$.*

The case ${}^- \neq \mathrm{id}_K$ is trivial, so assume ${}^-$ is the identity on K. Then the proposition follows from (1.35) and Th. 1.3.6(b), p. 30, and the fact that transpose is an orthogonal involution. $\square$

If f is reflexive but not skew symmetric, the space V has an *orthogonal basis*, i.e. a basis $v_1, \ldots, v_n$ such that $f(v_i, v_j) = 0$ if $i \neq j$. This means that the discriminant matrix mat f is diagonal with respect to this basis, with the diagonal entries $f(v_1, v_1) = d_1, \ldots, f(v_n, v_n) = d_n$. In this case we write $f = \langle d_1, d_2, \ldots, d_n \rangle$.

More generally, if $\underline{f}_1, \ldots, \underline{f}_k$ are square matrices over D, then the sesquilinear form whose discriminant matrix (with respect to some basis) is the matrix with the $\underline{f}_1, \ldots, \underline{f}_k$ as block diagonal matrices (and 0s elsewhere) is denoted by $\langle \underline{f}_1, \ldots, \underline{f}_k \rangle$.

If f is skew symmetric and nonsingular, then $f(v, v) = 0$ for all $v \in V$, and V is even dimensional and has a "symplectic" basis $v_1, \ldots, v_{2m}$ with the property that $f(v_{2i-1}, v_{2i}) = 1 = -f(v_{2i}, v_{2i-1})$ for all i and all other "inner products" $f(v_i, v_j)$ are $= 0$. In other words V is the orthogonal direct sum of m planes, each with discriminant matrix

$$\begin{pmatrix} 0 & 1 \\ -1 & 0 \end{pmatrix}. \tag{1.36}$$

Suppose again that $f : V \times V \to D$ is a sesquilinear form. A vector $v \in V$ is *isotropic* if $v \neq 0$ and $f(v, v) = 0$. V (or f) is called *isotropic* if it contains an isotropic vector, and is *totally isotropic* if f is identically 0.

A *hyperbolic plane* is a plane with an ε-reflexive form f with discriminant matrix

$$\mathrm{mat}\, f = \begin{pmatrix} 0 & 1 \\ \varepsilon & 0 \end{pmatrix} \qquad (\varepsilon = \pm 1)$$

with respect to some basis. A *hyperbolic space* is, by definition, an orthogonal direct sum of one or more such planes (with the same ε). It is of course isotropic.

The following theorem is easily checked.

Theorem 1.5.1 *Let (V, f) be a plane with an ε-reflexive form f which is not skew symmetric. The following are equivalent.*

(a) V is a hyperbolic plane.

(b) $f = \langle d, -d \rangle$ where $d \in \dot{D}$ and $\bar{d} = \varepsilon d$.

(c) V is nonsingular and isotropic.

If $D = K$ and $\varepsilon = 1$, then these conditions are also equivalent to

(d) $\det f = -1$. $\square$

Theorem 1.5.2 *Let (V, f) be a nonsingular ε-reflexive space, and let $U \subset V$ be a totally isotropic subspace of V with basis $u_1, \ldots, u_r$. Then there are vectors $w_1, \ldots, w_r$ in V such that if W is the space spanned by them,*

$$\operatorname{mat} f|_{U \oplus W} = \operatorname{diag}\left(\begin{pmatrix} 0 & 1 \\ \varepsilon & 0 \end{pmatrix}, \begin{pmatrix} 0 & 1 \\ \varepsilon & 0 \end{pmatrix}, \ldots, \begin{pmatrix} 0 & 1 \\ \varepsilon & 0 \end{pmatrix}\right)$$

in the basis $u_1, w_1, u_2, w_2, \ldots, u_r, w_r$. □

A corollary is that a nonsingular alternating form has a symplectic basis, as stated earlier.

Let (V, f) be a sesquilinear space over $(K, ^-)$, and $(L, ^-)/(K, ^-)$ an extension of fields with involution. Then there is a unique extension of f to a sesquilinear form f^L on the L-vector space $V^L = L \otimes_K V$ (which we assume contains V). This process is called "extension of the scalars". (The notation V_L and f_L is sometimes used, but we wish to reserve this for "restriction of scalars").

A little more generally, if $\varphi : (K, ^-) \to (L, ^-)$ is a homomorphism of fields with involution, then we can also "extend the scalars" to L. The resulting extension is a sesquilinear form

$$f^L : V^L \times V^L \to (L, ^-)$$

on $V^L = L \otimes_K V$ (where K operates on L via $\alpha.\lambda = \varphi(\alpha)\lambda$, $\alpha \in K, \lambda \in L$), satisfying

$$f^L(\lambda \otimes u, \lambda' \otimes v) = \lambda\varphi(f(u, v))\overline{\lambda'} \qquad (\lambda, \lambda' \in L,\ u, v \in V). \tag{1.37}$$

Since a K-basis of V is also an L-basis of V^L, f and f^L have the same discriminant matrix and so f is nonsingular if and only if f^L is nonsingular. Also f is symmetric if and only if f^L is symmetric, and a similar statement holds when f is skew symmetric, Hermitian or skew Hermitian.

One can also extend the scalars when f is a sesquilinear form over a division algebra D; this process sometimes leads to a sesquilinear form over an algebra, and is discussed later (p. 110*ff*).

1.5.2 Isometries. Let (V, f) and (W, g) be sesquilinear forms over the division algebra $(D, ^-)$ with involution. A *morphism* $\varphi : (W, g) \to (V, f)$ is a D-linear map $\varphi : W \to V$ such that $f(\varphi w, \varphi w') = g(w, w')$ for all $w, w' \in W$. It is necessarily injective if W is nonsingular, and is called an *isometry* if it is bijective. Two sesquilinear forms are *equivalent* if there is an isometry between them. In this case we write $(V, f) \cong (W, g), V \cong W$, or $f \cong g$, depending on the context.

One of the principal questions in the theory of sesquilinear spaces is the determination of invariants which characterize their equivalence classes. The nature of these invariants depends on whether the forms are Hermitian, symmetric, etc., and on the nature of the field K. The simplest invariants are the dimension of V and the determinant $\det f$.

The case of skew symmetric forms is trivial: since any nonsingular skew symmetric form has a symplectic basis, the dimension of a nonsingular skew symmetric form completely determines its equivalence class. The equivalence theory for other reflexive forms is much more complicated in general, and a summary of results for them is given in §1.5.7. The equivalence question for arbitrary sesquilinear spaces "reduces" to the reflexive case (cf. [57] and [61]), but we shall not need this theory.

The equivalence of reflexive forms over a field $(K, {}^-)$ with involution, which are not skew symmetric, can also be reduced to that of symmetric bilinear forms. First of all if h is skew Hermitian and $K = K_0(\sqrt{\delta})$, then $g = \sqrt{\delta}h$ is Hermitian. Furthermore any isometry between two skew Hermitian forms is also an isometry between the Hermitian forms associated to them in this way, and conversely. Thus when $K \neq K_0$, it suffices to consider the case of Hermitian forms. If h_1 and h_2 are Hermitian forms, they are equivalent if and only if their "trace forms" $h_i(u, v) + \overline{h_i(u, v)}$ are equivalent as symmetric bilinear forms over K_0. This is a theorem of Jacobson [30], and a proof is also given in Th. 10.1.1, p. 348, [65]. This also holds over a quaternion division algebra D with conjugation as involution, using the canonical trace of the quaternion algebra,

$$\mathrm{Tr}_{D/K}(\alpha + \beta i + \gamma j + \delta k) = \alpha,$$

or equivalently the reduced trace $\mathrm{trd}_{D/K}$ – see Th. 10.1.7, p. 352, *ibid.*

Exercises. 1. Show that if f and g are equivalent nonsingular reflexive forms over $(D, {}^-)$, then their adjoint involutions (on the algebras of endomorphisms of their vector spaces) are equivalent.

2. If ${}^-$ and ${}^\sim$ are symplectic involutions on $\mathrm{M}(2n, K)$, show that $(\mathrm{M}(2n, K), {}^-) \cong (\mathrm{M}(2n, K), {}^\sim)$. (Cf. the exercise on p. 46.) □

If we use bases for V and W over D to express $\varphi : (W, g) \to (V, f)$ as a matrix $\underline{\varphi}$ with entries in D, then the fact that φ is a morphism of sesquilinear spaces is expressed in terms of matrices by

$$\underline{g} = \underline{\varphi}\,\underline{f}\,\underline{\varphi}^{t-}.$$

The set of isometries $\varphi : V \to V$ of a sesquilinear space with itself is the *isometry group* of $f : V \times V \to (D, {}^-)$. It is denoted by $\mathbf{I}(V, f)$, $\mathbf{I}(V)$, or $\mathbf{I}(f)$, depending on the context. If f is known to be symmetric, it is called the *orthogonal group* and is denoted by $\mathbf{O}(V, f)$, etc. In the skew symmetric case, it is called the *symplectic group* and is denoted by $\mathbf{Sp}(V, f)$, and is the *unitary group* $\mathbf{U}(V, f)$ when f is ε-Hermitian.

The isometry group can also be viewed as the group of (invertible) matrices satisfying

$$\underline{\varphi}\,\underline{f}\,\underline{\varphi}^{t-} = \underline{f}. \tag{1.38}$$

Suppose that f is a nonsingular reflexive form over a field. Then (1.38) implies that $\det \underline{\varphi}\, \overline{\det \underline{\varphi}} = 1$, i.e. the norm of the determinant of an isometry is 1. In the skew symmetric case, it can be shown that $\det \underline{\varphi} = 1$. Suppose then that f is *not* skew symmetric. Then there is a vector $v \in V$ such that $f(v, v) \neq 0$ and we can split $V = Kv \perp W$. If $\lambda \in K$ has norm 1, the map $\varphi := \lambda(\mathrm{id}_{Kv}) \perp \mathrm{id}_W$ is an isometry with determinant λ. Thus if we define the "special" unitary group $\mathbf{SU}(V)$ respectively special orthogonal group $\mathbf{SO}(V)$ to be the subgroup of isometries of determinant 1, the sequences

$$1 \to \mathbf{SU}(V) \to \mathbf{U}(V) \to {}_N\dot{K} \to 1, \quad 1 \to \mathbf{SO}(V) \to \mathbf{O}(V) \to \pm 1 \to 1$$

are exact, where ${}_N\dot{K}$ is the subgroup of $\dot{K}$ of elements of norm 1.

One of the most important theorems for reflexive forms is

Theorem 1.5.3 Witt's Theorem. *Let $f : V \times V \to (D, ^-)$ be a nonsingular reflexive form over $(D, ^-)$, and suppose that $V = U_1 \perp W_1 = U_2 \perp W_2$ are splittings of V with $U_1 \cong U_2$. Then $W_1 \cong W_2$.* □

More generally, if U_1 and U_2 are equivalent subspaces of V which are not necessarily nonsingular, then $U_1^\perp \cong U_2^\perp$ (if V is nonsingular). And if $\varphi : (U, f|_{U\times U}) \to (V, f)$ is an injective morphism, it can be extended to an isometry of V. (This is proved later in a more general setting – see p. 100).

1.5.3 Witt rings. It follows from Theorem 1.5.2 (p. 48) that any ε-reflexive space V over $(D, ^-)$ has a "Witt decomposition"

$$V = H \perp W$$

where H is hyperbolic and W is *anisotropic*, that is to say, W contains no isotropic vectors. By Witt's theorem, H and W are both determined up to an isometry. W is called the *anisotropic kernel* of V. The equivalence classes of anisotropic forms $[W]$ can be made into an associative commutative monoid by defining $[W_1]+[W_2] = [W_3]$ where $W_1 \perp W_2 = H \perp W_3$ is a Witt decomposition. This monoid can be imbedded in a group in which every element is of the form $[W] - [W']$ (cf. §1.6); it is called the *Witt group* and is denoted by $\mathrm{W}_\varepsilon(D)$, or more precisely by $\mathrm{W}_\varepsilon(D, ^-)$. Of course since every nonsingular skew symmetric form is a sum of hyperbolic forms, the Witt group $\mathrm{W}_{-1}(K, \mathrm{id}_K)$ is trivial, so we ignore it in the rest of this discussion.

Since $\langle\alpha, -\alpha\rangle$ (where $\bar\alpha = \varepsilon\alpha \in \dot D$) is hyperbolic and hence $= 0$ in $\mathrm{W}_\varepsilon(D)$,

$$[-\alpha] = -[\alpha]$$

where we have written $[\alpha]$ for $[\langle\alpha\rangle]$.

In the case of symmetric forms or Hermitian forms over the field K, $\mathrm{W}_1(K) = \mathrm{W}_1(K, ^-)$ is also a ring. Namely we can define the product $(V \otimes W, fg)$ of 2 forms (V, f) and (W, g) where

$$(fg)(v_1 \otimes w_1, v_2 \otimes w_2) = f(v_1, v_2)g(w_1, w_2), \tag{1.39}$$

which in terms of diagonalizations is

$$\langle\alpha_1, \alpha_2, \ldots, \alpha_n\rangle\langle\beta_1, \beta_2, \ldots, \beta_m\rangle = \langle\alpha_1\beta_1, \alpha_1\beta_2, \ldots, \alpha_n\beta_m\rangle. \tag{1.40}$$

In general, the discriminant matrix of the product fg is the tensor product $\underline{f} \otimes \underline{g}$ of discriminant matrices (p. 17, [39]).

The product of two general sesquilinear forms f and g over $(K, ^-)$ can also be defined by (1.39), and again has discriminant matrix $\underline{f} \otimes \underline{g}$ – if the forms happen to be ε-reflexive and δ-reflexive respectively, then the product fg is $\varepsilon\delta$-reflexive.

Since the product is distributive and the product of a hyperbolic space and an arbitrary nonsingular space is hyperbolic, $\mathrm{W}_1(K, ^-)$ inherits a multiplication which makes it into a ring, the *Witt ring* of symmetric bilinear or Hermitian forms over $(K, ^-)$. In fact multiplication can be extended to the "graded Witt ring"

$$\mathrm{W}(K, ^-) := \mathrm{W}_1(K, ^-) \oplus \mathrm{W}_{-1}(K, ^-)$$

which is a $\mathbb{Z}/2$-graded ring, although as pointed out above, $\mathrm{W}_{-1}(K, \mathrm{id}) = 0$.

The product of forms is considered in greater generality in Ch. 3.

The Witt group $\mathrm{W}_\varepsilon(K, ^-)$ can also be defined in another way via the Grothendieck group – this is discussed in Ch. 3 in the more general context of forms over an involution algebra. One begins with the monoid of equivalence classes $[h]$ of nonsingular ε-reflexive forms, which is imbedded in the corresponding Grothendieck group $\hat{\mathrm{W}}_\varepsilon(K,{}^-)$ – the "*Witt-Grothendieck group*" – whose elements are the differences $[h_1]-[h_2]$ (cf. §1.6). Then the Witt group $\mathrm{W}_\varepsilon(K,{}^-)$ is the quotient $\hat{\mathrm{W}}_\varepsilon(K,{}^-)/\mathbb{Z}[H]$ where H is an ε-reflexive hyperbolic plane.

In fact $\hat{\mathrm{W}}(K,{}^-) := \hat{\mathrm{W}}_1(K,{}^-) \oplus \hat{W}_{-1}(K,{}^-)$ is also a graded ring, and the subgroup generated by 1-reflexive and -1-reflexive hyperbolic forms is a graded ideal $\hat{\mathrm{H}}(K,{}^-)$ such that $\hat{\mathrm{W}}(K,{}^-)/\hat{\mathrm{H}}(K,{}^-) \cong \mathrm{W}(K,{}^-)$.

Now consider the Witt group $\mathrm{W}_1(K) = \mathrm{W}_1(K,{}^-)$ of symmetric or Hermitian forms. The dimension gives a homomorphism $\dim : \hat{\mathrm{W}}_1(K) \to \mathbb{Z}$ whose kernel is denoted by $\hat{I}$. Thus $\hat{\mathrm{W}}_1/\hat{I} \cong \mathbb{Z}$, and $\hat{I}$ is generated, as an Abelian group, by the differences of (equivalence classes of) unary (one dimensional) forms. Since $\hat{I} \cap \mathbb{Z}[H] = 0$, $\hat{I}$ is isomorphic to its image $I \subset \mathrm{W}_1(K)$, the *fundamental ideal* of $\mathrm{W}_1(K)$, consisting of the classes of all even dimensional forms. The kernel of the composite homomorphism $\hat{\mathrm{W}}_1(K) \to \mathbb{Z} \to \mathbb{Z}/2$ contains the classes of all even dimensional forms, hence the homomorphism factors through $\mathrm{W}_1(K)$, so we get a (surjective) homomorphism "dimension mod 2"

$$\dim_{/2} : \mathrm{W}_1(K) \to \mathbb{Z}/2$$

with kernel I which assigns to the Witt class $[h]$ the parity of its dimension, i.e. $\dim_{/2}[h] = \dim h + 2\mathbb{Z}$. In particular,

$$\mathrm{W}_1(K)/I \cong \mathbb{Z}/2.$$

Recall that the discriminant of h is

$$\operatorname{disc} h = (-1)^{\frac{1}{2}n(n-1)} \det h \in \dot{K}_0/\mathrm{N}\dot{K}$$

(p. 45) where $n = \dim h$ ($= \dim V$ if V is the space of h), and that $\mathrm{N}\dot{K} = \dot{K}^2$ if ${}^-$ is the identity on K, $= \mathrm{N}_{K/K_0}\dot{K}$ otherwise. The discriminant is well-defined on Witt classes, in contrast to the determinant, so we can consider it to be a map on $\mathrm{W}_1(K)$. Consider the map

$$(\operatorname{disc}, \dim_{/2}) : \mathrm{W}_1(K) \to \dot{K}_0/\mathrm{N}\dot{K} \times \mathbb{Z}/2. \tag{1.41}$$

If we define multiplication on the (set) product $\dot{K}_0/\mathrm{N}\dot{K} \times \mathbb{Z}/2$ by

$$(\alpha\mathrm{N}\dot{K}, i)(\beta\mathrm{N}\dot{K}, j) = ((-1)^{ij}\alpha\beta\mathrm{N}\dot{K}, i+j),$$

then $\dot{K}_0/\mathrm{N}\dot{K} \times \mathbb{Z}/2$ is a group, the *central extension of* $\mathbb{Z}/2$ *by* $\dot{K}_0/\mathrm{N}\dot{K}$ *with cocycle* $\mathrm{z}(i,j) = (-1)^{ij} \in \mathrm{Z}^2(\mathbb{Z}/2, \dot{K}_0)$,[8] which we denote by $\dot{K}_0/\mathrm{N}\dot{K} \times_{\mathrm{z}} \mathbb{Z}/2$. Moreover (1.41) is then an epimorphism with kernel I^2 and

$$\mathrm{W}_1(K)/I^2 \cong \dot{K}_0/\mathrm{N}\dot{K} \times_{\mathrm{z}} \mathbb{Z}/2, \quad I/I^2 \cong \dot{K}_0/\mathrm{N}\dot{K} \tag{1.42}$$

– cf. Prop. 2.1 and Cor. 2.2, pp. 31,32, [39] – it is proved there only for symmetric bilinear forms but the proof applies, *mutatis mutandis*, to the Hermitian case as well.

The Witt ring can be considered as a (covariant) functor W_1 from the category of involution fields of characteristic $\neq 2$ to the category of commutative rings since a homomorphism $(K,{}^-) \to (L,{}^-)$ gives rise, by scalar extension, to a ring homomorphism $\mathrm{W}_1(K,{}^-) \to \mathrm{W}_1(L,{}^-)$.

[8] where $\mathbb{Z}/2$ operates trivially on $\dot{K}_0$.

1.5.4 K real closed or algebraically closed.

Theorem 1.5.4 *If K is an algebraically closed field $\mathbf{C}$, two nonsingular symmetric bilinear forms are equivalent if and only if they have the same dimension. Furthermore*

$$\dim : \hat{\mathrm{W}}_1(\mathbf{C}, \mathrm{id}) \to \mathbb{Z} \quad \textit{and} \quad \dim_{/2} : \mathrm{W}_1(\mathbf{C}, \mathrm{id}) \to \mathbb{Z}/2$$

are both ring isomorphisms. □

Let $n\langle\alpha\rangle$ denote the n-fold sum $\langle\alpha\rangle \perp \langle\alpha\rangle \perp \cdots \perp \langle\alpha\rangle$.

Theorem 1.5.5 *Let $(\mathbf{C},\ ^*\ )$ be an algebraically closed field with an involution* $^* \neq \mathrm{id}$.

(a) The subfield of $\mathbf{C}$ of elements fixed by * is a real closed field $\mathbf{R}$.
(b) Up to equivalence, there are exactly two anisotropic symmetric forms over $(\mathbf{R}, \mathrm{id})$ or Hermitian forms over $(\mathbf{C},\ ^*\ )$ of dimension n, namely $n\langle 1\rangle$ and $n\langle -1\rangle$.
(c) $\mathrm{W}_1(\mathbf{R}, \mathrm{id}) \cong \mathrm{W}_1(\mathbf{C},\ ^*\ ) \cong \mathbb{Z}$ *(as commutative rings) via the signature map.*
(d) (Sylvester's Law of Inertia) *Two nonsingular symmetric forms over $(\mathbf{R}, \mathrm{id})$ or Hermitian forms over $(\mathbf{C},\ ^*\ )$ are equivalent if and only if they have the same dimension and the same signature.*
(e) The identity of the Witt-Grothendieck ring is $\langle 1\rangle$, and if we identify $\mathbb{Z}\langle 1\rangle$ with $\mathbb{Z}$, then $\hat{\mathrm{W}}_1(K) = \mathbb{Z}[\langle -1\rangle]$ as a ring (note that $\langle -1\rangle^2 = \langle 1\rangle = 1$) and is $\cong \mathbb{Z} \oplus \mathbb{Z}$ as a group.

The subfield of $\mathbf{C}$ of elements fixed by * is a real closed field $\mathbf{R}$ by Th. 1.4.5 (p. 38).

Any nonsingular symmetric or Hermitian form is of the form $h = \langle\varepsilon_1, \varepsilon_2, \ldots, \varepsilon_n\rangle$ where for each i, $\varepsilon_i = \pm 1$, and the difference between the number of $1s$ and $-1s$ is the signature $\mathrm{sgn}(h) = r^+(h) - r^-(h)$ of h. It is additive and is $= 0$ on any hyperbolic space, so it induces a homomorphism $\mathrm{sgn} : \mathrm{W}_1(K) \to \mathbb{Z}$. The rest of the theorem follows without difficulty. □

Of course this theorem applies to an arbitrary real closed field $\mathbf{R}$ since $(\mathbf{C} : \mathbf{R}) = 2$ where $\mathbf{C} = \mathbf{R}(i)$.

We mention also *Pfister's Local-Global Principle* for symmetric bilinear forms (Th. 3.6.2, [65]): *The kernel of the canonical map*

$$\mathrm{W}_1(K) \to \prod_P \mathrm{W}_1(K_P)$$

is the torsion subgroup of $\mathrm{W}_1(K)$ *and is a 2-primary Abelian group.* Here P runs over the orderings of K, and K_P is a real closure of K for P. In particular if K is *not* formally real – for example if the characteristic of K is > 0 – the Witt group itself is a 2-primary torsion Abelian group.

1.5.5 Local and global fields. References for this section are [54], §12 and §13 of Ch. 2 and Ch. 5 in [65], as well as Ch. VI, [39].

For the moment, K continues to be an arbitrary field of characteristic $\neq 2$.

Let (V, b) be a symmetric space. If $\langle\alpha_1, \alpha_2, \ldots, \alpha_n\rangle$ is any diagonalization of b, the *Hasse algebra* of b is defined to be the tensor product of quaternion algebras

$$\mathrm{S}(b) = \bigotimes_{i<j} (\alpha_i, \alpha_j)_K.$$

It is independent of the chosen diagonalization (up to isomorphism), and is a central simple algebra since it is the tensor product of central simple (quaternion) algebras (cf. Th. (1.2.9), p. 9). Its Brauer class (cf. §1.2)

$$\mathrm{s}(b) = [\mathrm{S}(b)] \in \mathrm{Br}_2(K)$$

is called the *Hasse invariant* of b. Here $\mathrm{Br}_2(K)$ denotes the group of Brauer classes of order 1 or 2.

We will usually denote the Brauer class $[(\alpha, \beta)_K]$ by the same symbol $(\alpha, \beta)_K$ as the quaternion algebra itself – or even by (α, β) if the base field is obvious. It should be clear from the context which is meant. If it is necessary to make the field K explicit, we write s_K instead of s.

The quaternion Brauer classes satisfy the following identities (cf. [54], pp. 146, 147):

1. $(\alpha, \beta) = 1$ if and only if the algebra (α, β) is not a division algebra, if and only if the "norm form" $xx^* = x_0^2 - \alpha x_1^2 - \beta x_2^2 + \alpha\beta x_3^2$ is isotropic, if and only if $\alpha \in \mathrm{N}_{K(\sqrt{\beta})/K}$.
2. $(\alpha, \beta) = (\beta, \alpha)$.
3. $(\alpha^2, \beta) = 1 = (\alpha, -\alpha) = (\alpha, 1 - \alpha)$.
4. $(\alpha, \alpha\beta) = (\alpha, -\beta)$.
5. $(\alpha, \beta)(\alpha, \gamma) = (\alpha, \beta\gamma), \qquad (\alpha, \beta)(\gamma, \beta) = (\alpha\gamma, \beta)$.
6. If L/K is a finite extension of local fields, $\alpha \in \dot{K}$ and $\lambda \in \dot{L}$, then
$$(\alpha, \lambda)_L = (\alpha, \mathrm{N}_{L/K}\lambda)_K.$$

See p. 102, [37] for the last property.

The following lemma follows easily (and tediously!) from the definition of s and the foregoing properties of the Brauer classes $(\alpha, \beta)_K$.

Lemma 1.5.1 *Let b be a nonsingular symmetric form of dimension n over K.*

(a) If $\alpha \in \dot{K}$,

$$\mathrm{s}(\alpha b) = \mathrm{s}(b)(\alpha, -1)_K^{n(n-1)/2}(\alpha, \det b)_K^{n-1}.$$

(b) If $b_1, \ldots, b_r$ are nonsingular symmetric forms over K,

$$\mathrm{s}(b_1 \perp \ldots \perp b_r) = \prod_i \mathrm{s}(b_i) \prod_{i<j} (\det b_i, \det b_j)_K.$$

(c) If $\alpha_1, \ldots, \alpha_m \in \dot{K}$,

$$\begin{aligned} &\mathrm{s}(\langle \det \alpha_1 b, \ldots, \det \alpha_m b\rangle) \\ = \ &\mathrm{s}(\langle \alpha_1^n, \ldots, \alpha_m^n\rangle)(\det b, -1)_K^{m(m-1)/2}(\det b, \alpha_1 \cdots \alpha_m)_K^{n(m-1)}. \end{aligned}$$

(d) *Let f be another nonsingular symmetric form, say of rank m. Then*

$$\begin{aligned}&\mathrm{s}(b\otimes f)\\ &= \mathrm{s}(b)^m\mathrm{s}(f)^n(\det b,-1)_K^{m(m-1)/2}(\det b,\det f)_K^{mn-1}(\det f,-1)_K^{n(n-1)/2}.\end{aligned}$$

(e) *If b is hyperbolic of dimension $n=2k$, then*

$$\mathrm{s}(b)=(-1,-1)_K^{k(k-1)/2}.$$

□

Theorem 1.5.6 *Let (V,b) and (V',b') be symmetric spaces over K.*

(a) *If they have dimension ≤ 3, they are equivalent if and only if*

$$\dim b=\dim b', \quad \det b=\det b', \quad \mathrm{s}(b)=\mathrm{s}(b'). \tag{1.43}$$

(b) *If V and V' have dimension 4 and (1.43) holds, the two forms are equivalent to scalar multiples of each other, i.e. $b'\cong\alpha b$ for some $\alpha\in\dot{K}$.*

(c) *If every symmetric space over K of dimension ≥ 5 is isotropic, two symmetric spaces V and V' (of any dimension) are equivalent if and only if (1.43) holds.*

In Scharlau's book [65], (a) is 13.5, p. 86, (b) is 14.1, p. 88, and (c) is 14.5, p. 91. □

Every symmetric space of dimension ≥ 5 over a local field is isotropic, and so by (c), we get

Theorem 1.5.7 *Two symmetric spaces over a local field are equivalent if and only if (1.43) holds.* □

In the case of a local field K, $\mathrm{Br}_2(K)$ is the group of order 2 (cf. p. 17), and we usually denote its elements by ± 1; in particular the Hasse invariant $\mathrm{s}(b)=\pm 1$.

There are two variations on the theme of the Hasse invariant. The first is usually called the *Witt invariant* $\mathrm{c}(b)$ – cf. p. 119, [39]) – and arises in a natural way from the theory of Clifford algebras (cf. p. 259). It is defined by

$$\begin{aligned}\mathrm{c}(b) &= \mathrm{s}(b) && \text{if } n\equiv 1,2 \bmod 8\\ \mathrm{c}(b) &= \mathrm{s}(b)(-1,-\det b)_K && \text{if } n\equiv 3,4 \bmod 8\\ \mathrm{c}(b) &= \mathrm{s}(b)(-1,-1)_K && \text{if } n\equiv 5,6 \bmod 8\\ \mathrm{c}(b) &= \mathrm{s}(b)(-1,\det b)_K && \text{if } n\equiv 7,8 \bmod 8.\end{aligned}$$

Unlike the Hasse invariant, it is well-defined on Witt classes, and in fact provides an isomorphism

$$I^2/I^3\to\mathrm{Br}_2(K)$$

for an arbitrary field K (of characteristic $\neq 2$ of course). It is characterized by the fact that it is well-defined on the Witt group $\mathrm{W}_1(K)$ and coincides with the Hasse invariant when $n\equiv 1$ or $2 \bmod 8$.

The other variant of the Hasse invariant is the *Hasse-Witt invariant* which we shall denote by $\mathrm{w}(b)$ (or $\mathrm{w}_K(b)$ if the field K is not clear from the context). It also takes its values in $\mathrm{Br}_2(K)$, and can be defined by the property that it too is well-defined on the Witt group and coincides with the Hasse invariant when $n \equiv 0$ or 1 mod 8. One can show that this means that

$$\begin{array}{rcll} \mathrm{w}(b) & = & \mathrm{s}(b) & \text{if } n \equiv 0, 1 \bmod 8 \\ \mathrm{w}(b) & = & \mathrm{s}(b)(-1, -\det b)_K & \text{if } n \equiv 2, 3 \bmod 8 \\ \mathrm{w}(b) & = & \mathrm{s}(b)(-1, -1)_K & \text{if } n \equiv 4, 5 \bmod 8 \\ \mathrm{w}(b) & = & \mathrm{s}(b)(-1, \det b)_K & \text{if } n \equiv 6, 7 \bmod 8. \end{array}$$

It is clear that Ths. 1.5.6 and 1.5.7 continue to hold if s is replaced by c or w.

Exercise. Show that the Witt and Hasse-Witt invariants are well-defined on Witt classes, and that they are characterized by this fact and the fact that they agree with the Hasse invariant when $n \equiv 1, 2 \bmod 8$ respectively $n \equiv 0, 1 \bmod 8$. □

Another result needed later is:

Lemma 1.5.2 *If b is a nonsingular symmetric form of rank n over K and $\alpha \in \dot{K}$, then*

$$\mathrm{w}(\alpha b) = \mathrm{w}(b)(\alpha, \operatorname{disc} b)_K^{n+1}.$$

For example, if $n \equiv 3 \bmod 8$,

$$\begin{aligned} \mathrm{w}(\alpha b) &= \mathrm{s}(\alpha b)(-1, -\det \alpha b)_K \\ &= \mathrm{s}(b)(\alpha, -1)_K^{(n-1)/2}(-1, -\alpha \det b)_K \quad \text{by Lem. 1.5.1(a) since } n \text{ is odd} \\ &= \mathrm{s}(b)(-1, -\det b)_K(\alpha, -1)_K^{(n+1)/2} = \mathrm{w}(b). \end{aligned}$$

The other 7 cases can be proved in a similar fashion, and are left to the reader. □

Theorem 1.5.8 The Hasse-Minkowski Theorem. *Two symmetric spaces over a global field K of characteristic $\neq 2$ are equivalent if and only if for each prime $\mathfrak{p}$ of K, their completions ("scalar extensions") over $K_{\mathfrak{p}}$ are equivalent .* □

Furthermore one need only check the real primes and those nonArchimedean primes $\mathfrak{p}$ at which 2 or the discriminant of the forms is not a unit.

Using Jacobson's theorem (cf. p. 49), one can use Ths. 1.5.5, 1.5.7 and 1.5.8 to solve the Hermitian case as well:

Theorem 1.5.9 *Suppose that h and g are nonsingular Hermitian forms over $(K, ^-)$, $^- \neq \mathrm{id}$.*

(a) If K_0 is a real closed field, h and g are equivalent if and only if they have the same dimension and signature.

(b) If K is a local field, h and g are equivalent if and only if they have the same dimension and discriminant.

(c) If K is a global field, $K = K_0(\sqrt{\delta})$, then h and g are equivalent if and only if they have the same dimension, discriminant and, at each real prime at which $\delta < 0$, the same signature. □

The next three lemmas are needed in Ch. 5.

Lemma 1.5.3 *The quadratic form $q = \langle \alpha, \beta, -\alpha\beta \rangle$, $\alpha\beta \neq 0$, over the global field K represents an infinite number of square classes of K.*

The quadratic form $\langle 1\rangle \perp -q$ is the norm form of the quaternion algebra $D = (\alpha,\beta)_K$. If D is split, it is isomorphic to $(\gamma,1)_K$ for any $\gamma \in \dot{K}$ (p. 16, (b)) and since the norm form of this algebra is $\langle 1,-\gamma,-1,\gamma\rangle$, $q \cong \langle \gamma,1,-\gamma\rangle$ and so represents γ.

Therefore we may assume that D is *not* split over K. Let S consist of the (positive even number of) primes of K at which $D_{\mathfrak{p}}$ is not split (cf. Ex. 5, p. 18). If $\mathfrak{p} \notin S$, $q_{\mathfrak{p}}$ represents all elements of $K_{\mathfrak{p}}$ by the argument used above for K.

Let $\mathfrak{p}_1,\ldots,\mathfrak{p}_n$ be primes not in S. For each i, choose a nonsquare γ_i of $K_{\mathfrak{p}_i}$. Let M be an arbitrary subset of $\mathrm{N} = \{1,\ldots,n\}$. By the approximation theorem for valuations (p. 18), we can find $\gamma \in \dot{K}$ which is arbitrarily close to γ_i at $\mathfrak{p}_i$ for each $i \in \mathrm{M}$, arbitrarily close to 1 for each $i \in \mathrm{N}\setminus\mathrm{M}$, and arbitrarily close to α at each prime of S. If the approximation is close enough,

$$\gamma \in \begin{cases} \gamma_i \dot{K}^2_{\mathfrak{p}_i} & \text{for all } i \in \mathrm{M}, \\ \dot{K}^2_{\mathfrak{p}_i} & \text{for all } i \in \mathrm{N}\setminus\mathrm{M}, \\ \alpha \dot{K}^2_{\mathfrak{p}} & \text{for each prime in } S, \end{cases}$$

by an obvious argument for Archimedean primes, and by the Local Square Theorem in the case of finite primes. It follows that γ is represented by q at *every* prime $\mathfrak{p}$ of K, and so also over K by the Hasse Principle for "representations" of one quadratic form by another (p. 189, [54]).

There are 2^n choices for M, leading to 2^n distinct square classes of K. Since n is arbitrary, the lemma follows. □

Lemma 1.5.4 *If K is a global field and L/K is a quadratic extension, then $[\dot{K} : \mathrm{N}_{L/K}\dot{L}]$ is infinite.*

We note first that the norms – and therefore also the non-norms – in a quadratic extension $L_{\mathfrak{P}}/K_{\mathfrak{p}}$ of local fields form an open subset of $K_{\mathfrak{p}}$. This follows easily from the Local Square Theorem according to which an element of $K_{\mathfrak{p}}$ sufficiently close to 1 is a square in $K_{\mathfrak{p}}$ – and hence a norm in $L_{\mathfrak{P}}/K_{\mathfrak{p}}$. Let $\mathfrak{p}_1,\mathfrak{p}_2,\ldots$ be a sequence of distinct primes of K which do not split in L/K (cf. 3.7, p. 378, [53]). Let N be a positive integer, and for each $i, 1 \le i \le N$, choose a nonnorm $\alpha_i \in \dot{K}_{\mathfrak{p}_i}$. Apply the approximation theorem (p. 18) to find $\beta_i \in \dot{K}$ which is close to α_i at $\mathfrak{p}_i$ and close to 1 at the other $\mathfrak{p}_j$. If the approximations are good enough, for each $i \neq j$, $\beta_i\beta_j^{-1}$ is not a norm in $K_{\mathfrak{p}_i}$ (nor in $K_{\mathfrak{p}_j}$ for that matter) and hence is certainly not a norm in the extension L/K. Thus $\beta_1,\ldots,\beta_N$ are in distinct cosets of $\dot{K}/\mathrm{N}_{L/K}\dot{L}$ and the lemma follows. □

Let L/K be a finite Abelian extension of local fields, $\mathcal{G} = \mathrm{Gal}(L/K)$ its Galois group, and $\bar{\mathrm{N}} : \dot{L}/\dot{L}^2 \to \dot{K}/\dot{K}^2$ the homomorphism induced by $\mathrm{N}_{L/K}$. The image of $\bar{\mathrm{N}}$ is $(\mathrm{N}_{L/K}\dot{L})\dot{K}^2/\dot{K}^2$.

Lemma 1.5.5 $|\ker \bar{\mathrm{N}}| = \frac{[\dot{L}:\dot{L}^2]}{|\mathrm{im}\,\bar{\mathrm{N}}|} = \frac{[\dot{L}:\dot{L}^2]}{[\dot{K}:\dot{K}^2]}[\mathcal{G} : \mathcal{G}^2]$, *which is* $= [\mathcal{G} : \mathcal{G}^2]$ *if K is nondyadic since $[\dot{K} : \dot{K}^2] = 4$ for any nondyadic field ((p. 217, [65]).*

By local class field theory (p. 18), there is a (canonical) isomorphism $\dot{K}/\mathrm{N}_{L/K}\dot{L} \cong \mathcal{G}$, so there is an isomorphism between their subgroups generated by squares:

$\dot{K}^2 \mathrm{N}_{L/K}\dot{L}/\mathrm{N}_{L/K}\dot{L} \cong \mathcal{G}^2$. Thus

$$\begin{aligned} |\mathrm{im}\,\bar{N}| &= [(\mathrm{N}_{L/K}\dot{L})\dot{K}^2 : \dot{K}^2] = \frac{[\dot{K} : \dot{K}^2]}{[\dot{K} : (\mathrm{N}_{L/K}\dot{L})\dot{K}^2]} \\ &= \frac{[\dot{K} : \dot{K}^2][(\mathrm{N}_{L/K}\dot{L})\dot{K}^2 : \mathrm{N}_{L/K}\dot{L}]}{[\dot{K} : \mathrm{N}_{L/K}\dot{L}]} \\ &= \frac{[\dot{K} : \dot{K}^2]|\mathcal{G}^2|}{|\mathcal{G}|} = \frac{[\dot{K} : \dot{K}^2]}{[\mathcal{G} : \mathcal{G}^2]}. \end{aligned}$$ □

1.5.6 Trace forms over local fields. If $(L, ^-)/(K, ^-)$ is a finite extension of fields with involution, an (involution) trace (cf. p. 35)

$$\mathrm{t} : (L, ^-) \to (K, ^-)$$

is simply a nonzero K-linear map $\mathrm{t} : L \to K$ which commutes with the involutions,

$$\mathrm{t}(\bar{\alpha}) = \overline{\mathrm{t}(\alpha)} \text{ for all } \alpha \in L. \tag{1.44}$$

For example if L/K is separable, the field trace $\mathrm{Tr}_{L/K}$ is such a trace (Lem. 1.3.5, p. 33). If h is a nonsingular Hermitian form over $(L, ^-)$, it is easy to see that

$$(\mathrm{t}h)(u, v) = \mathrm{t}(h(u, v)) : V \times V \to (K, ^-) \tag{1.45}$$

is a nonsingular symmetric or Hermitian form over $(K, ^-)$. It is called the *transfer* of h with respect to (or via) t.

Transfers are of central importance in representation theory and will be studied systematically later in §3.2.3. Here we wish to deal with a special case which is important for representations over local and global fields (cf. Ch. 5).

The next lemma uses the notion of "field discriminant". If L/K is a finite separable extension, the usual notion of field discriminant is the determinant of the matrix $(\mathrm{Tr}_{L/K}\, \lambda_i\lambda_j)$ where $\lambda_1, \ldots, \lambda_l$ is a basis of L/K – this is the same thing as the determinant of the transfer of the form $\langle 1 \rangle$ over L to K – but we shall call this the *determinant of the extension* L/K, and denote it by $\det L/K$, since this is more consistent with the terminology of reflexive forms. We define the *discriminant of* L/K to be

$$\mathrm{disc}\, L/K = \mathrm{disc}\,(\mathrm{Tr}_{L/K}\, \langle 1 \rangle) = (-1)^{l(l-1)/2} \det L/K$$

where $l = (L : K)$.

Lemma 1.5.6 *Let $(L, ^-)/(K, ^-)$ be a finite extension of involution fields of the same kind*[9] *such that L/K is separable. Suppose that $h : V \times V \to (L, ^-)$ is either a symmetric bilinear form or an Hermitian or skew Hermitian form of rank $\dim_L V = n$. Then $\det(L/K) \in K_0$, the transfer $\mathrm{Tr}_{L/K} h : V \times V \to (K, ^-)$ has determinant*

$$\det \mathrm{Tr}_{L/K} h = (\det L/K)^n \mathrm{N}_{L/K}(\det h),$$

and, if h is not skew Hermitian of odd rank, discriminant

$$\mathrm{disc}\, \mathrm{Tr}_{L/K} h = (\mathrm{disc}\, L/K)^n \mathrm{N}_{L/K}(\mathrm{disc}\, h).$$

If h is Hermitian, we also have

$$\det \mathrm{Tr}_{L/K} h = (\det L_0/K_0)^n \mathrm{N}_{L_0/K_0}(\det h).$$

[9]That is to say, if the involution on L is $\neq$ the identity, then it is also $\neq$ the identity on K.

Assume first that $\dim_L V = 1$, say with basis vector v. Let $\lambda_1, \ldots, \lambda_l$ be a basis of L_0/K_0; it is also a basis of L/K and $\lambda_1 v, \ldots, \lambda_l v$ is a basis of V over K. (Note that if $K \neq K_0$, say $K = K_0(\sqrt{\delta})$, then $L \neq L_0$ and $L = L_0(\sqrt{\delta})$.) If $h(v,v)\lambda_i = \sum_k \alpha_{ik}\lambda_k$ where $\alpha_{ik} \in K_0$ for all i and k, then the matrix of the transfer $\mathrm{Tr}_{L/K}h$ is

$$\begin{aligned}\left(\mathrm{Tr}_{L/K}h(\lambda_i v, \lambda_j v)\right) &= \left(\mathrm{Tr}_{L/K}(h(v,v)\lambda_i\lambda_j)\right) = \left(\mathrm{Tr}_{L/K}\sum_k \alpha_{ik}\lambda_k\lambda_j\right) \\ &= \left(\sum_k \alpha_{ik}\mathrm{Tr}_{L/K}\lambda_k\lambda_j\right) = (\alpha_{ik})\left(\mathrm{Tr}_{L/K}\lambda_k\lambda_j\right).\end{aligned}$$

Now $\det(\alpha_{ik}) = \mathrm{N}_{L/K}h(v,v)$ and $\det\left(\mathrm{Tr}_{L/K}\lambda_k\lambda_j\right) = \det L/K$ – these are standard facts of the theory of fields – and so the proof of the determinant formula in this case is complete, and the proof for general n follows easily by choosing an orthogonal basis $v_1, \ldots, v_n$ of V. (We note as well that $\det L/K = \det L_0/K_0$, $\operatorname{disc} L/K = \operatorname{disc} L_0/K_0$, and, if h is Hermitian, $\mathrm{N}_{L/K}h(v,v) = \mathrm{N}_{L_0/K_0}h(v,v)$ since $\lambda_1, \ldots, \lambda_l$ is a basis of L_0/K_0).

The formula for discriminants follows easily from that for determinants. □

This lemma and its proof are adaptations of those for the bilinear case in Lemma 2.2, [50].

Corollary 1.5.1 (Dedekind's discriminant formula) *Let M be a midfield of the finite separable extension L/K. Then*

$$\operatorname{disc} L/K = (\operatorname{disc} M/K)^{(L:M)}\mathrm{N}_{M/K}\operatorname{disc} L/M.$$

This is proved by applying the lemma to the formula

$$\mathrm{Tr}_{L/K}\langle 1\rangle = \mathrm{Tr}_{M/K}(\mathrm{Tr}_{L/M}\langle 1\rangle).$$ □

Let $\mathrm{t} : (L, ^-) \to (K, ^-)$ be an involution trace (p. 35).

Theorem 1.5.10 *Suppose that L/K is an Abelian extension of local fields and that the involution $^-$ on L is $\neq \mathrm{id}$. If h_1 and h_2 are inequivalent nonsingular Hermitian forms over $(L, ^-)$, then their traces $\mathrm{t}h_1$ and $\mathrm{t}h_2$ are also inequivalent.*

We first prove the following special case.

Lemma 1.5.7 *Suppose that the fixed field L_0 of the involution on L is $= K$. If h_1 and h_2 have the same rank but different determinants, then the symmetric bilinear forms $\mathrm{t}h_1$ and $\mathrm{t}h_2$ have the same rank and determinant but different Hasse invariants.*

Note that $\mathrm{t}(h_1 \perp h_2) = \mathrm{t}h_1 \perp \mathrm{t}h_2$. Because Hermitian forms over L are classified by rank and discriminant (cf. Table 1, p. 61), $h_i \cong \langle 1, \ldots, 1, \det h_i\rangle$ and so by Lemma 1.5.1(b),

$$\mathrm{s}(\mathrm{t}h_i) = \mathrm{s}(\mathrm{t}\langle 1, \ldots, 1\rangle) \cdot \mathrm{s}(\mathrm{t}\langle \det h_i\rangle) \cdot (\det \mathrm{t}\langle 1, \ldots, 1\rangle,\ \det \mathrm{t}\langle \det h_i\rangle)_K.$$

It follows that it suffices to prove the lemma in the case that h_1 has rank 1.

So suppose $h_1 = \langle \alpha\rangle$, say $h_1(v,v) = \alpha \in K$. Write $L = K(\sqrt{\delta})$. By (1.44), $\mathrm{t}(\sqrt{\delta}) = 0$ and so $\mathrm{t}(1) \neq 0$. Then $\mathrm{t}h_1 = \langle \alpha\mathrm{t}(1), -\alpha\delta\mathrm{t}(1)\rangle$ in the K-basis $v, \sqrt{\delta}v$ of

V. Thus $\det \mathrm{t}h_1 = -\delta$ which is independent of α. So $\mathrm{t}h_1$ and $\mathrm{t}h_2$ have the same determinant. Furthermore the Hasse invariant of $\mathrm{t}h_1$ is

$$\mathrm{s}(\mathrm{t}h_1) = (\alpha\mathrm{t}(1), -\alpha\delta\mathrm{t}(1))_K = (\alpha\mathrm{t}(1), \delta)_K$$

by identity (c) on p. 16. This *does* depend on the choice of α since δ is a nonsquare in K (cf. Ex. 4, p. 17). It follows that $\mathrm{t}h_1$ and $\mathrm{t}h_2$ have different Hasse invariants since there are only two possible determinants because $[\dot{K} : \mathrm{N}_{L/K}\dot{L}] = 2$ (cf. (1.23)). □

For the moment we assume the following theorem.

Theorem 1.5.11 *Suppose that L/K is a finite Abelian extension of local fields and $\mathrm{t} : L \to K$ is a nonzero K-linear map. If $b, b' : V \times V \to L$ are nonsingular symmetric bilinear forms with the same discriminant but different Hasse invariants, then their traces $\mathrm{t}b$ and $\mathrm{t}b'$ are not equivalent.*

This theorem is due to J. Milnor [50] in the more general case when L/K is finite and separable.

We now show that Theorem 1.5.10 follows from this theorem and Lemma 1.5.7.

Suppose first that the (nonidentity) involution $^-$ on L is the identity on K, and that L_0 is the subfield of L of elements fixed by it. We have $L = L_0 \oplus L_0\sqrt{\delta}$ for some $\delta \in \dot{L}_0$. Thus if $\alpha_0 \in L_0$, $\mathrm{t}\left(\overline{\alpha_0\sqrt{\delta}}\right) = -\mathrm{t}(\alpha_0\sqrt{\delta}) = \overline{\mathrm{t}(\alpha_0\sqrt{\delta})} = \mathrm{t}(\alpha_0\sqrt{\delta})$. Therefore $\mathrm{t}(L_0\sqrt{\delta}) = 0$ and we can factor t through the projection $\pi : L = L_0 \oplus L_0\sqrt{\delta} \to L_0$: $\mathrm{t} = \mathrm{s}\pi$ where $\mathrm{s} = \mathrm{t}|_{L_0}$. Note that $\pi(\bar{\alpha}) = \pi(\alpha)$ for all $\alpha \in L$, so π is an involution trace.

By the lemma, the symmetric bilinear forms πh_1 and πh_2 over L_0 have the same rank and discriminant but different Hasse invariants. Therefore by Theorem 1.5.11 the symmetric forms $\mathrm{s}\pi h_1 = \mathrm{t}h_1$ and $\mathrm{s}\pi h_2 = \mathrm{t}h_2$ are not equivalent.

Suppose now that the involution is *not* the identity on K. In this case we may assume that $\delta \in K_0$, and that $K = K_0(\sqrt{\delta})$. Let b_i be the nonsingular symmetric bilinear form $b_i(u,v) = h_i(u,v) + \overline{h_i(u,v)} = \mathrm{Tr}_{L/L_0}(h_i(u,v))$ over L_0. By Lem. 1.5.7, b_1 and b_2 have the same rank and determinant but different Hasse invariants. Let $\mathrm{t}_0 : L_0 \to K_0$ denote the restriction of t to L_0. By Theorem 1.5.11 $\mathrm{t}_0 b_1$ and $\mathrm{t}_0 b_2$ are not equivalent as symmetric forms over K_0. But

$$\mathrm{t}h_i(u,v) + \overline{\mathrm{t}h_i(u,v)} = \mathrm{t}b_i(u,v) = \mathrm{t}_0 b_i(u,v),$$

and so $\mathrm{t}h_1$ and $\mathrm{t}h_2$ are not equivalent. This proves Theorem 1.5.10. □

We now proceed to the proof of Theorem 1.5.11.

The determinant of L/K is the determinant of the symmetric bilinear form $\mathrm{Tr}_{L/K}(\alpha\beta) : L \times L \to K$. For our purposes we need to consider the "t-determinant" of L/K, replacing $\mathrm{Tr}_{L/K}(\alpha\beta)$ by $\mathrm{t}(\alpha\beta)$:

$$\det{}_{\mathrm{t}} L/K = \det(\mathrm{t}(\alpha\beta)) = \det(\mathrm{t}(\lambda_i\lambda_j))$$

where $\lambda_1, \ldots, \lambda_l$ is a basis of L/K. It is well-defined mod $\dot{K}^2$. The proof of the following lemma is identical to that of Lem. 1.5.6 when the trace $\mathrm{Tr}_{L/K}$ is replaced by t.

Lemma 1.5.8 *If b is a symmetric form over L of dimension n, the determinant of the transfer $\mathrm{t}b$ over K is $(\det_{\mathrm{t}} L/K)^n \mathrm{N}_{L/K}(\det b)$.* □

Lemma 1.5.9 *In order to prove Theorem 1.5.11 it suffices to find two nonsingular symmetric forms b_1 and b_2 over L with the same rank and discriminant but with* $\mathrm{s}(\mathrm{t}b_1) \neq \mathrm{s}(\mathrm{t}b_2)$ ($\mathrm{s} =$ *the Hasse invariant*).

Since $\mathrm{t}b_1$ is not equivalent with $\mathrm{t}b_2$, b_1 and b_2 are also nonequivalent, and so their Hasse invariants are different (cf. Table 1, p. 61). Now we claim that $b \perp b_1$ and $b' \perp b_2$ are equivalent. They certainly have the same rank and discriminant. As for their Hasse invariants, by Lemma 1.5.1(b),

$$\begin{aligned} \mathrm{s}(b \perp b_1) &= \mathrm{s}(b)\mathrm{s}(b_1)(\det b, \det b_1)_L \\ &= (-\mathrm{s}(b'))(-\mathrm{s}(b_1))(\det b', \det b_2)_L = \mathrm{s}(b' \perp b_2). \end{aligned}$$

Thus we also have $\mathrm{t}b \perp \mathrm{t}b_1 \cong \mathrm{t}b' \perp \mathrm{t}b_2$, so

$$\mathrm{s}(\mathrm{t}b)\mathrm{s}(\mathrm{t}b_1)(\det \mathrm{t}b, \det \mathrm{t}b_1)_K = \mathrm{s}(\mathrm{t}b')\mathrm{s}(\mathrm{t}b_2)(\det \mathrm{t}b', \det \mathrm{t}b_2)_K.$$

Since $(\det \mathrm{t}b, \det \mathrm{t}b_1)_K = (\det \mathrm{t}b', \det \mathrm{t}b_2)_K$ by Lem. 1.5.6, and $\mathrm{s}(\mathrm{t}(b_1) = -\mathrm{s}(\mathrm{t}(b_1)$, $\mathrm{s}(\mathrm{t}b) = -\mathrm{s}(\mathrm{t}b')$. □

We now proceed with the proof of Theorem 1.5.11. The proof is divided into 4 cases.

Case 1. L contains an element α_0 whose norm in K is nonsquare. By §1.2.5 there is $\beta \in K$ such that $(\beta, \mathrm{N}_{L/K}\alpha_0)_K = -1$. Consider the binary space V_0 over L with symmetric form $b_0 = \langle \alpha, -\alpha\beta \rangle$ with respect to the orthogonal basis u_0, v_0, where $\alpha \in \dot{L}$ is to be determined. For any choice of α its determinant is $-\beta$. We now calculate the Hasse invariant of $\mathrm{t}b_0$, freely using Lem. 1.5.1 and the properties of $(\cdot,\cdot)_K$ on p. 53. Let b_0' respectively b_0'' be the restriction of b_0 to the subspace Lu_0 respectively Lv_0.

Choose an orthogonal basis for the space $(Lu_0, \mathrm{t}b_0')$ over K. Then $\mathrm{t}b_0' = \langle \alpha_1, \ldots, \alpha_d \rangle$ (where $d = (L : K)$) for certain $\alpha_i \in \dot{K}$. And there is an orthogonal basis for $(Lv_0, \mathrm{t}b_0'')$ in which $\mathrm{t}b_0'' = \langle -\alpha_1\beta, \ldots, -\alpha_d\beta \rangle$. If s' is the Hasse invariant of $\mathrm{t}b_0'$, that of $\mathrm{t}b_0''$ is

$$\begin{aligned} \mathrm{s}'' &= \prod_{i<j}(-\alpha_i\beta, -\alpha_j\beta)_K \\ &= \mathrm{s}'(-\beta, \alpha_1 \ldots \alpha_d)_K^{d-1}(-\beta, -\beta)_K^{d(d-1)/2}. \end{aligned}$$

Thus the Hasse invariant of $\mathrm{t}b_0$ is

$$\begin{aligned} \mathrm{s} &= \mathrm{s}'\mathrm{s}''(\alpha_1 \ldots \alpha_d, \alpha_1 \ldots \alpha_d(-\beta)^d)_K \\ &= (-\beta, \alpha_1 \ldots \alpha_d)_K^{2d-1}(-1, \alpha_1 \ldots \alpha_d)_K(-\beta, -\beta)_K^{d(d-1)/2} \\ &= (\beta, \alpha_1 \ldots \alpha_d)_K(-\beta, -\beta)_K^{d(d-1)/2}. \end{aligned}$$

By Lemma 1.5.8, $\alpha_1 \ldots \alpha_d = (\det_{\mathrm{t}} L/K)\mathrm{N}_{L/K}\alpha$. It follows that s is the product of $(\beta, \mathrm{N}_{L/K}\alpha)_K$ and a factor which is independent of α, so we get different values for s if we choose $\alpha = \alpha_0$ and $\alpha = 1$. By the foregoing lemma, this proves Theorem 1.5.11 in this case.

Case 2. L has degree 2 over K. Then $\mathrm{N}_{L/K}\dot{L}$ has index 2 in $\dot{K}$, while $\dot{K}^2$ has index ≥ 4. (cf. pp. 163, 167, [54]). Thus we can apply Case 1.

Case 3. L has odd degree over K. The norm of any nonsquare α_0 in K is an odd power of α_0 and so is not a square. Apply Case 1 again.

Case 4. L/K Abelian. There is a tower of fields

$$K = K_0 \subset K_1 \subset K_2 \subset \cdots \subset K_r = L$$

in which each successive extension K_{i+1}/K_i has prime degree.

Since $\mathrm{Hom}_K(L, K)$ is a one dimensional vector space over L (with the action $\lambda_1\psi(\lambda_2) = \psi(\lambda_1\lambda_2)$), there is a $\lambda_1 \in L$ such that $\mathrm{t}(\lambda) = \mathrm{Tr}_{L/K}(\lambda_1\lambda)$ for all $\lambda \in L$. Now if $b \not\cong b'$, then certainly $\lambda_1 b \not\cong \lambda_1 b'$. Apply Case 2 or 3 to get $\mathrm{Tr}_{L/K_{r-1}}(\lambda_1 b) \not\cong \mathrm{Tr}_{L/K_{r-1}}(\lambda_1 b')$. Again apply Case 2 or 3 to get $\mathrm{Tr}_{L/K_{r-2}}(\lambda_1 b) \not\cong \mathrm{Tr}_{L/K_{r-2}}(\lambda_1 b')$, and so on. Eventually we get $\mathrm{Tr}_{L/K}(\lambda_1 b) \not\cong \mathrm{Tr}_{L/K}(\lambda_1 b')$, i.e. $\mathrm{t}b \not\cong \mathrm{t}b'$. □

There are some sharper results for trace forms of local fields, due mainly to V.P. Gallagher [29] and M. Epkenhans [22].

1.5.7 Equivalence of reflexive forms over division algebras. Although forms over division algebras are not of direct interest to us as far as representations are concerned, they do nevertheless arise in the consideration of representations over fields. In particular it is important to determine the equivalence classes of Hermitian and skew Hermitian forms over a central division K-algebra with involution, when K is real closed, local or global. The following table gives invariants in all the "classical" cases, and is derived mainly from the table on p. 347, [65].

Table 1

Type	finite	R	local	global
symmetric	dim, det	dim, sgn	dim, det, s	Hasse Prin.
skew-symmetric	dim	dim	dim	dim
Herm/field	dim	dim, sgn	dim, det	dim, det, $\mathrm{sgn}_\mathfrak{p}$
Herm/quat	$\nexists$	dim, sgn	dim	dim, $\mathrm{sgn}_\mathfrak{p}$
skew-Herm/quat	$\nexists$	dim	dim, det	dim, det, b
Herm/skew(2nd kd)	$\nexists$	$\nexists$	$\nexists$	dim, det, $\mathrm{sgn}_\mathfrak{p}$

Explanations. 1. R stands for "real closed", s for the Hasse invariant, skew(2nd kd) for a division algebra with an involution of the second kind, b for the Bartels invariant (discussed below), and Hasse Prin. for the Hasse Principle (p. 55).

2. It is understood that in the case of forms over the quaternions, an involution of the 1st kind is conjugation * .

3. $\mathrm{sgn}_\mathfrak{p}$ (K global). The signature is only defined when $\mathrm{char}\, K = 0$, i.e. when K is a number field. In the above table it can be ignored when the characteristic is $\neq 0$.

Assume that $\mathrm{char}\, K = 0$. There are three cases:

(i) *Quadratic extension K/K_0.* In this case the signature $\mathrm{sgn}_\mathfrak{p} h$ of an Hermitian form $h : V \times V \to (K, ^-)$ is defined for real primes $\mathfrak{p}$ of K_0 (or the corresponding orders P of K_0 – cf. Prop. 1.4.1, p. 42) for which K/K_0 is definite, i.e. $K = K_0(\sqrt{\delta})$ with $\delta < 0$ at P. For such $\mathfrak{p}$, $K_{0\mathfrak{p}} \cong \mathbb{R}$, $K_{0\mathfrak{p}} \otimes_{K_0} K \cong \mathbb{C}$, and $\mathrm{sgn}_\mathfrak{p} h$ is the signature of the "classical" Hermitian form

$$h_\mathfrak{p} = h^{\mathbb{C}} : V^{\mathbb{C}} \times V^{\mathbb{C}} \to (\mathbb{C}, ^*).$$

(ii) *Quaternion division algebra $(D, ^*)$.* Here the signature $\mathrm{sgn}_\mathfrak{p} h$ of an Hermitian form $h : V \times V \to (D, ^*)$ is defined for the real primes $\mathfrak{p}$ of the center K (or the corresponding orderings of K) for which the quaternion algebra $K_\mathfrak{p} \otimes_K D$ is not split (and so is $\cong \mathbb{H}$). It is the signature of the Hermitian form

$$h_\mathfrak{p} = h^\mathbb{R} : V^\mathbb{R} \times V^\mathbb{R} \to (\mathbb{H}, {}^*).$$

See p. 41.

(iii) *Noncommutative division algebra* $(D, {}^-)$ *with an involution of the second kind.* The signature $\operatorname{sgn}_\mathfrak{p} h$ of an Hermitian form $h : V \times V \to (D, {}^-)$ is defined for the real primes $\mathfrak{p}$ of K_0 for which $(K, {}^-)$ is definite. If $\mathfrak{p}$ is such a prime,

$$K_{0\mathfrak{p}} \otimes_{K_0} D \cong (K_{0\mathfrak{p}} \otimes_{K_0} K) \otimes_K D = K_\mathfrak{P} \otimes_K D = D_\mathfrak{P}$$

where $\mathfrak{P}$ is the unique (complex) prime of K lying above $\mathfrak{p}$. The involution $\mathrm{id} \otimes {}^-$ on $K_{0\mathfrak{p}} \otimes_{K_0} K = K_\mathfrak{P} \cong \mathbb{C}$ is complex conjugation * and it follows that the involution algebra $(D_\mathfrak{P}, {}^* \otimes {}^-)$ is $\cong (\mathrm{M}(d, \mathbb{C}), {}^\sim)$ where

$$\widetilde{a} = b^{-1} a^{t*} b, \quad \text{for some } b \text{ satisfying } b^{t*} = b$$

(Th. 1.3.6(a), p. 30). Suppose that $h : V \times V \to (D, {}^-)$ is a nonsingular Hermitian form, and that

$$h_\mathfrak{P} : V_\mathfrak{P} \times V_\mathfrak{P} \to (D_\mathfrak{P}, {}^* \otimes {}^-) \cong (\mathrm{M}(d, \mathbb{C}), {}^\sim) \tag{1.46}$$

is its "completion" (extension of scalars from K to $\mathbb{C}$). If $\dim_D V = m$, by Th. 3.2.9 (p. 120) there is a matrix $\underline{h} \in \mathbf{GL}(md, \mathbb{C})$ such that

$$h_\mathfrak{P}(u, v) = u \underline{h} v^{t*} b \quad \text{for all } u, v \in \mathbb{C}^{d \times md}, \qquad \underline{h}^{t*} = \underline{h},$$

where we have identified $V_\mathfrak{P} = \mathbb{C} \otimes_K V$ with $\mathbb{C}^{d \times md}$. (The reader is advised to take this on faith at this point and return to it later when it becomes necessary in subsequent developments). Thus $\underline{h}$ is a nonsingular Hermitian matrix over $\mathbb{C}$; the signature $\operatorname{sgn}_\mathfrak{p} h$ is defined to be the signature of the corresponding Hermitian form

$$\check{h} : \mathbb{C}^{md} \times \mathbb{C}^{md} \to (\mathbb{C}, {}^*), \qquad \check{h}(u, v) = u \underline{h} v^{t*}.$$

This definition depends on the identification of $D_\mathfrak{P}$ with $\mathrm{M}(d, \mathbb{C})$ and the choice of b. As we have defined it, a different identification and choice of b would lead to the matrix $\alpha c_0 \underline{h} c_0^{t*}$ instead of $\underline{h}$, where $\alpha \in \dot{\mathbb{R}}$ and $c_0 \in \mathbf{GL}(md, \mathbb{C})$, by Th. 3.2.10 (p. 125). Thus with a change in the identification and the choice of b, $\operatorname{sgn} h_\mathfrak{p}$ will either be the same (if $\alpha > 0$) or will change sign. Therefore if h' is another Hermitian form of the same rank, $\operatorname{sgn} h_\mathfrak{p}$ and $\operatorname{sgn} h'_\mathfrak{p}$ will be equal with a particular choice of an identification and b if and only if they are equal with any other choice.

The definition of $\operatorname{sgn}_\mathfrak{p} h$ given by Scharlau on pp. 375-376 in [65] is the same as that given here. He defines the form $h_\mathfrak{P}$ as above, and then "scales" it by $b^{-1} : (h_\mathfrak{P} b^{-1})(u, v) = h_\mathfrak{P}(u, v) b^{-1}$. This scaled form is an Hermitian form over $(\mathrm{M}(d, \mathbb{C}), {}^{t*})$. It is then noted that it can be expressed by an Hermitian matrix in $\mathbb{C}^{md \times md}$, which in fact is the matrix $\underline{h}$ defined above.

4. The results on Hermitian forms over $(K, {}^-)$ can be applied to the skew Hermitian case since if h is skew Hermitian over $(K, {}^-)$, $\sqrt{\delta} h$ is Hermitian ($K = K_0(\sqrt{\delta})$). Similarly skew Hermitian forms over a central division algebra $(D, {}^-)$ with an involution of the 2nd kind over a global field K, can be scaled by $\sqrt{\delta}$ to get Hermitian forms over $(D, {}^-)$.

5. Suppose $(D, ^-)$ is a quaternion division algebra with an orthogonal involution. The pure quaternions D_0 are stable under any involution since

$$\dot{D}_0 = \{d \in D : \ d \notin K, \ d^2 \in K\},$$

and so the 1-eigenspace $(D_0)_1$ of $^-$ has dimension 2 and $(D_0)_{-1}$ has dimension 1. This means that i, j and k can be chosen so that $\bar{i} = -i$, $\bar{j} = j$ and $\bar{k} = k$, and this in turn implies that $\bar{d} = i^{-1}d^*i$ for all $d \in D$.

Now suppose h is an Hermitian form over $(D, ^-)$. Then hi (defined like hb^{-1} in 3.) is a skew Hermitian form over $(D, ^*)$ and it is straightforward to check that the correspondence $h \leftrightarrow hi$ is bijective between Hermitian forms over $(D, ^-)$ and skew Hermitian forms over $(D, ^*)$, and also between skew Hermitian forms over $(D, ^-)$ and Hermitian forms over $(D, ^*)$. These are in fact category isomorphisms.

Thus statements concerning equivalence of ε-Hermitian forms over a quaternion algebra with conjugation as involution apply equally well to $-\varepsilon$-Hermitian forms over the same algebra with an orthogonal involution.

6. In the case of (nonsingular) skew Hermitian forms over quaternion algebras over local fields, we note that the determinant takes its values in $\dot{K}/\dot{K}^2$ and that furthermore it can take on *any* value if the dimension is > 1, while it misses the single value $(-1)\dot{K}^2$ when the dimension is 1. This, and the fact that dim and det characterize equivalence in this case, constitute "Tsukamoto's Theorem" (p. 363, [65]).

7. The Hasse Principle and the Bartels Invariant. The Hasse Principle applies in all cases in the above table except skew Hermitian forms over a quaternion division algebra over a global field with involution = conjugation, that is to say, it is possible for forms in this case to be locally equivalent at all primes but not globally equivalent. In fact if D is the quaternion algebra in question and s is the number of primes $\mathfrak{p}$ of K at which $D_\mathfrak{p}$ is not split, then the number of global equivalence classes which correspond to a single local equivalence class is 2^{s-2} – thus the Hasse Principle holds if and only if $s = 2$ (the minimum number if D is not split). See [65], §§10.4.5 – 10.4.7, pp. 370-371. If h and g are skew Hermitian forms over $(D, ^*)$, then ("Bartels theorem") it can be shown that they are equivalent if and only if they have the same rank and determinant (or are locally equivalent at each prime $\mathfrak{p}$ of K) and have the same Bartels invariant $\mathrm{b}(h, g) \in \mathrm{Br}(K)/\langle [D] \rangle$. See [3] or [65] for its definition. We give its properties later on p. 203.

The remaining cases, in which the Hasse Principle *does* hold, require a little explanation. The symmetric case is of course the classical case, while the skew symmetric case is trivial since nonsingular forms of the same dimension over any field are equivalent. Hermitian forms over a division algebra with an involution of the second kind are covered in 10.6.1, [65].

Now consider the case of Hermitian forms over a nonsplit quaternion algebra $(D, ^*)$ with conjugation as involution, say $D = (\alpha, \beta)_K$. Let h and g be nonsingular Hermitian forms over $(D, ^*)$, and suppose that[10] $h_\mathfrak{p} \cong g_\mathfrak{p}$ for all primes $\mathfrak{p}$ of K.

[10] If $D_\mathfrak{p}$ is split, so $\cong \mathrm{M}(2, K_\mathfrak{p})$, then $h_\mathfrak{p}$ is an Hermitian form over a matrix algebra – such forms are dealt with in detail in Ch. 3.

Since the diagram

$$\begin{array}{ccc} D & \longrightarrow & D_{\mathfrak{p}} \\ \downarrow{\scriptstyle \mathrm{trd}} & & \downarrow{\scriptstyle \mathrm{trd}} \\ K & \longrightarrow & K_{\mathfrak{p}} \end{array}$$

is commutative, the trace forms satisfy $(\mathrm{trd}\, h)_{\mathfrak{p}} \cong \mathrm{trd}\, h_{\mathfrak{p}}$ and so $(\mathrm{trd}\, h)_{\mathfrak{p}} \cong (\mathrm{trd}\, g)_{\mathfrak{p}}$ for all $\mathfrak{p}$ which, since these are symmetric bilinear forms, implies that $\mathrm{trd}\, h \cong \mathrm{trd}\, g$ and so $h \cong g$ by Jacobson's Theorem (p. 49).

An entirely similar proof works in the case of Hermitian forms over $(K, ^-)$. □

Existence theorems. We record here conditions for the existence of symmetric forms over local and global fields.

Theorem 1.5.12 *Let V be a vector space of dimension ≥ 2 over the local field K. Suppose that $(\alpha\dot{K}^2, \varepsilon) \in (\dot{K}/\dot{K}^2) \times \{\pm 1\}$ is given. Then there is a symmetric bilinear form $b : V \times V \to K$ with $\det b = \alpha$ and $\mathrm{s}(b) = \varepsilon$ if and only if $\varepsilon = 1$ when $\dim V = 2$ and $\alpha =_2 -1$.*

This is essentially 63:23, p. 171, [54]. (Of course b is unique up to equivalence by Table 1, p. 61). □

Theorem 1.5.13 *Suppose that for each prime $\mathfrak{p}$ of the global field K, $(U_{\mathfrak{p}}, f_{\mathfrak{p}})$ is a symmetric space over $K_{\mathfrak{p}}$ of the same dimension for each $\mathfrak{p}$. In order that there exist a symmetric space (V, b) over K such that $(V_{\mathfrak{p}}, b_{\mathfrak{p}}) \cong (U_{\mathfrak{p}}, f_{\mathfrak{p}})$ for each $\mathfrak{p}$, it is necessary and sufficient that there exists $d \in K$ such that $\det f_{\mathfrak{p}} \equiv d \bmod \dot{K}_{\mathfrak{p}}^2$ for each $\mathfrak{p}$, that $\mathrm{s}_{\mathfrak{p}} f_{\mathfrak{p}} = 1$ for all but finitely many $\mathfrak{p}$, and that $\prod_{\mathfrak{p}} \mathrm{s}_{\mathfrak{p}} f_{\mathfrak{p}} = 1$.*

This is proved for algebraic number fields (of finite degree over $\mathbb{Q}$) in 72:1, p. 203, [54], and the proof there can be easily seen to extend to global fields (of characteristic $\neq 2$, of course). We leave it as an exercise to show that the condition $\prod_{\mathfrak{p}} \mathrm{s}_{\mathfrak{p}} f_{\mathfrak{p}} = 1$, using the slightly different definition of the Hasse invariant there, is equivalent to the same relation using our definition. □

1.6 Grothendieck groups

Suppose that $(\mathbf{M}, +)$ is an Abelian monoid – that is to say $\mathbf{M}$ has a binary law of composition $x+y$ which is commutative and associative. Then we can construct a group $\mathrm{K}(\mathbf{M})$ called the *Grothendieck group of* $\mathbf{M}$ as follows (an alternative approach can be found in §7, Ch.1, [41]).

Take the set of ordered pairs $\mathbf{M} \times \mathbf{M}$ and say that two pairs (x_1, y_1) and (x_2, y_2) are related, $(x_1, y_1) \sim (x_2, y_2)$, if there is $z \in \mathbf{M}$ such that $x_1 + y_2 + z = x_2 + y_1 + z$. This is easily seen to be an equivalence relationship.

Denote the equivalence class of (x_1, y_1) by $[x_1, y_1]$, and let $\mathrm{K}(\mathbf{M})$ be the set of equivalence classes. It can be made into an Abelian group as follows. (One should intuitively think of $[x_1, y_1]$ as the difference $x_1 - y_1$). Addition in $\mathrm{K}(\mathbf{M})$ is defined by

$$[x_1, y_1] + [x_2, y_2] = [x_1 + x_2, y_1 + y_2],$$

and is easily seen to be well-defined, associative and commutative. The equivalence class $[x, x]$ for any $x \in \mathbf{M}$ is a 0 for addition – note that $[x, x] = [y, y]$ for all $x, y \in \mathbf{M}$. And $[y, x]$ is the negative of $[x, y]$ since $[x, y] + [y, x] = [x + y, x + y] = 0$. Thus $\mathrm{K}(\mathbf{M})$ *is* an Abelian group.

Since $[x+y, y] = [x+z, z]$ for all $x, y, z \in M$, the map $\iota : \mathbf{M} \to \mathrm{K}(\mathbf{M})$ given by $\iota(x) = [x+y, y]$ for any $y \in \mathbf{M}$ is well-defined. It is a homomorphism, and

$$[x_1, y_1] = [x_1 + 2z, y_1 + 2z] = [x_1 + z, z] + [z, y_1 + z] = \iota(x_1) - \iota(y_1).$$

Thus the image of $\mathbf{M}$ generates $\mathrm{K}(\mathbf{M})$ as an Abelian group.

We note that

$$\iota \text{ is injective if and only if } \mathbf{M} \text{ has cancellation.} \tag{1.47}$$

Indeed $\iota(x_1) = \iota(x_2)$ if and only if $x_1 + 2y + z = x_2 + 2y + z$ for some y and z, which implies that $x_1 = x_2$ if $\mathbf{M}$ has cancellation. Conversely suppose $x_1 + z = x_2 + z$. Then $\iota(x_1) = [x_1 + z, z] = [x_2 + z, z] = \iota(x_2)$, so $x_1 = x_2$ if ι is injective.

$\mathrm{K}(\mathbf{M})$ has a universal mapping property:

Proposition 1.6.1 *Let $\varphi : \mathbf{M} \to A$ be a homomorphism of $\mathbf{M}$ into an Abelian group A. Then there is a unique homomorphism $\widehat{\varphi} : \mathrm{K}(\mathbf{M}) \to A$ making*

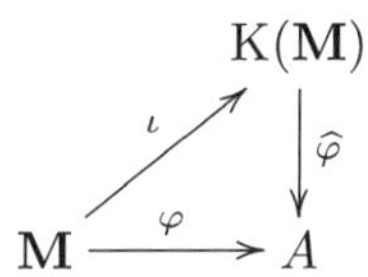

(1.48)

commutative.

If $[x_1, y_1] = [x_2, y_2]$, say $x_1 + y_2 + z = x_2 + y_1 + z$, then applying φ to this equation and cancelling $\varphi(z)$ leads to $\varphi(x_1) - \varphi(y_1) = \varphi(x_2) - \varphi(y_2)$. Thus we may define

$$\hat{\varphi}[x_1, y_1] = \varphi(x_1) - \varphi(y_1).$$

It is straightforward to check that this *is* a homomorphism $\mathrm{K}(\mathbf{M}) \to A$ and that it makes (1.48) commutative. Furthermore if $\psi : \mathrm{K}(\mathbf{M}) \to A$ is another homomorphism such that $\psi\iota = \varphi$, then $\psi[x_1, y_1] = \psi(\iota(x_1)) - \psi(\iota(y_1)) = \hat{\varphi}[x_1, y_1]$, i.e. $\psi = \hat{\varphi}$. □

In the usual way, one gets:

Corollary 1.6.1 K *is a functor from the category of Abelian monoids to the category of Abelian groups.* □

A commutative monoid is a *commutative semiring* if it has, in addition to the sum $x + y$, a product xy which is associative, commutative and distributive over addition.

Proposition 1.6.2 *Let $\mathbf{S}$ be a commutative semiring.*

(a) The product

$$[x_1, y_1] \cdot [x_2, y_2] = [x_1 x_2 + y_1 y_2,\ x_1 y_2 + y_1 x_2] \tag{1.49}$$

makes $\mathrm{K}(\mathbf{S})$ into a commutative ring (without a multiplicative identity, in general), and $\iota : \mathbf{S} \to \mathrm{K}(\mathbf{S})$ is a semiring homomorphism. If $1_{\mathbf{S}}$ is a multiplicative identity of $\mathbf{S}$, then $\iota(1_{\mathbf{S}})$ is a multiplicative identity for $\mathrm{K}(\mathbf{S})$. If in addition $\mathbf{H}$ is a nonempty subset of $\mathbf{S}$ with the property $\mathbf{SH} \subset \mathbf{H}$, then the subgroup $\langle \iota\mathbf{H} \rangle$ of $\mathrm{K}(\mathbf{S})$ generated by $\iota\mathbf{H}$ is an ideal of $\mathrm{K}(\mathbf{S})$.

(b) Let $\mathbf{M}$ be an Abelian monoid, and $\mathbf{S} \times \mathbf{M} \to \mathbf{M}$, $(s, x) \to sx$, a map satisfying

$$s(x+y) = sx + sy, \quad (s+t)x = sx + tx, \quad s(tx) = (st)x$$

for all $x, y \in \mathbf{M}$ *and all* $s, t \in \mathbf{S}$. *Then*

$$[s,t][x,y] = [sx + ty, tx + sy] \tag{1.50}$$

makes $\mathrm{K}(\mathbf{M})$ *into a* $\mathrm{K}(\mathbf{S})$*-module. If* $1_{\mathbf{S}}$ *is a multiplicative identity of* $\mathbf{S}$ *satisfying* $1_{\mathbf{S}}x = x$ *for all* $x \in \mathbf{M}$, *then the multiplicative identity* $\iota 1_{\mathbf{S}}$ *of* $\mathrm{K}(\mathbf{S})$ *satisfies* $\iota 1_{\mathbf{S}}[x,y] = [x,y]$ *for all* $[x,y] \in \mathrm{K}(\mathbf{M})$.

The proof is long but straightforward. We simply sketch the proof of (b) and leave the details (and (a)) as an exercise.

One first shows that (1.50) is well-defined by showing that if $[x,y] = [x',y']$ respectively $[s,t] = [s',t']$, then $[sx' + ty', tx' + sy'] = [sx + ty, tx + sy]$ respectively $[s'x + t'y, t'x + s'y] = [sx + ty, tx + sy]$. One then easily verifies the module axioms

$$\begin{aligned} [s,t]\left([x,y] + [x',y']\right) &= [s,t][x,y] + [s,t][x',y'], \\ \left([s,t] + [s',t']\right)[x,y] &= [s,t][x,y] + [s',t'][x,y], \\ [s,t]\left([s',t'][x,y]\right) &= \left([s,t][s',t']\right)[x,y] \end{aligned}$$

and the fact that $\iota 1_{\mathbf{S}}$ has the stated property. □

We also need a "graded" version of the Grothendieck group. Suppose that $\mathbf{M}$ is a direct sum $\mathbf{M} = \bigoplus_\nu \mathbf{M}_\nu$ of submonoids, where ν runs over an indexing set N. The direct sum here means that every element of $\mathbf{M}$ is a sum of a finite number of elements, each from a different "homogeneous component" $\mathbf{M}_\nu$, in one and only one way. (If $\mathbf{M}$ contains a zero element, the requirement is that every *nonzero* element of $\mathbf{M}$ is a sum of a finite number of nonzero elements, each from a different homogeneous component, in one and only one way).

Proposition 1.6.3 *(a)* $\mathrm{K}(\mathbf{M})$ *is also graded by* N, $\mathrm{K}(\mathbf{M}) = \bigoplus \mathrm{K}(\mathbf{M})_\mu$, *in such a way that for all* ν, $\mathrm{K}(\mathbf{M})_\nu$ *is the Grothendieck group* $\mathrm{K}(\mathbf{M}_\nu)$, *and the canonical homomorphism* $\iota : \mathbf{M} \to \mathrm{K}(\mathbf{M})$ *is graded – i.e.* $\iota(\mathbf{M}_\nu) \subset \mathrm{K}(\mathbf{M})_\nu$. *In fact* $\iota|_{\mathbf{M}_\nu}$ *is the canonical map* $\mathbf{M}_\nu \to \mathrm{K}(\mathbf{M}_\nu)$.

(b) Suppose in addition that $\mathbf{H}$ *is a graded subset of* $\mathbf{M}$ *in the sense that every element of* $\mathbf{H}$ *is a finite sum of elements, each from a different* $\mathbf{H}_\nu := \mathbf{H} \cap \mathbf{M}_\nu$. *The subgroups* $\langle \iota \mathbf{H} \rangle$ *of* $\mathrm{K}(\mathbf{M})$ *and* $\langle \iota \mathbf{H}_\nu \rangle$ *of* $\mathrm{K}(\mathbf{M})_\nu$ *satisfy*

$$\langle \iota \mathbf{H} \rangle = \oplus_\nu \langle \iota \mathbf{H}_\nu \rangle$$

and so

$$\mathrm{K}(\mathbf{M}) / \langle \iota \mathbf{H} \rangle \cong \oplus_\nu \left(\mathrm{K}(\mathbf{M}_\nu) / \langle \iota \mathbf{H}_\nu \rangle \right).$$

Define $\mathrm{K}'(\mathbf{M}) = \oplus_\nu \mathrm{K}(\mathbf{M}_\nu)$. Let $\varphi : \mathbf{M} \to A$ be a homomorphism into an Abelian group A, and put $\varphi_\nu = \varphi|_{\mathbf{M}_\nu}$. By the universal mapping property of $\mathrm{K}(\mathbf{M}_\nu)$, there is a unique homomorphism $\widehat{\varphi}_\nu : \mathrm{K}(\mathbf{M}_\nu) \to A$ making

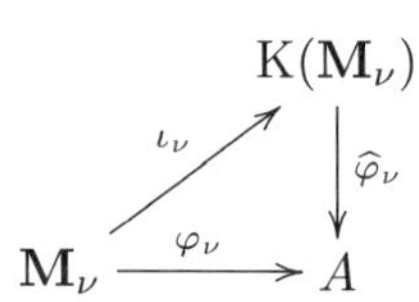

commutative, where ι_ν is the canonical map. Define $\widehat{\varphi} = \oplus_\nu \widehat{\varphi}_\nu : \mathrm{K}'(\mathbf{M}) \to A$ and $\iota' = \oplus_\nu \iota_\nu : \mathbf{M} \to \mathrm{K}'(\mathbf{M})$. Since $\iota' \mathbf{M}$ generates $\mathrm{K}'(\mathbf{M})$, it is clear that $\widehat{\varphi}$ is the

unique map making

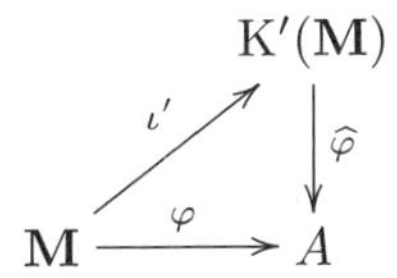

commutative, so $\mathbf{M} \xrightarrow{\iota'} \mathrm{K}'(\mathbf{M})$ has the same universal mapping property as $\mathbf{M} \to \mathrm{K}(\mathbf{M})$. By the usual argument involving universal mapping properties in the diagrams

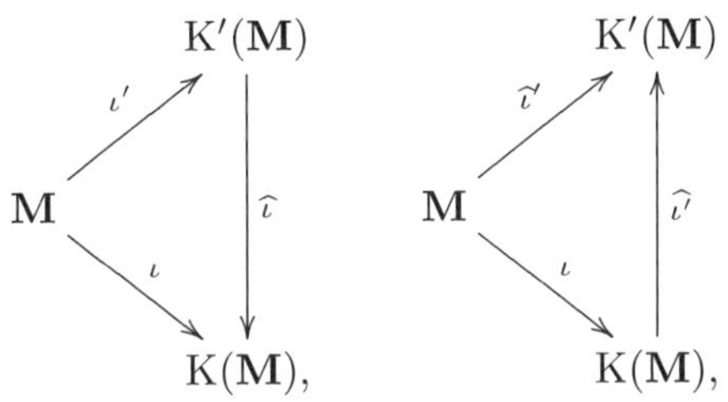

$\widehat{\iota}$ and $\widehat{\iota'}$ are mutually inverse isomorphisms. It follows that

$$\mathrm{K}(\mathbf{M}) = \oplus_\nu \widehat{\iota}\mathrm{K}(\mathbf{M}_\nu) = \oplus_\nu \mathrm{K}(\mathbf{M}_\nu)$$

(after identification) and that $\iota|_{\mathbf{M}_\nu}$ is the canonical map $\mathbf{M}_\nu \to \mathrm{K}(\mathbf{M}_\nu)$.

Now (b). Since $\iota\mathbf{H} \subset \sum_\nu \iota\mathbf{H}_\nu$, $\langle\iota\mathbf{H}\rangle \subset \sum_\nu \langle\iota\mathbf{H}_\nu\rangle$, and therefore $\langle\iota\mathbf{H}\rangle = \sum_\nu \langle\iota\mathbf{H}_\nu\rangle$. Since $\langle\iota\mathbf{H}_\nu\rangle \subset \langle\iota\mathbf{M}_\nu\rangle = \mathrm{K}(\mathbf{M})_\nu$, we get $\sum_\nu \langle\iota\mathbf{H}_\nu\rangle = \oplus_\nu \langle\iota\mathbf{H}_\nu\rangle$. □

Now assume further that $\mathbf{S}$ is a commutative semiring with a multiplicative identity, and that $(\mathsf{N}, +)$ is an Abelian monoid such that $\mathbf{S} = \oplus_\nu \mathbf{S}_\nu$ and $\mathbf{S}_\nu\mathbf{S}_\mu \subset \mathbf{S}_{\nu+\mu}$ for all $\nu, \mu \in \mathsf{N}$. It is straightforward to check the following graded version of Prop. 1.6.2.

Proposition 1.6.4 *The graded Abelian group* $\mathrm{K}(\mathbf{S}) = \oplus_\nu \mathrm{K}(\mathbf{S}_\nu)$ *is a graded commutative ring, satisfying* $\mathrm{K}(\mathbf{S}_\nu)\mathrm{K}(\mathbf{S}_\mu) \subset \mathrm{K}(\mathbf{S}_{\nu+\mu})$ *for all* $\nu, \mu \in \mathsf{N}$. *Furthermore if the graded subset* $\mathbf{H}$ *satisfies* $\mathbf{SH} \subset \mathbf{H}$, *then* $\langle\iota\mathbf{H}\rangle$ *is a graded ideal* $\oplus_\nu\langle\iota\mathbf{H}_\nu\rangle$ *of* $\mathrm{K}(\mathbf{S})$, *and* $\mathrm{K}(\mathbf{S})/\langle\iota\mathbf{H}\rangle \cong \oplus_\nu (\mathrm{K}(\mathbf{S}_\nu)/\langle\iota\mathbf{H}_\nu\rangle)$ *is also a graded commutative ring.* □

Example. An important example of the Grothendieck group is $\mathrm{K}_0(R)$ where R is a ring. It is by definition the Grothendieck group $\mathrm{K}(\mathbf{P})$ where $\mathbf{P}$ is the Abelian monoid of isomorphism classes of finitely generated projective R-modules, with addition $[P_1] + [P_2] = [P_1 \oplus P_2]$.

Of special interest to us later is $\mathrm{K}_0(A)$ where A is a separable algebra over a field K. In this case every A-module is projective – see Th. 1.2.3, p. 6, and the definition of semisimple ring on the same page. By Th. 1.2.7, p. 8, $\mathbf{P}$ is graded by the set S of isomorphism classes of simple A-modules, and so $\mathrm{K}_0(A)$ is similarly graded by S. In fact by Th. 1.2.2(c) $\mathrm{K}_0(A)$ is the free Abelian group on these isomorphism classes, so is $\cong \mathbb{Z}^{|\mathsf{S}|}$. □

Exercises. 1. If R is the direct sum $R = R_1 \oplus R_2$ of two rings, show that there is a canonical isomorphism $K_0(R) \to K_0(R_1) \oplus K_0(R_2)$, taking $[P]$ to $([R_1P], [R_2P])$. (We identify $K_0(R)$ with $K_0(R_1) \oplus K_0(R_2)$ via this isomorphism).

2. Let $(R, ^-)$ be a ring with involution. If $P \in \mathbf{P}$, then the map $[P] \to [\overline{P^*}]$ where $\overline{P^*}$ is the twisted dual, gives an involution, also denoted $^-$, of the Abelian group $\mathrm{K}_0(R)$.

3. Suppose $(A, ^-)$ is a hyperbolic involution algebra, $A = A_1 \oplus A_2$ where $\overline{A_1} = A_2$. Show that
(i) $\overline{\mathrm{K}_0(A_1)} = K_0(A_2)$,
(ii) if A_1 is a simple algebra and V_1 and V_2 are A_1 respectively A_2-modules of the same length, $\overline{[V_1]} = [V_2]$.

4. Let $(A, ^-)$ be a $(K, ^-)$-involution algebra and $(L, ^-)/(K, ^-)$ an extension of fields with involution. Then there is a canonical map $\mathrm{K}_0(A) \to \mathrm{K}_0(L \otimes_K A)$ taking $[P] \to [L \otimes_K P]$ which commutes with the involutions.

5. (This exercise requires knowledge of §3.1.2). Assume A is a semisimple K-algebra with a $(K, ^-)$-involution and $\varepsilon = \pm 1$. If $h : (V \oplus W) \times (V \oplus W) \to (A, ^-)$ is an ε-hyperbolic form with V and W totally isotropic, then $\overline{[V]} = [W]$, and conversely, if $\overline{[V]} = [W]$, then there is an ε-hyperbolic form on $V \oplus W$ under which V and W are totally isotropic. □

REMARK. The canonical map in 4. is injective if A is separable – cf. Exercise 6., p. 417, [19]. □

1.7 Linear representations of finite groups

References for this section are [68], [18], and [19]. The basic facts, with which we begin, can be found in §§1-3 of [68] and Chs. II and V of [18].

We assume throughout that G is a finite group of order $g = |G|$. All vector spaces over the field K are assumed to be finite dimensional.

1.7.1 Basic facts about linear representations. A *linear representation* – or simply *representation* – of G is a homomorphism

$$\rho : G \to \mathbf{GL}(V) \tag{1.51}$$

of G into the "general linear group" of the K-vector space V, the group of invertible K-linear transformations of V into itself. More precisely, ρ is a *K-linear representation of G*, or a *linear representation of G over K*. The dimension of V over K is called the *degree* of ρ.

Another linear representation $\rho' : G \to \mathbf{GL}(V')$ over K is *equivalent* to ρ if there is a K-linear isomorphism $\varphi : V \to V'$ making

$$\begin{array}{ccc} V & \xrightarrow{\varphi} & V' \\ {\scriptstyle \rho(s)}\downarrow & & \downarrow{\scriptstyle \rho'(s)} \\ V & \xrightarrow[\varphi]{} & V' \end{array} \tag{1.52}$$

commutative for all $s \in G$. Such an isomorphism is sometimes called an *equivariant* isomorphism, or more precisely a G-equivariant isomorphism, and we write $\rho \simeq \rho'$.

If a basis of V is chosen and P_s is the matrix of $\rho(s)$ with respect to the basis, the map $s \to \mathsf{P}_s$ is a *matrix representation* of G over K, $\mathsf{P} : G \to \mathbf{GL}(n, K)$.

The notion of representation of G can also be stated in terms of modules over the group algebra KG – V is a (finitely generated) KG-module via the action $sv = \rho(s)v$ of G on V, extended to KG by linearity:

$$(\sum_s \alpha_s s)v = \sum_s \alpha_s \rho(s)v.$$

We then refer to V as the KG-module *corresponding to* or *arising from* the representation ρ.

Conversely if V is a KG-module, of finite dimension over K (or equivalently, finitely generated over KG), we can define a map $\rho : G \to \mathrm{End}(V)$ by decreeing that $\rho(s)v = sv$. Then

$$\rho(s)(\rho(t)v) = s(tv) = (st)v = \rho(st)v$$

so $\rho(st) = \rho(s)\rho(t)$. Furthermore its image is in $\mathbf{GL}(V)$ since $\rho(s)\rho(s^{-1}) = \rho(1) = \mathrm{id}_V$, so ρ is a representation – we say that it is the representation *afforded* by the KG-module V.

The commutativity of the diagram (1.52) is equivalent to $s\varphi = \varphi s$. This means that φ is a KG-homomorphism, and so we see that *the representations ρ and ρ' are equivalent if and only if the KG-modules corresponding to them are isomorphic as KG-modules.*

Formally, there is an isomorphism between the category of K-representations of G of finite dimension and the category of finitely generated KG-modules.

In terms of the corresponding matrix representations, P and P' are equivalent if and only if there is an invertible matrix $\underline{\varphi}$ such that

$$\mathsf{P}'_s = \underline{\varphi}\mathsf{P}_s\underline{\varphi}^{-1} \quad \text{for all } s \in G.$$

Note that if $V = V'$, then $\varphi \in \mathbf{GL}(V)$ and (1.52) is equivalent to $\rho'(s) = \varphi\rho(s)\varphi^{-1}$, which is expressed as "ρ and ρ' are conjugate representations".

The connection between representations and modules allows us to introduce the notions and theory of modules into representation theory. Thus for example we get the notion of *subrepresentation* of ρ as the representation afforded by a KG-submodule of V, and the notion of the *direct sum* $\rho_1 \oplus \rho_2$ of two representations as the representation afforded by the direct sum $V_1 \oplus V_2$ of their modules. When V is simple as a KG-module, ρ is called *irreducible* (over K). If ρ is reducible over K, it means that there is a basis of V such that

$$\mathsf{P}_s = \begin{pmatrix} \mathsf{P}_{1,s} & * \\ 0 & \mathsf{P}_{2,s} \end{pmatrix},$$

where $\mathsf{P}_{1,s}$ and $\mathsf{P}_{2,s}$ are matrix representations of G afforded by a KG-submodule U of V and the factor module V/U respectively, 0 is a zero matrix of the appropriate size and $*$ is also a matrix of the appropriate size. Such a basis of V is obtained by taking a basis of the KG-submodule U and augmenting it to a basis of V.

If $V = V_1 \oplus V_2$ is the space of the direct sum $\rho_1 \oplus \rho_2$, then in a basis of V which is the union of bases of V_1 and V_2, the matrix P_s is the "direct sum" $\mathsf{P}_{1,s} \oplus \mathsf{P}_{2,s}$ of the matrices of $\rho_1(s)$ and $\rho_2(s)$:

$$\mathsf{P}_s = \begin{pmatrix} \mathsf{P}_{1,s} & 0 \\ 0 & \mathsf{P}_{2,s} \end{pmatrix}.$$

In this case one says that ρ or P is *decomposable* (assuming, of course, that V_1 and V_2 are $\neq 0$).

And when V is a semisimple KG-module, ρ is referred to as *completely reducible*, and there is a basis of V such that

$$\mathsf{P}_s = \begin{pmatrix} \mathsf{P}_{1,s} & 0 & \cdots & 0 \\ 0 & \mathsf{P}_{2,s} & \cdots & 0 \\ 0 & 0 & \ddots & 0 \\ 0 & 0 & \cdots & \mathsf{P}_{k,s} \end{pmatrix},$$

where the P_i are irreducible matrix representations.

Of particular importance is *Maschke's Theorem*:

Theorem 1.7.1 *The algebra KG is separable if and only if the order g of G is not divisible by the characteristic of the field K.*

The sufficiency (the most important part of the theorem and the only part we will use) is proved as follows. It is enough to prove that KG is semisimple, since then every scalar extension $L \otimes_K KG \cong LG$ (given by $\sum_s \alpha_s \otimes s \to \sum_s \alpha_s s$) would also be semisimple. By Th. 1.2.3 it is enough to show that any KG-submodule U of V splits V as a KG-module. It certainly splits V as a vector space; let $\pi : V \to U$ be the corresponding projection of such a splitting. Then one can check that the map

$$\widetilde{\pi} : V \to U, \quad \widetilde{\pi}(v) = \tfrac{1}{g} \sum_s s\pi(s^{-1}v)$$

is a KG-linear projection of V onto U, and so $V = U \oplus \ker \widetilde{\pi}$ is a splitting of V into KG-submodules. □

Thus when the characteristic does not divide g, or more generally when the representation ρ is completely reducible, the KG-module V is semisimple and so (Th. 1.2.2, p. 6) is the direct sum of its homogeneous components V_S and each of these homogeneous components is a direct sum of a uniquely determined (finite) number of simple submodules isomorphic to S. In representation-theoretic terms, this means that ρ is the direct sum of irreducible representations, and that the multiplicity of any particular equivalence class of irreducible representations in this direct sum is invariant.

The *tensor product* $\rho_1 \otimes \rho_2$ of two representations of G is defined as the representation afforded by the tensor product of their spaces $V_1 \otimes_K V_2$, with the action $s(v_1 \otimes v_2) = sv_1 \otimes sv_2$.

To put it another way, the tensor product $V_1 \otimes_K V_2$ of two KG-modules is, in a natural way, also a KG-module. It is easy to see that this implies a commutative semiring structure on the monoid $\mathbf{P}$ of isomorphism classes of KG-modules, and so by Prop. 1.6.2, p. 65, $\mathrm{K}_0(KG)$ is a ring – in fact with a 1 equal to the equivalence class $[K]$ of the trivial representation.

If L/K is any field extension, the inclusion $\mathbf{GL}(V) \overset{\iota}{\hookrightarrow} \mathbf{GL}(L \otimes_K V)$ leads to an L-representation

$$\rho^L = \iota \circ \rho : G \to \mathbf{GL}(L \otimes_K V) \tag{1.53}$$

– the representation obtained from ρ by *extending the scalars to L*. The LG-module corresponding to ρ^L is simply the module $L \otimes_K V = V^L$ on which LG operates naturally via the canonical isomorphism $LG \cong L \otimes_K KG$. (We usually identify LG and $L \otimes_K KG$ via this isomorphism.) Slightly more generally, using a homomorphism $\varphi : K \to L$ of fields, one obtains an L representation (1.53) by making L into a K-space via $\alpha.\lambda = \varphi(\alpha)\lambda$, $\alpha \in K, \lambda \in L$. This is also referred to as extending the scalars.

On the other hand if V is a finite dimensional vector space over L, a representation $\rho : G \to \mathbf{GL}(V)$ over L is said to be *defined over* the subfield K if an L-basis for V can be chosen so that the corresponding matrix representation $\mathsf{P} : G \to \mathbf{GL}(n, L)$ is over K, that is to say the entries of the matrix $\mathsf{P}(s)$ are in K for all $s \in G$, or equivalently that $\mathsf{P} : G \to \mathbf{GL}(n, K)$. This is also equivalent to the existence of a vector space V_0 over K and a representation $\rho_0 : G \to \mathbf{GL}(V_0)$ over K such that $\rho_0^L \simeq \rho$.

1.7.2 Characters. The *character* of a representation ρ over K is the map $\chi = \chi_\rho : G \to K$ which assigns to each $s \in G$ the trace of the K-linear transformation $\rho(s)$:

$$\chi_\rho(s) = \operatorname{tr}(\rho(s)).$$

χ is in the K-algebra of K-valued functions on G – in fact it is in the K-algebra of K-valued "class functions" of G – the functions constant on the conjugacy classes of G – since

$$\chi(tst^{-1}) = \chi(s)$$

for all $s, t \in G$. It is occasionally convenient to write $\chi_V = \chi_\rho$.

Characters satisfy

$$\chi_{\rho_1 \oplus \rho_2} = \chi_{\rho_1} + \chi_{\rho_2}, \quad \chi_{\rho_1 \otimes \rho_2} = \chi_{\rho_1}\chi_{\rho_2}, \quad \chi_V = \chi_U + \chi_W$$

if

$$0 \to U \to V \to W \to 0$$

is an exact sequence of KG-modules. Also

$$\chi(1) = (\dim V)1_K, \quad \chi(s^{-1}) = \chi(s)^*,$$

where $\chi(s)^*$ is the "complex conjugate" of $\chi(s)$. By this we mean the following: over an algebraically closed extension field, the linear transformation $\rho(s)$ has $n = \dim V$ eigenvalues (including multiplicities) all of which are g^{th} roots of unity, and $\operatorname{tr}(\rho(s))$ is their sum. Further, $\operatorname{tr}(\rho(s^{-1}))$ is the sum of the inverses of these eigenvalues and we refer to the process of taking the sum of their inverses as "complex conjugation".

REMARK. If the field K has characteristic 0 and we identify, in any way, the maximal cyclotomic subextension of the prime field ("$\mathbb{Q}$") of K with the corresponding subfield of $\mathbb{C}$, this process *is* complex conjugation in the ordinary sense. But if K has characteristic $p \neq 0$, this process is not always induced by an element of the Galois group of some extension of K which contains the eigenvalues, and in fact even depends on the actual eigenvalues of $\rho(s)$ rather than just their sum. For example if K has characteristic 5, the elements $1, 2, 3, 4$ of the prime field are all 4th roots of 1, and $2+4 = 3+3$ but $2^{-1}+4^{-1} = 3+4 \neq 3^{-1}+3^{-1} = 2+2$. However since the roots of unity which appear in the trace of $\rho(s)$ are uniquely determined as the eigenvalues of $\rho(s)$, the sum of their inverses is also uniquely determined by $\rho(s)$. □

It is clear also that if L/K is an extension field, then

$$\chi_{\rho^L} = \chi_\rho.$$

The central theorem of character theory is

Theorem 1.7.2 *If* char $K = 0$, *two representations of* G *over* K *are equivalent if and only if their characters are equal.* □

This is a very powerful theorem, and especially so since the character of a representation is often easily calculated.

The subring $\mathsf{X}_K(G)$ of the ring of K-valued functions on G, generated by the characters of representations over K, is called the *ring of virtual characters*, and it is clear that each virtual character is the difference $\chi_1 - \chi_2$ of two characters. When the characteristic is 0, the ring of virtual characters is canonically isomorphic to the Grothendieck group of K-representations of G, which is by definition the Grothendieck group of the monoid of equivalence classes $[\rho]$ of representations, with addition defined in the obvious way, $[\rho_1] + [\rho_2] = [\rho_1 \oplus \rho_2]$. The isomorphism is of course $[\rho] \to \chi_\rho$, and it respects the multiplication defined in both groups – namely $\chi_1\chi_2$ and $[\rho_1 \otimes \rho_2]$ – and so is an isomorphism of rings as well.

These rings are isomorphic to a third ring, namely $\mathrm{K}_0(KG)$, via the categorical isomorphism between representations of G over K and finitely generated KG-modules. The ring structure on $\mathrm{K}_0(KG)$ arises from the KG-action $s.(v_1 \otimes_K v_2) = sv_1 \otimes_K sv_2$ on the tensor product $V_1 \otimes_K V_2$ of two KG-modules. Of course it is essentially the same as the Grothendieck group of K-representations of G.

Many of these same properties are enjoyed by *Brauer characters*, which are defined in the case that K is a field of prime characteristic p, as follows. An element s of G is called *p-regular* if p does *not* divide the order of s, otherwise s is called *p-singular*. Assume that there is a field $\widetilde{K}$ of characteristic 0 with a discrete valuation whose ring of integers $\mathfrak{o}$ and prime ideal $\mathfrak{p}$ satisfy $\mathfrak{o}/\mathfrak{p} \cong K$ – there *is* always such a field if K is perfect – cf. p. 45, [66]. The triple $(\widetilde{K}, \mathfrak{o}, K)$ is called a *p-modular system* (cf. p. 419*ff*, [19]). We shall also assume that $\widetilde{K}$ is "sufficiently large" relative to G, meaning that it contains the group of all g^{th} roots of unity. Then K also is sufficiently large relative to G. Let $g = p^r m$ where $p \nmid m$. The residue class map $\mathfrak{o} \to K$ induces an isomorphism between the groups of mth roots of unity in $\widetilde{K}$ and K; if $\zeta \in K$ is one of them, we denote by $\widetilde{\zeta}$ the corresponding root of unity in $\mathfrak{o}$.

Suppose now that ρ is a K-representation (1.51) of degree n and let $s \in G$ be a p-regular element. Since $\rho(s)^m = 1$, the eigenvalues $\zeta_1, \ldots, \zeta_n$ (with multiplicities) of $\rho(s)$ are m^{th} roots of unity and $\rho(s)$ is diagonalizable. The *Brauer character*[11] is defined for each p-regular element s of G by

$$\chi_\rho(s) = \widetilde{\zeta}_1 + \cdots + \widetilde{\zeta}_n \in \mathfrak{o}.$$

The *complex conjugate* $\chi_\rho(s)^*$ can be defined for a Brauer character as it was for an ordinary character (p. 71):

$$\chi_\rho(s)^* = \widetilde{\zeta}_1^{-1} + \cdots + \widetilde{\zeta}_n^{-1} \in \mathfrak{o}.$$

Let $\mathbb{Q}'$ be the prime subfield of $\widetilde{K}$ and $\mathbb{Q}'(\zeta')$ any cyclotomic subfield of $\widetilde{K}$ containing $\widetilde{\zeta}_1, \ldots, \widetilde{\zeta}_n$. Then under any imbedding $\varphi : \mathbb{Q}'(\zeta') \to \mathbb{C}$, it is clear that

[11] A full treatment of Brauer characters can be found in §17B, p. 419*ff*, [19] or in ch. 18, p. 147*ff*, [68].

$\varphi(\chi_\rho(s)^*) = (\varphi(\chi_\rho(s))^*$ where * on the right side really *is* complex conjugation. Thus if $\chi_\rho(s)^* = \chi_\rho(s)$, its image under any imbedding $\mathbb{Q}'(\zeta') \to \mathbb{C}$ is contained in $\mathbb{R}$, and we accordingly say that $\chi_\rho(s)$ is *real*.

Let σ be an automorphism of K – for example an involution $^-$ – then $\chi_\rho(s)^\sigma$ is defined as follows: Express $\rho(s)$ as a matrix in $\mathrm{M}(n, K)$ with respect to some basis of V, and apply σ to its entries; then the eigenvalues $\eta_1, \ldots, \eta_n$ of the transformed matrix are again m^{th} roots of unity and do not depend on the basis chosen; we define $\chi_\rho(s)^\sigma$ to be the sum of their lifts $\widetilde{\eta}_1, \ldots, \widetilde{\eta}_n$ to K.

The following is a basic theorem of Brauer (Corollary 17.10, p. 424, [19]):

Theorem 1.7.3 *Two completely reducible representations of G over K are equivalent if and only if their Brauer characters*[12] *are equal.* □

When the characteristic of K does not divide g, then of course *every* representation of G is completely reducible by Maschke's Theorem.

Suppose for the rest of this section that the characteristic of K does not divide the order g of G, and let ρ and ρ' be representations which are *absolutely irreducible* – this means that they are irreducible over *every* extension L/K, and is equivalent to being irreducible over an algebraic closure of K. Choose bases for their vector spaces and consider the corresponding matrix representations:

$$\mathsf{P}(s) = (\alpha(s)_{ij}), \quad \mathsf{P}'(s) = (\beta(s)_{kl}), \quad \text{for all } s \in G.$$

If ρ and ρ' are not equivalent, then according to p. 14, [68],

$$\sum_{s\in G} \alpha(s)_{ij}\beta(s^{-1})_{kl} = 0 \quad \text{for any } i, j, k, l, \tag{1.54}$$

while if they are equivalent, we assume they are actually equal and then

$$\sum_{s\in G} \alpha(s)_{ij}\, \alpha(s^{-1})_{kl} = \begin{cases} g/n & \text{if } i = l \text{ and } j = k, \\ 0 & \text{otherwise,} \end{cases} \tag{1.55}$$

since the degree of ρ is not divisible by the characteristic of K (Cor. 1.2.2, p. 16).

If φ and ψ are complex valued functions on G, define

$$(\varphi|\psi) = \tfrac{1}{g}\sum_{s\in G} \varphi(s)\psi(s)^*.$$

This is a nonsingular positive definite Hermitian form on the complex vector space $\mathbb{C}^G$ of dimension g. Now suppose that χ and χ' are the characters of *absolutely* irreducible representations ρ and ρ'. Then the *(first) orthogonality relations* are

$$(\chi|\chi') = \begin{cases} 1 & \text{if } \rho \text{ and } \rho' \text{ are equivalent,} \\ 0 & \text{otherwise.} \end{cases} \tag{1.56}$$

In fact if $\mathbf{H}$ is the subspace of $\mathbb{C}^G$ of class functions and if K is a *splitting field* of G – that is to say all K-irreducible representations over K are actually absolutely irreducible – the characters of the K-irreducible representations form an orthonormal basis of $\mathbf{H}$. Thus the number of absolutely irreducible representations of G is the class number of G. By another theorem of Brauer, a field K of characteristic 0 *is* a splitting field if K contains the mth roots of unity, where m is the *exponent* of G – the smallest positive integer m such that $s^m = 1$ for all $s \in G$.

---

[12] Defined using the same p-modular system.

In particular χ is the character of an absolutely irreducible representation if and only if $(\chi|\chi) = 1$.

1.7.3 The twisted contragredient representation. Let $\rho : G \to \mathbf{GL}(V)$ be a representation. Since V is a left KG-module (via ρ), $V^* = \operatorname{Hom}_K(V, K)$ is a *right* KG-module determined by

$$\langle v, xa\rangle = \langle av, x\rangle.$$

Now twist the action of KG on V^* via the standard $(K, ^-)$-involution $^-$ on KG to make it into a *left* KG-module $\overline{V^*}$: thus, if we write $\bar{x}$ for $x \in V^*$ when it is considered as an element of $\overline{V^*}$,

$$\langle v, a\bar{x}\rangle = \langle v, x\bar{a}\rangle = \langle \bar{a}v, x\rangle.$$

If $\bar{\rho}^*$ denotes the representation thus afforded by $\overline{V^*}$,

$$\langle v, \bar{\rho}^*(s)\bar{x}\rangle = \langle v, s\bar{x}\rangle = \langle s^{-1}v, x\rangle = \langle \rho(s^{-1})v, x\rangle. \tag{1.57}$$

If $^-$ is the identity on K, this means that $\bar{\rho}^*(s) = \rho(s^{-1})^t$. This is the "usual" definition of the contragredient representation.

Suppose $^-$ is *not* necessarily the identity on K. In order to calculate the matrix of $\bar{\rho}^*$, the action of K on $\overline{V^*}$ must agree with that implicit in the action of KG, so $\alpha x = x\bar{\alpha}$ where the right side represents the "usual" action of $\bar{\alpha} \in K$ on V^*. Let $\{v_i\}$ be a K-basis of V and $\{x_i\}$ the dual basis of V^*. Then if $\rho(s^{-1})v_j = \sum_i \alpha_{ij}v_i$ and $\bar{\rho}^*(s)x_k = \sum_l \beta_{lk}x_l$ $(\alpha_{ij}, \beta_{ij} \in K)$,

$$\begin{aligned}
\langle \rho(s^{-1})v_j, x_k\rangle &= \alpha_{kj} \\
&= \langle v_j,\ \bar{\rho}^*(s)x_k\rangle = \langle v_j, \sum_l \beta_{lk}x_l\rangle = \langle v_j, \sum_l x_l\bar{\beta}_{lk}\rangle \\
&= \bar{\beta}_{jk}.
\end{aligned}$$

Thus

$$\operatorname{mat}\bar{\rho}^*(s) = (\operatorname{mat}\rho(s^{-1}))^{t-}.$$

The representation $\bar{\rho}^* : G \to \mathbf{GL}(\overline{V^*})$ is called the *twisted contragredient* representation of ρ. If χ_ρ is the character of ρ, then the character of $\bar{\rho}^*$ is

$$\chi_{\bar{\rho}^*}(s) = \overline{\chi_\rho(s)^*}, \tag{1.58}$$

where $\chi_\rho(s)^*$ is the "complex conjugate" of $\chi_\rho(s)$ (cf. p. 71).

Now suppose that $\varphi : V \to \overline{V^*}$ is a KG-homomorphism. Define

$$f : V \times V \to K \quad \text{by} \quad f(u, v) = \langle u, \varphi v\rangle.$$

Then f is a sesquilinear form over $(K, ^-)$ – note in particular that

$$f(u, \alpha v) = \langle u, \varphi(\alpha v)\rangle = \langle u, \alpha\varphi(v)\rangle = \langle u, \varphi(v)\bar{\alpha}\rangle = f(u, v)\bar{\alpha}$$

since $\varphi(v) \in \overline{V^*}$. Its right adjoint is $f_r = \varphi$, so f is nonsingular if and only if φ is an isomorphism. Furthermore if $s \in G$,

$$f(su, sv) = \langle su, \varphi(sv)\rangle = \langle su, s\varphi(v)\rangle = \langle s^{-1}su, \varphi(v)\rangle = \langle u, \varphi(v)\rangle = f(u, v),$$

so that f is G-*invariant*. Conversely much the same computation shows that if f is a G-invariant sesquilinear form on V, then f_r is a KG-homomorphism $V \to \overline{V^*}$.

Theorem 1.7.4 *Let ρ be a representation of G over K. Then ρ is equivalent to its twisted contragredient $\bar{\rho}^*$ if and only if there is a nonsingular G-invariant sesquilinear form over $(K,\,^-)$ on the space V of ρ.*

More precisely the G-invariant sesquilinear forms on V correspond bijectively to the KG-homomorphisms $V \to \overline{V^*}$ via $f \rightsquigarrow f_r$, with nonsingular forms corresponding to isomorphisms. □

We now consider what happens to the contragredient under restriction of the scalars. Let$(L,\,^-)/(K,\,^-)$ be a finite separable extension of fields with involution, and let $\mathrm{Tr} = \mathrm{Tr}_{L/K}$ be the trace in L/K – it is $\neq 0$ since L/K is separable. Let V be a finite dimensional vector space over L, and denote by V_K the vector space V viewed as a vector space over K. Let $V^* = \mathrm{Hom}_L(V, L)$ and $(V_K)^* = \mathrm{Hom}_K(V_K, K)$. Then the map

$$\tau : (V^*)_K \to (V_K)^*, \quad \tau(x) = \mathrm{Tr} \circ x,$$

is an isomorphism of K-vector spaces. It suffices to show it is injective, and that is clear since if $x \in V^*$, $x \neq 0$, we choose $v \in V$ such that $x(v) \neq 0$, and then for some $\lambda \in L$, $\mathrm{Tr}(x(\lambda v)) = \mathrm{Tr}(\lambda x(v)) \neq 0$. (The isomorphism also follows from Th. 1.2.10, p. 9.)

Let $\langle\,\cdot\,,\cdot\,\rangle_L$ denote the pairing between V and V^* and $\langle\,\cdot\,,\cdot\,\rangle_K$ that between V_K and $(V_K)^*$, and let the L-representation $\rho : G \to \mathbf{GL}(V)$ be denoted by ρ_K when considered as a K-representation $G \to \mathbf{GL}(V_K)$, with twisted contragredients $\bar{\rho}^* : G \to \mathbf{GL}(\overline{V^*})$ and $\bar{\rho}_K^* : G \to \mathbf{GL}(\overline{(V_K)^*})$ respectively. If $v \in V$, $x \in \overline{V^*}$ and $s \in G$,

$$\begin{aligned}\langle v, \bar{\rho}_K^*(s)\tau(x)\rangle_K &= \langle \rho_K(s^{-1})v, \tau(x)\rangle_K \\ &= \mathrm{Tr}\langle \rho(s^{-1})v, x\rangle_L = \mathrm{Tr}\langle v, \bar{\rho}^*(s)x\rangle_L \\ &= \langle v, \tau(\bar{\rho}^*(s)x)\rangle_K.\end{aligned}$$

Thus $\bar{\rho}_K^*(s)\tau(x) = \tau(\bar{\rho}^*(s)x)$ for all $x \in V^*$ and $s \in G$, so

Theorem 1.7.5 *The canonical K-isomorphism $\tau : (V^*)_K \to (V_K)^*$ establishes an equivalence between the representation $\bar{\rho}^*$ of G on $\overline{V^*}$, considered as K-space $(\overline{V^*})_K$, and the representation $\bar{\rho}_K^*$ of G on the space $\overline{(V_K)^*}$.* □

To put it another way, if we identify $(V^*)_K$ and $(V_K)^*$ via τ, the twisted contragredient of ρ as an L-representation, when considered as a K-representation, is the same as the twisted contragredient of ρ considered as a K-representation.

Exercise. If U and V are finite dimensional K-spaces, there is a canonical isomorphism

$$\varphi : U^* \otimes_K V^* \to (U \otimes_K V)^*$$

determined by

$$\langle u \otimes v, \varphi(u^* \otimes v^*)\rangle = \langle u, u^*\rangle\langle v, v^*\rangle.$$

Suppose that U and V are KG-modules. Then $\overline{U^*}$ and $\overline{V^*}$ are (left) KG-modules, and so therefore is $\overline{U^*} \otimes_K \overline{V^*}$ (p. 70). Also $U \otimes_K V$ is a KG-module, and so its twisted contragredient $\overline{(U \otimes_K V)^*}$ is also. Show that

$$\bar{\varphi} : \overline{U^*} \otimes_K \overline{V^*} \to \overline{(U \otimes_K V)^*},$$

defined in the obvious way, is a KG-isomorphism. □

1.7.4 Induced representations. References are §3.3, pp. 28*ff*, [68] and §43, pp. 314*ff*, [18]). See also pp. 161*ff* later in this book.

Let H be a subgroup of the finite group G, and let W be a (finite dimensional) KH-module, affording the representation σ of H. If we view KG as a (KG, KH)-bimodule, the tensor product

$$V = \mathrm{Ind}_H^G W = KG \otimes_{KH} W$$

is a KG-module; the representation of G it affords, the *representation induced from* σ, is denoted by $\mathrm{Ind}_H^G \sigma$. Since V is the vector space direct sum of the $s \otimes W$ as $s \in G$ runs over a system of (left coset) representatives of G/H, the dimension of V is $[G : H] \dim W$.

When there is no danger of confusion, we write $\mathrm{Ind}\, W$ for $\mathrm{Ind}_H^G W$ and $\mathrm{Ind}\, \sigma$ for $\mathrm{Ind}_H^G \sigma$.

Induction of representations is transitive in the sense that if G is in turn a subgroup of a finite group F, then

$$\mathrm{Ind}_G^F(\mathrm{Ind}_H^G W) \cong \mathrm{Ind}_H^F W \quad (\text{ isomorphism of } KF\text{-modules}).$$

One can also start with a representation $\rho : G \to \mathbf{GL}(V)$ of G, and restrict it to a representation $\rho|_H : H \to \mathbf{GL}(V)$ of H. This representation is denoted by $\mathrm{Res}_H^G\, \rho$ or simply $\mathrm{Res}\, \rho$, and its space (which is V of course) by $\mathrm{Res}_H^G V$ or $\mathrm{Res}\, V$.

Now suppose that $\rho : G \to \mathbf{GL}(V)$ and $\sigma : H \to \mathbf{GL}(W)$ are representations. Then there is a KG-isomorphism

$$\varphi : \mathrm{Ind}(W \otimes_K \mathrm{Res}\, V) \to (\mathrm{Ind}\, W) \otimes_K V, \tag{1.59}$$

usually referred to as "Frobenius Reciprocity". Thus φ is an isomorphism

$$\varphi : KG \otimes_{KH} (W \otimes_K \mathrm{Res}\, V) \to (KG \otimes_{KH} W) \otimes_K V, \tag{1.60}$$

and we shall show that φ can be chosen as

$$\varphi((\sum_s \alpha_s s) \otimes (w \otimes v)) = \sum_s \alpha_s((s \otimes w) \otimes sv). \tag{1.61}$$

First note that the map

$$KG \times W \times \mathrm{Res}\, V \to (KG \otimes_{KH} W) \otimes_K V, \quad (\sum_s \alpha_s s, w, v) \to \sum_s \alpha_s((s \otimes w) \otimes sv)$$

is K-trilinear, and so the map

$$KG \times (W \otimes_K V) \to (KG \otimes_{KH} W) \otimes_K V$$

determined by

$$(\sum_s \alpha_s s, w \otimes v) \to \sum_s \alpha_s((s \otimes w) \otimes sv)$$

is K-bilinear. This map is balanced with respect to KH – where $h \in H$ operates on $W \otimes_K V$ via $w \otimes v \to hw \otimes hv$ – and so the map φ in (1.61) is well-defined. Moreover it is a KG-homomorphism since $\varphi(sx) = s\varphi(x)$ for all s and x.

To show that it is an isomorphism, we construct the inverse homomorphism ψ. Consider the map

$$KG \times W \times V \to KG \otimes_{KH} (W \otimes_K \mathrm{Res} V)$$
$$(\sum_s \alpha_s s, w, v) \to \sum_s \big(\alpha_s s \otimes_{KH} (w \otimes_K s^{-1} v)\big).$$

It is K-trilinear and the triples $(\sum_s \alpha_s sh, w, v)$ and $(\sum_s \alpha_s s, hw, v)$ have the same image, and so gives a K-linear map

$$(KG \otimes_{KH} W) \otimes_K V \to KG \otimes_{KH} (W \otimes_K \operatorname{Res} V)$$

which is the inverse of φ. □

In terms of representations, Frobenius Reciprocity is

$$\operatorname{Ind}_H^G(\sigma \otimes \rho|_H) \cong \operatorname{Ind}_H^G(\sigma) \otimes \rho.$$

1.7.5 Simple components of KG and irreducible representations. Reference: §70, pp. 463*ff*, [18]

Since the characteristic of K does not divide the order g of G, the group algebra KG is separable over K by Maschke's Theorem (p. 70) and so is a direct sum of simple separable K-algebras

$$KG = A_1 \oplus A_2 \oplus \cdots \oplus A_h.$$

Each simple component A_i corresponds to a uniquely determined equivalence class of K-irreducible representations: if $\rho : G \to \mathbf{GL}(V)$ is an irreducible representation, then there is a unique i, $1 \leq i \leq h$, such that A_i acts faithfully on the KG-module V (cf. p. 69) and ρ is given by

$$G \xrightarrow{\text{incl}} KG \xrightarrow{\text{proj}} A_i \to \operatorname{End}_K V,$$

and $A_j V = 0$ for all $j \neq i$. Conversely if V is a simple A_i-module, this sequence yields a K-irreducible representation. By Ths. 1.2.5 and 1.2.6 (p. 8), A_i is isomorphic to a matrix algebra $\mathrm{M}(n_i, D_i)$ for some K-division algebra D_i, and V is the unique (up to isomorphism) simple A_i-module and is isomorphic to the A_i-module of column vectors $D_i^{n_i \times 1}$. If we identify $A_i = \rho(KG) \subset \operatorname{End}_K V$, then $A_i = \operatorname{End}_{D_i} V$ where $D_i = \operatorname{End}_{KG} V = \operatorname{End}_{A_i} V$ (with D_i acting on the right of V). We let $\rho_i : G \to \mathbf{GL}(V_i)$ be the irreducible representation which corresponds in this way to A_i.

There is a finite separable extension L/K in which ρ_i splits into absolutely irreducible representations. If char $K = 0$, we know by Brauer's theorem (p. 73) that if m is the exponent of G, the field L obtained by adjoining the mth roots of unity to K is such a splitting field for *any* representation of G.

The index $d_i = \sqrt{\dim_{\mathrm{Z}(D_i)} D_i}$ of D_i (or A_i) is significant from a representation-theoretic point of view: if we extend the K-representation ρ_i to the algebraic closure $\widetilde{K}$ of K (or to any splitting field of G), it will in general no longer be irreducible, but the $\widetilde{K}G$-module $V_i^{\widetilde{K}}$ (which affords $\rho_i^{\widetilde{K}}$ over $\widetilde{K}$) splits into a direct sum of simple modules over $A_i^{\widetilde{K}}$, in such a way that the multiplicity (of the isomorphism class) of each simple module in the direct sum is exactly d_i. Furthermore the number of distinct isomorphism classes in this splitting of $V_i^{\widetilde{K}}$ is $z_i = (\mathrm{Z}(D_i) : K)$.

In particular ρ_i is absolutely irreducible (or "split") over K if and only if $d_i = z_i = 1$, which means that $A_i \cong \mathrm{M}(n_i, K)$, i.e. A_i is split. More generally we say that a (possibly reducible) representation is split (over K) if it is the direct sum of absolutely irreducible representations. This is the same as saying that if V is the space of the representation, then the simple summands A_i such that $A_i V \neq 0$ are full matrix algebras $\mathrm{M}(n_i, K)$ over K.

The algebra A_i (or the representation ρ_i) is called "quasisplit" if d_i (but not necessarily z_i) is $= 1$. This means that $A_i \cong \mathrm{M}(n_i, L_i)$ where $L_i = \mathrm{Z}(A_i)$.

Theorem 1.7.6 *Suppose that A is a simple direct summand of KG. If the characteristic of K is $\neq 0$, A is quasisplit.*

Suppose the characteristic is 0 and that $\bar{A} = A$ where $^-$ is the standard (K, id)-involution of KG. Then if $(A, ^-)$ is symplectic or orthogonal, either A is quasisplit or A is similar (over its center) to a quaternion algebra.

Let Q be the prime subfield of K, and identify $KG = K \otimes_Q QG$. Let A_1 be the direct summand of QG such that A is a direct summand of $K \otimes_Q A_1 = KA_1 \subset KG$. Suppose first that $\operatorname{char} K \neq 0$. By Ex. 3, p. 17, A_1 is quasisplit, say $A_1 \cong \mathrm{M}(n, L_1)$ where L_1/Q is a finite field extension. Now by the exercise on p. 14, $K \otimes_Q L_1 \cong K_1 \oplus \cdots \oplus K_l$ for fields $K_1, \ldots, K_l$, and so

$$K \otimes_Q A_1 \cong (K \otimes_Q L_1) \otimes_{L_1} A_1 \cong \big(K_1 \otimes_{L_1} A_1\big) \oplus \cdots \oplus \big(K_l \otimes_{L_1} A_1\big), \qquad (1.62)$$

and since one of these summands is A, the first statement follows. This argument also shows that, in the case of characteristic 0 (which we now assume), A is quasisplit if A_1 is quasisplit.

Assume A is not quasisplit. Since $\bar{A} = A$ and the involution is the identity on the center of A, A is isomorphic to its opposite algebra and so has order 2 in the Brauer group (over its center). Since A is one of the summands in (1.62), A_1 is not quasisplit either and furthermore $\bar{A}_1 = A_1$. So it has order 2 in the Brauer group of its center which, as a finite extension of $Q = \mathbb{Q}$, is a global field. Thus A_1 is similar to a quaternion algebra (p. 19) and so therefore is A. □

We note that the degree (cf. p. 68) of the irreducible representation ρ_i over K corresponding to $A_i \cong \mathrm{M}(n_i, D_i)$ is $\dim_K D_i^{n_i \times 1} = n_i z_i d_i^2$, while the degree (cf. p. 12) of the K-algebra A_i is $n_i d_i$, which is the degree of any irreducible component of $\rho_i^{\widetilde{K}}$.

An arbitrary linear representation over K is said to be *of odd type* (with respect to $(K, ^-)$) if $AV = 0$ for each simple summand A of KG which is stable under the standard $(K, ^-)$-involution of KG and has *even* degree.

Suppose that the characteristic of K is 0, and that G has a normal Abelian subgroup H. By the corollary on p. 61 in [68], the degrees of all irreducible representations of G over an algebraic closure $\hat{K}$ of K divide $[G : H]$. Thus if G has such a subgroup H of odd index – e.g. if the order of G is odd – the degrees of the simple algebra components of $\hat{K}G$ are all odd. Since these degrees are identical with the degrees of the simple algebra components of KG (cf. p. 13), the latter are odd as well. Thus the representations of G over K are all of odd type (if G has a normal Abelian subgroup of odd index and $\operatorname{char} K = 0$).

In general, since the characteristic of K does not divide the degree $d_i n_i$ of A_i (Cor. 1.2.2, p. 16), the center $L_i = \mathrm{Z}(D_i)$ is generated over K by the character values of any character of a representation which is an irreducible constituent of ρ_i over an algebraic closure (cf. (70.8), p. 468, [18]). Since character values are sums of roots of unity, this means that L_i/K is an Abelian extension, i.e. a Galois extension with Abelian Galois group. In fact the character values are sums of mth roots of unity where m is the exponent of G, so L_i is a subfield of the cyclotomic extension of K generated by the mth roots of unity.

We need to generalize these facts slightly:

Lemma 1.7.1 *If A is a simple direct summand of KG then L/K_0 is Abelian.*

We may suppose that $K \neq K_0$. There is a simple direct summand A_1 of K_0G such that A is a direct summand of $K \otimes_{K_0} A_1$. The center L_1 of A_1 is an Abelian extension of K_0, and $K \otimes_{K_0} L_1$ is the center of $K \otimes_{K_0} A_1$.

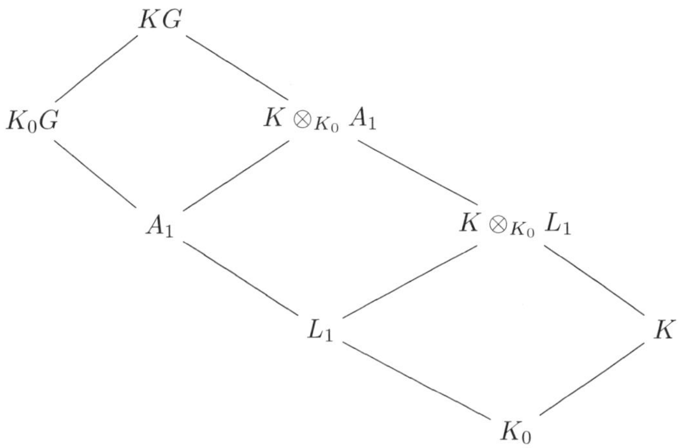

Since K/K_0 is a separable extension, $K \otimes_{K_0} L_1$ is a separable L_1-algebra and so is either a separable quadratic extension of L_1 or is $\cong L_1 \oplus L_1$.

In the first case, $K \otimes_{K_0} A_1$ is a simple algebra, so $= A$ and $K \otimes_{K_0} L_1 = L$. Thus L is the compositum of two Abelian extensions of K_0 and so is Abelian over K_0 – it is elementary that the Galois group of L/K_0 is the product of the Galois groups of K/K_0 and L_1/K_0 or one can apply Cor. 1.2.6, p. 22. In the second case,

$$K \otimes_{K_0} A_1 \cong (K \otimes_{K_0} L_1) \otimes_{L_1} A_1 \cong (L_1 \oplus L_1) \otimes_{L_1} A_1 \cong A_1 \oplus A_1.$$

These two copies of A_1 are K-algebras via the two implicit maps $K \to L_1$, but are not in general isomorphic as K-algebras. In any case A must be isomorphic to one of them, and so $L = L_1$ is Abelian over K_0. □

It is often useful to view the process of going up from K to a splitting field $\hat{K}$ of the representation ρ_i – e.g. to an algebraic closure $\widetilde{K}$ – as occurring in two stages: from K to $L_i = \mathrm{Z}(D_i)$, and then from L_i to $\hat{K}$. In the first stage ρ_i splits into its "conjugate representations". They are most easily understood from the point of view of matrix representations: if the matrix version of one of the L_i-irreducible components of $\rho_i^{L_i}$ is P_{ij} with character χ_{ij}, then the other irreducible components are the $(L_i : K)$ irreducible and mutually inequivalent representations $s \to (\mathsf{P}_{ij}(s))^\sigma$, one for each $\sigma \in \mathrm{Gal}(L_i/K)$. The character of this conjugate representation is simply χ_{ij}^σ, that is to say $s \to \chi_{ij}(s)^\sigma$. Then in the second stage, going from L_i to $\hat{K}$, each of these conjugate representations splits into d_i mutually equivalent absolutely irreducible representations. In the context of representation theory, the index d_i is often referred to as the "Schur index" of the absolutely irreducible representation or character belonging to $\rho_i^{L_i}$ or any of its conjugates (or to ρ_i itself).

1.7.6 Representations over the real and complex numbers. Reference: 13.2, pp. 106*ff*, [68].

Theorem 1.7.7 *(Frobenius–Schur). Let $\rho : G \to \mathbf{GL}(V)$ be a linear representation over the complex numbers $\mathbb{C}$ with character χ. In order that χ have values in $\mathbb{R}$ (resp. that ρ be realizable over $\mathbb{R}$), it is necessary and sufficient that V have a nonsingular bilinear form (resp. nonsingular symmetric bilinear form) invariant under G.* □

A slightly more general version of this theorem, when $\mathbb{C}$ is replaced by the algebraic closure of a real closed field, is given later (Theorem 4.5.4, p. 176).

Assume now that ρ is an irreducible representation over $\mathbb{C}$ of degree $\dim_{\mathbb{C}} V = n$. So the corresponding simple component of $\mathbb{C}G$ is (isomorphic to) $\mathrm{M}(n, \mathbb{C})$. There is a unique (up to equivalence) irreducible representation ρ_0 over $\mathbb{R}$ from which ρ arises by scalar extension from $\mathbb{R}$ to $\mathbb{C}$ – ρ is either $= \rho_0^{\mathbb{C}}$ or is an irreducible component of it.

There are three possible cases:

(a) At least one value $\chi(s)$, $s \in G$, is not real. Then $\rho_0^{\mathbb{C}} = \rho \oplus \rho^*$ where ρ^* is the conjugate representation and is not equivalent to ρ – in terms of the corresponding matrix representation, $\mathsf{P}^*(s) = \mathsf{P}(s)^*$ for all $s \in G$. The character of ρ_0 is $\chi + \chi^*$ and the corresponding simple summand of $\mathbb{R}G$ is $\mathrm{M}(n, \mathbb{C})$. In this case V has no G-invariant nonsingular bilinear form, and ρ is called a *type* 0 representation, or a representation of *unitary type.*[13]

(b) If ρ can be realized over $\mathbb{R}$ – in particular all values of χ are real – ρ is a *type* 1 representation or is *of orthogonal type.* This means that $\rho = \rho_0^{\mathbb{C}}$. The character of ρ_0 is χ and the corresponding simple summand of $\mathbb{R}G$ is $\mathrm{M}(n, \mathbb{R})$. Such a representation ρ has a nonsingular invariant bilinear form which is unique up to a nonzero scalar multiple; the form is symmetric. (See Th. 1.7.8 below).

(c) ρ is a *type* -1 representation or is of *symplectic type* if the values of χ are real but ρ is not realizable over $\mathbb{R}$. In this case $\rho_0^{\mathbb{C}} = \rho \oplus \rho$ and the character of ρ_0 is 2χ. The corresponding direct summand of $\mathbb{R}G$ is $\mathrm{M}(\frac{n}{2}, \mathbb{H})$ where $\mathbb{H}$ is the Hamiltonian quaternion algebra $(-1, -1)_{\mathbb{R}}$. In this case there is again a nonsingular invariant bilinear form which is unique up to a nonzero scalar multiple, but this time it is skew symmetric. (See Th. 1.7.8 below).

This trichotomy is verified in Corollary 4.5.2 (p. 175) for a formally real field $\mathbf{R}$ and its algebraic closure $\mathbf{C} = \mathbf{R}(i)$. In fact it holds in the following form for an arbitrary field K of characteristic 0:

Theorem 1.7.8 *Let $\rho : G \to \mathbf{GL}(V)$ be an irreducible representation over an arbitrary field K of characteristic 0, with character χ, and suppose that A is the simple summand of KG such that $AV \neq 0$. If $h : V \times V \to (K, ^-)$ is a nonzero invariant sesquilinear form, it is nonsingular and, if ρ is absolutely irreducible, is unique up to a scalar.*

Now suppose ρ is absolutely irreducible. Then the "Frobenius-Schur indicator"

$$\frac{1}{|G|} \sum_{s \in G} \chi(s^2) \;=\; 0, \; \pm 1. \tag{1.63}$$

---

[13]Our terminology is a little unorthodox here – a type 0 representation is often called type 1 in the literature, a type 1 is called a type 2, and a type -1 is called a type 3.

Moreover if $^{-}$ *is the standard* (K, id)*-involution on* KG,

(a) ρ *has a nonzero invariant bilinear form if and only if the indicator is* $\neq 0$, *and in this case* $\bar{A} = A$.

(b) *If the indicator is* 1, *the unique invariant nonzero bilinear form is a nonsingular symmetric form and* $(A, ^{-})$ *is orthogonal.*

(c) *If the indicator is* -1, *the unique invariant nonzero bilinear form is a nonsingular skew symmetric form and* $(A, ^{-})$ *is symplectic.*

Suppose $h : V \times V \to K$ is a nonzero G-invariant sesquilinear form. If $u \in \ker h_r = \{u \in V : h(V, u) = 0\}$, then for all $s \in G$

$$h(V, su) = h(sV, su) = h(V, u) = 0,$$

so $\ker h_r$ is a proper invariant subspace of V and therefore $= 0$. Thus h is nonsingular, and h_r is a KG-isomorphism (Th. 1.7.4, p. 75). If ρ is absolutely irreducible and h' is another G-invariant sesquilinear form, then $h_r^{-1}h'_r \in \mathrm{End}_{KG}V = K$, so h' is a scalar multiple of h.

Now suppose that ρ is absolutely irreducible – so $A \cong \mathrm{M}(n, K)$ – and let χ be its character. View ρ as a matrix representation $\mathsf{P} : G \to \mathbf{GL}(n, K)$ by choosing a basis of V. Then $\mathsf{P}G \subset \mathbf{GL}(n, K')$ for some subfield K' of K which is finitely generated over the prime subfield and is therefore isomorphic to a subfield K'' of $\mathbb{C}$ – say $\varphi : K' \to K''$. We denote by $\mathsf{P}^{K''}$ the matrix representation over K'' obtained from P via φ, by X $(= \varphi\chi)$ its character, and by $\mathsf{P}^{\mathbb{C}}$ this representation viewed as a representation over $\mathbb{C}$ and $\mathsf{X}^{\mathbb{C}} = \mathsf{X}$ its character. By Prop. 39, p. 109, [68], the relation (1.63) holds over $\mathbb{C}$ (with χ replaced by $\mathsf{X}^{\mathbb{C}}$ of course).

Since $\varphi\chi = \mathsf{X}^{\mathbb{C}}$, we immediately get (1.63).

Suppose that the indicator is -1, for example. By Props. 38 and 39, [68], there is a nonsingular invariant skew symmetric form f, which we view as a polynomial over $\mathbb{C}$ in the $2n$ indeterminates $\vec{x} = (x_1, \dots, x_n)$ and $\vec{y} = (y_1, \dots, y_n)$. Let $\{\omega_i\}$ be a basis of $\mathbb{C}$ over K'', and write,

$$f(\vec{x}, \vec{y}) = \sum_i \omega_i f_i(\vec{x}, \vec{y}), \quad f_i \in K''[\vec{x}, \vec{y}] \text{ for all } i.$$

Then for all $s \in G$

$$\begin{aligned} f(\mathsf{P}^{K''}(s)\vec{x},\ \mathsf{P}^{K''}(s)\vec{y}) &= \sum_i \omega_i f_i(\mathsf{P}^{K''}(s)\vec{x}, \mathsf{P}^{K''}(s)\vec{y}) \\ &= f(\vec{x}, \vec{y}) = \sum_i \omega_i f_i(\vec{x}, \vec{y}), \end{aligned}$$

so $f_i(\mathsf{P}^{K''}(s)\vec{x},\ \mathsf{P}^{K''}(s)\vec{y}) = f_i(\vec{x}, \vec{y})$ for all i and all s, that is to say, the f_i are invariant. Similarly

$$\begin{aligned} f(\vec{x}, \vec{y}) &= \sum_i \omega_i f_i(\vec{x}, \vec{y}) \\ &= -f(\vec{y}, \vec{x}) = \sum_i \omega_i(-f_i(\vec{y}, \vec{x})) \end{aligned}$$

implies that all f_i are skew symmetric. At least one of the f_i is $\neq 0$, say f_j, and so $\varphi^{-1}f_j$ is a nonzero (and so nonsingular) invariant skew symmetric form for P, and so also for ρ. Thus $\mathsf{P} : G \to \mathbf{Sp}(n, K')$ and $\rho : G \to \mathbf{Sp}(n, K)$, and

$(\varphi^{-1}f_j)(au, v) = (\varphi^{-1}f_j)(u, \bar{a}v)$ for all $a \in KG$. This implies that $\bar{A} = A$, and that

$$(A, ^-) \xrightarrow{\rho} (\mathrm{End}V, ^\sim), \quad \text{(where } ^\sim \text{ is the adjoint involution of } \varphi^{-1}f_j),$$

is an isomorphism. Thus $(A, ^-)$ is symplectic (Prop. 1.5.1, p. 47).

Similarly if the indicator is 1, there is a nonsingular invariant symmetric form for ρ, and $(A, ^-)$ is orthogonal.

If the indicator for χ is 0, it is easy to see that the indicator for $\mathsf{X}^{\mathbb{C}}$ is also 0, and so $\mathsf{P}^{\mathbb{C}}$ has no nonzero invariant forms. But if ρ has a nonzero invariant form, an argument similar to the above shows that $\mathsf{P}^{\mathbb{C}}$ must also have one, which is a contradiction. □

If K can be imbedded in $\mathbb{C}$, as is "usually" the case, this theorem is more easily proved by simply replacing K by an imbedding into $\mathbb{C}$.

CHAPTER 2

Isometry representations of finite groups

From now on, G is a finite group of order $g = |G|$ and K is a field of characteristic $\neq 2$ and relatively prime to g, with an involution $^-$ which is possibly the identity. All vector spaces over K are assumed to be finite dimensional, and bilinear and sesquilinear forms will be assumed to be nonsingular unless mentioned otherwise.

2.1 Introduction

An *orthogonal representation* of G is a homomorphism

$$\rho : G \to \mathbf{O}(V, b) \tag{2.1}$$

of G into the orthogonal group of a symmetric bilinear form $b : V \times V \to K$. Similarly a *symplectic* representation is a homomorphism $\rho : G \to \mathbf{Sp}(V, b)$ where (V, b) is a skew symmetric space, and a *unitary* representation is a homomorphism $\rho : G \to \mathbf{U}(V, h)$ where (V, h) is an Hermitian space. Equivalently, we can say that b or h is a *G-invariant* form, or if the group G is clear from the context, simply that b or h is *invariant*.

We shall generally ignore the case of skew Hermitian forms since any such form is a scalar multiple of an Hermitian form; thus their respective unitary groups are identical, and the theory of skew Hermitian representations of G is equivalent to the theory of Hermitian representations. A similar remark applies to "μ-Hermitian" forms, where $\mu\bar{\mu} = 1$ – by "Hilbert's Theorem 90" (cf. p. 281, [53]), there is $\alpha \in K$ such that $\mu = \alpha\bar{\alpha}^{-1}$ and then αh is Hermitian if h is μ-Hermitian.

Recall that a form $h : V \times V \to (K, ^-)$ which is symmetric bilinear, skew symmetric bilinear, Hermitian or skew Hermitian, is called a *reflexive form* over $(K, ^-)$, and (V, h) is called a *reflexive space* over $(K, ^-)$ (if h is nonsingular). Its group of isometries is denoted by $\mathbf{I}(h)$, $\mathbf{I}(V, h)$ or $\mathbf{I}(V)$. A homomorphism $\rho : G \to \mathbf{I}(V, h)$ will be called an *isometry representation* (over $(K, ^-)$) if h is nonsingular. Thus an isometry representation is an orthogonal, symplectic, or unitary representation. When such a representation is given, we sometimes refer to (V, h) as an *equivariant space*, or more precisely as an *equivariant reflexive space*.

With an obvious meaning, we often refer to the *type* of a reflexive form h as symmetric, skew symmetric, Hermitian or skew Hermitian, and similarly the (isometry) type of an isometry representation is orthogonal, symplectic or unitary. The form h is ε-reflexive with $\varepsilon = 1$ if h is symmetric or Hermitian, $= -1$ if it is skew symmetric or skew Hermitian.

As we saw in §1.7, the theory of linear representations of G over K is equivalent to the theory of modules over the group algebra KG. We now discuss an analogue of this which is critical for the theory of isometry representations.

Let $\rho : G \to \mathbf{I}(V, h)$ be an isometry representation. Then $\rho(s) \in \mathbf{I}(V, h)$ means that $h(\rho(s)u, \rho(s)v) = h(u, v)$ for all $u, v \in V$. This can also be written $h(su, sv) = h(u, v)$ in terms of the KG-module structure on V, and this is in turn equivalent to $h(su, v) = h(u, s^{-1}v) = h(u, \bar{s}v)$ where $^-$ is the standard $(K, ^-)$-involution on KG. It follows that $\rho(s) \in \mathbf{I}(V, h)$ is equivalent to

$$h(au, v) = h(u, \bar{a}v) \text{ for all } a \in KG. \tag{2.2}$$

Thus the adjoint involution with respect to h (cf. (1.34)) on $\mathrm{End}_K V$ induces the standard involution $^-$ on the image of KG. Or to put it another way, if $^-$ also denotes the adjoint involution with respect to h on $\mathrm{End}_K V$, the homomorphism $KG \to \mathrm{End}_K V$ given by the KG-module structure of V is a homomorphism

$$(KG, ^-) \to (\mathrm{End}_K V, ^-)$$

of involution algebras. One also says that h *is invariant under the involution algebra* $(KG, ^-)$ (or simply under KG) if (2.2) holds – of course this is the same as saying that h is G-invariant.

More generally, if $(A, ^-)$ is an involution algebra which acts on V in such a way that $h(au, v) = h(u, \bar{a}v)$ for all $a \in A$ and $u, v \in V$, then h is said to be A-invariant or more precisely $(A, ^-)$-invariant. This occurs, for example, when $(A, ^-)$ is a direct summand of $(KG, ^-)$ and $\rho : G \to \mathbf{I}(V, h)$ is an isometry representation which is A-isotypic as a linear representation.

If $(A, ^-)$ is an involution algebra which is a direct summand of $(KG, ^-)$ – e.g. a simple involution algebra – and if V is an A-module with a form h invariant under $(A, ^-)$, then of course V can also be viewed as a KG-module, and h is $(KG, ^-)$-invariant; thus we again get an isometry representation $\rho : G \to \mathbf{I}(h)$.

Equivalence. Two isometry representations

$$\rho_1 : G \to \mathbf{I}(V_1, h_1) \text{ and } \rho_2 : G \to \mathbf{I}(V_2, h_2) \tag{2.3}$$

are *equivalent* if there is an isometry

$$\varphi : (V_1, h_1) \to (V_2, h_2) \tag{2.4}$$

such that the diagram

$$\begin{array}{ccc} V_1 & \xrightarrow{\varphi} & V_2 \\ {\scriptstyle \rho_1(s)}\downarrow & & \downarrow{\scriptstyle \rho_2(s)} \\ V_1 & \xrightarrow{\varphi} & V_2 \end{array}$$

is commutative for all $s \in G$ – that is to say, φ is also a KG-module homomorphism. In this case we write $\rho_1 \cong \rho_2$. We can also express this by saying that the invariant forms h_1 and h_2 are (equivariantly) equivalent. More generally if $(A, ^-)$ is a $(K, ^-)$-involution algebra, two $(A, ^-)$-invariant forms (V_i, h_i) $(i = 1, 2)$ are said to be (equivariantly) equivalent if there is an A-isomorphism $\varphi : V_1 \to V_2$ such that $h_2(\varphi u, \varphi v) = h_1(u, v)$ for all $u, v \in V_1$.

If ρ_1 and ρ_2 are equivalent isometry representations, then of course they are also equivalent as linear representations.

Examples. 1. Let $V = KG$; the *regular representation* $\rho : G \to \mathbf{GL}(V)$ is the (linear) representation given by $\rho(s)v = sv$. Let $^-$ be the standard $(K, ^-)$-involution on KG. If t is the trace form $\mathrm{Tr}_{KG/K}(u\bar{v})$, it is easy to see that the regular representation is an isometry representation $\rho : G \to \mathbf{I}(V, \mathrm{t})$, called the *regular* isometry representation of G. It is orthogonal if $(K, {}^-)$ is of the first kind, unitary if $(K, {}^-)$ is of the second kind.

2. Let $(L, {}^-)/(K, {}^-)$ be a finite extension of fields with involution, and suppose that L/K is a Galois extension, with Galois group G, and that ${}^-$ commutes with all elements of G (which is of course true if $(L, {}^-)$ is of the first kind). If $V = L$ and t is the trace form of the extension L/K, then the inclusion $G \subset \mathbf{GL}(V)$ is an isometry representation $G \to \mathbf{I}(V, \mathrm{t})$, orthogonal if $(K, {}^-)$ is of the first kind, unitary otherwise. □

Exercise. If $(L, {}^-)$ is of the first kind, show that the representation in Example 2 is equivalent to that in Example 1 if and only if L/K has a "self-dual normal basis", i.e. there is an element $\lambda \in L$ such that $\{\sigma\lambda |\ \sigma \in G\}$ is a basis of L/K which satisfies $\mathrm{Tr}_{L/K}(\sigma\lambda \cdot \tau\lambda) = \delta_{\sigma,\tau}$ for all $\sigma, \tau \in G$. □

We shall now see how the notions of isometry representation (or invariant form) and their equivalence are related to ε-Hermitian forms defined over KG itself, that is to say to forms

$$\hat{h} : V \times V \to (KG, {}^-)$$

with properties analogous to those of ε-reflexive forms over $(K, {}^-)$. A thorough and systematic study of such forms is given later in Ch. 3, but for the moment, the definition will suffice: if $(A, {}^-)$ is a $(K, {}^-)$-involution algebra, a map

$$f : V \times V \to (A, {}^-)$$

is a *sesquilinear form* over $(A, {}^-)$ if it is biadditive and satisfies $f(au, bv) = af(u, v)\bar{b}$ for all $u, v \in V$ and $a, b \in A$. The sesquilinear form $\hat{h}$ is *ε-Hermitian* ($\varepsilon = \pm 1$) if also[1] $\overline{\hat{h}(u, v)} = \varepsilon\hat{h}(v, u)$ for all u and v. The ε-Hermitian forms over $(A, {}^-)$ are denoted by ${}_{(A,{}^-)}\mathcal{H}_\varepsilon$ or simply ${}_A\mathcal{H}_\varepsilon$, and those that are nonsingular by ${}_{(A,{}^-)}\mathcal{H}^*_\varepsilon$ – when A is semisimple (the case of most interest to us), the ε-Hermitian form $\hat{h}$ is *nonsingular* if $\hat{h}(u, v) = 0$ for all $v \in V$ implies that $u = 0$ (cf. Lem. 3.1.1).

Let

$$\mathrm{Tr} = \mathrm{Tr}_{KG/K} : KG \to K$$

be the canonical trace $\mathrm{Tr}(\sum_s \alpha_s s) = \alpha_1$ – it is an involution trace (cf. p. 35).

If $\hat{h} : V \times V \to (KG, {}^-)$ is a sesquilinear form, it is easy to check that

$$h := \mathrm{Tr} \circ \hat{h} : V \times V \to (K, {}^-)$$

is a $(KG, {}^-)$-invariant form over K, and that it is ε-reflexive if $\hat{h}$ is ε-Hermitian. If $\hat{h}(u, v) \neq 0$, then there exists $s \in G$ such that $\mathrm{Tr}(\hat{h}(u, v)s) \neq 0$, i.e. $\mathrm{Tr}(\hat{h}(u, s^{-1}v)) \neq 0$. It follows that h is nonsingular if $\hat{h}$ is, and the converse is trivial.

Conversely, if we are given the $(KG, {}^-)$-invariant form $h : V \times V \to (K, {}^-)$, it turns out that there is a *unique* sesquilinear form $\hat{h} : V \times V \to (KG, {}^-)$ such that $h = \mathrm{Tr} \circ \hat{h}$, namely

$$\hat{h}(u, v) = \sum_s h(u, sv)s. \tag{2.5}$$

[1]Strictly speaking, it is of course symmetric or skew symmetric when $(A, {}^-) = (K, \mathrm{id})$, or more generally, when A is commutative and ${}^- = \mathrm{id}_A$.

Indeed it is easy to check that $\hat{h}$ is such a form, and if $\breve{h}$ is another and $\breve{h}(u,v) = \sum_s \alpha_s s$, then

$$\alpha_s = \mathrm{Tr}(\breve{h}(u,v)s^{-1}) = \mathrm{Tr}(\breve{h}(u,sv)) = h(u,sv).$$

If the form h is ε-reflexive, then $\hat{h}$ is in fact ε-Hermitian[2], and conversely.

$\hat{h}$ is called the *$(KG,^-)$-form (or KG-form) associated to the invariant form h or to the isometry representation $\rho : G \to \mathbf{I}(h)$*. It is nonsingular respectively totally isotropic if and only if h is nonsingular respectively totally isotropic.

Now suppose that the isometry representations ρ_1 and ρ_2 in (2.3) are equivalent via the KG-linear isometry φ. Then for any $u, v \in V$

$$\begin{aligned}\hat{h}_2(\varphi u, \varphi v) &= \sum_s h_2(\varphi u, s\varphi v)s \\ &= \sum_s h_2(\varphi u, \varphi sv)s \\ &= \sum_s h_1(u, sv)s = \hat{h}_1(u,v),\end{aligned}$$

i.e. the forms $\hat{h}_2$ and $\hat{h}_1$ are equivalent via φ.

Conversely, if $\hat{h}_2 \cong \hat{h}_1$ via the KG-linear isometry φ,

$$h_2(\varphi u, \varphi v) = \mathrm{Tr}(\hat{h}_2(\varphi u, \varphi v)) = \mathrm{Tr}(\hat{h}_1(u,v)) = h_1(u,v),$$

and so $\rho_1 \cong \rho_2$. Therefore

Theorem 2.1.1 *For each $\varepsilon = \pm 1$, the association of the form $\hat{h}$ given by (2.5) with an isometry representation $\rho : G \to \mathbf{I}(V,h)$, provides an isomorphism between the category of isometry representations of G on nonsingular ε-reflexive spaces (V,h) (which is essentially the same category as the category ${}^{(KG,^-)}_{(K,^-)}\mathcal{H}^*_\varepsilon$ of nonsingular ε-reflexive $(KG,^-)$-invariant forms over $(K,^-)$) and the category ${}_{(KG,^-)}\mathcal{H}^*_\varepsilon = {}_{KG}\mathcal{H}^*_\varepsilon$ of nonsingular ε-Hermitian forms over $(KG,^-)$. The inverse isomorphism associates the isometry representation $s \to (v \to sv)$ on the space $(V, \mathrm{Tr}_{KG/K} \circ \hat{h})$ with the space $(V,\hat{h}) \in {}_{(KG,^-)}\mathcal{H}^*_\varepsilon$.*

In particular, two isometry representations $\rho_1 : G \to \mathbf{I}(h_1)$ and $\rho_2 : G \to \mathbf{I}(h_2)$ are equivalent if and only if the associated $(KG,^-)$-forms $\hat{h}_1$ and $\hat{h}_2$ are equivalent. □

There is a useful generalization of this theorem to forms invariant under a separable involution algebra – see Th. 3.2.4, p. 113.

Since we are assuming that the characteristic of K does not divide the order g of G, the group algebra KG is separable and is therefore the direct sum

$$KG = A_1 \oplus A_2 \oplus \cdots \oplus A_h$$

of simple separable algebras, in a unique way. As we saw in §1.3, any A_i is either fixed by the standard involution of KG or is interchanged with another simple summand. And so we can (with a change in notation) write

$$KG = A_1 \oplus \cdots \oplus A_k \oplus A_{k+1} \oplus \cdots \oplus A_l$$

---

[2]There is a small inconsistency in the terminology here (which we ignore) – if G is the trivial group $\{1\}$ and h is symmetric, say, then $\hat{h} = h$ and is symmetric, not Hermitian.

where each A_i is a simple involution algebra, simple and separable as an algebra for $1 \le i \le k$ and, for $k < i \le l$, a direct sum of a simple separable component A of KG and its image $\bar{A}$ under the involution of KG – the latter are the "(simple) hyperbolic involution algebras". Thus the involution algebra $(KG, ^-)$ is the direct sum

$$(KG, ^-) = (A_1, ^-) \oplus (A_2, ^-) \oplus \cdots \oplus (A_l, ^-)$$

of simple involution algebras, which are separable and uniquely determined up to order.

For any KG-module V we get a corresponding decomposition

$$V = A_1V \oplus \cdots \oplus A_kV \oplus A_{k+1}V \oplus \cdots \oplus A_lV \tag{2.6}$$

(cf. Th. 1.2.7) where the A_iV for $1 \le i \le k$ are isotypic components of V, and for $i > k$ are the direct sum of two isotypic components. We shall refer to the A_iV as the *involution isotypic* components of V.

Now let $h : V \times V \to (K, ^-)$ be a nonsingular reflexive form and $\rho : G \to \mathbf{I}(V, h)$ an isometry representation. Since $h(au, bv) = h(u, \bar{a}bv) = 0$ if a and b are elements of different A_i's, the decomposition in (2.6) is orthogonal,

$$V = A_1V \perp A_2V \perp \ldots \perp A_lV. \tag{2.7}$$

We denote by h_i the restriction of the form h to A_iV.

Since any KG-homomorphism $\varphi : V \to V'$ takes A_iV to A_iV', two isometry representations

$$\rho : G \to \mathbf{I}(V, h) \quad \text{and} \quad \rho' : G \to \mathbf{I}(V', h') \tag{2.8}$$

are equivalent if and only if the subrepresentations $G \to \mathbf{I}(A_iV, h_i)$ and $G \to \mathbf{I}(A_iV', h_i')$ are equivalent for all i.

Thus the equivalence class of ρ is determined by the equivalence classes of the subrepresentations $G \to \mathbf{I}(A_iV, h_i)$, $1 \le i \le l$. The form h_i is of course determined by h and the canonical decomposition (2.7), and is called an *isotypic component of* h or more precisely the $(A_i, ^-)$*-isotypic component* of h. It is also G-invariant. If $h = h_i$, i.e. $V = A_iV$, then ρ is called an $(A_i, ^-)$*-isotypic (isometry) representation.*

The category of $(A_i, ^-)$-isotypic isometry representations of G on ε-reflexive spaces (V, h) (which is essentially the same category as the category ${}^{(A_i,^-)}_{(K,^-)}\mathcal{H}^*_\varepsilon$ of nonsingular ε-reflexive $(A_i, ^-)$-invariant forms over $(K\, ^-)$) is a full subcategory of the category ${}^{(KG,^-)}_{(K,^-)}\mathcal{H}^*_\varepsilon$. The category isomorphism of Th. 2.1.1 provides an isomorphism of this subcategory with the full subcategory ${}_{(A_i,^-)}\mathcal{H}^*_\varepsilon$ of nonsingular ε-Hermitian forms over $(A_i, ^-)$, of the category ${}_{(KG,^-)}\mathcal{H}^*_\varepsilon$. Indeed, if $(V, h) \in {}^{(A_i,^-)}_{(K,^-)}\mathcal{H}^*_\varepsilon$ and 1_i is the identity element of A_i, $\hat{h}(u, v) = \hat{h}(1_iu, v) = 1_i\hat{h}(u, v)$ for all $u, v \in V$, so $\hat{h}$ takes its values in $(A_i, ^-)$. And it is clear that the inverse isomorphism has image in the category ${}^{(A_i,^-)}_{(K,^-)}\mathcal{H}^*_\varepsilon$. Thus:

Corollary 2.1.1 *Let* $1 \le i \le l$ *and* $\varepsilon = \pm 1$. *The association of the form* $\hat{h} : V \times V \to (A_i, ^-)$ *with an* A_i*-isotypic isometry representation* $\rho : G \to \mathbf{I}(V, h)$, *provides an isomorphism between the category* ${}^{(A_i,^-)}_{(K,^-)}\mathcal{H}^*_\varepsilon$ *of* $(A_i, ^-)$*-isotypic isometry representations of* G *and the category* ${}_{(A_i,^-)}\mathcal{H}^*_\varepsilon$ *of nonsingular* ε*-Hermitian forms over* $(A_i, ^-)$.

The inverse isomorphism associates with a space $(V, \hat{h}) \in {}_{(A_i,^-)}\mathcal{H}^*_\varepsilon$ *the* $(A_i, ^-)$*-isotypic isometry representation* $v \to sv$ *on the space* $(V, \mathrm{Tr}_{KG/K} \circ \hat{h})$*, where* G *acts on* V *via* $s \in KG \to A_i$. □

REMARK. It follows from the definition of $\hat{h}$ that (2.7) is also an orthogonal decomposition of V with respect to the Hermitian form $\hat{h}$. Let $\hat{h}_i \in {}_{(A_i,^-)}\mathcal{H}^*_\varepsilon$ be the restriction of $\hat{h}$ to A_iV. Then we say that $\hat{h}$ is the *vector sum* of the $\hat{h}_i$ – cf. p. 95. □

We shall see later (Corollary 3.1.6, p. 98) that if $(A_i, ^-)$ is a hyperbolic involution algebra, the equivalence class of $\hat{h}_i$ is completely determined by the length of the A_i-module A_iV . This implies the following theorem:

Theorem 2.1.2 *Let* $(A_1, ^-), \ldots, (A_k, ^-)$ *be the nonhyperbolic simple involution components of* $(KG, ^-)$*. Then two isometry representations (2.8) are equivalent if and only if*

(a) *the* ε*-Hermitian forms* $\hat{h}_i$ *and* $\hat{h}'_i$ *are equivalent for* $1 \leq i \leq k$*, and*
(b) $\mathrm{len}_{KG} A_iV = \mathrm{len}_{KG} A_iV'$ *(or* $\dim_K A_iV = \dim_K A_iV'$*) for* $k < i \leq l$*.*

Condition (a) can be replaced by the condition that the subrepresentations of G *on* (A_iV, h_i) *and* (A_iV', h'_i) *be equivalent for* $1 \leq i \leq k$*, or that the isotypic components* h_i *and* h'_i *are equivariantly equivalent for* $1 \leq i \leq k$*. Condition (b) can be replaced by the requirement that* ρ *and* ρ' *be equivalent as linear representations.* □

Corollary 2.1.2 *The following conditions are necessary for the equivalence of two isometry representations (2.3) :*

(a) ρ_1 *and* ρ_2 *are equivalent as linear representations.*
(b) h_1 *and* h_2 *are equivalent forms.*
(c) h_1 *and* h_2 *are isotypically equivalent.*
(d) $(\hat{h}_1)_i$ *and* $(\hat{h}_2)_i$ *are equivalent for* $1 \leq i \leq k$*.* □

Of course (c) implies (b) but in general the converse is not true.

It is apparent from the above theorem that determining the equivalence of ε-Hermitian forms over a simple algebra A is critical. And the solution of *that* problem will be effected by reducing it to a more familiar one, namely that of the equivalence of reflexive forms over the division algebra D Brauer equivalent to A. See Th. 3.2.14, p. 132.

When $(K, ^-)$ is of the first kind, the division algebras D which occur in this context are very restricted in nature, as we have already seen in Th. 1.7.6 (p. 78), namely D is either quasisplit or, if $\mathrm{char}\, K = 0$ and $(A, ^-)$ is orthogonal or symplectic, D is a quaternion division algebra.

An important special case of the equivalence of two isometry representations (2.8) is when $V = V'$ and $h = h'$. One can in fact always assume this by (b) of Corollary 2.1.2.

In this case, ρ and ρ' are equivalent if and only if there is an automorphism φ of V which commutes with the actions of G arising from ρ and ρ',

$$\varphi\rho'(s) = \rho(s)\varphi \quad \text{for all } s \in G,$$

and which is an isometry of h with itself, that is to say $\varphi \in \mathbf{I}(V,h)$. Thus $\rho \cong \rho'$ if and only if there exists $\varphi \in \mathbf{I}(V,h)$ such that $\rho'(s) = \varphi^{-1}\rho(s)\varphi$ for all $s \in G$ – in other words, ρ and ρ' are "conjugate representations" of G into $\mathbf{I}(V,h)$.

Conjugacy in $\mathbf{I}(V,h)$. Suppose that G and G' are finite subgroups of $\mathbf{I}(V,h)$. If they are conjugate in $\mathbf{I}(V,h)$, say

$$\tau^{-1}G\tau = G' \tag{2.9}$$

for $\tau \in \mathbf{I}(V,h)$, then the equation

$$s\tau = \tau \cdot \tau^{-1}s\tau \quad \text{for all } s \in G$$

can be interpreted as meaning that the representation given by inclusion $G \hookrightarrow \mathbf{I}(V,h)$ and the representation $\text{inn}_{\tau^{-1}}$ are equivalent as isometry representations via the map $\tau : V \to V$.

Similarly if G and G' are conjugate in $\mathbf{GL}(V)$, then the same representations (into $\mathbf{GL}(V)$) are linearly equivalent.

Theorem 2.1.3 *Suppose $h \in {}_{(K,^-)}\mathcal{H}^*_\varepsilon$ and that the quadruple $(G,(K,^-),\varepsilon,h)$ has the property that any two isometry representations $G \to \mathbf{I}(V,h)$ are equivalent if they are linearly equivalent. If G and G' are subgroups of $\mathbf{I}(V,h)$ which are conjugate in $\mathbf{GL}(V)$, then they are conjugate in $\mathbf{I}(V,h)$.*

We know that (2.9) holds for some $\tau \in \mathbf{GL}(V)$. Then the inclusion $G \hookrightarrow \mathbf{I}(V,h)$ and conjugation $\text{inn}_{\tau^{-1}} : G \to \mathbf{I}(V,h)$ are linearly equivalent and so these two representations are equivalent as isometry representations, say via $\varphi \in \mathbf{I}(V,h)$,

$$\varphi s = \tau^{-1}s\tau\varphi,$$

from which it follows that $\varphi G\varphi^{-1} = \tau^{-1}G\tau = G'$. □

We shall see later on that the condition in Th. 2.1.3 on the isometry representations $G \to \mathbf{I}(V,h)$ does arise – see for example Corollary 5.2.1, p. 224.

Extension of the scalars. We have already seen that if $\rho : G \to \mathbf{GL}(V)$ is a linear representation and $K \to L$ is a homomorphism of fields – e.g. the inclusion of a subfield K into L, then there is a representation $\rho^L : G \to \mathbf{GL}(V^L)$ obtained by "extension of the scalars" (p. 70). If $\rho : G \to \mathbf{I}(V,h)$ is an *isometry* representation, we can also extend the scalars to L if $(K,^-) \to (L,^-)$ is a homomorphism of fields with involution, such as the inclusion $(K,^-) \to (L,^-)$ of a subfield with involution $(K,^-)$ into the field $(L,^-)$ with involution.[3] When this is the case, the linear extension ρ^L of ρ is also an isometry extension since $\rho^L(G) = \rho(G) \subset \mathbf{I}(V,h) \subset \mathbf{I}(V^L,h^L)$. (See p. 48).

Hyperbolic representations. Suppose that $\rho : G \to \mathbf{GL}(V)$ is a linear representation and that $\varepsilon = \pm 1$. The *hyperbolic space based on V* is

$$\mathrm{H}_\varepsilon(V) = \mathrm{H}(V) = V \oplus \overline{V^*},$$

the associated *ε-hyperbolic form* is

$$h = h_{\mathrm{H}(V)} : \mathrm{H}(V) \times \mathrm{H}(V) \to (K,^-), \quad h(v_1 + v_1^*, v_2 + v_2^*) = \langle v_1, v_2^* \rangle + \varepsilon\overline{\langle v_2, v_1^* \rangle},$$

[3]The possibility that the identity involution on K is extended by a *non-identity* involution on L is allowed.

and the ε-*hyperbolic representation based on* ρ is

$$\mathrm{H}_\varepsilon(\rho) = \mathrm{H}(\rho) = \rho \oplus \rho^*,$$

where $\rho^* : G \to \mathbf{GL}(\overline{V^*})$ is the twisted contragredient representation (p. 74) of ρ. It is easy to see that h is an ε-reflexive form – for example

$$\begin{aligned} h(v_1 + v_1^*, \alpha(v_2 + v_2^*)) &= \langle v_1, \alpha v_2^* \rangle + \varepsilon\overline{\langle \alpha v_2, v_1^* \rangle} = \langle v_1, v_2^* \bar{\alpha} \rangle + \varepsilon\overline{\alpha \langle v_2, v_1^* \rangle} \\ &= \langle v_1, v_2^* \rangle \bar{\alpha} + \varepsilon\overline{\langle v_2, v_1^* \rangle} \bar{\alpha} = h(v_1 + v_1^*, v_2 + v_2^*)\bar{\alpha} \end{aligned}$$

by the definition of $\overline{V^*}$. The subspaces V and $\overline{V^*}$ are totally isotropic and the right adjoint

$$h_r : \mathrm{V} \oplus \overline{V^*} \to (\mathrm{V} \oplus \overline{V^*})^* = V^* \oplus (\overline{V^*})^*$$

is injective on both V and $\overline{V^*}$ (since $V \times V^* \to K$ is a perfect pairing), taking V into $(\overline{V^*})^*$ and $\overline{V^*}$ into V^*; therefore h_r is an isomorphism and h is nonsingular. Finally h is G-invariant since $\langle sv, sv^* \rangle = \langle s^{-1}sv, v^* \rangle = \langle v, v^* \rangle$ by (1.57), p. 74, and so the hyperbolic representation is an isometry representation

$$\mathrm{H}(\rho) : G \to \mathbf{I}(\mathrm{H}(V), h_{\mathrm{H}(V)}).$$

Of course an isometry representation $\sigma : G \to \mathbf{I}(W, g)$ is called hyperbolic if it is isomorphic to a hyperbolic representation $\mathrm{H}(\rho)$.

Exercise. Let G be the cyclic group $\langle s \rangle$ of order 2.
(a) Consider the hyperbolic plane K^2 whose form h has discriminant matrix

$$\begin{pmatrix} 0 & 1 \\ 1 & 0 \end{pmatrix}$$

with respect to the canonical basis of K^2. Let $\rho : G \to \mathbf{GL}(2, K)$ be the linear representation such that $\rho(s)$ interchanges the two vectors in the canonical basis. Show that ρ is an orthogonal representation $\rho : G \to \mathbf{O}(K^2, h)$ but that it is not (equivalent to) a hyperbolic representation in the above sense.
(b) Generalize to K^{2n}. □

2.2 Notes on Chapter 2

1. The existence of self-dual normal bases, alluded to in the exercise on p. 85, has received a great deal of attention – see the first note on Ch. 4, p. 216.

CHAPTER 3

Hermitian forms over semisimple algebras

3.1 Introduction

The central importance of Hermitian forms over semisimple algebras in this book arises from the fact (cf. Th. 2.1.1, p. 86) that the category of orthogonal (or, more generally, isometry) representations of a finite group G over K is equivalent to the category of nonsingular Hermitian forms (respectively ε-Hermitian forms) over the group algebra KG. The point is that many facts about isometry representations are more easily dealt with from the point of view of ε-Hermitian forms over semisimple algebras, in much the same way that certain properties of linear representations are most easily verified (and interpreted) by using the language and theory of KG-modules.

From another point of view, there is some virtue in considering Hermitian forms over semisimple algebras *ab initio* since their category is closed under scalar extension, unlike the category of Hermitian forms over division algebras.

We assume that A is a finite dimensional semisimple algebra with a $(K, ^-)$-involution, also denoted by $^-$, over the field K of characteristic $\neq 2$, and that all A-modules are finitely generated.

$V^* = \mathrm{Hom}_A(V, A)$ will stand for the dual of a left A-module V. V^* is naturally a right A-module, and the twisted dual $\overline{V^*}$ is a left A-module, via $a\bar{x} = \overline{x\bar{a}}$. Here $\bar{x}$ denotes the linear functional $x \in V^*$ when considered as an element of $\overline{V^*}$. (It is tempting, though sometimes a little confusing, to write this relationship as $\overline{xa} = \bar{a}\bar{x}$). Finally $f : V \times V \to (A, ^-)$ will generally denote a sesquilinear form and h an ε-Hermitian form ($\varepsilon = \pm 1$). (In the special case when $A = K = K_0$, such a form is *not* ε-Hermitian in the sense of forms over a field but rather symmetric or skew symmetric, but we will ignore this inconsistency since it will be clear from the context what is meant).

3.1.1 Basic properties of ε-Hermitian forms over semisimple algebras. The definition of sesquilinear forms and ε-Hermitian forms ($\varepsilon = \pm 1$) is given on p. 85. The sesquilinear forms over $(A, ^-)$ form a category ${}_{(A,^-)}\mathcal{S}$, which we usually abbreviate as ${}_A\mathcal{S}$, whose morphisms $\varphi : (V, f) \to (U, g)$ are A-homomorphisms $\varphi : V \to U$ which satisfy

$$g(\varphi u, \varphi v) = f(u, v) \quad \text{for all } u, v \in V. \tag{3.1}$$

In particular the forms f and g are *equivalent*, $f \cong g$, if they are isomorphic, i.e. there is a morphism φ satisfying (3.1) which is an isomorphism of A-modules. We usually refer to such a morphism φ as an *isometry*.

The isometries $(V, f) \to (V, f)$ form a group, called the *unitary group* $\mathbf{U}(V, f) = \mathbf{U}(V) = \mathbf{U}(f)$ if f is an ε-Hermitian form.

The ε-Hermitian forms over $(A, ^-)$ constitute a full subcategory ${}_{(A,^-)}\mathcal{H}_\varepsilon = {}_A\mathcal{H}_\varepsilon$ of ${}_A\mathcal{S}$.

There is an involutory functor $^-$ on ${}_{(A,^-)}\mathcal{S}$ and the subcategory ${}_{(A,^-)}\mathcal{H}_\varepsilon = {}_A\mathcal{H}_\varepsilon$, given by $f \rightsquigarrow \bar{f}$, the "reflection" of f defined by $\bar{f}(u,v) = \overline{f(v,u)}$.

Let t: $(A, ^-) \to (K, ^-)$ be an involution trace (Def. 1.3.1, p. 35). The examples which are important in the present context are the reduced trace $\mathrm{trd}_{A/K}$ when A is separable over K and, if $A = KG$, the canonical trace $\mathrm{Tr}_{KG/K}(\sum_s \alpha_s s) = \alpha_1$ – cf. p. 35.

Composition with t induces an isomorphism

$$\mathrm{t}_* : \mathrm{Hom}_A(V, A) \to \mathrm{Hom}_K(V, K) \tag{3.2}$$

of K-vector spaces (Th. 1.2.10, p. 9).

Now let U be a left A-module and W a right A-module, and let φ: $U \to W$ be an $(A, ^-)$-*homomorphism* – this is an additive map satisfying

$$\varphi(au) = \varphi(u)\bar{a} \text{ for all } u \in U \text{ and } a \in A.$$

(A similar definition applies to a map from a *right* module to a *left* module.) Equivalently $\varphi : U \to \bar{W}$ is an A-homomorphism to the *twisted* (left) A-module $\bar{W}$, or $\varphi : \bar{U} \to W$ is an A-homomorphism from the twisted (right) module $\bar{U}$ to W.

The transpose $\varphi^t : W^* \to U^*$ of φ is an $(A, ^-)$-homomorphism defined by

$$\langle u, \varphi^t(x)\rangle = \overline{\langle x, \varphi(u)\rangle} \quad (u \in U, x \in W^*). \tag{3.3}$$

Since this equation can be written as $(\varphi^t(x))(u) = \overline{x(\varphi(u))}$, φ^t is well-defined and $\varphi^t(x)$ *is* in U^*. Moreover it is easy to see that φ^t is an $(A, ^-)$-homomorphism $W^* \to U^*$ from the left A-module W^* to the right A-module U^*. Thus we get a group homomorphism between groups of $(A, ^-)$-homomorphisms:

$$\mathrm{Hom}_{(A,^-)}(U, W) \xrightarrow{t} \mathrm{Hom}_{(A,^-)}(W^*, U^*). \tag{3.4}$$

If we identify U with U^{**} and W with W^{**} (Th. 1.2.4(b)), then $\varphi^{tt} : U \to W$ and, applying (3.3) twice,

$$\langle x, \varphi^{tt}(u)\rangle = \overline{\langle u, \varphi^t(x)\rangle} = \overline{\overline{\langle x, \varphi(u)\rangle}} = \langle x, \varphi(u)\rangle,$$

so

$$\varphi^{tt} = \varphi. \tag{3.5}$$

It follows from consideration of

$$\mathrm{Hom}_{(A,^-)}(U, W) \xrightarrow{t} \mathrm{Hom}_{(A,^-)}(W^*, U^*) \xrightarrow{t} \mathrm{Hom}_{(A,^-)}(U, W) \xrightarrow{t} \mathrm{Hom}_{(A,^-)}(W^*, U^*)$$

that (3.4) is an isomorphism.

Similarly the transpose $\psi^t : U^* \to W^*$ of an $(A, ^-)$-homomorphism $\psi : W \to U$ is defined by

$$\langle \psi^t(y), w\rangle = \overline{\langle \psi(w), y\rangle} \quad (w \in W, y \in U^*), \tag{3.6}$$

leading to the isomorphism

$$\mathrm{Hom}_{(A,^-)}(W, U) \xrightarrow{t} \mathrm{Hom}_{(A,^-)}(U^*, W^*).$$

Let $\theta : W \to V$ be an $(A, ^-)$-homomorphism from the right module W to the left module V, so $\theta\varphi : U \to V$ is an A-homomorphism. If $u \in U$ and $x \in V^*$,

$$\begin{aligned}\langle(\theta\varphi)(u), x\rangle &= \overline{\langle\theta^t(x), \varphi(u)\rangle} && \text{by (3.6)}\\ &= \overline{\overline{\langle u, \varphi^t\theta^t(x)\rangle}} = \langle u, \varphi^t\theta^t(x)\rangle && \text{by (3.3).}\end{aligned}$$

Since $\langle(\theta\varphi)(u), x\rangle = \langle u, (\theta\varphi)^t(x)\rangle$ by (1.8), it follows that

$$(\theta\varphi)^t = \varphi^t\theta^t$$

as in the usual case when both θ and φ are A-homomorphisms. And from this we see that φ^t is an isomorphism if and only if φ is an isomorphism (in the category of A-modules and $(A, ^-)$-homomorphisms).

Now suppose that $f : V \times V \to (A, ^-)$ is a sesquilinear form. The *right adjoint* respectively *left adjoint* $f_r, f_l : V \to V^*$ is the $(A, ^-)$-homomorphism defined by $f_r(v) = f(\cdot, v)$ respectively $f_l(v) = \overline{f(v, \cdot)}$. It is easy to see that

$$\bar{f}_r = f_l \text{ and } \bar{f}_l = f_r. \tag{3.7}$$

Theorem 3.1.1 *If we identify V^{**} with V via the canonical isomorphism (Th. 1.2.4, p. 7), then $f_r^t = f_l$ and $f_l^t = f_r$.*

$f_r^t = f_l$ follows from $\langle u, f_r^t(v)\rangle = \overline{\langle v, f_r(u)\rangle} = \overline{f(v, u)} = \langle u, f_l(v)\rangle$, and $f_l^t = f_r$ follows from this and (3.5). □

The form f is *nonsingular* if its right adjoint is an isomorphism. Thus the 0 form on the 0 module is nonsingular, but we shall generally assume that a nonsingular form is $\neq 0$ unless mentioned otherwise.

Corollary 3.1.1 *The sesquilinear form $f : V \times V \to (A, ^-)$ is nonsingular if and only if the* left *adjoint f_l is an isomorphism.* □

The category of nonsingular forms is denoted by ${}_{(A,^-)}\mathcal{S}^* = {}_A\mathcal{S}^*$ and the category of nonsingular ε-Hermitian forms by ${}_{(A,^-)}\mathcal{H}_\varepsilon^* = {}_A\mathcal{H}_\varepsilon^*$. If $(V, f) \in {}_{(A,^-)}\mathcal{H}_\varepsilon^*$, (V, f), or V itself, is called an *ε-Hermitian space*, or sometimes simply an *Hermitian space*.

It is often easier to check nonsingularity using the following lemma:

Lemma 3.1.1 *If f_r or f_l is injective ("f is nondegenerate"), then f is nonsingular.*

This follows immediately from $\text{len}_A V = \text{len}_A V^*$ (Th. 1.2.4, p. 7). □

Since $f(u, v) = \langle u, f_r(v)\rangle$ and $f(v, u) = \overline{\langle u, f_l(v)\rangle} = \overline{\langle u, f_r^t(v)\rangle}$,

$$f \text{ is } \varepsilon\text{-Hermitian if and only if } f_r^t = \varepsilon f_r.$$

When f is nonsingular, the *adjoint* (antiautomorphism) on $\text{End}_A V$, $\varphi \to \bar{\varphi}$ is defined by

$$f(\varphi u, v) = f(u, \bar{\varphi} v) \quad \text{for all } u, v \in V. \tag{3.8}$$

Equivalently

$$\bar{\varphi} = f_r^{-1}\varphi^t f_r,$$

and it is an involution when f is ε-Hermitian.

Let $(B, ^-)$ be another K-algebra with $(K, ^-)$-involution and suppose that V is an (A, B)-bimodule. (It is understood that $\alpha v = v\alpha$ for all $\alpha \in K$ and $v \in V$.) A sesquilinear form $f : V \times V \to (A, ^-)$ is *B-invariant* (or more precisely $(B, ^-)$-*invariant*) if $f(ub, v) = f(u, v\bar{b})$ for all $b \in B$.

Proposition 3.1.1 *Let $g : V \times V \to (A, ^-)$ be a nonsingular sesquilinear form.*

(a) If $\varphi \in \mathrm{End}_A V$, then $g_\varphi(u, v) := g(\varphi u, v)$ is also a sesquilinear form $V \times V \to (A, ^-)$. Conversely if $f : V \times V \to (A, ^-)$ is an arbitrary sesquilinear form on V, then there is a uniquely determined $\varphi \in \mathrm{End}_A V$ such that $f = g_\varphi$.

(b) g_φ is nonsingular if and only if $\varphi \in \mathbf{GL}_A V$.

(c) If g is invariant under $(B, ^-)$, then g_φ is $(B, ^-)$-invariant if and only if $\varphi \in \mathrm{End}_{(A,B)} V$ (the bimodule endomorphisms of V).

(d) Let $\delta, \varepsilon = \pm 1$. If g is δ-Hermitian, then g_φ is ε-Hermitian if and only if $\bar{\varphi} = \delta\varepsilon\varphi$ where $\bar{\varphi}$ is the adjoint of φ with respect to g.

(e) g_φ is equivalent to g_ψ if and only if there is $\theta \in \mathbf{GL}_A V$ such that $\psi = \widetilde{\theta}\varphi\theta$ where $^\sim$ is the inverse of $^-$; thus if g is δ-Hermitian, g_φ is equivalent to g_ψ if and only if there is $\theta \in \mathbf{GL}_A V$ such that $\psi = \bar{\theta}\varphi\theta$.

(f) Suppose that g is δ-Hermitian and $(B, ^-)$-invariant, and that φ and $\psi \in \mathrm{End}_{(A,B)} V$. Then g_φ is equivariantly equivalent to g_ψ if and only if there is $\theta \in \mathbf{GL}_{(A,B)} V$ such that $\psi = \bar{\theta}\varphi\theta$.

(a) It is clear that g_φ is sesquilinear. Suppose that $f : V \times V \to (A, ^-)$ is sesquilinear. Let $f_l, g_l : V \to \overline{V^*} = \overline{\mathrm{Hom}_A(V, A)}$ be the left adjoints. g_l is an isomorphism since g is nonsingular. If $\varphi = g_l^{-1} f_l \in \mathrm{End}_A V$,

$$g_\varphi(u, v) = \overline{\langle v, g_l g_l^{-1} f_l u\rangle} = \overline{\langle v, f_l u\rangle} = f(u, v).$$

φ is uniquely determined because g is nonsingular.

(b) is a consequence of $(g_\varphi)_l = g_l\varphi$, (c) follows from $g_\varphi(ub, v) = g(\varphi(ub), v)$ and $g_\varphi(u, v\bar{b}) = g((\varphi u)b, v)$, and (d), (e) and (f) are easily checked as well. □

Now consider an ε-Hermitian form $h : V \times V \to (A, ^-)$. If U is any submodule of V, the orthogonal complement of U is

$$U^\perp = \{v \in V : \ h(U, v) = 0\} = \{v \in V : \ h(v, U) = 0\}.$$

Its length satisfies

$$\mathrm{len}\, U + \mathrm{len}\, U^\perp = \mathrm{len}\, V \tag{3.9}$$

if either h or $h|_{U \times U}$ is nonsingular. To prove this, we first note that if we choose W so that $V = U \oplus W$, then $V^* = U^o \oplus W^o$ where U^o and W^o are the annihilators of U and W, and furthermore $U^o \cong W^*$ and $W^o \cong U^*$ (cf. p. 3). Since we are assuming that A is semisimple, U^* and W^* have the same length as U and W respectively (Th. 1.2.4, p. 7). Now $U^\perp$ is the annihilator of $h_r U$, and so has length $\mathrm{len}\, V^* - \mathrm{len}\, h_r U$ (*ibid*), $= \mathrm{len}\, V - \mathrm{len}\, U$ since $h_r|_U$ is injective.

Thus when U is nonsingular (that is to say $h|_{U \times U}$ is nonsingular), $U \cap U^\perp = 0$ and it follows that V is the orthogonal direct sum of U and $U^\perp$ which we write as

$$V = U \perp U^\perp. \tag{3.10}$$

Now suppose that (V, h) is nonsingular. Then there is a converse, namely that if $V = U \perp W$, then U (and W) are nonsingular subspaces of V. This results from the fact that the kernel of $(h|_{U\times U})_r$ is contained in the kernel of h_r which is of course 0. If the restriction of h to U is h_1 and to W is h_2, we also write

$$h = h_1 \perp h_2.$$

Note also that

$$\ker h_r = \ker h_l = V^\perp.$$

By p. 36, A has a splitting into simple involution components

$$(A, {}^-) = (A_1, {}^-) \oplus \cdots \oplus (A_k, {}^-) \oplus (A_{k+1}, {}^-) \oplus \cdots \oplus (A_l, {}^-),$$

where A_i is a simple algebra for $1 \le i \le k$ (separable if A is separable) and is a simple hyperbolic involution algebra for $k+1 \le i \le l$ – that is to say $A_i = B_i \oplus \bar{B}_i$ where B_i is a simple algebra – again separable if A is so. This decomposition implies a direct sum splitting of V – in fact an orthogonal splitting

$$V = A_1V \perp \cdots \perp A_kV \perp A_{k+1}V \perp \cdots \perp A_lV$$

since $h(av, a'v') = ah(v, v')\overline{a'} = 0$ if a and a' are in distinct simple involution components. We note too that the value of h on any of these components of V lies in the corresponding simple involution component of A, and so the restriction of h to such a component can, if we wish, be considered as an ε-Hermitian form over the corresponding component of $(A, {}^-)$:

$$h_i = h|_{A_iV\times A_iV} : A_iV \times A_iV \to (A_i, {}^-).$$

When viewed in this way, $h : V \times V \to (A, {}^-)$ is called the *vector sum*

$$\mathsf{X}_{1\le i\le l}\, h_i = h_1 \times \cdots \times h_l$$

of the forms h_i, $1 \le i \le l$, (or spaces $\mathsf{X}_{1\le i\le l}\,(V_i, h_i)$).

This implies in particular that if h is unsplittable, i.e. V is not the orthogonal direct sum of proper subspaces, then h is a form over one of the simple involution components of A. We shall see shortly what shape these unsplittable forms over the simple involution components have.

Since the existence of such a vector sum depends only on the fact that $(A, {}^-)$ is a direct sum of involution algebras, we can talk more generally about a vector sum even if the involution components of $(A, {}^-)$ are not simple – in fact if the ring $(R, {}^-)$ with involution is a direct sum $(R, {}^-) = (R_1, {}^-) \oplus \cdots \oplus (R_l, {}^-)$, then:

Lemma 3.1.2 *If (M, f) is a sesquilinear space over $(R, {}^-)$, it is the vector sum of the forms $(R_iM, f|_{R_iM\times R_iM})$ over $(R_i, {}^-)$, and if (N, g) is another, $(M, f) \cong (N, g)$ if and only if $(R_iM, f|_{R_iM\times R_iM}) \cong (R_iN, g|_{R_iN\times R_iN})$ for all i.* □

Exercises. Let $(A, {}^-)$ be an involution algebra over $(K, {}^-)$ of dimension n, $\mathrm{t} : (A, {}^-) \to (K, {}^-)$ an involution trace, and $\mathrm{t} : A \times A \to (K, {}^-)$ the trace form $\mathrm{t}(a, b) = \mathrm{t}(a\bar{b})$. Let G be a finite group of K-linear automorphisms of $(A, {}^-)$ with the property

$$\mathrm{t}(sa) = \mathrm{t}(a) \quad \text{for all } s \in G \text{ and } a \in A.$$

Then the (nonsingular) reflexive form t over $(K, {}^-)$ is G-invariant.

The associated Hermitian form on A over $(KG, {}^-)$ is

$$\hat{\mathrm{t}}(a, b) = \sum_{s\in G} \mathrm{t}(a, sb)s.$$

If $a \in A$ has the property that the orbit $\{sa : s \in G\}$ is a basis of A (over K), then a is said to *generate a normal basis of* A (with respect to G), and $\{sa : s \in G\}$ is called a *normal basis* (and of course then $|G| = n$, assuming that the sa are distinct). It generates an *orthonormal* basis if it generates a normal basis which is orthonormal, i.e. $\mathrm{t}(sa, ta) = \delta_{s,t}$. We shall assume that $|G| = n$ in what follows.

Let $(L, ^-)/(K, ^-)$ be an extension of involution fields.

1. G is a group of automorphisms of the involution algebra $(A, ^-)^L = (L, ^-) \otimes_{(K,^-)} (A, ^-)$ over $(L, ^-)$, $1 \otimes a$ generates a normal basis of A^L if a generates one of A, and $\hat{\mathrm{t}}^L = \widehat{\mathrm{t}^L}$.

2. Show that a generates a normal basis if $\hat{\mathrm{t}}(a, a) \in \dot{K}$, and that it generates an orthonormal basis if and only if $\hat{\mathrm{t}}(a, a) = 1$.

3. Show that A has a normal orthonormal basis if and only if $\hat{\mathrm{t}} \cong \langle 1 \rangle$.

4. Let L/K be a finite Galois extension with Galois group G with the field trace $\mathrm{Tr}_{L/K}$ as involution trace. Show that the Hermitian form $\widehat{\mathrm{id} \otimes \mathrm{Tr}_{L/K}}$ over LG on the "extended algebra" $L^L = L \otimes_K L$ is $\cong \langle 1 \rangle$. (Use the exercise on p. 23). □

3.1.2 Hyperbolic forms. Let U be a left A-module and W a right A-module, and

$$U \times W \to A, \quad (u, w) \to \langle u, w \rangle, \tag{3.11}$$

an A-pairing (cf. p. 3) where A is a semisimple K-algebra. The left and right kernels of the pairing are respectively

$$W^o = \{u \in U : \ \langle u, W \rangle = 0\}, \quad U^o = \{w \in W : \ \langle U, w \rangle = 0\}.$$

They are submodules of U and W respectively and are the kernels of the "left and right adjoints" of the pairing,

$$\langle \cdot, \cdot \rangle_l : U \to W^*, \ \ u \to \langle u, \cdot \rangle, \qquad \langle \cdot, \cdot \rangle_r : W \to U^*, \ \ w \to \langle \cdot, w \rangle,$$

which are A-homomorphisms. We shall assume that the pairing is *perfect* in the sense that both kernels are 0, i.e. that the left and right adjoints are injective. Then $\mathrm{len}\, U \leq \mathrm{len}\, W^* = \mathrm{len}\, W$, and so by symmetry

$$\mathrm{len}\, U = \mathrm{len}\, W. \tag{3.12}$$

Define a form g on $U \oplus \overline{W}$ by

$$g((u_1, \bar{w}_1), (u_2, \bar{w}_2)) = \langle u_1, w_2 \rangle + \varepsilon \overline{\langle u_2, w_1 \rangle}. \tag{3.13}$$

It is an ε-Hermitian form over $(A, ^-)$. Now

$$g_l : U \oplus \overline{W} \to (U \oplus \overline{W})^* = U^* \oplus \overline{W}^*$$

and $g_l|_U = \overline{\langle \cdot, \cdot \rangle_l}$ and so is a bijection $U \to \overline{W}^*$. Similarly $g_l|_{\overline{W}}$ is equal to $\varepsilon \langle \cdot, \cdot \rangle_r$ and so is a bijection of W with U^*. Thus g_l is an isomorphism, and so g is nonsingular and is called the *ε-hyperbolic form based on the pairing* (1.4). Note that U and $\overline{W}$ are totally isotropic in the sense that $g(U, U) = 0 = g(\overline{W}, \overline{W})$, and that the 0 form on the 0 module is hyperbolic.

We say that a sesquilinear form $h : V \times V \to (A, ^-)$ is hyperbolic if it is equivalent to a hyperbolic form g in the above sense.

The following useful lemma is easy to prove.

Lemma 3.1.3 *Let $U' \times W' \to A$ be another perfect pairing and g' the ε-hyperbolic form based on it. If U' is isomorphic to U, g' is equivalent to g. In fact if $\varphi : U' \to U$ is an isomorphism and $\psi : W' \to W$ is defined by $\langle \varphi u', \psi w' \rangle = \langle u', w' \rangle$, the map $\varphi \oplus \psi$ is an isometry $g' \to g$. Similarly the two hyperbolic forms are equivalent if W' is isomorphic to W.* □

Consider the canonical pairing $U \times U^* \to A$. The right adjoint is the identity on U^*, and the left adjoint is the canonical map $\gamma_U : U \to U^{**}$ which is an isomorphism (Th. 1.2.4, p. 7). Thus the pairing $U \times U^* \to A$ is perfect, and the resulting form is called the *(ε-Hermitian) hyperbolic form* or *ε-hyperbolic form based on U*. Its module $U \oplus \overline{U^*}$ is denoted by $\mathrm{H}(U)$ and its form by $h_{\mathrm{H}(U)}$, or by $\mathrm{H}_\varepsilon(U)$ and $h_{\mathrm{H}_\varepsilon(U)}$ respectively if it is necessary to make ε explicit. By the lemma, it is equivalent to the form based on $\overline{U^*}$ since the isotypic components of U and $\overline{U^*}$ have the same length (Ths. 1.2.7(d) and 1.2.4) and are therefore isomorphic. Furthermore, any hyperbolic form over A is equivalent to one of the form $h_{\mathrm{H}(U)}$ since the hyperbolic form based on the pairing (1.4) is isomorphic to $h_{\mathrm{H}(U)}$ because W is isomorphic to U^*. The hyperbolic space $\mathrm{H}(U)$ is called the *hyperbolic space or module based on U*.

As a converse to this, we have the theorem:

Theorem 3.1.2 *If h is a nonsingular ε-Hermitian form over $(A, ^-)$ on $V = U \oplus W$ with U and W totally isotropic A-modules, the pairing $U \times \overline{W} \to A$ defined by $\langle u, \bar{w} \rangle = h(u, w)$ is perfect and the resulting hyperbolic form on $U \oplus W$ is h. Thus V is an ε-Hermitian hyperbolic space, isometric to both $\mathrm{H}_\varepsilon(U)$ and $\mathrm{H}_\varepsilon(W)$.* □

Corollary 3.1.2 *Let $h : V \times V \to (K, ^-)$ be a nonsingular ε-reflexive form and $\rho : G \to \mathbf{I}(V, h)$ an isometry representation, with associated ε-Hermitian form $\hat{h}$ over $(KG, ^-)$. Then*

(a) ρ is a hyperbolic representation if and only if $\hat{h}$ is hyperbolic.
(b) If $V = U \oplus W$ with U and W both totally isotropic with respect to h and stable under G, then ρ is hyperbolic, $\rho \cong \mathrm{H}_\varepsilon(\rho|_U)$ (cf. p. 90).

If ρ is hyperbolic, then $V = U \oplus W$ where both U and W are totally isotropic with respect to h and stable under the action of G. The definition of $\hat{h}$ implies that they are also totally isotropic with respect to $\hat{h}$, so by the theorem, $\hat{h}$ is hyperbolic. Conversely if $\hat{h}$ is hyperbolic, say based on the pairing $U \times W \to KG$, let $\rho' : G \to \mathbf{I}(\mathrm{H}(U), h')$ be the hyperbolic representation based on the linear representation of G on U arising from the action of KG on U. By Lem. 3.1.3, $\widehat{h'}$ is equivalent to $\hat{h}$, and so h' is equivariantly equivalent to h. This proves (a), and (b) follows at once from (a) since both U and W are totally isotropic with respect to $\hat{h}$, and so $\hat{h}$ is hyperbolic by the theorem. □

Corollary 3.1.3 *The map which associates to a hyperbolic representation $\rho : G \to \mathbf{I}(V, h)$ the hyperbolic form $\hat{h} : V \times V \to (KG, ^-)$, is an isomorphism of categories.*

This follows easily from (a) of the foregoing corollary, taking into account that the inverse functor is $(g : V \times V \to (KG, ^-)) \rightsquigarrow (\rho : G \to \mathbf{I}(V, \mathrm{Tr}_{KG/K} g))$. The details are omitted. □

Proposition 3.1.2 *Let $h : V \times V \to (A, ^-)$ be a nonsingular ε-Hermitian form. Then $h \perp -h$ is equivalent to the ε-hyperbolic form $h_{\mathrm{H}_\varepsilon(V)}$ based on V.*

Since the characteristic of K is $\neq 2$, the space $V \perp V$ of $h \perp -h$ is the direct sum of the submodules $U = \{(v,v) : v \in V\}$ and $W = \{(v,-v) : v \in V\}$. Both are totally isotropic and isomorphic as A-modules to V, so the proposition follows from Th. 3.1.2. □

Note that $\mathrm{H}_\varepsilon(U_1 \oplus U_2) \cong \mathrm{H}_\varepsilon(U_1) \perp \mathrm{H}_\varepsilon(U_2)$ (isometry), and by induction

$$\mathrm{H}_\varepsilon(\oplus_i U_i) \cong \perp_i \mathrm{H}_\varepsilon(U_i). \tag{3.14}$$

Theorem 3.1.3 *Two ε-Hermitian hyperbolic spaces are isometric if and only if they are isomorphic (as A-modules).*

The necessity is trivial so assume that $\mathrm{H}_\varepsilon(U)$ and $\mathrm{H}_\varepsilon(W)$ are isomorphic A-modules. Since their forms are vector sums (p. 95) of forms over the simple involution components $(A_i, ^-)$ of $(A, ^-)$, we can assume that $(A, ^-)$ is a simple involution algebra. If A is a simple algebra, the twisted dual $\overline{V^*}$ of an A-module V is also an A-module, of the same length as V. Since $\mathrm{H}_\varepsilon(U)$ and $\mathrm{H}_\varepsilon(W)$ have the same length, they are both an orthogonal direct sum of the same number of ε-hyperbolic spaces based on a simple module and so are equivalent. If $(A, ^-)$ is hyperbolic, $A = B \oplus \bar{B}$ where B is simple. If V is a simple left A-module, it is a module over one of the components of A, say B', so its dual is a simple right B'-module and the twisted dual $\overline{V^*}$ is a simple left $\overline{B'}$-module. The theorem now follows as in the previous case. □

Corollary 3.1.4 *If $\rho : G \to \mathbf{I}(V,h)$ and $\rho' : G \to \mathbf{I}(V',h')$ are hyperbolic representations, they are equivalent if and only if V and V' are isomorphic KG-modules – in other words, ρ and ρ' are equivalent as linear representations.*

This is clear from the above theorem and Cor. 3.1.3. □

If (V,h) is an ε-Hermitian space over a hyperbolic algebra $\mathrm{H}(A)$ on the semisimple algebra A, $\mathrm{H}(A) = A \oplus \bar{A}$ (cf. Ex. 7, p. 25), then $V = AV \oplus \bar{A}V$ and both AV and $\bar{A}V$ are totally isotropic since, for example, $h(AV, AV) = Ah(V,V)\bar{A} = 0$. Thus by Th. 3.1.2, (V,h) is a hyperbolic space and so we get

Corollary 3.1.5 *An ε-Hermitian space over the semisimple hyperbolic involution algebra $\mathrm{H}(A)$ is hyperbolic. Two such spaces are equivalent if and only if they are isomorphic as $\mathrm{H}(A)$-modules.* □

Corollary 3.1.6 *If A is a simple algebra, an ε-Hermitian space over $\mathrm{H}(A)$ is hyperbolic, and two such spaces (V,h) and (W,g) are equivalent if and only if they have the same length over $\mathrm{H}(A)$, or if AV and AW have the same length over A or the same K-dimension.* □

Lemma 3.1.4 *Suppose that V is an A-module of length 2 with an ε-Hermitian form h over $(A, ^-)$ and a totally isotropic simple submodule $S_1 \not\subset V^\perp$. Then V is (equivalent to) a hyperbolic "plane" over $(A, ^-)$ based on S_1.*

If $V^\perp \neq 0$, then $V = S_1 \oplus V^\perp$, which is impossible since that would imply that $S_1 \subset V^\perp$. Thus $V^\perp = 0$ and h is nonsingular.

Let S_2 be any (simple) complement of S_1, $V = S_1 \oplus S_2$, and suppose that the isotypic component of A of type S_1 is A_1 and that of type S_2 is A_2. If $A_1 \neq A_2$, then A_1 and A_2 are switched by the involution on A since otherwise $(A_1, ^-)$ and $(A_2, ^-)$ would be distinct simple involution components of $(A, ^-)$ and so S_1 and S_2

would be orthogonal (and $S_1 \subset V^\perp$). Thus $(A_1 \oplus A_2, \bar{\ })$ is a hyperbolic component of $(A, \bar{\ })$ in this case, and so h is hyperbolic by Cor. 3.1.5, based on either S_1 or S_2.

We can therefore assume that $A_1 = A_2$. If $\bar{A}_1 \neq A_1$, then h is a nonsingular form over the hyperbolic algebra $A_1 \oplus \bar{A}_1$ and so would be hyperbolic, which implies that V also contains a nonzero isotypic component of type different than S_1 – contradiction. Therefore $\bar{A}_1 = A_1$. We can also assume that S_2 is not totally isotropic by Th. 3.1.2. Choose generators v_1 and v_2 for S_1 and S_2 respectively. Then $h(v_2, v_1)$ and $h(v_2, v_2)$ are nonzero elements of A_1, with the same left annihilator as v_2, a maximal left ideal of A_1. Therefore the left ideals $Ah(v_2, v_1)$ and $Ah(v_2, v_2)$ are isomorphic as A-modules via the map $ah(v_2, v_1) \to ah(v_2, v_2)$, and since they both split the semisimple left A-module A, the isomorphism can be extended to an automorphism of A as a left A-module. Such an automorphism is given by right multiplication by a unit $b \in A$, so $h(v_2, v_2) = h(v_2, v_1)b = h(v_2, \bar{b}v_1)$. Define $v_3 = v_2 - \frac{1}{2}\bar{b}v_1$. Then $v_3 \notin S_1$ so $V = S_1 + Av_3$, and

$$h(v_3, v_3) = h(v_2, v_2) - \tfrac{1}{2}h(v_2, \bar{b}v_1) - \tfrac{1}{2}h(\bar{b}v_1, v_2)$$

since $h(v_1, v_1) = 0$. Now $\overline{h(v_2, v_2)} = \varepsilon h(v_2, v_2)$ and $h(v_2, v_2) = h(v_2, \bar{b}v_1)$ imply that

$$h(\bar{b}v_1, v_2) = h(v_2, \bar{b}v_1)$$

and so $h(v_3, v_3) = 0$. Therefore Av_3 is totally isotropic, so is $\neq V$ and $V = S_1 \oplus Av_3$ and the lemma follows from Th. 3.1.2. □

Proposition 3.1.3 *If A is a simple algebra with involution $\bar{\ }$, and $h : V \times V \to (A, \bar{\ })$ is a nonsingular ε-Hermitian form, then*
(a) if a simple module supports at least one nonsingular ε-Hermitian form, h is an orthogonal direct sum of (nonsingular) forms on simple modules;
(b) otherwise h is an orthogonal direct sum of hyperbolic planes.

It is clear from Lem. 3.1.4 that (V, h) is an orthogonal direct sum of ε-hyperbolic planes and nonsingular ε-Hermitian spaces of length 1. If a simple module does not support any nonsingular ε-Hermitian form, (V, h) is therefore an orthogonal direct sum of hyperbolic planes. If a simple module *does* support a nonsingular ε-Hermitian form, then by Prop. 3.1.2 (V, h) is an orthogonal direct sum of length 1 spaces. □

Theorem 3.1.4 *Let $(A, \bar{\ }) = \oplus_i (A_i, \bar{\ })$ be the decomposition of $(A, \bar{\ })$ into simple involution algebras. Then a nonsingular form $h : V \times V \to (A, \bar{\ })$ is hyperbolic if and only if all of the involution isotypic components*

$$h|_{A_iV \times A_iV} : A_iV \times A_iV \to (A_i, \bar{\ })$$

are hyperbolic.

Since $V = \perp_i A_iV$, the sufficiency follows from (3.14).

Conversely, since (V, h) is a hyperbolic space based on a direct sum of simple modules, it is an orthogonal direct sum of hyperbolic planes by (3.14). Each of these planes, as a sum of two nonorthogonal simple modules, is in some A_iV. Thus each A_iV is an orthogonal direct sum of hyperbolic planes, and so is itself hyperbolic. □

Lemma 3.1.5 *Let $h : V \times V \to (A, \bar{\ })$ be an ε-Hermitian form, and U a totally isotropic submodule such that $U \cap V^\perp = 0$. If $U = U_1 \oplus \cdots \oplus U_r$ is a* *decomposition of U into simple modules, then there are totally isotropic simple submodules $W_1, \ldots, W_r$ of V such that*

$$U + \sum W_i = (U_1 \oplus W_1) \perp \cdots \perp (U_r \oplus W_r)$$

where each $U_i \oplus W_i$ is a hyperbolic plane.

Let V' be a submodule of V which contains U and satisfies $V' \oplus V^\perp = V$. It suffices to prove the lemma with V replaced by the nonsingular subspace V', in effect allowing us to assume that $V^\perp = 0$, i.e. that h is nonsingular.

The orthogonal complement $W' = (U_2 \oplus \cdots \oplus U_r)^\perp$ contains U_1 and has length $\operatorname{len} V - (r-1)$ (cf. p. 94), while $(U_1 \perp U_2 \oplus \cdots \oplus U_r)^\perp$ has length $\operatorname{len} V - r$. Therefore we can choose a simple submodule W_1' of W' satisfying $h(U_1, W_1') \neq 0$. Then $H := U_1 \oplus W_1'$ is a hyperbolic plane by Lem. 3.1.4; let W_1 be a totally isotropic simple module such that $H = U_1 \oplus W_1$. Then $H^\perp$ contains $U_2, \ldots, U_r$, and so we can proceed by induction. □

3.1.3 Witt's Theorem.

Theorem 3.1.5 (Witt's Theorem) *Suppose that A is a semisimple K-algebra with a $(K, {}^-)$-involution ${}^-$ and that $h : V \times V \to (A, {}^-)$ is an ε-Hermitian form. If U is a submodule of V, an injective isometry $\varphi : U \to V$ can be extended to an element of the unitary group $\mathbf{U}(h)$ if and only if*

$$\varphi(U \cap V^\perp) = \varphi(U) \cap V^\perp. \tag{3.15}$$

In particular φ can be extended if either (V, h) or $(U, h|_{U \times U})$ is nonsingular.

If $\hat{\varphi} \in \mathbf{U}(h)$ is an extension of φ, then $\hat{\varphi}(V^\perp) = V^\perp$, so $\varphi(U \cap V^\perp) = \hat{\varphi}(U \cap V^\perp) = \hat{\varphi}(U) \cap \hat{\varphi}(V^\perp) = \varphi(U) \cap V^\perp$. Thus (3.15) is necessary.

Conversely suppose that (3.15) holds. Let

$$(A, {}^-) = (A_1, {}^-) \oplus \cdots \oplus (A_l, {}^-)$$

be the decomposition of A into simple involution algebras. Then there are corresponding decompositions $V = A_1 V \perp \cdots \perp A_l V$ and $U = A_1 U \perp \cdots \perp A_l U$, and since any A-homomorphism $U \to V$ takes $A_i U$ into $A_i V$, it is apparent that φ has an extension in $\mathbf{U}(h)$ if and only if each $\varphi|_{A_i U}$ has an extension to an element of $\mathbf{U}(h|_{A_i V \times A_i V})$. Furthermore

$$U \cap V^\perp = \left(A_1 U \cap A_1 V^\perp\right) \perp \cdots \perp \left(A_h U \cap A_h V^\perp\right)$$

and since $A_i \varphi(U) = \varphi(A_i U)$ for all i,

$$\varphi(U) \cap V^\perp = \left(\varphi(A_1 U) \cap A_1 V^\perp\right) \perp \cdots \perp \left(\varphi(A_h U) \cap A_h V^\perp\right),$$

so (3.15) implies that $\varphi(A_i U \cap A_i V^\perp) = \varphi(A_i U) \cap A_i V^\perp$ for all i. And since $A_i V^\perp = (A_i V)^\perp$ – the orthogonal complement of $A_i V$ in itself – it follows that it suffices to prove the theorem in the case when $V = A_i V$ for some i. Thus we may assume that A is a simple involution algebra.

Suppose first that A is a hyperbolic involution algebra. Then two ε-Hermitian spaces are isometric if and only if their underlying modules are isomorphic (Cor. 3.1.5, p. 98). Choose U' so that

$$U = U' \oplus (U \cap V^\perp), \quad = U' \perp (U \cap V^\perp),$$

and let $U_0' = (U', h|_{U'\times U'})^\perp$. Choose U'' so that

$$U' = U'' \oplus U_0', \quad = U'' \perp U_0'.$$

Then U'' is a nonsingular subspace, and we can split

$$V = U'' \perp V'$$

(cf. (3.10)). Since $U_0' \subset V'$ and is totally isotropic (or 0), and $U_0' \cap V^\perp = 0$, by Lem. 3.1.5 there is a hyperbolic space $\mathrm{H}(U_0') \subset V'$ based on U_0'. Thus

$$\begin{aligned} V &= U'' \perp \mathrm{H}(U_0') \perp V'' \\ &= U'' \perp \mathrm{H}(U_0') \perp V''' \perp (U \cap V^\perp) \perp (V^\perp)' \end{aligned} \tag{3.16}$$

where $V^\perp = (U \cap V^\perp) \perp (V^\perp)'$, and V''' is nonsingular.

Now

$$\varphi U = \varphi U' \perp \varphi(U \cap V^\perp) = \varphi U' \perp \big((\varphi U) \cap V^\perp\big), \quad \text{and} \quad \varphi U' = \varphi U'' \perp \varphi U_0'$$

and we can find a similar splitting

$$V = \varphi U'' \perp \mathrm{H}(\varphi U_0') \perp V'''' \perp \big((\varphi U) \cap V^\perp\big) \perp (V^\perp)'' \tag{3.17}$$

where $V^\perp = \big((\varphi U) \cap V^\perp\big) \perp (V^\perp)''$ – so $\operatorname{len}(V^\perp)'' = \operatorname{len}(V^\perp)'$ – and V'''' is nonsingular and has the same length as V''', hence there is an isometry $V''' \to V''''$. By Lem. 3.1.3 there is an isometry $\mathrm{H}(U_0') \to \mathrm{H}(\varphi U_0')$ extending $\varphi|_{U_0'}$. It is now clear how to extend φ to an element of $\mathbf{U}(h)$ since $(V^\perp)''$ and $(V^\perp)'$ are totally isotropic and easily seen to be A-isomorphic since the other summands in (3.16) and (3.17) can be paired off into isomorphic pairs.

We may assume from now on that A is a simple algebra. The proof is divided into 7 steps.

1. We first show that the theorem follows if we assume the special case:
φ can be extended to an element of $\mathbf{U}(h)$ when there is a hyperplane (submodule of colength 1) of U containing $V^\perp$ and on which φ is the identity map.

By (3.15) and Lem. 1.2.1, p. 6, there is a submodule W of $V^\perp$ such that

$$V^\perp = (U \cap V^\perp) \oplus W = ((\varphi U) \cap V^\perp) \oplus W.$$

Then $U + V^\perp = U \oplus W$ and $\varphi U + V^\perp = \varphi(U) \oplus W$. Extend φ to an isometry $\widehat{\varphi} : U + V^\perp \to \varphi U + V^\perp$ by defining it to be the identity on W. Then $\widehat{\varphi}(V^\perp) = V^\perp$, so

$$\widehat{\varphi}((U + V^\perp) \cap V^\perp) = \widehat{\varphi}(V^\perp) = V^\perp = \widehat{\varphi}(U + V^\perp) \cap V^\perp.$$

This is condition (3.15) with U and φ replaced by $U + V^\perp$ and $\widehat{\varphi}$, and so we can assume that $V^\perp \subset U$.

In this case, (3.15) becomes $\varphi(V^\perp) = \varphi(U) \cap V^\perp$, whence $\varphi(V^\perp) = V^\perp$ and $V^\perp \subset \varphi(U)$. Now we use induction on $\operatorname{len} U \geq \operatorname{len} V^\perp$. If $U = V^\perp$, then $\varphi U = U$, and so φ can be extended by defining it to be the identity on a submodule W for which $V = U \oplus W$. Suppose then that $\operatorname{len} U > \operatorname{len} V^\perp$, and let H be a hyperplane of U containing $V^\perp$. Then $\varphi(H \cap V^\perp) = \varphi(V^\perp) = \varphi(H) \cap V^\perp$ since $V^\perp = \varphi(V^\perp) \subset \varphi(H)$. By the induction hypothesis, there is an extension φ_1 of $\varphi|_H$ to an element of $\mathbf{U}(V)$, and

$$\varphi_1^{-1}\varphi(U \cap V^\perp) = \varphi_1^{-1}(\varphi(U) \cap V^\perp) = \big(\varphi_1^{-1}\varphi(U)\big) \cap V^\perp$$

since $\varphi_1^{-1}V^\perp = V^\perp$. Moreover $\varphi_1^{-1}\varphi : U \to V$ is the identity on H, and therefore has an extension $\varphi_2 \in \mathbf{U}(V)$ by our assumption. The map $\hat{\varphi} = \varphi_1\varphi_2$ is the desired extension of φ.

2. The proof is now reduced to that of the special case in 1: φ is the identity on a hyperplane H of U which contains $V^\perp$. Define $S = (\varphi - 1)U$. If $S = 0$, φ is the identity map of U, and the theorem is trivial. So assume $S \neq 0$ – then $S \cong U/H$ is simple.

3. Suppose that $S \subset V^\perp$. Then it follows that $\varphi(U) = U$ since

$$\varphi U \subset U + S \subset U + V^\perp \subset U.$$

Moreover $h(v, \varphi u) = h(v, u)$ for all $v \in V$ and $u \in U$ since $\varphi u \equiv u \bmod V^\perp$. And φ can be extended to an element in $\mathbf{U}(V)$ by defining it to be the identity on any complement W of U, $V = U \oplus W$:

$$\begin{aligned} h(\varphi u + w, \varphi u' + w') &= h(\varphi u, \varphi u') + h(\varphi u, w') + h(w, \varphi u') + h(w, w') \\ &= h(u, u') + h(u, w') + h(w, u') + h(w, w') \\ &= h(u + w, u' + w'). \end{aligned}$$

4. We may therefore suppose that

$$S \not\subset V^\perp, \quad S \text{ simple.}$$

Then the composite map $V \xrightarrow{h_r} \overline{V^*} \xrightarrow{|_S} \overline{S^*}$ (where the second map is restriction) is not zero, hence is onto and so the kernel $S^\perp$ is a "hyperplane" of V. Furthermore

$$h(\varphi(u), \varphi(v) - v) = -h(\varphi(u) - u, v) \text{ for all } u, v \in U \tag{3.18}$$

and it follows by taking $v \in H$ that

$$H \subset S^\perp. \tag{3.19}$$

Thus we have this configuration:

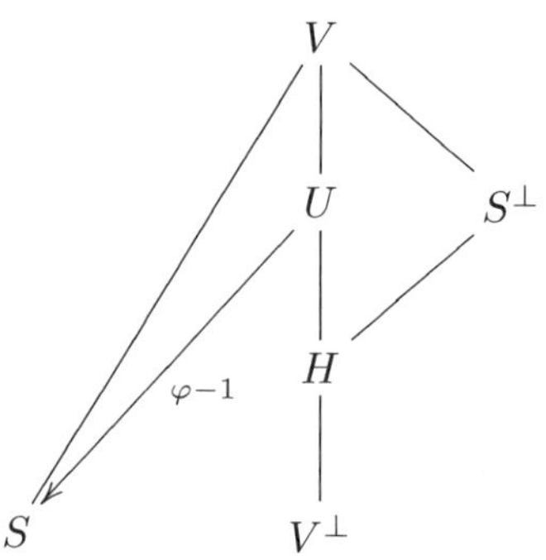

5. Suppose next that $U \not\subset S^\perp$, so $V = S^\perp + U$ and $U \cap S^\perp = H$. If $v \in V$, say $v = s + u = s' + u'$ with $s, s' \in S^\perp$ and $u, u' \in U$, then $u - u' = s' - s \in S^\perp \cap U = H$, which implies that

$$s + \varphi u - (s' + \varphi u') = s - s' + \varphi(u - u') = s - s' + u - u' = 0.$$

Therefore if $\hat{\varphi}(v) := s + \varphi(u)$, $\hat{\varphi}$ is a well-defined A-endomorphism of V which extends φ. Since $S = (\varphi - 1)U$ and $U \not\subset S^\perp$, $h(\varphi(u) - u, v) \neq 0$ for some $u, v \in U$, so by (3.18) $\varphi(U) \not\subset S^\perp$. Therefore $V = S^\perp + \varphi(U)$, and so $\hat{\varphi}$ is onto and therefore is an A-automorphism of V.

For any $u \in U$ and $s' \in S^\perp$, $h(\varphi(u) - u, s') = 0$, so $h(\varphi(u), s') = h(u, s')$. Therefore

$$\begin{aligned} & h(\varphi(u) + s, \varphi(u') + s') \\ = \;& h(\varphi(u), \varphi(u')) + h(\varphi(u), s') + h(s, \varphi(u')) + h(s, s') \\ = \;& h(u, u') + h(u, s') + h(s, u') + h(s, s') = h(u + s, u' + s'), \end{aligned}$$

and so $\hat{\varphi} \in \mathbf{U}(h)$.

6. We may now assume that $U \subset S^\perp$. It follows easily from (3.18) that $\varphi(U) \subset S^\perp$ also, and that S is totally isotropic since if $u, v \in U$,

$$h(\varphi u - u, \varphi v - v) = h(\varphi u - u, \varphi v) - h(\varphi u - u, v) = 0.$$

We can in fact assume that

$$U = S^\perp$$

by defining φ to be the identity on a submodule W such that $S^\perp = U \oplus W = \varphi(U) \oplus W$ (Lem. 1.2.1, p. 6). For then

$$\begin{aligned} & h(\varphi(u) + w, \varphi(u') + w') \\ = \;& h(\varphi(u), \varphi(u')) + h(\varphi(u), w') + h(w, \varphi(u')) + h(w, w') \\ = \;& h(u, u') + h(u, w') + h(w, u') + h(w, w') = h(u + w, u' + w') \end{aligned}$$

since, e.g., $h(\varphi(u), w') = h(u, w')$ because $w' \in S^\perp$. The new hyperplane H is (the old one)$\oplus W$, and S is the same as before. The relationship (3.15) still holds since $\varphi|_H = \mathrm{id}$ and $V^\perp \subset H \subset U$ implies that both sides are $V^\perp$.

7. Let $\widehat{\varphi}$ be any A-module automorphism of V which extends φ. The form

$$g(u + V^\perp, v + V^\perp) = h(u, v)$$

is a nonsingular ε-Hermitian form on $V/V^\perp$. Since $\widehat{\varphi}(V^\perp) = \varphi(V^\perp) = V^\perp$, $\widehat{\varphi}$ induces an automorphism on $V/V^\perp$, which we also denote by $\widehat{\varphi}$. Its inverse has an adjoint ψ with respect to g, satisfying $g(\widehat{\varphi}(u+V^\perp), \psi(v+V^\perp)) = g(u+V^\perp, v+V^\perp)$. If we also denote by ψ a lift of ψ to an automorphism of V, then

$$h(\widehat{\varphi}(u), \psi(v)) = g(\widehat{\varphi}(u + V^\perp), \psi(v + V^\perp)) = g(u + V^\perp, v + V^\perp) = h(u, v) \quad (3.20)$$

for all $u, v \in V$.

Now let W_1 be a (simple) module such that $V = S^\perp \oplus W_1$ – then $h(S, W_1) \neq 0$. Since S is totally isotropic (cf. 6.), $S \subset S^\perp$ and so the module $S + W_1$ has length 2. By Lem. 3.1.4 (p. 98), we can assume that W_1 is totally isotropic since S is. Now $S \subset S^\perp = U$ implies that $(\varphi - 1)S^\perp = S$ so

$$\varphi S^\perp = S^\perp, \quad \varphi S = S. \quad (3.21)$$

Moreover $\psi S^\perp = S^\perp$ since $\psi S^\perp \subset S^\perp$ by (3.20) (with $v \in S^\perp$ and $u \in S$). It follows that $V = S^\perp \oplus \psi W_1$.

Suppose $W_1 = Aw_1$. By Lem. 3.1.4, $S \oplus \psi W_1$ is a hyperbolic plane and so there is an isotropic vector

$$w_2 \equiv \psi w_1 \mod S \quad (3.22)$$

such that $S \oplus \psi W_1 = S \oplus Aw_2$. Let $\mathfrak{a} \subset A$ be the annihilator of w_2. Then $\mathfrak{a}w_2 = 0$ implies that $\mathfrak{a}\psi w_1 \subset S$ which implies that $\mathfrak{a}\psi w_1 = 0$. Thus $\mathfrak{a} \subset$ the annihilator of ψw_1 which equals the annihilator of w_1, so $\mathfrak{a}$ is equal to the annihilator of w_1 since both are maximal left ideals of A. Therefore there is an isomorphism $\psi_1 : W_1 \to W_2$

which takes w_1 to w_2. Now define $\widehat{\varphi} = \varphi \oplus \psi_1$. It is an automorphism of V, and is in $\mathbf{U}(h)$ since

$$\begin{aligned} & h(\varphi u + a w_2, \varphi u' + a' w_2) \\ = \quad & h(\varphi u, \varphi u') + h(\varphi u, w_2)\overline{a'} + a h(w_2, \varphi u') \\ = \quad & h(u, u') + h(\varphi u, \psi w_1)\overline{a'} + a h(\psi w_1, \varphi u') \text{ by (3.21) and (3.22)} \\ = \quad & h(u, u') + h(u, w_1)\overline{a'} + a h(w_1, u') \quad \text{by (3.20)} \\ = \quad & h(u + a w_1, u' + a' w_1). \end{aligned}$$

□

The slightly weaker "cancellation" form of Witt's Theorem follows easily:

Corollary 3.1.7 *Suppose that A is a semisimple K-algebra with a $(K, {}^-)$-involution ${}^-$ and that $h_i : V_i \times V_i \to (A, {}^-)$, $i = 1, 2, 3$, are nonsingular ε-Hermitian spaces. Then*

$$h_1 \perp h_2 \cong h_1 \perp h_3 \quad \textit{implies that} \quad h_2 \cong h_3.$$

□

Corollary 3.1.8 *Suppose that $\rho_i : G \to \mathbf{I}(V_i, h_i)$, $i = 1, 2, 3$, are isometry representations where h_1, h_2 and h_3 are nonsingular ε-Hermitian forms over $(K, {}^-)$. Then*

$$\rho_1 \perp \rho_2 \cong \rho_1 \perp \rho_3$$

implies that $\rho_2 \cong \rho_3$.

The map implied by $\rho_1 \perp \rho_2 \cong \rho_1 \perp \rho_3$ is an isometry $\widehat{h_1 \perp h_2} \cong \widehat{h_1 \perp h_3}$ between the associated forms over $(KG, {}^-)$. Since $\widehat{h_1 \perp h_2} = \hat{h}_1 \perp \hat{h}_2$ and $\widehat{h_1 \perp h_3} = \hat{h}_1 \perp \hat{h}_3$, this corollary follows from the previous one. □

One can also prove Witt's Theorem for forms over a semisimple algebra by using Morita equivalence and Witt's Theorem for ε-reflexive forms over division algebras – see §3.6.

3.2 The algebra of forms

In this section, the basic properties of products and sums of sesquilinear forms over algebras with involution are developed and then applied to the determination of indecomposable forms and to the twisted contragredient representation. An excellent reference is [27].

3.2.1 Sums and products of forms. Let $(A, {}^-)$ be a $(K, {}^-)$-involution algebra as usual. As before, ${}_{(A,{}^-)}\mathcal{S}$ – or simply ${}_A\mathcal{S}$ – is the category of sesquilinear forms $f : V \times V \to (A, {}^-)$ on finitely generated left A-modules V, and $\bar{U}$ is the A-module U with the action of A twisted by the involution to make it into an A-module of the opposite hand. An element $u \in U$, when considered as an element of $\bar{U}$, is usually denoted by $\bar{u}$.

Let $(V, f) \in {}_A\mathcal{S}$. Since $f(v, \overline{v'}a) = f(v, \bar{a}v') = f(v, v')a$ for $a \in A$, f can be interpreted as a pairing

$$V \times \bar{V} \to A,$$

or, since $f(\alpha v, v') = f(v, \bar{\alpha} v') = f(v, \overline{v'}\alpha)$ for $\alpha \in K$, equally well as a homomorphism

$$\widetilde{f} : V \otimes_K \bar{V} \to A$$

of (A,A)-bimodules satisfying

$$\widetilde{f}(v\otimes\overline{v'})=f(v,v').$$

We shall not generally distinguish notationally between these two points of view, and so we usually write $\widetilde{f}$ simply as f.

For both the definition of product of forms and later applications, it is necessary to generalize these notions slightly to (A,B)-bimodules, where $(B,^{-})$ is another involution algebra over $(K,^{-})$, with a $(B,^{-})$-invariant sesquilinear form over $(A,^{-})$, that is to say

$$f(vb,v')=f(v,v'\bar{b})\ \text{ for all } v,v'\in V \text{ and } b\in B.$$

The action of A and B on V can also be twisted so that it becomes a (B,A)-bimodule $\bar{V}$ and $f:\ V\otimes_B\bar{V}\to A$ is an (A,A)-bimodule homomorphism. The previous case of V an A-module over the K-algebra A is the special case $B=K$.

If $g:U\times U\to(A,^{-})$ is another $(B,^{-})$-invariant sesquilinear form, a *bivariant morphism* φ from h to g is an (A,B)-bimodule homomorphism $\varphi:V\to U$ satisfying $g(\varphi v,\varphi v')=h(v,v')$ for all $v,v'\in V$. A bivariant isomorphism is referred to as a *bivariant* or *equivariant isometry* or as an *isometry of invariant forms.*

As usual we shall restrict our attention to the special case when A and B are semisimple algebras over K. In this case the underlying modules are reflexive by Th. 1.2.4. ${}_{(A,^{-})}\mathcal{S}^{(B,^{-})}={}_{A}\mathcal{S}^{B}$ stands for the category of $(B,^{-})$-invariant sesquilinear forms over $(A,^{-})$, and ${}_{(A,^{-})}\mathcal{H}_{\varepsilon}^{(B,^{-})}={}_{A}\mathcal{H}_{\varepsilon}^{B}$ for the full subcategory of $(B,^{-})$-invariant ε-Hermitian forms. We often write ${}_{A}\mathcal{H}_{\varepsilon}^{B}={}_{A}\mathcal{H}^{B}$ if ε is clear from the context, and also ${}_{(A,^{-})}\mathcal{H}_{\varepsilon}^{(B,^{-})}$ if the involutions are *not* clear from the context. As earlier, a superscript * on any of these symbols will signify the subcategory of nonsingular forms – for example ${}_{(A,^{-})}\mathcal{H}_{\varepsilon}^{(B,^{-})*}$

Occasionally we shall consider sesquilinear forms $f:V\times V\to(A,^{-})$ where V is a *right* A-module. Such a form is biadditive and satisfies the sesquilinearity condition

$$1'.\ f(va,v'a')=\bar{a}f(v,v')a'.$$

Of course if V is a (B,A)-bimodule, such an f is $(B,^{-})$-invariant if $f(bv,v')=f(v,\bar{b}v')$. The category of these sesquilinear forms will be denoted by ${}^{(B,^{-})}\mathcal{S}_{(A,^{-})}={}^{B}\mathcal{S}_{A}$ and the subcategory of ε-Hermitian forms by ${}^{(B,^{-})}\mathcal{H}_{\varepsilon,(A,^{-})}={}^{B}\mathcal{H}_{\varepsilon,A}={}^{B}\mathcal{H}_{A}$. Most results for "lefthanded" forms will also apply with obvious changes to these "righthanded" forms, and we will generally not mention them explicitly.

There are however simple useful category isomorphisms

$${}_{(A,^{-})}\mathcal{S}^{(B,^{-})}\leftrightarrow{}^{(B,^{-})}\mathcal{S}_{(A,^{-})},\qquad {}_{(A,^{-})}\mathcal{H}_{\varepsilon}^{(B,^{-})}\leftrightarrow{}^{(B,^{-})}\mathcal{H}_{\varepsilon,(A,^{-})}\tag{3.23}$$

given by taking "opposite forms"[1]: $(V,f)\rightsquigarrow(\bar{V},f^{o})$, where if $(V,f)\in{}^{(B,^{-})}\mathcal{S}_{(A,^{-})}$ for example,

$$f^{o}(\bar{u},\bar{v})=\overline{f(v,u)},$$

and $\bar{V}$ is the (B,A)-bimodule obtained by twisting the (A,B)-bimodule V: $b\bar{v}a=\overline{\bar{a}v\bar{b}}$.

Exercise. Show that these *are* well-defined isomorphisms and that they preserve nonsingularity and hyperbolicity. □

[1]This is different from the "reflection" $\bar{f}$ defined on p. 92.

We now consider several operations on invariant forms – these generally apply as well to "noninvariant" forms by taking $B = K$.

The first of these operations is *scaling*: if one multiplies the form f by $\delta \in \dot{K}$, the scaled form δf is B-invariant if f is. If h is ε-Hermitian ($\varepsilon = \pm 1$) and $\bar{\delta} = \pm\delta$, then δh is $\delta\bar{\delta}^{-1}\varepsilon$-Hermitian.

The categories of $(B, ^-)$-invariant sesquilinear and ε-Hermitian forms over $(A, ^-)$ admit a sum – the usual "orthogonal" sum $(V, f) \perp (U, g)$.

A product can also be defined, between a form $(V, f) \in {}_A\mathcal{S}^B$ and a form $(U, g) \in {}_B\mathcal{S}^C$ where $(C, ^-)$ is a third $(K, ^-)$-involution algebra. The product fg is a form in ${}_A\mathcal{S}^C$, namely the form on $V \otimes_B U$ determined by

$$\begin{aligned}&(fg)((v_1 \otimes_B u_1) \otimes_C (\bar{u}_2 \otimes_B \bar{v}_2))\\ &\quad = f(v_1 g(u_1 \otimes_C \bar{u}_2) \otimes_B \bar{v}_2) = f(v_1 \otimes_B g(u_1 \otimes_C \bar{u}_2)\bar{v}_2),\end{aligned}$$

or, in more conventional notation,

$$(fg)(v_1 \otimes u_1, v_2 \otimes u_2) = f(v_1 g(u_1, u_2), v_2) = f(v_1, v_2\overline{g(u_1, u_2)}). \tag{3.24}$$

It is straightforward to check that this definition is legitimate by applying standard properties of the tensor product to $(V \otimes_B U) \otimes_C (\bar{U} \otimes_B \bar{V})$; here we have identified $\overline{V \otimes_B U}$ with $\bar{U} \otimes_B \bar{V}$ as a (C, A)-bimodule via the canonical map $\overline{v \otimes u} \to \bar{u} \otimes \bar{v}$.

Thus the product is a map

$${}_A\mathcal{S}^B \times {}_B\mathcal{S}^C \to {}_A\mathcal{S}^C.$$

Moreover if f is ε-Hermitian and g is δ-Hermitian, then fg is $\varepsilon\delta$-Hermitian, that is the product restricts to a map

$${}_A\mathcal{H}_\varepsilon^B \times {}_B\mathcal{H}_\delta^C \to {}_A\mathcal{H}_{\varepsilon\delta}^C.$$

It is easy to check that the distributive laws

$$f(g_1 \perp g_2) = fg_1 \perp fg_2 \text{ and } (f_1 \perp f_2)g = f_1 g \perp f_2 g$$

both hold under the usual identifications $V \otimes (U_1 \oplus U_2) = (V \otimes U_1) \oplus (V \otimes U_2)$ and $(V_1 \oplus V_2) \otimes U = (V_1 \otimes U) \oplus (V_2 \otimes U)$.

If L/K is a finite extension of fields and $A = B = C = L$ then

$$(fg)(v_1 \otimes u_1, v_2 \otimes u_2) = f(v_1 g(u_1, u_2), v_2) = f(v_1, v_2) g(u_1, u_2),$$

in particular if we identify $V \otimes U$ with $U \otimes V$ as usual, then

$$gf = fg.$$

When $L = K$, the involution is the identity and $\varepsilon = 1$, the product is the "classical" one defined for symmetric forms over K (cf. p. 50).

If $(h, W) \in {}_C\mathcal{S}^D$, the associative law

$$(fg)h = f(gh)$$

is easy to check, once one identifies the (A,D)-bimodules $(V \otimes_B U) \otimes_C W$ and $V \otimes_B (U \otimes_C W)$.

We now consider how the nonsingularity of f and g carries over to $f \perp g$ and fg. Recall that $V^* = \operatorname{Hom}_A(V, A)$ is a (B, A)-bimodule via

$$\langle v, bxa \rangle = \langle vb, x \rangle a,$$

so $\overline{\mathrm{Hom}_A(V,A)}$ is an (A,B)-bimodule. The left and right adjoints

$$f_r : V \to \overline{\mathrm{Hom}_A(V,A)}, \quad v \to f(\cdot, v)$$

and

$$f_l : V \to \overline{\mathrm{Hom}_A(V,A)}, \quad v \to \overline{f(v,\cdot)}$$

are both A-linear. Furthermore the B-linearity of f_l or f_r is equivalent to f being B-invariant. For example, if f is B-invariant,

$$f_l(vb) = \overline{f(vb,\cdot)} = \overline{f(v,\cdot\bar{b})} = \langle \cdot \bar{b}, f_l(v)\rangle = \langle \cdot, \bar{b} f_l(v)\rangle = \bar{b} f_l(v) = f_l(v)b.$$

If we identify $\mathrm{Hom}_A(V \oplus U, A)$ with $\mathrm{Hom}_A(V,A) \oplus \mathrm{Hom}_A(U,A)$, it is apparent that

$$(f \perp g)_r : \ V \perp U \to \mathrm{Hom}_A(V,A) \oplus \mathrm{Hom}_A(U,A)$$

is the direct sum $f_r \oplus g_r$. Thus $f \perp g$ is nonsingular if and only if f and g are both nonsingular.

Consider next the nonsingularity of the product fg where, as before, $(V,f) \in {}_A\mathcal{S}^B$ and $(U,g) \in {}_B\mathcal{S}^C$. We need a lemma:

Lemma 3.2.1 *There is a canonical isomorphism*

$$\varphi : \ \overline{\mathrm{Hom}_A(V,A)} \otimes_B \overline{\mathrm{Hom}_B(U,B)} \to \overline{\mathrm{Hom}_A(V \otimes_B U, A)} \tag{3.25}$$

of (A,C)-bimodules determined by

$$\bar{\theta} \otimes_B \bar{\eta} \to \big(v \otimes_B u \to \theta(v\eta(u))\big).$$

The map $(v,u) \mapsto \theta(v\eta(u))$ is biadditive and admits B, and so provides a homomorphism $V \otimes_B U \to A$ which is clearly A-linear. We therefore get an element of $\mathrm{Hom}_A(V \otimes_B U, A)$ for each pair $(\bar{\theta}, \bar{\eta})$, namely

$$\sum_i v_i \otimes_B u_i \to \sum_i \bar{\theta}(v_i \bar{\eta}(u_i)).$$

This map from $\overline{\mathrm{Hom}_A(V,A)} \times \overline{\mathrm{Hom}_B(U,B)}$ is biadditive and admits B, and so provides a homomorphism (3.25) of Abelian groups. A straightforward check shows that it is in fact an (A,C)-bimodule homomorphism.

Suppose next that $U = B$ (and, say, $C = B$). The right B-module $\mathrm{Hom}_B(B,B)$ consists of the right multiplications $r_b(b') = b'b$, and the map $B \to \overline{\mathrm{Hom}_B(B,B)}$, $b \to r_{\bar{b}}$, is an isomorphism of left B-modules. The diagram

$$\begin{array}{ccc}
\overline{\mathrm{Hom}_A(V,A)} \otimes_B \overline{\mathrm{Hom}_B(B,B)} & \longrightarrow & \overline{\mathrm{Hom}_A(V,A)} \otimes_B B \\
\varphi \downarrow & & \downarrow \\
\overline{\mathrm{Hom}_A(V \otimes_B B, A)} & \longleftarrow & \overline{\mathrm{Hom}_A(V,A)}
\end{array}$$

is commutative (where the maps are the obvious ones). Since the three unlabelled maps are isomorphisms, so is φ.

It now follows from the additivity of the functors

$$U \rightsquigarrow \mathrm{Hom}_A(V,A) \otimes_B \mathrm{Hom}_B(U,B) \ \text{ and } \ U \rightsquigarrow \mathrm{Hom}_A(V \otimes_B U, A)$$

from left B-modules to Abelian groups that φ is an isomorphism first for free B-modules B^n of finite rank, and then for finitely generated projective modules U. But a finitely generated module over a semisimple algebra *is* a finitely generated projective since B is semisimple (Th. 1.2.3, p. 6), and so the lemma follows. $\square$

Now suppose that f and g are nonsingular forms as before. The diagram

$$\begin{array}{ccc}
V \otimes_B U & \xrightarrow{\;(fg)_r\;} & \mathrm{Hom}_A(V \otimes_B U, A) \\
{\scriptstyle f_r \otimes g_r}\big\downarrow & \nearrow{\scriptstyle \varphi} & \\
\mathrm{Hom}_A(V, A) \otimes_B \mathrm{Hom}_B(U, B) & &
\end{array}$$

is commutative, and so $(fg)_r$ is an isomorphism since φ and $f_r \otimes g_r$ both are. Thus fg is nonsingular.

Conversely suppose that fg is nonsingular. Then it is not always true that f and g must be too (see the exercise below). But a sufficient condition is that the "types" of the nonzero isotypic B-components of U and V be the same, that is to say that for each simple summand B_i of B, $VB_i \neq 0$ if and only if $B_iU \neq 0$. For suppose this condition holds. If g were not nonsingular, g_r would have a nonzero kernel, say $W \subset U$. By the lemma following, $V \otimes_B W \neq 0$ and is in the kernel of $(fg)_r$ by the commutativity of the above diagram (or by (3.24)). This is a contradiction, so g is nonsingular. A similar argument works for f – note that $\ker f_r$ is an (A, B)-bimodule.

Alternatively if f is nonsingular and for each simple summand B_i of B, $B_iU \neq 0$ implies that $VB_i \neq 0$, then fg nonsingular implies that g is nonsingular. Similarly if g is nonsingular and $VB_i \neq 0$ implies $B_iU \neq 0$ for each i, then fg nonsingular implies f nonsingular.

Lemma 3.2.2 *Let W be a left B-module and V a right B-module such that $B'W \neq 0 \neq VB'$ for some simple algebra component B' of B. Then $V \otimes_B W \neq 0$.*

By the biadditivity of the tensor product $V \otimes_B W$, it suffices to prove it in the case that V and W are simple B'-modules, and therefore also when $V = B' = W$. It is easy to see, using the universal mapping property of the tensor product (p. 1), that $V \otimes_B W$ and $V \otimes_{B'} W$ are canonically isomorphic if V and W are (isotypic) B'-modules. Since $B' \otimes_{B'} B' \cong B'$, the lemma follows. □

Exercise. Let $(B_1, ^-) = \mathrm{M}(n_1, (K, ^-))$, $(B_2, ^-) = \mathrm{M}(n_2, (K, ^-))$ and $(B, ^-) = (B_1, ^-) \oplus (B_2, ^-)$. Let $V = B_1$, viewed as a (B_1, B)-bimodule, and $U = B$, viewed as a (B, B)-bimodule, with the forms

$$f : V \times V \to (B_1, ^-), \quad f(b_1, b_1') = b_1\overline{b_1'}$$

and

$$g : U \times U \to (B, ^-), \quad g(b, b') = b1_{B_1}\overline{b'}.$$

(a) Show that $f \in {}_{(B_1,^-)}\mathcal{H}_1^{(B,^-)}$ and $g \in {}_{(B,^-)}\mathcal{H}_1^{(B,^-)}$.
(b) Show that $fg : (V \otimes_B U) \times (V \otimes_B U) \to (B_1, ^-)$ is equivalent to f via an isometry $V \otimes_B U \to V$, namely the canonical isomorphism $B_1 \otimes_B B \to B_1$, given by $b_1 \otimes_B b \to b_1 b = b_1 b_{(1)}$ where $b = (b_{(1)}, b_{(2)})$.
(c) Show that fg is a nonsingular form but $g : U \times U \to (B, ^-)$ is not. □

Now suppose again that $(V, f) \in {}_A\mathcal{H}_\varepsilon^B$ and that g is δ-hyperbolic on the (B, C)-bimodule U_1; thus $U = U_1 \oplus \overline{U_1^*}$ is also a (B, C)-bimodule and so $(U, g) \in {}_B\mathcal{H}_\delta^C$. Then the form fg on

$$V \otimes_B U = (V \otimes_B U_1) \oplus (V \otimes_B \overline{U_1^*})$$

is also hyperbolic by Th. 3.1.2 since the summands are both totally isotropic. A similar proof shows that fg is hyperbolic if f is hyperbolic.

We summarize our results so far on sums and products of forms:

Theorem 3.2.1 *(a) The sum $f \perp g$ is a composition law*

$$ {}_A\mathcal{S}^B \times {}_A\mathcal{S}^B \to {}_A\mathcal{S}^B $$

which is associative and commutative and, for each $\varepsilon = \pm 1$, induces a composition law

$$ {}_A\mathcal{H}^B_\varepsilon \times {}_A\mathcal{H}^B_\varepsilon \to {}_A\mathcal{H}^B_\varepsilon $$

by restriction. The sum $f \perp g$ is nonsingular if and only if f and g are both nonsingular.

(b) The product fg is a composition law

$$ {}_A\mathcal{S}^B \times {}_B\mathcal{S}^C \to {}_A\mathcal{S}^C $$

which is associative, distributive with respect to $\perp$, and for any $\varepsilon, \delta = \pm 1$, induces a composition law

$$ {}_A\mathcal{H}^B_\varepsilon \times {}_B\mathcal{H}^C_\delta \to {}_A\mathcal{H}^C_{\varepsilon\delta}. $$

It is commutative when $A = B = C = L$, a finite extension field of K.

(c) fg is nonsingular if both f and g are nonsingular, and conversely f and g are nonsingular if fg is nonsingular and either

(i) the types of their nonzero isotypic components are the same, or

(ii) f is nonsingular and for each simple summand B_i of B, $B_iU \neq 0$ implies that $VB_i \neq 0$, or

(iii) g is nonsingular and for each simple summand B_i of B, $VB_i \neq 0$ implies $B_iU \neq 0$.

(d) fg is hyperbolic if $f \in {}_A\mathcal{H}^{*B}_\varepsilon$, $g \in {}_B\mathcal{H}^{*C}_\delta$ and at least one of f and g is hyperbolic. □

3.2.2 Going up. We consider the extension of a sesquilinear form $f : V \times V \to (A, ^-)$ to an overring.

Let A' be another K-algebra with $(K, ^-)$-involution and $\varphi : (A, ^-) \to (A', ^-)$ a K-homomorphism of involution algebras – for example the inclusion map if $(A, ^-)$ is an involution subalgebra of $(A', ^-)$. Let A act on the right of A' via φ, namely $a'.a = a'\varphi(a)$, and put $V^{A'} = A' \otimes_A V$. Let g be the Hermitian form $g(a'_1, a'_2) = a'_1\overline{a'_2}$ on A'. It is nonsingular since the dual of A' as a left A'-module is the ring of right multiplications. It is clear that g is $(A, ^-)$-invariant: $g \in {}_{A'}\mathcal{H}^A_1$. Thus if $f \in {}_A\mathcal{S}^B$ is a $(B, ^-)$-invariant sesquilinear form over $(A, ^-)$, the product $gf \in {}_{A'}\mathcal{S}^B$ and we denote it by $f^{(A', ^-)}$ (or simply $f^{A'}$ if the involution is clear from the context). Explicitly

$$ f^{A'} : V^{A'} \times V^{A'} \to (A', ^-) $$

where, by (3.24),

$$ f^{A'}(a'_1 \otimes v_1, a'_2 \otimes v_2) = (a'_1 f(v_1, v_2))\overline{a'_2} = a'_1\varphi(f(v_1, v_2))\overline{a'_2}. \qquad (3.26) $$

If f is ε-Hermitian (or nonsingular or ε-hyperbolic), then so is $f^{A'}$ by Th. 3.2.1.

Theorem 3.2.2 *(a) The map* $f \rightsquigarrow f^{A'}$ *is a covariant functor* ${}_A\mathcal{S}^B \to {}_{A'}\mathcal{S}^B$ *which preserves orthogonal sums, totally isotropic subspaces, hyperbolic forms, and nonsingular forms. It restricts to a functor* ${}_A\mathcal{H}^B_\varepsilon \to {}_{A'}\mathcal{H}^B_\varepsilon$ *for* $\varepsilon = \pm 1$.
(b) f *is nonsingular if* $f^{A'}$ *is nonsingular and, for each simple component* A_i *of* A,

$$A_i V \neq 0 \Rightarrow A' A_i \neq 0$$

or, equivalently,

$$A_i V \neq 0 \Rightarrow \varphi A_i \neq 0.$$

If $\psi : (V_1, f_1) \to (V_2, f_2)$ is a morphism in ${}_A\mathcal{S}^B$, it is apparent from (3.26) that $\mathrm{id}_{A'} \otimes \psi$ is a morphism $(A' \otimes V_1, f_1^{A'}) \to (A' \otimes V_2, f_2^{A'})$ in ${}_{A'}\mathcal{S}^B$ and that

$$f \rightsquigarrow f^{A'}, \quad \psi \rightsquigarrow \mathrm{id}_{A'} \otimes \psi$$

is a functor. The fact that it preserves orthogonal sums is simply the distributive law $g(f_1 \perp f_2) = gf_1 \perp gf_2$ of Th. 3.2.1(b). By the same theorem, nonsingular forms and hyperbolic forms are preserved, and the last statement of (a) follows too. That totally isotropic subspaces are preserved is trivial by (3.26).

Part (b) follows from Th. 3.2.1(c)(ii). □

There is a special case of "going up" which is especially useful, using a homomorphism $\psi : (K, {}^-) \to (L, {}^-)$ of commutative rings with involution. (Of course the case of L a field is particularly important, and $(K, {}^-)$ is an involution field as usual). There are two ways of doing this.

The first is the most direct: Our naive notation causes problems here since if $\lambda \in \bar{L}$, $\bar{\lambda}$ is ambiguous - it is either the element λ of L, considered as an element of the twisted L-module $\bar{L}$, or it is the transform of λ by the involution of L. We ask the reader's forbearance – we temporarily indicate twisting by $\widehat{\ }$. We first note that

$$\widehat{L} \otimes_K \widehat{V} \to \widehat{L \otimes_K V}, \quad \widehat{\lambda} \otimes_K \widehat{v} \to \widehat{\lambda \otimes_K v}$$

is an isomorphism of right $L \otimes_K A$-modules:

$$(\widehat{\lambda} \otimes \widehat{v})(\mu \otimes a) = (\widehat{\lambda . \mu}) \otimes \widehat{v a} = \widehat{\bar{\mu}\lambda} \otimes \widehat{\bar{a} v} \quad \to \quad \widehat{\bar{\mu}\lambda \otimes \bar{a} v} = \widehat{\lambda \otimes v}(\mu \otimes a).$$

Next, the map

$$L \times V \times \hat{L} \times \hat{V} \to L \otimes_K A, \quad (\lambda_1, v_1, \hat{\lambda}_2, \hat{v}_2) \to (\lambda_1 \bar{\lambda}_2) \otimes_K f(v_1, v_2) = \lambda_1 \bar{\lambda}_2 f(v_1, v_2),$$

is multilinear and balanced in K (where L and A are identified in the usual way with subalgebras of $L \otimes_K A$) and so we get a map

$$(L \otimes_K V) \otimes_K \widehat{L \otimes_K V} \to L \otimes A, \quad (\lambda_1 \otimes_K v_1) \otimes_K (\hat{\lambda}_2 \otimes_K \hat{v}_2) \to \lambda_1 \bar{\lambda}_2 f(v_1, v_2),$$

that is to say, a sesquilinear form

$$f^L : (L \otimes_K V) \times (L \otimes_K V) \to L \otimes_K A$$

given by

$$f^L(\lambda_1 \otimes_K v_1, \lambda_2 \otimes_K v_2) = \lambda_1 \bar{\lambda}_2 f(v_1, v_2). \tag{3.27}$$

We say that f^L *is obtained from* f *by extension of scalars.* It is easy to see that if f is $(B, {}^-)$-invariant, then f^L is $(L, {}^-) \otimes_{(K,{}^-)} (B, {}^-)$-invariant.

Alternatively, we first extend the scalars in the K-algebras with involution $(A, {}^-)$, $(B, {}^-)$, obtaining L-algebras with involution

$$(A', {}^-) = (A^L, {}^-) = (L, {}^-) \otimes_{(K,{}^-)} (A, {}^-), \quad (B', {}^-) = (B^L, {}^-) = (L, {}^-) \otimes_{(K,{}^-)} (B, {}^-).$$

The L-module V^L can be identified with $V^{A'}$ via the (A^L, B^L)-bimodule isomorphism

$$V^{A'} = (L \otimes_K A) \otimes_A V \to L \otimes_K V = V^L, \quad (\lambda \otimes_K a) \otimes_A v \to \lambda \otimes_K av. \tag{3.28}$$

The canonical homomorphism $A \to A^L$, $a \to 1 \otimes_K a$, is a homomorphism of $(K, {}^-)$-involution algebras, so if $(V, f) \in {}_A\mathcal{S}^B$, the sesquilinear space $(V^{A^L}, f^{A^L}) \in {}_{A^L}\mathcal{S}^{B^L}$ is defined and can be identified with (V^L, f^L) since the isomorphism (3.28) is an isometry:

$$\begin{aligned} f^L(\lambda_1 \otimes a_1 v_1, \lambda_2 \otimes a_2 v_2) &= \lambda_1 \bar{\lambda}_2 f(a_1 v_1, a_2 v_2) \quad \text{by (3.27)} \\ &= ((\lambda_1 \otimes 1)(1 \otimes a_1) f(v_1, v_2)(1 \otimes \bar{a}_2)(\bar{\lambda}_2 \otimes 1) \\ &= (\lambda_1 \otimes a_1) f(v_1, v_2) \overline{(\lambda_2 \otimes a_2)} \\ &= f^{A^L}((\lambda_1 \otimes a_1) \otimes v_1, (\lambda_2 \otimes a_2) \otimes v_2) \quad \text{by (3.26)} \end{aligned} \tag{3.29}$$

where we have again freely used the identifications of L and A into $L \otimes_K A$.

Theorem 3.2.3 *(a) Scalar extension $f \rightsquigarrow f^L$ is a functor ${}_A\mathcal{S}^B \to {}_{A^L}\mathcal{S}^{B^L}$ which preserves orthogonal sums, totally isotropic subspaces, and hyperbolic forms. For $\varepsilon = \pm 1$, it restricts to a functor ${}_A\mathcal{H}_\varepsilon^B \to {}_{A^L}\mathcal{H}_\varepsilon^{B^L}$.*
(b) f^L is nonsingular if and only if f is nonsingular.
(c) If $(C, {}^-)$ is another $(K, {}^-)$-involution algebra and $(U, g) \in {}_B\mathcal{S}^C$, then $(fg)^L = f^L g^L$.

(a) It is easy to check that f^L is B^L-invariant, and the rest of (a), and (b), follow from Th. 3.2.2. Now (c). The form $f^L g^L$ is defined on

$$(L \otimes_K V) \otimes_{B^L} (L \otimes_K U),$$

and $(fg)^L$ is defined on

$$L \otimes_K V \otimes_B U$$

where the (A^L, C^L)-bimodule structure on $L \otimes_K V \otimes_B U$ is given by

$$(\lambda_1 \otimes a)(\lambda \otimes v \otimes u)(\lambda_2 \otimes c) = \lambda_1 \lambda \lambda_2 \otimes av \otimes uc.$$

These two (A^L, C^L)-bimodules are isomorphic via $(\lambda_1 \otimes v) \otimes (\lambda_2 \otimes u) \to \lambda_1 \lambda_2 \otimes v \otimes u$. If we identify them via this isomorphism and use (3.29) several times, we get $f^L g^L = (fg)^L$. □

3.2.3 Going down. Now we come to "going down", namely going from forms over the $(K, {}^-)$-involution algebra $(A, {}^-)$ to left $(A, {}^-)$-invariant forms over $(K, {}^-)$:

$$_{(A,^-)}\mathcal{S} \to {}^{(A,^-)}_{(K,^-)}\mathcal{S}, \quad {}_{(A,^-)}\mathcal{H}_\varepsilon \to {}^{(A,^-)}_{(K,^-)}\mathcal{H}_\varepsilon. \tag{3.30}$$

In fact when A is a separable K-algebra, we shall see that the maps in (3.30) can be chosen to be category isomorphisms, as in the special case of $(KG, {}^-)$ (cf. Th. 2.1.1, p. 86).

In order to do this, we assume that we have an involution trace

$$\mathrm{t}: (A, {}^-) \to (K, {}^-),$$

(cf. p. 35). The symmetric or Hermitian form $\mathrm{t}(a, b) = \mathrm{t}(a\bar{b})$ is left $(A, {}^-)$-invariant – i.e. $\mathrm{t} \in {}^{(A,^-)}_{(K,^-)}\mathcal{H}$ – since if $a, b, c \in A$,

$$\mathrm{t}(ca, b) = \mathrm{t}(ca\bar{b}) = \mathrm{t}(a\bar{b}c) = \mathrm{t}(a\overline{\bar{c}b}) = \mathrm{t}(a, \bar{c}b).$$

We begin with a sesquilinear form $f : V \times V \to (A, ^-)$ and define

$$\mathrm{t}_*f : V \times V \to (K, ^-), \quad (\mathrm{t}_*f)(u, v) = \mathrm{t}(f(u, v)), \tag{3.31}$$

the *transfer* of the form f (with respect to the involution trace t). It is a sesquilinear form over $(K, ^-)$ and is left $(A, ^-)$-invariant:

$$\mathrm{t}_*f(au, v) = \mathrm{t}(af(u, v)) = \mathrm{t}(f(u, v)a) = \mathrm{t}_*f(u, \bar{a}v).$$

If f is ε-Hermitian, then t_*f is ε-reflexive, and we have maps

$$\mathrm{t}_* : {}_A\mathcal{S} \to {}_K^A\mathcal{S}, \quad \mathrm{t}_* : {}_A\mathcal{H}_\varepsilon \to {}_K^A\mathcal{H}_\varepsilon. \tag{3.32}$$

Since the form $\mathrm{t}(a, b)$ is nonsingular (by definition of an involution trace), every linear form in $\mathrm{Hom}_K(A, K)$ is $= \mathrm{t}(\cdot\ b)$ for a uniquely determined $b \in A$. Thus if $(V, g) \in {}_K^A\mathcal{S}$ and $u, v \in V$, the linear form $a \to g(au, v)$ is given by

$$g(au, v) = \mathrm{t}(af(u, v)) = \mathrm{t}(f(u, v)a)$$

for a uniquely determined $f(u, v) \in A$. It is easy to see that $f : V \times V \to (A, ^-)$ is sesquilinear: first of all, for all $a \in A$,

$$\begin{aligned} \mathrm{t}(af(u, v_1 + v_2)) &= g(au, v_1 + v_2) = g(au, v_1) + g(au, v_2) \\ &= \mathrm{t}(af(u, v_1)) + \mathrm{t}(af(u, v_2)) = \mathrm{t}(a(f(u, v_1) + f(u, v_2))) \end{aligned}$$

and it follows that $f(u, v_1 + v_2) = f(u, v_1) + f(u, v_2)$ by uniqueness. Semilinearity in v follows from

$$\mathrm{t}(af(u, bv)) = g(au, bv) = g(\bar{b}au, v) = \mathrm{t}(\bar{b}af(u, v)) = \mathrm{t}(af(u, v)\bar{b}).$$

Similarly $f(u, v)$ is A-linear in u. If g is ε-Hermitian, then

$$\begin{aligned} \mathrm{t}(af(u, v)) &= g(au, v) = \varepsilon\,\overline{g(v, au)} = \varepsilon\,\overline{g(\bar{a}v, u)} \\ &= \varepsilon\,\overline{\mathrm{t}(\bar{a}f(v, u))} = \varepsilon\,\mathrm{t}(\overline{f(v, u)}a) = \mathrm{t}(a\varepsilon\,\overline{f(v, u)}), \end{aligned}$$

so f is also ε-Hermitian. Thus both transfers in (3.32) are onto, and by the uniqueness property, injective as well.

Consider next the commutative diagram (of Abelian groups)

$$\begin{array}{ccc} V & \xrightarrow{f_r} & \mathrm{Hom}_A(V, A) \\ & \searrow^{g_r} & \downarrow \mathrm{t}_* \\ & & \mathrm{Hom}_K(V, K) \end{array}$$

where $\mathrm{t}_*(x) = \mathrm{t} \circ x$ and f_r and g_r are the right adjoints. Since t_* is an isomorphism (Th. 1.2.10, p. 9), $g = \mathrm{t}_*f$ is nonsingular if and only if f is.

Thus the transfers

$$\mathrm{t}_* : {}_A\mathcal{S}^* \to {}_K^A\mathcal{S}^* \quad \text{and} \quad \mathrm{t}_* : {}_A\mathcal{H}_\varepsilon^* \to {}_K^A\mathcal{H}_\varepsilon^* \quad (\varepsilon = \pm 1)$$

are also bijections.

As one would expect, the transfers are isomorphisms of categories. Of course it is trivial that if $\varphi : (V_1, f_1) \to (V_2, f_2)$ is a morphism in ${}_A\mathcal{S}$, then φ is also a morphism $(V_1, \mathrm{t}_*f_1) \to (V_2, \mathrm{t}_*f_2)$ in ${}_K^A\mathcal{S}$. Conversely if $\varphi : (V_1, \mathrm{t}_*f_1) \to (V_2, \mathrm{t}_*f_2)$ is a morphism in ${}_K^A\mathcal{S}$ (which we require to be A-linear), then

$$\mathrm{t}(af_2(\varphi u, \varphi v)) = \mathrm{t}(f_2(\varphi au, \varphi v)) = \mathrm{t}(f_1(au, v)) = \mathrm{t}(af_1(u, v))$$

so $f_2(\varphi u, \varphi v) = f_1(u, v)$ for all u, v, and φ is also a morphism in ${}_A\mathcal{S}$.

The categories ${}_A\mathcal{H}_\varepsilon$ and ${}^A_K\mathcal{H}_\varepsilon$ are full subcategories of ${}_A\mathcal{S}$ and ${}^A_K\mathcal{S}$, and it is clear that the isomorphism ${}_A\mathcal{S} \to {}^A_K\mathcal{S}$ restricts to an isomorphism ${}_A\mathcal{H}_\varepsilon \to {}^A_K\mathcal{H}$.

Before stating the next theorem, we introduce one more notion, that of an ε-hyperbolic representation of an involution algebra $(A, ^-)$: this is an ε-hyperbolic form $h : V \times V \to (K, ^-)$ on a left A-module V, which is $(A, ^-)$-invariant and which has a "hyperbolic decomposition" $V = U \oplus W$ where U and W are totally isotropic A-submodules of V, rather than simply totally isotropic K-subspaces. We also say in this case that h is a strongly ε-hyperbolic $(A, ^-)$-invariant form. We note that if $\rho : G \to \mathbf{I}(V, h)$ is an ε-hyperbolic representation (p. 90), then h is a strongly ε-hyperbolic $(A, ^-)$-invariant form where A is the summand of KG such that $AV = V$.

Theorem 3.2.4 . *Let $(A, ^-)$ be a (semisimple) algebra with $(K, ^-)$-involution and $\mathrm{t} : (A, ^-) \to (K, ^-)$ an involution trace. If $(V, f) \in {}_A\mathcal{S}$, the transfer $\mathrm{t}_*f : V \times V \to (K, ^-)$ provides a functor $(V, f) \rightsquigarrow (V, \mathrm{t}_*f)$ which is a category isomorphism ${}_A\mathcal{S} \to {}^A_K\mathcal{S}$ (defined as the identity on morphisms) and induces, by restriction, a category isomorphism between the full subcategories ${}_A\mathcal{H}_\varepsilon$ and ${}^A_K\mathcal{H}_\varepsilon$ for $\varepsilon = \pm 1$. Furthermore this functor preserves orthogonal decompositions, and is an isomorphism between the following subcategories of nonsingular forms:*
(i) ${}_A\mathcal{S}^*$ and ${}^A_K\mathcal{S}^*$,

(ii) ${}_A\mathcal{H}^*_\varepsilon$ and ${}^A_K\mathcal{H}^*_\varepsilon$, and

(iii) the ε-hyperbolic forms over $(A, ^-)$ and the strongly ε-hyperbolic $(A, ^-)$-invariant forms over $(K, ^-)$.

It is clear that the functor preserves orthogonality. The rest of the theorem has already been proved above except for (iii). Certainly if h is a hyperbolic form over $(A, ^-)$, it is clear that t_*h is a strongly hyperbolic invariant form. Conversely suppose t_*h is a strongly hyperbolic invariant form, say $V = U \oplus W$ with A-submodules U and W which are totally isotropic under t_*h. Since $\mathrm{t}_* : {}_A\mathcal{S} \to {}^A_K\mathcal{S}$ is a category isomorphism, the forms over $(A, ^-)$ whose transfers are the restrictions of h to U and to W must be 0-forms, i.e. the form whose transfer is h itself is a hyperbolic form over $(A, ^-)$ (Th. 3.1.2, p. 97). □

Corollary 3.2.1 *Same as the theorem, except we replace left A-modules by right A-modules.*

This can be proved in a similar manner, or one can use the commutativity of the diagram of functors:

$$\begin{array}{ccc} {}_A\mathcal{S} & \longrightarrow & \mathcal{S}_A \\ \downarrow{\scriptstyle \mathrm{t}} & & \downarrow{\scriptstyle \mathrm{t}} \\ {}^A_K\mathcal{S} & \longrightarrow & {}_K\mathcal{S}^A \end{array}$$

where the horizontal maps are both "taking the opposite form" (cf. the exercise, p. 105). The three functors, aside from that on the right, are isomorphisms, and so therefore is the fourth. □

Exercise. Check the commutativity of the diagram. □

3.2.4 Indecomposable forms. Recall that an involution algebra $(A, ^-)$ is a *simple* involution algebra if it is semisimple and has no two-sided ideals, other than 0 and A, which are stable under the involution. On the other hand it is a simple algebra with involution if it has an involution and is simple as an algebra.

A simple involution algebra is either a simple algebra with involution or is a hyperbolic involution algebra $\mathrm{H}(B)$ based on a simple algebra B (cf. p. 36).

An ε-Hermitian form h is called *(orthogonally) indecomposable* if it is not the orthogonal direct sum of two nonzero ε-Hermitian forms.

Theorem 3.2.5 *Let $(A, ^-)$ be a simple involution algebra and $h : V \times V \to (A, ^-)$ an indecomposable nonsingular ε-Hermitian form.*

(a) If $(A, ^-)$ is hyperbolic, then h is the hyperbolic form based on a simple A-module and is uniquely determined up to isometry.

(b) If A is a simple algebra, then either

(i) V is simple and the underlying module of every *indecomposable form over A is simple, or*

(ii) h is the hyperbolic form based on a simple A-module and is uniquely determined up to isometry.

(a) If $(A, ^-)$ is hyperbolic, then h must be hyperbolic (Cor. 3.1.5, p. 98) and is hyperbolic on a *simple* module by (3.14), p. 98, since it is indecomposable. It therefore has length 2, and so is uniquely determined up to isometry by Cor. 3.1.6 (p. 98). This proves (a).

(b) This follows from Prop. 3.1.3 (p. 99), Prop. 3.1.2, and Th. 3.1.3. □

It is possible to characterize exactly the circumstances under which a simple module supports a nonsingular ε-Hermitian form:

Theorem 3.2.6 *Let $(A, ^-)$ be a simple algebra with involution and V a simple A-module. Then V supports both a nonsingular Hermitian form and a nonsingular skew Hermitian form over $(A, ^-)$ unless A is quasisplit and $(A, ^-)$ is of the first kind as an involution algebra over its center. In the latter case, V supports an Hermitian form if and only if $(A, ^-)$ is orthogonal, and it supports a skew Hermitian form if and only if $(A, ^-)$ is symplectic.*

Suppose there is no nonsingular ε-Hermitian form on the simple module V for some particular choice of $\varepsilon = \pm 1$. This means that 0 is the only ε-Hermitian form on V since an A-homomorphism $V \to \overline{V^*}$ is either 0 or an isomorphism. Since $\overline{V^*}$ is also a simple A-module, there *is* an A-isomorphism $\varphi : V \to \overline{V^*}$. The form $\langle u, \varphi(v)\rangle + \varepsilon\overline{\langle v, \varphi(u)\rangle}$ on V is ε-Hermitian and so must be 0. This means that the nonsingular form $g(u, v) := \langle u, \varphi(v)\rangle$ is $-\varepsilon$-Hermitian.

Let $^\sim$ be the adjoint involution with respect to g on the division algebra $D = \mathrm{End}_A V$. Since it is an involution, its eigenvalues as a K_0-linear transformation of D have square 1. Suppose that $d \in \dot{D}$ is an eigenvector with eigenvalue -1, i.e. $\widetilde{d} = -d$. Then the form $h(u, v) := g(ud, v)$ is ε-Hermitian and nonzero, contrary to our assumptions. Thus all eigenvalues of $^\sim$ are 1, so $^\sim = \mathrm{id}_D$. In particular D is a (commutative) field L, $A \cong \mathrm{M}(n, L)$, and the involution on A is of the first kind.

Thus we have shown that V supports both a nonsingular Hermitian form and a nonsingular skew Hermitian form over $(A, ^-)$ unless A is a split algebra over its center and the involution on A is of the first kind, and in this case it supports at least one kind of form. It remains to be shown that in this exceptional case, it supports exactly *one* kind of form, and that the type of the involution determines whether it is Hermitian or skew Hermitian.

So suppose that $A = \mathrm{M}(n, L)$ with an involution $^-$ of the first kind, and that h is a nonsingular ε-Hermitian form over A on the simple A-module V. By Th. 3.2.4, p. 113, and Ex. 1, p. 35, the form $b(u, v) := \mathrm{tr}(h(u, v))$, where tr is the matrix trace, is a nonsingular ε-symmetric bilinear form over L which is $(A, ^-)$-invariant. Thus $^-$ is the adjoint involution of b, and so $\varepsilon = 1$ if $^-$ is of orthogonal type and $= -1$ if $^-$ is of symplectic type (Prop. 1.5.1, p. 47). That is to say, h is Hermitian if $^-$ is of orthogonal type and is skew Hermitian if $^-$ is of symplectic type. □

Corollary 3.2.2 *When $(A, ^-)$ is quasisplit and orthogonal, any nonsingular skew Hermitian form is hyperbolic; similarly when $(A, ^-)$ is quasisplit and symplectic, any nonsingular Hermitian form is hyperbolic. In all other cases (when A is a simple algebra), a nonsingular Hermitian or skew Hermitian form over $(A, ^-)$ is the orthogonal direct sum of "unary" forms, that is to say of forms on simple modules.*

This follows easily from the last two theorems. However this important fact is perhaps best proved using Morita theory (Th. 3.2.14, p. 132). With the notation of that theorem, the form $\breve{h}$ is skew-symmetric over a field in the first two cases mentioned in the corollary since $\breve{\varepsilon} = \varepsilon\varepsilon_0 = -1$, and so an indecomposable form must be hyperbolic of length 2. In all other cases, every space $(\breve{V}, \breve{h})$ has an orthogonal basis, and so an indecomposable "space" over A must be simple. □

These results are equivalent to existence theorems for $(A, ^-)$-invariant forms over $(K, ^-)$ – see pp. 146*ff*.

3.2.5 The twisted contragredient representation revisited. The twisted contragredient representation $\bar{\rho}^* : G \to \mathbf{GL}(\overline{V^*})$ of a linear representation $\rho : G \to \mathbf{GL}(V)$ is defined by $\langle \rho(s^{-1})v, x\rangle = \langle v, \bar{\rho}^*(s)x\rangle$ where $x \in \overline{V^*}$ (cf. p. 74). It differs from the usual contragredient representation when $(K, ^-)$ is of the second kind, since $\alpha x = x\bar{\alpha}$ for $\alpha \in K$.

As we saw there, if the matrix of $\rho(s)$ with respect to some basis of V over K is $(\rho(s)_{ij})$, then the matrix of $\bar{\rho}^*(s)$ with respect to the twisted dual basis of $\overline{V^*}$ is

$$\left(\overline{\rho(s^{-1})_{ij}}\,\right)^t = \left(\overline{\rho(s)_{ji}}\,\right)^{-1}.$$

So if the character of ρ is χ, the character of ρ^* is

$$\overline{\mathrm{tr}(\rho(s)^{-1})} = \overline{\chi(s)^*}$$

where $\chi(s)^*$ is the "complex conjugate" of $\chi(s)$ (cf. p. 71) and tr is the matrix trace. We write this

$$\chi_{\bar{\rho}^*} = \overline{\chi_\rho^*}. \tag{3.33}$$

If $\mathrm{char}\, K = p \neq 0$ and χ is the Brauer character of ρ arising from a p-modular system $(\widetilde{K}, \mathfrak{o}, K)$ in which $\widetilde{K}$ is sufficiently large for G (cf. p. 72), then a similar argument shows that (3.33) again holds – see p. 73 for the definition of $\overline{\chi(s)^*}$ in this case.

Theorem 3.2.7 *Let A be a simple direct summand of KG, and $\rho : G \to \mathbf{GL}_K(V)$ a linear K-representation of G which is A-isotypic ($AV = V$). Let χ be the ordinary character of ρ if* char $K = 0$ *or the Brauer character if* char $K \neq 0$.

(a) The following are equivalent:

(i) ρ *is equivalent to its twisted contragredient* $\bar{\rho}^*$.
(ii) $\bar{A} = A$ *where* $^-$ *is the standard* $(K, ^-)$*-involution on* KG.
(iii) There is a nonsingular sesquilinear form $h : V \times V \to (K, ^-)$ *which is* G*-invariant.*

If char $K = 0$, *or if the Brauer character exists, they are equivalent to*
(iv) $\overline{\chi^*} = \chi$.

If, in addition, $^-$ *is the identity on* K, *then they are equivalent to*
(v) χ *is real (cf. p. 73).*

(b) For either choice of $\varepsilon = \pm 1$, *each of (i) – (iii) is sufficient for the existence of an invariant* ε*-reflexive form* $h : V \times V \to (K, ^-)$ *except in the case*

$$A \text{ is quasisplit,} \quad (A, ^-) \text{ is of the first kind,} \quad \text{and } \operatorname{len}_A V \text{ is odd.} \tag{3.34}$$

(c) In the exceptional case (3.34), there is either a nonsingular invariant symmetric form b *on* V *or a nonsingular invariant skew symmetric form* b *on* V, *but not both. The first case occurs when* $(A, ^-)$ *is orthogonal and the second when it is symplectic.*

REMARK. The exceptional case corresponds to the exceptional case of Th. 3.2.6. □

The dual $V^* = \operatorname{Hom}_K(V, K)$ is a right A-module, hence a left $\bar{A}$-module via twisting by the standard involution $^-$ on KG. Furthermore $\bar{\rho}^*$ is an $\bar{A}$-isotypic representation. Since V and V^* have the same length (Th. 1.2.4, p. 7), it follows at once that (i) and (ii) are equivalent, and they are equivalent to (iii) by Th. 1.7.4 (p. 75), and to (iv) by Ths. 1.7.2 and 1.7.3 (p. 73). If $^-$ is the identity on K, then $\overline{\chi^*} = \chi^*$, and so the equivalence of (iv) and (v) follows.

Now assume $\bar{A} = A$. Since A is a separable (simple) K-algebra by Maschke's Theorem, the reduced trace $\operatorname{trd}_{A/K}$ is an involution trace by Ex. 1, p. 35. Therefore by Th. 3.2.4(ii), the existence of a nonsingular $(A, ^-)$-invariant ε-reflexive form on an A-module V is equivalent to the existence of a nonsingular ε-Hermitian form over $(A, ^-)$, and so we may consider the latter.

Since an A-module supports ε-hyperbolic forms over $(A, ^-)$ for each choice of $\varepsilon = \pm 1$ if it has even length and nonsingular ε-Hermitian forms of *any* length if A is *not* quasisplit with an involution of the first kind (Cor. 3.2.2), (b) follows at once.

Assume therefore that V has odd length, $A \cong \mathrm{M}(n, L)$ where L/K is a finite extension of fields, and that $(A, ^-)$ is of the first kind. Since the A-modules which support nonsingular indecomposable ε-Hermitian forms over $(A, ^-)$ have length 1 or 2 by Th. 3.2.5(b), V supports nonsingular ε-Hermitian forms if and only if a simple A-module supports such forms. Thus (c) follows from Th. 3.2.6. □

3.2.6 Adjoint involutions. Let $h : V \times V \to (K, ^-)$ be a nonsingular ε-reflexive form and $\rho : G \to \mathbf{I}(V, h)$ an isometry representation, *irreducible* as a linear representation over K. We also denote by ρ its linear extension to $KG \to \operatorname{End}_K V$. Let A be the corresponding simple direct summand of KG, $^\sim$ the adjoint involution adj_h on $\operatorname{End}_K V$ with respect to h (a $(K, ^-)$-involution), and $^-$ the standard $(K, ^-)$-involution on KG. By Schur's Lemma (p. 1), the centralizer $\operatorname{End}_{KG} V$ of ρA in $\operatorname{End}_K V$ is a division algebra D over K, and the centralizer $\operatorname{End}_D V$ of D is $= \rho A$ by the Centralizer Theorem (p. 11). We will write the elements of D on the *right* of vectors in V, so that its multiplication is the "opposite" of that in $\operatorname{End}_K V$ –

thus the opposite algebra D^o is a subalgebra of $\mathrm{End}_K V$ but D itself is only a vector subspace (unless D is commutative, of course). Thus

$$A \overset{\rho}{\cong} \mathrm{End}_D V \cong \mathrm{M}(n, D)$$

and D is separable over K (since KG is – and therefore also A and D – by Maschke's Theorem).

Theorem 3.2.8 *(a) A is stable under the standard $(K, {}^{-})$-involution ${}^{-}$ of KG, and*

$$\rho : (A, {}^{-}) \to (\mathrm{End}_K V, {}^{\sim})$$

is a homomorphism of $(K, {}^{-})$-involution algebras (where ${}^{-}$ also denotes the restriction of the standard involution on KG to A), and so $\rho(A) = \mathrm{End}_D V$ is stable under ${}^{\sim}$ and

$$\rho : (A, {}^{-}) \to (\mathrm{End}_D V, {}^{\sim})$$

is an isomorphism of $(K, {}^{-})$-involution algebras.

(b) D is stable under ${}^{\sim}$, so h is (right) $(D, {}^{\sim})$-invariant.

(c) Let $h_0 : V \times V \to (D, {}^{\sim})$ be the unique ε-reflexive form such that $h = \mathrm{trd}_{D/K} \circ h_0$ (cf. Cor. 3.2.1). Then the adjoint involution adj_{h_0} with respect to h_0 on $\mathrm{End}_D V = \rho A$ is the restriction of $\mathrm{adj}_h = {}^{\sim}$.

(d) Choose a basis of V over D, and use it to identify $\mathrm{End}_D V$ with $\mathrm{M}(n, D)$. If $a \in \mathrm{M}(n, D)$, then

$$\widetilde{a} = \underline{h}_0^{-1} a^{t\sim} \underline{h}_0, \tag{3.35}$$

where $\underline{h}_0$ is the matrix of h_0 with respect to the same basis. In particular if the basis is orthonormal with respect to h_0, then

$$\widetilde{a} = a^{t\sim}.$$

(e) The G-invariant forms on V over $(K, {}^{-})$ are the forms $h_d(u, v) = h(ud, v)$, $d \in D$, which are nonsingular if $d \neq 0$. The form h_d is ε-reflexive if and only if $d \in (D, {}^{\sim})_+$ (the symmetric elements of $(D, {}^{\sim})$), $-\varepsilon$-reflexive if and only if $d \in (D, {}^{\sim})_-$.

(f) More generally if V is not necessarily a simple KG-module, then $B := \mathrm{End}_{KG} V$ is stable under ${}^{\sim} = \mathrm{adj}_h$, the G-invariant forms on V are the forms $h_b := h(ub, v)$ where $b \in B$, and h_b is

(i) ε-reflexive if and only if $b \in (B, {}^{\sim})_+$,
(ii) $-\varepsilon$-reflexive if and only if $b \in (B, {}^{\sim})_-$, and
(iii) nonsingular if and only if $b \in B^\times$.

Remarks. 1. The following diagrams show the relationships described in the theorem.

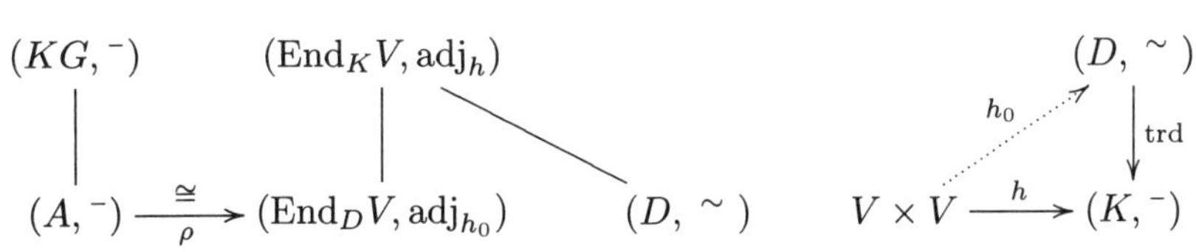

In the left diagram the lines indicate containment of involution subalgebras[2], and the arrow an isomorphism of involution algebras.

2. If A is any simple summand of KG which is stable under the standard involution, we know by Theorem 3.2.7 that, for at least one choice of $\varepsilon = \pm 1$, there *is* an ε-reflexive form h over $(K, ^-)$ on a simple A-module V such that the associated representation $\rho : G \to \mathbf{GL}_K V$ is an isometry representation with respect to h. Thus by (a) and (c), the involution on A "is" the adjoint of the ε-Hermitian form h_0. □

The first statement in (a) is a consequence of the equivalence of (ii) and (iii) in Th. 3.2.7(a). The fact that h is invariant under $(KG, ^-)$ means that it is invariant under $(A, ^-)$ – which is equivalent to $(A, ^-) \to (\mathrm{End}_K V, \mathrm{adj}_h)$ being a homomorphism of involution algebras. This proves (a), and (b) follows at once since D is the centralizer of ρA in $\mathrm{End}_K V$.

(c) For all $d \in D$ and $a \in A$,

$$0 = h(au, vd) - h(u, \widetilde{a}vd) = \mathrm{trd}_{D/K}((h_0(au, v) - h_0(u, \widetilde{a}v))d),$$

so $h_0(au, v) = h_0(u, \widetilde{a}v)$ since D is a separable division K-algebra.

(d) In terms of matrices, $h_0(\underline{u}, \underline{v}) = \underline{u}^{t\sim}\underline{h}_0\underline{v}$ for all $\underline{u}, \underline{v} \in D^{(\dim_D V)\times 1}$ and the adjoint equation $h_0(au, v) = h_0(u, \widetilde{a}v)$ is

$$\underline{u}^{t\sim} a^{t\sim} \underline{h}_0 \underline{v} = \underline{u}^{t\sim} \underline{h}_0 \widetilde{a} \underline{v}$$

which implies (3.35).

(e) follows immediately from (c) and (d) of Prop. 3.1.1, p. 94, taking into account that $D = \mathrm{End}_{KG} V = \mathrm{End}_A V$.

(f) If $b \in B$ and $a \in KG$,

$$h(u, \widetilde{b}av) = h(bu, av) = h(\bar{a}bu, v) = h(b\bar{a}u, v) = h(u, a\widetilde{b}v),$$

so $\widetilde{b} \in B$ and $\widetilde{B} = B$. The G-invariant forms on V are of the form h_b by Prop. 3.1.1(c) since $\mathrm{End}_{(K,KG)} V = B$, and then (i), (ii) and (iii) follow easily. □

3.2.7 Forms over simple algebras in terms of matrices. It is important for later use to be able express sesquilinear forms over a simple K-algebra $(A, ^-)$ with $(K, ^-)$-involution in terms of matrices as one does in the "classical" case over a field (cf. p. 44), and we will see how this can be done in this section.

We begin by identifying A with a matrix algebra, $A = \mathrm{M}(n, D)$, for some division algebra D. By Th. 1.3.7(c), p. 34 (applied to the center of A), D also has a $(K, ^-)$-involution $^-$ which agrees on its center with the involution on the center of A; if it is of the first kind and D is a quaternion algebra, we assume that it is conjugation *. By Th. 1.3.6,

$$\bar{a} = \underline{h}_0^{-1} a^{t-} \underline{h}_0$$

for some $\underline{h}_0 \in \mathbf{GL}(n, D)$ such that $\underline{h}_0^{t-} = \varepsilon_0 \underline{h}_0$ where $\varepsilon_0 = \pm 1$, and $= 1$ if and only if $(A, ^-)$ and $(D, ^-)$ have the same type.

A (left) simple A-module is isomorphic to the column vectors $D^{n\times 1}$ of length n, with A acting by left multiplication. Since any (finitely generated, as always)

[2]Strictly speaking, $(A, ^-)$ is not a subalgebra of $(KG, ^-)$ in the usual sense, since in general $1_A \neq 1_{KG}$.

left A-module is a direct sum of (say) m simple A-modules, it is isomorphic to the A-module $D^{n\times m}$ of n by m matrices.

Lemma 3.2.3 *(a) The map* $D^{m\times k} \to \mathrm{Hom}_A(D^{n\times m}, D^{n\times k})$ *given by right multiplication, is an isomorphism of* $(\mathrm{M}(m,D),\mathrm{M}(k,D))$*-bimodules.*
(b) Multiplication $D^{m\times n}\otimes_A D^{n\times k}\to D^{m\times k}$ *is an isomorphism of* $(\mathrm{M}(m,D),\mathrm{M}(k,D))$*-bimodules.*

(a) follows from Prop. 1.1.1, p. 3, with $M = D^{n\times 1}$.
(b) Every element of $D^{1\times n}\otimes_A D^{n\times 1}$ is of the form $(1,0,\ldots,0)\otimes(d_1,d_2,\ldots,d_n)^t$ where $d_i\in D$. The vector $(1,0,\ldots,0)$ is fixed under right multiplication by the matrix unit $e_{11}\in A$; since $e_{11}(d_1,d_2,\ldots,d_n)^t=(d_1,0,\ldots,0)^t$, it follows that every element of $D^{1\times n}\otimes_A D^{n\times 1}$ is of the form $(1,0,\ldots,0)\otimes(1,0,\ldots,0)^t d$ for some $d\in D$, and so $D^{1\times n}\otimes_A D^{n\times 1}\to D$ is an isomorphism. Then (b) follows from the biadditivity of $M\otimes_A N$ as a functor of M and N and the obvious fact that it is a homomorphism of bimodules. □

If $A\cong \mathrm{M}(n,D)$

Corollary 3.2.3 *If D is replaced by a matrix ring* $B=\mathrm{M}(l,D)$*, then (a) holds: the map* $B^{m\times k}\to\mathrm{Hom}_A(B^{n\times m},B^{n\times k})$ *given by right multiplication, is an isomorphism of* $(\mathrm{M}(m,B),\mathrm{M}(k,B))$*-bimodules.*

This follows easily from the lemma using the fact that matrices over $\mathrm{M}(l,D)$ are matrices over D with l-fold larger dimensions and that this correspondence preserves multiplication. □

Now suppose that $V=D^{n\times m}$. By the foregoing lemma, $V^*=\mathrm{Hom}_A(V,A)=D^{m\times n}$. Let $f: V\times V\to(A,{}^-)$ be a sesquilinear form, and consider its right adjoint $f_r: V\to\overline{V^*}$, $f_r(v)=f(\,\cdot\,,v)$. If we follow this map by the map $\overline{V^*}\to V$ given by $x\to \underline{h_0}^{-1}x^{t-}$, the result is an A-linear homomorphism $V\to V$, and so is given by right multiplication by a matrix $b\in D^{m\times m}$. It follows that $f(\cdot,v)=\varepsilon_0 b^{t-}v^{t-}\underline{h_0}$, and so if we set $\varepsilon_0 b^{t-}=\underline{f}\in\mathrm{M}(m,D)$, we get the important formula

$$f(u,v)=u\underline{f}v^{t-}\underline{h_0}\quad(u,v\in D^{n\times m}).\tag{3.36}$$

Thus $\underline{f}$ is the generalization of the notion of discriminant matrix of a sesquilinear form over a field. Naturally we call $\underline{f}$ the *(discriminant) matrix of f with respect to the given matrix realizations*[3] $A=\mathrm{M}(n,D)$ *and* $V=D^{n\times m}$, and

$$\mathrm{nrd}_{\mathrm{M}(m,D)/K}\underline{f}=:\det f\in K$$

the *determinant* of f. As in the "classical" case, f is nonsingular if and only if $\underline{f}$ is an invertible matrix, if and only if $\det f$ is nonzero.

The form f (with discriminant matrix $\underline{f}$) is ε-Hermitian if and only if, for all u and v, $\varepsilon\overline{f(v,u)}=f(u,v)$, i.e.

$$\varepsilon\underline{h_0}^{-1}(v\underline{f}u^{t-}\underline{h_0})^{t-}\underline{h_0}=\varepsilon\underline{h_0}^{-1}\underline{h_0}^{t-}u\underline{f}^{t-}v^{t-}\underline{h_0}=\varepsilon_0\varepsilon u\underline{f}^{t-}v^{t-}\underline{h_0}=u\underline{f}v^{t-}\underline{h_0}$$

since $\underline{h_0}^{t-}=\varepsilon_0 h_0$. This is equivalent to $\underline{f}=\varepsilon_0\varepsilon\underline{f}^{t-}$ i.e. $\underline{f}$ is an $\varepsilon_0\varepsilon$-Hermitian matrix with respect to the involution $b\to b^{t-}$ of $\mathrm{M}(m,D)$. Equivalently $\underline{f}\in\mathrm{M}(m,(D,{}^-))_{\varepsilon_0\varepsilon}$.

[3]The dependence of the discriminant matrix and determinant on the matrix realizations is discussed on pp. 122*ff*.

Now we consider the adjoint involution on $\mathrm{End}_A V$ with respect to a nonsingular ε-Hermitian form f: if $\varphi \in \mathrm{End}_A V$ is given by the matrix $\underline{\varphi} \in \mathrm{M}(m, D)$ and that of its adjoint $\mathrm{ad}_f = \widetilde{\varphi}$ by $\underline{\widetilde{\varphi}}$,

$$\begin{aligned} f(u\varphi, v) &= u\underline{\varphi}\,\underline{f}v^{t-}\underline{h}_0 \\ &= f(u, v\widetilde{\varphi}) \\ &= u\underline{f}(v\underline{\widetilde{\varphi}})^{t-}\underline{h}_0 = u\underline{f}\,\underline{\widetilde{\varphi}}^{t-}v^{t-}\underline{h}_0, \end{aligned}$$

which implies that $\underline{\varphi}\,\underline{f} = \underline{f}\,\underline{\widetilde{\varphi}}^{t-}$, i.e.

$$\underline{\widetilde{\varphi}} = (\underline{f}^{-1}\underline{\varphi}\,\underline{f})^{t-} = \underline{f}^{t-}\underline{\varphi}^{t-}(\underline{f}^{t-})^{-1} = \underline{f}\,\underline{\varphi}^{t-}\underline{f}^{-1}.$$

In other words, the adjoint involution on $\mathrm{M}(m, D)$ with respect to f is

$$\underline{\varphi} \to \underline{f}\,\underline{\varphi}^{t-}\underline{f}^{-1}. \tag{3.37}$$

Theorem 3.2.9 *Let $(A, ^-)$, $A = \mathrm{M}(n, D)$, be a simple algebra with a $(K, ^-)$-involution. Let $^-$ also denote a $(K, ^-)$-involution on D (cf. Th. 1.3.3). Then there is a matrix $\underline{h}_0 \in \mathbf{GL}(n, D)$ such that*

$$\bar{a} = \underline{h}_0^{-1}a^{t-}\underline{h}_0 \text{ for all } a \in A$$

where

$$\underline{h}_0^{t-} = \varepsilon_0\underline{h}_0, \quad \varepsilon_0 = \pm 1,$$
$$\underline{h}_0 = 1 \quad \text{if } n = 1,$$

and

$$\varepsilon_0 = 1 \text{ if and only if } (A, ^-) \text{ and } (D, ^-) \text{ have the same type.} \tag{3.38}$$

A sesquilinear form

$$f : D^{n\times m} \times D^{n\times m} \to (\mathrm{M}(n, D), ^-)$$

is of the form

$$f(u, v) = u\underline{f}v^{t-}\underline{h}_0$$

for a suitable matrix $\underline{f} \in \mathrm{M}(m, D)$. Furthermore f is a nonsingular form if and only if $\underline{f}$ is an invertible matrix, if and only if $\det f = \mathrm{nrd}_{A/K}\underline{f} \neq 0$.

The form f is ε-Hermitian if and only if $\underline{f}^{t-} = \varepsilon\varepsilon_0\underline{f}$. In this case the adjoint involution on $\mathrm{End}_A V = \mathrm{M}(m, D)$ with respect to f has type

$$\varepsilon(\text{type of } (A, ^-)) = \varepsilon\varepsilon_0(\text{type of } (D, ^-)), \tag{3.39}$$

and is given by

$$\underline{\widetilde{\varphi}} = \underline{f}\,\underline{\varphi}^{t-}\underline{f}^{-1}. \tag{3.40}$$

Furthermore f is hyperbolic if and only if it is equivalent to a form $g : D^{n\times m} \times D^{n\times m} \to (\mathrm{M}(n, D), ^-)$ with discriminant matrix of the form

$$\begin{pmatrix} 0 & M \\ \varepsilon\varepsilon_0 M^{t-} & 0 \end{pmatrix}. \tag{3.41}$$

REMARKS. The stipulation that $\underline{h}_0 = 1$ when $n = 1$ is made to ensure that the definition of $\det f$ given here is consistent with that given earlier on p. 45.

By identifying the space $D^{n\times m}$ of f with the space $D^{n\times m}$ of g by an equivalence isomorphism in the last statement of the theorem, one can also say that by choosing a new set of "column simple modules $D^{n\times 1}$" in the space $D^{n\times m}$ of

f, the discriminant matrix of f is given by (3.41). When $n = 1$, this is of course tantamount to choosing a new basis for the space of f. □

Most of the theorem has already been proved. (3.38) follows from Th. 1.3.6, p. 30, and Lem. 1.3.2, p. 28; (3.39) follows from them and (3.40).

Suppose f is hyperbolic. By definition of a hyperbolic form, $D^{n\times m} = U \oplus W$ with U and W totally isotropic and $\operatorname{len} U = q = \operatorname{len} W$ $(= \frac{m}{2})$, so there are isomorphisms $D^{n\times q} \to U$ and $D^{n\times q} \to W$ of A-modules, giving rise to an automorphism

$$\varphi : \left(D^{n\times q}\middle|D^{n\times q}\right) \to D^{n\times m}$$

of $D^{n\times m}$ where $\left(D^{n\times q}\middle|D^{n\times q}\right)$ represents $D^{n\times m}$ partitioned in the obvious way. The ϵ-Hermitian form

$$g(u,v) := f(\varphi u, \varphi v), \quad u,v \in \left(D^{n\times q}\middle|D^{n\times q}\right)$$

is obviously equivalent to f, and $g(u,v) = 0$ if $u,v \in \left(D^{n\times q}\middle|\,0^{n\times q}\right)$ or if $u,v \in \left(0^{n\times q}\middle|D^{n\times q}\right)$. Suppose the discriminant matrix of g is

$$\underline{g} = \begin{pmatrix} E & F \\ G & H \end{pmatrix}$$

where the matrices E, F, G, H are in $D^{q\times q}$, and let

$$u,v \in \left(D^{n\times q}\middle|\,0^{n\times q}\right), \quad u = (B,0), \quad v = (C,0).$$

Then

$$g(u,v) = (B,0)\begin{pmatrix} E & F \\ G & H \end{pmatrix}\begin{pmatrix} C^{t-} \\ 0 \end{pmatrix}\underline{h}_0 = (BEC^{t-})\underline{h}_0 = 0 \in \mathrm{M}(n,D).$$

Since this holds for all B and C, $E = 0$. Similarly $H = 0$, so $\underline{g}$ has the form in (3.41) since $\underline{g}^{t-} = \varepsilon\varepsilon_0\underline{g}$. Conversely if $\underline{g}$ has this form, then certainly g – and so also f – is hyperbolic by Th. 3.1.2, p. 97. □

By way of an example, we illustrate this approach under "scalar extension" of a form over a division ring with involution.

Example. We begin with a quaternion algebra $D = (\alpha\,,\,\beta)_K$, and its standard involution * (conjugation). Let L be the splitting field $K(\sqrt{\alpha})$ of D. Then $L \otimes_K D \cong \mathrm{M}(2,L)$ via the isomorphism defined by

$$i \to \begin{pmatrix} \sqrt{\alpha} & 0 \\ 0 & -\sqrt{\alpha} \end{pmatrix} \quad \text{and} \quad j \to \begin{pmatrix} 0 & \beta \\ 1 & 0 \end{pmatrix}.$$

This is an isomorphism $(L\otimes D, \mathrm{id}_L \otimes {}^*) \cong (\mathrm{M}(2,L), {}^-)$ of involution algebras where, if $X \in \mathrm{M}(2,L)$,

$$\bar{X} = J^{-1}X^tJ, \qquad J = \begin{pmatrix} 0 & -1 \\ 1 & 0 \end{pmatrix}.$$

We identify the two involution algebras by this isomorphism.

Now consider the Hermitian form $h = \langle\gamma,\delta\rangle : D^2 \times D^2 \to (D, {}^*)$,

$$h((x',y'),(x,y)) = x'\gamma x^* + y'\delta y^*$$

where $\gamma, \delta \in \dot{K}$. In terms of its discriminant matrix

$$h((x',y'),(x,y)) = (x',y')\begin{pmatrix} \gamma & 0 \\ 0 & \delta \end{pmatrix}\begin{pmatrix} x^* \\ y^* \end{pmatrix} = (x',y')\begin{pmatrix} \gamma & 0 \\ 0 & \delta \end{pmatrix}(x,y)^{t*}.$$

Now extend the scalars to L, to get the Hermitian form

$$h^L((x',y'),(x,y)) = (x',y')\begin{pmatrix} \gamma & 0 & 0 & 0 \\ 0 & \gamma & 0 & 0 \\ 0 & 0 & \delta & 0 \\ 0 & 0 & 0 & \delta \end{pmatrix}(x,y)^{t*} \in L\otimes D = \mathrm{M}(2,L)$$

where now $x',y',x,y \in \mathrm{M}(2,L)$, so $(x',y'),(x,y) \in L^{2\times 4}$. Since

$$(x,y)^{t*} = \begin{pmatrix} J^{-1}x^tJ \\ J^{-1}y^tJ \end{pmatrix} = \begin{pmatrix} J^{-1} & 0 \\ 0 & J^{-1} \end{pmatrix}\begin{pmatrix} x^t \\ y^t \end{pmatrix} J,$$

$$h^L((x',y'),(x,y)) = (x',y')\begin{pmatrix} \gamma & 0 & 0 & 0 \\ 0 & \gamma & 0 & 0 \\ 0 & 0 & \delta & 0 \\ 0 & 0 & 0 & \delta \end{pmatrix}\begin{pmatrix} J^{-1} & 0 \\ 0 & J^{-1} \end{pmatrix}\begin{pmatrix} x^t \\ y^t \end{pmatrix} J.$$

Thus the discriminant matrix of h^L is

$$\underline{h} = \begin{pmatrix} \gamma & 0 & 0 & 0 \\ 0 & \gamma & 0 & 0 \\ 0 & 0 & \delta & 0 \\ 0 & 0 & 0 & \delta \end{pmatrix}\begin{pmatrix} J^{-1} & 0 \\ 0 & J^{-1} \end{pmatrix} = \begin{pmatrix} 0 & \gamma & 0 & 0 \\ -\gamma & 0 & 0 & 0 \\ 0 & 0 & 0 & \delta \\ 0 & 0 & -\delta & 0 \end{pmatrix},$$

and

$$\underline{h}_0 = J.$$

Of course $\underline{h}_0$ is skew symmetric (so $\varepsilon_0 = -1$), and so is $\underline{h}$. These facts follow also from earlier theorems. First of all by Th. 1.3.1, conjugation on D is symplectic, so by Th. 1.3.6, (c), p. 30, and the fact that $(\mathrm{M}(2,L),\ ^t\ )$ is orthogonal, $\underline{h}_0$ must be a skew symmetric matrix, i.e. $\varepsilon_0 = -1$. On the other hand since h^L is Hermitian, $\varepsilon = 1$ and so $\varepsilon\varepsilon_0 = -1$, which means that $\underline{h}$ also must be skew symmetric.

This example is continued on p. 133. □

Dependence of the discriminant matrix on identifications. The definition of the discriminant matrix depends on certain identifications. The given data are a central[4] simple algebra A and its involution $^-$, along with an ε-Hermitian form $h: V\times V \to (A,^-)$. The identifications made in the above computations are $A = \mathrm{M}(n,D)$ and $V = D^{n\times m}$. There is also the question of the involution chosen on D.

We now look into the effect of these choices on the discriminant matrix $\underline{f}$ of a sesquilinear form f. In particular we wish to show the extent to which $\underline{f}$ is independent of the identifications.

We begin by considering the identifications $A = \mathrm{M}(n,D)$ and $V = D^{n\times m}$. Suppose they are based on the isomorphisms $\varphi_1 : A \to \mathrm{M}(n,D)$ and $\varphi_1' : V \to D^{n\times m}$, and that we have a second identification based on φ_2 and φ_2'. Each φ_i is

[4] If A is not central, we replace K by its center.

an isomorphism of K-algebras, and φ_i' is an isomorphism of Abelian groups (or K-vector spaces) satisfying $\varphi_i'(av) = \varphi_i(a)\varphi_i'v$. Thus the diagram

$$\begin{CD}
\mathrm{M}(n,D) \times D^{n\times m} @<(\varphi_2,\varphi_2')<< A \times V @>(\varphi_1,\varphi_1')>> \mathrm{M}(n,D) \times D^{n\times m}\\
@V\text{mult.}VV @VVV @VV\text{mult.}V\\
D^{n\times m} @<\varphi_2'<< V @>\varphi_1'>> D^{n\times m}
\end{CD}$$

is commutative.

Now $\varphi_1\varphi_2^{-1}$ is an automorphism of the central simple K-algebra $\mathrm{M}(n,D)$ and so is an inner automorphism inn_{a_0} by the Skolem-Noether Theorem. Let $\varphi' = \varphi_1'\varphi_2'^{-1}$. Then we have the commutative diagram

$$\begin{CD}
\mathrm{M}(n,D) \times D^{n\times m} @>(\mathrm{inn}_{a_0},\varphi')>> \mathrm{M}(n,D) \times D^{n\times m}\\
@VVV @VVV\\
D^{n\times m} @>>\varphi'> D^{n\times m}
\end{CD}$$

so if $a \in \mathrm{M}(n,D)$ and $v \in D^{n\times m}$,

$$\varphi'(av) = a_0 a a_0^{-1}\varphi'(v), \quad \text{i.e.} \quad a_0^{-1}\varphi'(av) = a a_0^{-1}\varphi'(v).$$

Therefore $a_0^{-1}\varphi'$ is an $\mathrm{M}(n,D)$-;automorphism of $D^{n\times m}$, and so is of the form $a_0^{-1}\varphi'(v) = vc_0$ for some $c_0 \in \mathbf{GL}(m,D)$. Thus

$$\varphi'(v) = a_0 v c_0 \quad \text{for all } v \in D^{n\times m}.$$

Conversely if a_0 and c_0 are arbitrary invertible matrices (of the appropriate size) and φ' is defined by this formula, then the above diagram is commutative, and so corresponds to a "difference" between identifications of V and A with $D^{n\times m}$ and $\mathrm{M}(n,D)$ respectively.

Since we are assuming that the involution $^-$ on D is of the same kind as that on A, for each $i = 1, 2$, there is a matrix $\underline{h}_{0i}$ such that

$$\varphi_i : (A, ^-) \to \left(\mathrm{M}(n,D),\ \underline{a} \to \underline{h}_{0i}^{-1}\underline{a}^{t-}\underline{h}_{0i}\right)$$

is an isomorphism of $(K, ^-)$-involution algebras. The two involutions on $\mathrm{M}(n,D)$ have the same type since they have the same type as that on A and so by Th. 1.3.6, p. 30, there is $\varepsilon_0 = \pm 1$ such that

$$\underline{h}_{01}^{t-} = \varepsilon_0\underline{h}_{01} \quad \text{and} \quad \underline{h}_{02}^{t-} = \varepsilon_0\underline{h}_{02}. \tag{3.42}$$

Let $\underline{f}_1$ be the discriminant matrix of f with respect to the first identification (φ_1, φ_1') and $\underline{f}_2$ the discriminant matrix with respect to the second. They are, by definition, the discriminant matrices of the sesquilinear forms f_1 and f_2 on $D^{n\times m}$ which make the diagram

$$\begin{CD}
D^{n\times m} \times D^{n\times m} @<(\varphi_2',\varphi_2')<< V \times V @>(\varphi_1',\varphi_1')>> D^{n\times m} \times D^{n\times m}\\
@VV f_2 V @VV f V @VV f_1 V\\
\mathrm{M}(n,D) @<\varphi_2<< A @>\varphi_1>> \mathrm{M}(n,D)
\end{CD} \tag{3.43}$$

commutative:

$$f_i(u,v) = u\underline{f}_i v^{t-}\underline{h}_{0i} \quad \text{for} \quad i = 1, 2.$$

The "difference" in the above diagram gives the commutative diagram

$$\begin{array}{ccc} D^{n\times m} \times D^{n\times m} & \xrightarrow{(\varphi',\varphi')} & D^{n\times m} \times D^{n\times m} \\ \downarrow{\scriptstyle f_2} & & \downarrow{\scriptstyle f_1} \\ \mathrm{M}(n,D) & \xrightarrow{\mathrm{inn}_{a_0}} & \mathrm{M}(n,D), \end{array}$$

so

$$a_0 u\underline{f}_2 v^{t-}\underline{h}_{02}a_0^{-1} = a_0 uc_0\underline{f}_1(a_0vc_0)^{t-}\underline{h}_{01} = a_0uc_0\underline{f}_1 c_0^{t-}v^{t-}a_0^{t-}\underline{h}_{01}$$

for all u and v in $D^{n\times m}$, which implies that

$$u\underline{f}_2 v^{t-}\underline{h}_{02}a_0^{-1} = uc_0\underline{f}_1 c_0^{t-}v^{t-}a_0^{t-}\underline{h}_{01} \text{ for all } u, v \in D^{n\times m}.$$

By Lem. 3.2.3(a),

$$\underline{f}_2 v^{t-}\underline{h}_{02}a_0^{-1} = c_0\underline{f}_1 c_0^{t-}v^{t-}a_0^{t-}\underline{h}_{01} \text{ for all } v \in D^{n\times m}$$

and, on replacing v^{t-} by w,

$$cw = wb \quad \text{for all } w \in D^{m\times n}$$

where

$$c = (c_0\underline{f}_1 c_0^{t-})^{-1}\underline{f}_2 \in \mathbf{GL}(m,D) \text{ and } b = a_0^{t-}\underline{h}_{01}a_0\underline{h}_{02}^{-1} \in \mathbf{GL}(n,D).$$

This means, for example, that left multiplication by c is an $\mathrm{M}(m,D)$-endomorphism of $D^{m\times n}$, and so c must be in the center $\dot{K}$ of $\mathbf{GL}(m,D)$. Thus $c = b \in \dot{K}$, say $= \alpha$, and

$$\underline{f}_2 = \alpha c_0\underline{f}_1 c_0^{t-}, \quad \underline{h}_{02} = \alpha^{-1}a_0^{t-}\underline{h}_{01}a_0. \tag{3.44}$$

If we apply ${}^{t-}$ to the second relation, we see that

$$\alpha \in K_0$$

by (3.42).

For the moment, suppose that we use the same identification $A = \mathrm{M}(n,D)$ in both cases, that is to say that $\varphi_1 = \varphi_2$. Then we can choose $a_0 = 1$ and $\underline{h}_{01} = \underline{h}_{02}$, so $\alpha = b = 1 = c$ and $\underline{f}_2 = c_0\underline{f}_1 c_0^{t-}$ which imply

$$f_1(uc_0, vc_0) = uc_0\underline{f}_1(vc_0)^{t-}\underline{h}_{01} = u\underline{f}_2 v^{t-}\underline{h}_{02} = f_2(u,v),$$

that is to say,

$$f_1 \cong f_2$$

and

$$\det f_2 = \mathrm{n}_{K/K_0}(\mathrm{nrd}_{\mathrm{M}(n,D)/K}c_0)\det f_1$$

since $\mathrm{nrd}\, c_0^{t-} = \overline{\mathrm{nrd}\, c_0}$ by Lem. 1.3.7 (p. 34), and where, if $\beta \in K$,

$$\mathrm{n}_{K/K_0}\beta = \beta\bar{\beta} = \begin{cases} \beta^2 & \text{if } (K, {}^-) \text{ is of the first kind,} \\ \mathrm{N}_{K/K_0}\beta & \text{if } (K, {}^-) \text{ is of the second kind.} \end{cases} \tag{3.45}$$

Thus in this case – that is to say when we fix an identification $A = \mathrm{M}(n,D)$ – the determinant of f is well-defined modulo

$$\mathrm{n}_{K/K_0}(\mathrm{nrd}_{D/K}D^\times) = \begin{cases} (\mathrm{nrd}_{D/K}D^\times)^2 & \text{if } (K, {}^-) \text{ is of the first kind,} \\ \mathrm{nrd}_{D/K_0}D^\times & \text{if } (K, {}^-) \text{ is of the second kind,} \end{cases}$$

since $\mathrm{nrd}_{\mathrm{M}(n,D)/K}\mathrm{M}(n,D) = \mathrm{nrd}_{D/K}D$ – cf. (1.15), p. 14.

On the other hand, if φ_1 is not necessarily $= \varphi_2$, then by (3.44), f_1 and f_2 are "similar" (cf. p. 188):

$$\alpha f_1 \cong f_2 \text{ for some } \alpha \in \dot{K}_0,$$

and the determinant is well-defined up to a factor in

$$\dot{K}_0^{md}\mathrm{n}_{K/K_0}(\mathrm{nrd}_{D/K}\dot{D})$$

where $d^2 = \dim_K D$. But

$$\dot{K}_0^{2d}\mathrm{n}_{K/K_0}(\mathrm{nrd}_{D/K}\dot{D}) = \mathrm{n}_{K/K_0}(\dot{K}_0^{d}\,\mathrm{nrd}_{D/K}\dot{D}) = \mathrm{n}_{K/K_0}(\mathrm{nrd}_{D/K}\dot{D}),$$

so the determinant is well-defined up to a factor in

$$\begin{array}{ll} \dot{K}_0^{d}\mathrm{n}_{K/K_0}(\mathrm{nrd}_{D/K}\dot{D}) & \text{if } m \text{ is odd,} \\ \mathrm{n}_{K/K_0}(\mathrm{nrd}_{D/K}\dot{D}) & \text{if } m \text{ is even.} \end{array}$$

(In particular, if $D = K$ and m is odd, the determinant carries no information.)

Theorem 3.2.10 *Suppose that $A \cong \mathrm{M}(n,D)$ has a $(K,^-)$-involution, and that a $(K,^-)$-involution on D is given. Let V be an A-module of length m.*

(i) The discriminant matrix $\underline{f} \in B = \mathrm{M}(m,D)$ of a sesquilinear form $f : V \times V \to (A,^-)$ is well-defined up to a transformation of the form

$$\underline{f} \rightsquigarrow \alpha c_0 \underline{f} c_0^{t-} \tag{3.46}$$

where $\alpha \in \dot{K}_0$, and $c_0 \in \mathbf{GL}(m,D)$. This is equivalent to saying that the sesquilinear forms $f_1, f_2 : D^{n\times m} \times D^{n\times m} \to \mathrm{M}(n,D)$ belonging to two different identifications (3.43) are similar.

The determinant of the sesquilinear form is well-defined up to a factor in

$$\begin{array}{ll} \dot{K}_0^{d}\,\mathrm{n}_{K/K_0}(\mathrm{nrd}_{D/K}\dot{D}) & \text{if } m \text{ is odd,} \\ \mathrm{n}_{K/K_0}(\mathrm{nrd}_{D/K}\dot{D}) & \text{if } m \text{ is even,} \end{array} \tag{3.47}$$

where $d^2 = \dim_K D$, and n_{K/K_0} is defined in (3.45).

(ii) If the same identification $(A,^-) = \big(\mathrm{M}(n,D),\ a \to \underline{h}_0^{-1}a^{t-}\underline{h}_0\big)$ is used in determining the discriminant matrix via two identifying maps $\varphi_1', \varphi_2' : V \to D^{n\times m}$, the resulting discriminant matrices satisfy $\underline{f}_2 = c_0\underline{f}_1c_0^{t-}$ for some $c_0 \in \mathbf{GL}(m,D)$, the corresponding forms on $D^{n\times m}$ are equivalent, $f_1 \cong f_2$, and the determinant is well-defined up to a factor in $\mathrm{n}_{K/K_0}(\mathrm{nrd}_{D/K}D^\times)$. □

REMARKS. 1. When $n = 1$ – that is to say, $A = D$ – the factor α in (3.46) and the factor $\dot{K}_0^d$ in (3.47) can be omitted if one takes the involution on D to be the given one on A – then the factor $\underline{h}_0$, which is only defined up to an element in $\dot{K}_0$, does not appear in the formula (3.36) for $\underline{f}$. This done, the definition of $\det f$ agrees with that on p. 45.

2. In practice, the assumption that the involution on D is known is often valid. This is the case if D is a field, for it then is the center of A and so inherits an involution from A. If D is a quaternion algebra and the involution is of the first kind, we will always choose conjugation as the involution on D. □

In any case we now consider the effect (on the discriminant matrix and determinant of an ε-Hermitian form h) of the choice of the involution on D. This will not be used elsewhere in this book.

Because of Remark 2 above, we will assume that D is a noncommutative central division algebra over K (with an involution $^-$ of course).

We assume as before that A and D have $(K, ^-)$-involutions, both denoted by $^-$, and that with some identification $A = \mathrm{M}(n, D)$, the given involution on A is $\bar{a} = \underline{h}_0^{-1} a^{t-} \underline{h}_0$, where $\underline{h}_0^{t-} = \varepsilon_0 \underline{h}_0$, $\varepsilon_0 = \pm 1, = 1$ if $^-$ is of the second kind.

Let $d \to \widetilde{d}$ be another $(K, ^-)$-involution on D, $\bar{a} = \underline{h}_1^{-1} a^{t\sim} \underline{h}_1$ for some $\underline{h}_1 \in A^\times$ such that $\underline{h}_1^{t\sim} = \varepsilon_1 \underline{h}_1$, $\varepsilon_1 = \pm 1$ and $= 1$ if $^-$ is of the second kind.

We shall also assume that $(D, ^\sim)$ has the same type as $(D, ^-)$; this implies that $\varepsilon_1 = \varepsilon_0$ (Th. 1.3.6, p. 30) and, if $d_1 \bar{d} d_1^{-1} = \widetilde{d}$ for all $d \in D$, that $\bar{d}_1 = d_1 = \widetilde{d}_1$.

Now if b is any matrix with entries in D,[5]

$$d_1^{-1} b^{t\sim} d_1 = b^{t-}.$$

If $a \in \mathrm{M}(n, D)$,

$$(d_1 \underline{h}_0)^{-1} a^{t\sim} (d_1 \underline{h}_0) = \underline{h}_0^{-1} a^{t-} \underline{h}_0 = \bar{a} = \underline{h}_1^{-1} a^{t\sim} \underline{h}_1,$$

so $\underline{h}_1 = \beta d_1 \underline{h}_0$ for some $\beta \in \dot{K}$, and it is easy to check that $\beta \in K_0$. Therefore by replacing d_1 by βd_1, we can assume that

$$\underline{h}_1 = d_1 \underline{h}_0.$$

Suppose that $f : D^{n\times m} \times D^{n\times m} \to (\mathrm{M}(n, D), ^-)$ is a sesquilinear form, so for suitable matrices $\underline{f}$ and $\underline{f}'$,

$$\begin{aligned} f(u,v) = u \underline{f} v^{t-} \underline{h}_0 &= u \underline{f}' v^{t\sim} \underline{h}_1 \\ &= u \underline{f}' d_1 v^{t-} d_1^{-1} d_1 \underline{h}_0 = u \underline{f}' d_1 v^{t-} \underline{h}_0. \end{aligned}$$

Thus the two discriminant matrices $\underline{f}$ and $\underline{f}'$ are related by

$$\underline{f} = \underline{f}' d_1,$$

and the determinant itself changes by the factor $(\mathrm{nrd}_{D/K} d_1)^m$.

Combining this with the previous theorem and using the same notation and the assumption that different choices of the involution on D are of the same type, we get (since α can also be absorbed into d_1)

Theorem 3.2.11 *Suppose that D is noncommutative. The discriminant matrix of a sesquilinear form $f : V \times V \to A \cong \mathrm{M}(n, D)$ is well-defined up to a transformation of $\mathrm{M}(m, D)$ of the form*

$$\underline{f} \rightsquigarrow c_0 \underline{f} d_1 c_0^{t-}$$

where m is the length of V as an A-module, $c_0 \in \mathbf{GL}(m, D)$ and

$$d_1 \in \dot{D}_+^{(-)},$$

where $^-$ is an arbitrary $(K, ^-)$-involution on D of a given type. The determinant of f is well-defined up to a factor in

$$(\mathrm{nrd}_{D/K} \dot{D}_+^{(-)})^m (\mathrm{n}_{K/K_0} \mathrm{nrd}_{D/K} \dot{D}).$$

which is

$$\begin{array}{ll} (\mathrm{nrd}_{D/K} \dot{D}_+^{(-)})(\mathrm{n}_{K/K_0} \mathrm{nrd}_{D/K} \dot{D}) & \text{if } m \text{ is odd,} \\ \mathrm{n}_{K/K_0} \mathrm{nrd}_{D/K} \dot{D} & \text{if } m \text{ is even.} \end{array}$$

[5] The left side, written as a matrix product, is $\mathrm{diag}(d_1^{-1}, \ldots, d_1^{-1}) b^{t\sim} \mathrm{diag}(d_1, \ldots, d_1)$.

Most of this has already been proved, and the remaining details are left to the reader. Note that, in comparing the factors up to which the determinant is defined in this theorem and the previous one, $\dot{K}_0^d \subset \mathrm{nrd}_{D/K}\dot{D}_+^{(-)}$. □

3.2.8 Transfers of forms over simple algebras. We continue to assume that $(D, ^-)$ is a division algebra with a $(K, ^-)$-involution, and that $A = \mathrm{M}(n, D)$ with center L and $(K, ^-)$-involution

$$\bar{a} = \underline{h}_0^{-1} a^{t-} \underline{h}_0, \quad \underline{h}_0^{t-} = \varepsilon_0 \underline{h}_0, \tag{3.48}$$

$$\varepsilon_0 = \begin{cases} \pm 1 & \text{if } (A, ^-) \text{ is symplectic or orthogonal,} \\ 1 & \text{if } (A, ^-) \text{ is unitary} \end{cases}$$

If the involution on A is of the first kind and D is a quaternion algebra, we choose $^-$ on D to be conjugation *.

Recall that if $a = (d_{ij}) \in A$,

$$\mathrm{trd}_{A/K}(a) = \sum_i \mathrm{trd}_{D/K}(d_{ii}) = \mathrm{trd}_{D/K} \sum_i d_{ii}$$

(cf. (1.13), p. 14). Define $\mathrm{trd}_{A/D}(a) = \sum_i d_{ii}$. Unless $D = L$, this depends on the choice of the identification $A = \mathrm{M}(n, D)$, but is nevertheless useful. The above formula becomes

$$\mathrm{trd}_{A/K} = \mathrm{trd}_{D/K} \circ \mathrm{trd}_{A/D}.$$

We shall assume for the rest of this section that A *is separable over* K as well as simple.

If $f : V \times V \to (A, ^-)$ is a sesquilinear form, the *transfer of* f (from A to K),

$$\mathrm{trd}_{A/K} \circ f : V \times V \to (K, ^-),$$

is a sesquilinear form over $(K, ^-)$ which is

(a) $(A, ^-)$-invariant,

(b) nonsingular if and only if f is nonsingular,

(c) ε-reflexive if and only if f is ε-Hermitian.

(cf. Th. 3.2.4, p. 113).

We will use the discriminant matrix to compute the (equivalence class of the) transfer of a nonsingular ε-Hermitian form $h : V \times V \to (A, ^-)$.

We can assume that $V = D^{n\times m}$, so that

$$h(u, v) = u\underline{h}v^{t-}\underline{h}_0$$

where $\underline{h} \in \mathrm{M}(m, D)$ is the discriminant matrix of h (Th. 3.2.9, p. 120), say $\underline{h} = (d_{ij})$. If $u = (u_{ij}), v = (v_{ij})$, and $\underline{h}_0 = (d_{0ij})$, the (i, i) entry of $h(u, v)$ is

$$h(u, v)_{ii} = \sum_{j,k,l} u_{ij} d_{jk} \bar{v}_{lk} d_{0li}, \tag{3.49}$$

and so

$$\mathrm{trd}_{A/K} h(u, v) = \sum_{i,j,k,l} \mathrm{trd}_{D/K}(u_{ij} d_{jk} \bar{v}_{lk} d_{0li}). \tag{3.50}$$

Now suppose that $\{e_{pq}\}$ is the standard basis (over D) of matrix units of $D^{n\times m}$. Then

$$\mathrm{trd}_{A/K} h(e_{pq}, e_{rs}) = \mathrm{trd}_{D/K}(d_{qs} d_{0rp}).$$

So if $D = K$, $\mathrm{trd}_{A/K}h(e_{pq}, e_{rs}) = d_{0rp}d_{qs}$. This proves:

Theorem 3.2.12 *Let $A = \mathrm{M}(n, K)$ have the involution (3.48) and let*

$$h : K^{n\times m} \times K^{n\times m} \to (A, ^-), \quad h(u, v) = u\underline{h}v^{t-}\underline{h}_0$$

be an ε-Hermitian form. Then the transfer $\mathrm{trd}_{A/K}h$ has discriminant matrix $\underline{h}_0^t \otimes \underline{h}$ (in a suitable basis), and determinant $(\det \underline{h}_0)^m(\det \underline{h})^n$. □

It is also possible to give an answer for some of the remaining cases (when A is not necessarily split):

Theorem 3.2.13 *(a) If $(K, ^-) = (K, \mathrm{id})$ and $h : V \times V \to (A, ^-)$ is skew Hermitian, the transfer $\mathrm{trd}_{A/K}h$ is skew symmetric, and so is determined, up to equivalence, by its rank $nm(D{:}K)$.*

Assume now that

$$\varepsilon = 1 \text{ if } (K, ^-) = (K, \mathrm{id}). \tag{3.51}$$

(b) If $D = L$, $(L, ^-) = (L, \mathrm{id})$ and $\varepsilon_0 = -1$ (i.e. $(A, ^-)$ is quasisplit and symplectic), then $\mathrm{trd}_{A/K}h$ is a hyperbolic symmetric form, and so is determined, up to equivalence, by its rank $nm(L{:}K)$.

Now assume also that

$$\text{if } D = L, \text{ then } \varepsilon_0 = 1 \text{ (i.e. } (A, ^-) \text{ is not quasisplit symplectic).} \tag{3.52}$$

(c) By choosing another identification of A with $\mathrm{M}(n, D)$, and replacing h by an equivalent form, we can assume that

$$\underline{h}_0 = \mathrm{diag}(d_{01}, \ldots, d_{0n}) \text{ with } \bar{d}_{0i} = \varepsilon_0 d_{0i} \in \dot{D} \text{ for all } i$$

and

$$\underline{h} = \mathrm{diag}(d_1, \ldots, d_m), \ \bar{d}_j = \varepsilon\varepsilon_0 d_j \in \dot{D} \text{ for all } j. \tag{3.53}$$

(d) The ε-reflexive form $\mathrm{trd}_{A/K}h : D^{n\times m} \times D^{n\times m} \to (K, ^-)$ is the orthogonal direct sum of the transfers from D to K of the nm one dimensional ε-Hermitian forms

$$h_{ij} : D \times D \to (D, \,^{\sim_i}), \quad h_{ij}(u, v) = ud_j\bar{v}d_{0i}, \quad 1 \le i \le n, \ 1 \le j \le m,$$

where $^{\sim_i}$ is the involution

$$\overset{\sim_i}{d} = d_{0i}^{-1}\bar{d}d_{0i}.$$

If $D = L$, then $\varepsilon_0 = 1$ and

$$\begin{aligned} \mathrm{trd}_{A/K}h &= \mathrm{Tr}_{L/K}\langle \underline{h}_0 \otimes \underline{h}\rangle \\ &= \mathrm{Tr}_{L/K}\left(\langle \underline{h}_0\rangle\langle \underline{h}\rangle\right), \end{aligned}$$

a symmetric form if $(K, ^-) = (K, \mathrm{id})$, otherwise an ε-Hermitian form over $(K, ^-)$.

(e) Suppose that $\underline{h}_0 = \mathrm{diag}(d_0, \ldots, d_0)$ (for example $\underline{h}_0 = \mathrm{I}_n$), so $\bar{d}_0 = \varepsilon_0 d_0$. If $a = (d_{ij}) \in \mathrm{M}(n, D) = A$, recall that $\mathrm{trd}_{A/D}(a) = \sum_i d_{ii}$ (cf. p. 127). Let $\sim$ be the involution

$$\widetilde{d} = d_0^{-1}\bar{d}d_0$$

on D. Then the transfer

$$h_D = \mathrm{trd}_{A/D}h : D^{n\times m} \times D^{n\times m} \to (D, \,^{\sim}\,)$$

is a nonsingular ε-Hermitian form whose transfer to K via $\mathrm{trd}_{D/K}$ is the transfer of h to K,

$$\mathrm{trd}_{D/K}h_D = \mathrm{trd}_{A/K}h.$$

Assume that $\underline{h} = \operatorname{diag}(d_1, \ldots, d_m)$ *(cf. (3.53)). The discriminant matrix* $\underline{h}_D \in D^{nm \times nm}$ *of* h_D*, as defined by* $h_D(u,v) = u\underline{h}_D v^{t\sim}$*,* $u, v \in D^{1 \times nm} = D^{nm}$*, is given by the diagonal matrix*

$$\underline{h}_D = \underline{h} \otimes \underline{h}_0$$

where the space $D^{n \times m}$ *of* h_D *has been identified with* D^{nm} *by "lining up the columns":*

$$(d_{ij}) \rightsquigarrow (d_{11}, \ldots, d_{m1}, d_{12}, \ldots, d_{m2}, \ldots, d_{1n}, \ldots, d_{mn}).$$

(a). The transfer is skew symmetric by Lem. 1.3.7, p. 34, and the rest is clear.

Consider (b). Since the involution is also the identity on K, $\varepsilon = 1$ by (3.51), so $\varepsilon\varepsilon_0 = -1$, i.e. $\underline{h} \in \mathrm{M}(n, L)$ is skew symmetric (Th. 3.2.9, p. 120). It follows from Th. 3.2.12 that the discriminant matrix of $\operatorname{trd}_{A/L} h$ is the tensor product of two skew symmetric matrices over L and so $\operatorname{trd}_{A/L} h$ is a hyperbolic symmetric form – this is easy to see directly and also follows from Th. 3.2.1(d), p. 109. Therefore its transfer $\operatorname{Tr}_{L/K}(\operatorname{trd}_{A/L} h)$ is a hyperbolic symmetric form over K.

(c). By (3.52), the reflexive form $\langle \underline{h}_0 \rangle$ is *not* skew symmetric, and so has an orthogonal basis. By using such an orthogonal basis in identifying A with $\mathrm{M}(n, D)$, $\underline{h}_0$ will be diagonal, as indicated.

And in fact we can assume that $\underline{h}$ also is diagonal. This follows by considering the $\varepsilon\varepsilon_0$-reflexive form $\breve{h} = \langle \underline{h} \rangle$,

$$\breve{h} : D^{1 \times m} \times D^{1 \times m} \to (D, {}^{-}), \quad \breve{h}(u,v) = u\underline{h}v^{t-},$$

which has an orthogonal basis if D is noncommutative. On the other hand if $D = L$, $\varepsilon\varepsilon_0 = \varepsilon$ since $\varepsilon_0 = 1$, so $\breve{h}$ is ε-Hermitian if $(A, {}^{-})$ is unitary and is symmetric if $(A, {}^{-})$ is of the first kind by (3.51); thus $\breve{h}$ has an orthogonal basis in both cases.

Therefore there is a matrix $T \in \mathbf{GL}(m, D)$ such that $T\underline{h}T^{t-}$ is a diagonal matrix $\operatorname{diag}(d_1, \ldots, d_m)$. The transformation of $D^{n \times m}$ by right multiplication by T is an A-automorphism, and

$$h(uT, vT) = uT\underline{h}T^{t-}v^{t-}\underline{h}_0 = u(\operatorname{diag}(d_1, \ldots, d_m))v^{t-}\underline{h}_0$$

shows that h is equivalent to the form over $(A, {}^{-})$ with discriminant matrix $\operatorname{diag}(d_1, \ldots, d_m)$, so we can assume[6] $\underline{h} = \operatorname{diag}(d_1, \ldots, d_m)$. (Note that $\underline{h}_0$ is not changed by this process).

(d) With these changes, (3.50) becomes

$$\operatorname{trd}_{A/K} h(u,v) = \sum_{i,j} \operatorname{trd}_{D/K}(u_{ij} d_j \bar{v}_{ij} d_{0i}),$$

which implies the first statement in (d). The second statement follows from Th. 3.2.12 and Th. 3.2.1(b) (p. 109) since $\underline{h}^{t-} = \varepsilon_0\varepsilon\underline{h} = \varepsilon\underline{h}$ and $\varepsilon = 1$ if $(K, {}^{-}) = (K, \mathrm{id})$.

(e) $A = \mathrm{M}(n, D)$ and $D^{n \times m}$ are left modules over both the rings $\mathrm{M}(n, D)$ and D, satisfying $dx = d^{(n)}x$ for all $d \in D$ and $x \in A$ or $D^{n \times m}$ where $d^{(n)} = \operatorname{diag}(d, \ldots, d) \in \mathrm{M}(n, D)$, and

$$\begin{aligned} h(du, d'v) &= h(d^{(n)}u, d'^{(n)}v) = d^{(n)}h(u,v)\overline{d'^{(n)}} \\ &= d^{(n)}h(u,v)(d_0^{-1}\bar{d}'d_0)^{(n)} = dh(u,v)\widetilde{d'}. \end{aligned}$$

[6]This is a special case of Morita equivalence – cf. §3.2.9.

It follows that $h_D = \mathrm{trd}_{A/D}h$ is a sesquilinear form over $(D, \,\tilde{}\,)$, and it is straightforward to check that it is in fact an ε-Hermitian form over $(D, \,\tilde{}\,)$.

Since the transfer $\mathrm{trd}_{A/K}h$ is a nonsingular form and $= \mathrm{trd}_{D/K}\mathrm{trd}_{A/D}h$ (p. 127), $\mathrm{trd}_{A/K}h(u,v) \neq 0$ implies that $\mathrm{trd}_{A/D}h(u,v) \neq 0$, whence h_D is nonsingular since $\mathrm{trd}_{A/K}h$ is.

If $\underline{h} = \mathrm{diag}(d_1, \dots, d_m)$, we get, from (3.49),

$$h_D(u,v) = \sum_{i,j,k} u_{ij} d_{jk} \bar{v}_{ik} d_0 = \sum_{i,j} u_{ij} d_j \bar{v}_{ij} d_0 = \sum_{i,j} u_{ij} d_j d_0 \widetilde{v}_{ij}.$$

Therefore the matrix units e_{ij} of $D^{n\times m}$ form an orthogonal basis of D^{nm} since

$$h_D(e_{ij}, e_{kl}) = \begin{cases} d_j d_0 & \text{if } i = k \text{ and } j = l, \\ 0 & \text{otherwise,} \end{cases}$$

and so the discriminant matrix of h_D is $\underline{h} \otimes \underline{h}_0$. □

REMARK. If K is a nonArchimedean local field and D is the unique quaternion division algebra with conjugation * as the involution, it is always possible to achieve $\underline{h}_0 = d_0 \mathrm{I}_n$ when $\underline{h}_0$ is Hermitian. This follows easily from the table on p. 61: the equivalence class of an Hermitian form in this situation is characterized up to equivalence by its dimension, so in fact $\underline{h}_0$ can be assumed to be the identity matrix I_n. □

Let K be an arbitrary field (of characteristic $\neq 2$), and let D be a central quaternion division algebra with an orthogonal involution $^-$. By Ths. 1.3.1 and 1.3.6 (pp. 28, 30) there is a pure quaternion j such that $\bar{d} = j^{-1} d^* j$ for all $d \in D$, where * denotes conjugation in D.

Thus if $h : D \times D \to (D, ^-)$ is a rank 1 Hermitian form, there is a pure quaternion l such that

$$h(d_1, d_2) = d_1 l d_2^* j = d_1 l j \bar{d}_2.$$

Lemma 3.2.4 *The quaternary quadratic form $\mathrm{trd}_{D/K}h$ over K has determinant 1; it is isotropic (hence hyperbolic) if and only if $(l^2, j^2)_K \cong D$.*

Since $K(j)$ is self-centralizing in D, it follows that the (-1)-eigenspace of the involutory linear transformation of the space of pure quaternions D_0 given by conjugation by j, is two dimensional, so we can find another pure quaternion $i \neq 0$ satisfying $l = \theta j + \delta i$ with $\theta \in K$, $ij = -ji$, and $\delta = 0$ or 1.

Had we chosen γj, $\gamma \in \dot{K}$, instead of j, we would have

$$h(d_1, d_2) = d_1 l j \bar{d}_2 = d_1 (l\gamma^{-1})(\gamma j)\bar{d}_2 \text{ and } l\gamma^{-1} = (\gamma^{-2}\theta)(\gamma j) + \delta(\gamma^{-1} i).$$

Thus i is uniquely determined up to nonzero scalars if $\delta = 1$.

Let $i^2 = \alpha, j^2 = \beta$, so $D = (\alpha, \beta)_K$. And

$$l^2 = \theta^2 \beta + \delta^2 \alpha.$$

Using the fact that $\mathrm{trd}_{D/L}(\alpha_0 1 + \alpha_1 i + \alpha_2 j + \alpha_3 ij) = 2\alpha_0$ (cf. p. 16), one can check that the discriminant matrix of $\mathrm{trd}_{D/L}h$ in the basis $1, ij, i, j$ of D is

$$2\beta \begin{pmatrix} \theta & -\alpha\delta & 0 & 0 \\ -\alpha\delta & -\theta\alpha\beta & 0 & 0 \\ 0 & 0 & \theta\alpha & -\alpha\delta \\ 0 & 0 & -\alpha\delta & -\theta\beta \end{pmatrix}.$$

Its determinant is $(2\beta)^4(-\theta^2\alpha\beta - \alpha^2\delta^2)^2 =_2 1$ (since $\mathrm{trd}_{D/L}h$ is nonsingular by Th. 3.2.4).

If $\theta = 0$, $\mathrm{trd}_{D/K}h$ is obviously isotropic; furthermore $l = i$ so $(l^2, j^2)_K = D$.

Suppose $\theta \neq 0$. The (nonsingular symmetric) forms b_1 and b_2 with matrices

$$\begin{pmatrix} \theta & -\alpha\delta \\ -\alpha\delta & -\theta\alpha\beta \end{pmatrix} \text{ and } \begin{pmatrix} \theta\alpha & -\alpha\delta \\ -\alpha\delta & -\theta\beta \end{pmatrix}$$

have the same determinant, and so their sum $b_1 \perp b_2$ is isotropic if and only if $b_2 \cong -b_1$ (see the exercise below). This is equivalent to the form $b_1 = \langle \theta, -\theta^3\alpha\beta - \theta\alpha^2\delta^2 \rangle$ representing $\theta\beta$, hence to $\langle 1, -\theta^2\alpha\beta - \alpha^2\delta^2 \rangle$ representing β. Since binary forms are classified by their determinant and Hasse invariant (Th. 1.5.6(a), p. 54), this in turn is equivalent to the Hasse invariants

$$(\beta, -\theta^2\alpha\beta^2 - \alpha^2\delta^2\beta)_K = (1, -\theta^2\alpha\beta - \alpha^2\delta^2)_K = 1$$

i.e. to $(\beta, \theta^2\alpha\beta + \alpha^2\delta^2)_K = 1$ (4., p. 53), thus to

$$1 = (\beta, \alpha)_K(\beta, \theta^2\beta + \alpha\delta^2)_K = (\beta, \alpha)_K(j^2, l^2)_K,$$

finally to $D = (\alpha, \beta)_K \cong (l^2, j^2)_K$. □

Exercise. Prove that if b_1 and b_2 are nonsingular binary symmetric forms with the same determinant, then $b_2 \cong -b_1$ if $b_1 \perp b_2$ is isotropic. (If the common determinant is $\neq_2 -1$, they are both anisotropic, and there is $\alpha \in \dot{K}$ such that α is represented by b_1 and $-\alpha$ by b_2). □

3.2.9 Morita theory for forms over simple algebras. In Th. 2.1.2 (p. 88), we saw that the equivalence of isometry representations can be "reduced" to the equivalence of ε-Hermitian forms over simple K-algebras, and here we shall see how this can in turn be reduced to the equivalence of certain reflexive forms over division algebras with involution.

As in the previous sections, we take a simple algebra $A = \mathrm{M}(n, D)$ over K, where D is a division algebra. The left A-module $D^{n\times m}$ has length m, and is a right B-module, $B = \mathrm{M}(m, D)$ – in fact $D^{n\times m}$ is an (A, B)-bimodule.

Now assume that D and A have $(K, {}^-)$-involutions, both also denoted ${}^-$. Then there is an $\underline{h}_0 \in \mathrm{M}(n, D)$ such that

$$\bar{a} = \underline{h}_0^{-1} a^{t-} \underline{h}_0,$$

where $\underline{h}_0^{t-} = \varepsilon_0 \underline{h}_0$ with $\varepsilon_0 = \pm 1$. We shall assume $\varepsilon_0 = 1$ if the involutions are of the second kind (cf. Th. 1.3.6, (a), p. 30).

An ε-Hermitian form $h : D^{n\times m} \times D^{n\times m} \to (A, {}^-)$ has the form

$$h(u, v) = u\underline{h}v^{t-}\underline{h}_0$$

where $\underline{h} \in \mathrm{M}(m, D)$ satisfies $\underline{h}^{t-} = \varepsilon_0\varepsilon\underline{h}$ (cf. §3.2.7, p. 118).

Now let $h' : D^{n\times m} \times D^{n\times m} \to (A, {}^-)$ be another ε-Hermitian form which is equivalent to h. Since $\mathrm{End}_A D^{n\times m} = \mathrm{M}(m, D)$ by Lem. 3.2.3, the relationship

$$h(\varphi u, \varphi v) = h'(u, v)$$

is equivalent to

$$u\underline{\varphi}\underline{h}\underline{\varphi}^{t-}v^{t-}\underline{h}_0 = u\underline{h}'v^{t-}\underline{h}_0$$

with $\underline{\varphi} \in \mathrm{M}(m, D)$, so also with

$$\underline{\varphi}\underline{h}\underline{\varphi}^{t-} = \underline{h}',$$

which is the relationship between the matrices of $\varepsilon_0\varepsilon$-reflexive forms $\breve{h}$ and $\breve{h}'$ over $(D, ^-)$ signifying their equivalence (p. 49). This sets up a bijection between the equivalence classes of ${}_{(A,^-)}\mathcal{H}_\varepsilon$ and those of ${}_{(D,^-)}\mathcal{H}_{\varepsilon_0\varepsilon}$, under which nonsingular forms correspond. (Of course the same statement holds for sesquilinear forms).

Thus we have the following theorem:

Theorem 3.2.14 "Morita Equivalence" *Two ε-Hermitian forms*

$$h, h' : D^{n\times m} \times D^{n\times m} \to (\mathrm{M}(n, D), ^-)$$

over $(A, ^-) = (\mathrm{M}(n, D), ^-)$,

$$h(u, v) = u\underline{h}v^{t-}\underline{h}_0, \quad h'(u, v) = u\underline{h}'v^{t-}\underline{h}_0,$$

are equivalent if and only if the $\varepsilon_0\varepsilon$-reflexive forms

$$\breve{h}, \breve{h}' : D^{1\times m} \times D^{1\times m} \to (D, ^-)$$

with the same discriminant matrices respectively as h and h',

$$\breve{h}(u, v) = u\underline{h}v^{t-}, \quad \breve{h}'(u, v) = u\underline{h}'v^{t-},$$

are equivalent.

Thus there is a bijection between the equivalence classes of ε-Hermitian forms over $(A, ^-)$ and the equivalence classes of $\varepsilon_0\varepsilon$-reflexive forms over $(D, ^-)$, under which both nonsingular and hyperbolic forms correspond and which preserves orthogonal direct sums.

The "Morita equivalent" forms h and $\breve{h}$ are said to be *Morita correspondents* of each other.

Only the statements about hyperbolic forms and orthogonal direct sums remain to be proved. By Th. 3.2.9, p. 120, the discriminant matrix of a hyperbolic form h over $(A, ^-)$ can be chosen to be of the shape

$$\begin{pmatrix} 0 & X \\ \varepsilon\varepsilon_0 X^{t-} & 0 \end{pmatrix},$$

so $\breve{h}$ is a hyperbolic form over $(D, ^-)$; the converse clearly holds as well. Similarly if h is an orthogonal sum $h = h_1 \perp h_2$, then $\breve{h} \cong \breve{h}_1 \perp \breve{h}_2$, and conversely. □

There is a slightly more general version of Morita equivalence which is occasionally useful. Let p be a positive integer, $\varepsilon_1 = \pm 1$, $\underline{h}_1 \in \mathbf{GL}(p, D)$ satisfying $\underline{h}_1^{-t} = \varepsilon_1 \underline{h}_1$, and the involution $^-$ on $\mathrm{M}(p, D)$ given by $\bar{b} = \underline{h}_1^{-1} b^{t-} \underline{h}_1$.

Corollary 3.2.4 *Two ε-Hermitian forms*

$$h, h' : D^{n\times m} \times D^{n\times m} \to (\mathrm{M}(n, D), ^-)$$

over $(A, ^-) = (\mathrm{M}(n, D), ^-)$,

$$h(u, v) = u\underline{h}v^{t-}\underline{h}_0, \quad h'(u, v) = u\underline{h}'v^{t-}\underline{h}_0,$$

are equivalent if and only if the $\varepsilon_0\varepsilon\varepsilon_1$-reflexive forms

$$\breve{h}, \breve{h}' : D^{p\times m} \times D^{p\times m} \to (\mathrm{M}(p, D), ^-)$$

over $(\mathrm{M}(p, D), ^-)$ *with the same discriminant matrices respectively as h and h',*

$$\breve{h}(u, v) = u\underline{h}v^{t-}\underline{h}_1, \quad \breve{h}'(u, v) = u\underline{h}'v^{t-}\underline{h}_1,$$

are equivalent.

Thus there is a bijection between the equivalence classes of ε-Hermitian forms over $(A, ^-)$ and the equivalence classes of $\varepsilon_0\varepsilon\epsilon_1$-reflexive forms over $(D, ^-)$, under which both nonsingular and hyperbolic forms correspond and which preserves orthogonal direct sums.

This follows easily from applying Th. 3.2.14 to both pairs $\mathrm{M}(n, D), D$ and $\mathrm{M}(p, D), D$. □

If we set $\varepsilon\varepsilon_0 = \breve{\varepsilon}$, then $\breve{h}$ is $\breve{\varepsilon}$-Hermitian and we have the important relationship

$$\varepsilon\varepsilon_0\breve{\varepsilon} = 1. \tag{3.54}$$

Corollary 3.2.5 *Let $\rho : G \to \mathbf{I}(V, h)$ be an A-isotypic isometry representation where $(A, ^-)$ is a component of $(KG, ^-)$ (and A is simple), and let $\hat{h}$ and $\breve{h}$ be the corresponding forms over $(A, ^-)$ and $(D, ^-)$. Then the following conditions are equivalent:*
(a) ρ is hyperbolic,
(b) $\hat{h}$ is hyperbolic,
(c) $\breve{h}$ is hyperbolic.

The equivalence of (a) and (b) is Cor. 3.1.2, p. 97, and that of (b) and (c) is the last part of the above theorem. □

Note that if ρ is ε-hyperbolic, then so is $\hat{h}$ but $\breve{h}$ is $\breve{\varepsilon}$-hyperbolic.

We can also give a sharper form of Th. 2.1.2 on p. 88. We use the notation of that theorem. Suppose that $1 \leq i \leq k$ and that the simple algebra A_i is quasi-split, say $A_i = \mathrm{M}(n_i, L_i)$, and that the involution on A_i is of the first kind. The form $\breve{h}_i$ is skew symmetric if $\varepsilon\varepsilon_0 = -1$. This happens when ρ is orthogonal ($\varepsilon = 1$) and $(A_i, ^-)$ is symplectic ($\varepsilon_0 = -1$), or when ρ is symplectic and $(A_i, ^-)$ is orthogonal. Since the equivalence class of a nonsingular skew symmetric form depends only on its rank, we get:

Theorem 3.2.15 *Suppose that for all $i > k_0$, $(A_i, ^-)$ is either hyperbolic, or quasisplit with $^-$ orthogonal and $\varepsilon = -1$, or quasisplit with $^-$ symplectic and $\varepsilon = 1$. Then two isometry representations on ε-reflexive spaces are equivalent if and only if they are equivalent as linear representations, and for all i, $1 \leq i \leq k_0$, either*
(a) the ε-Hermitian forms $\hat{h}_i$ and $\hat{h}'_i$ are equivalent,
or
(b) the $\breve{\varepsilon}$-reflexive forms $\breve{h}_i$ and $\breve{h}'_i$ are equivalent. □

Thus in principle at least, the equivalence problem for isometry representations over K is solved as soon as the equivalence problems for reflexive forms over finite dimensional division algebras over K are solved. Since this is the case for finite, algebraically and real closed, local and global fields, we will be able to give far reaching results for isometry representations over these fields (cf. Chs. 4 and 5).

Example. We can interpret the result in the example on p. 121 as saying that the Hermitian form $h^L : L^{2\times 4} \times L^{2\times 4} \to \mathrm{M}(2, L)$, arising from the Hermitian form $h : D^2 \times D^2 \to (D, ^*)$ where D is a quaternion algebra with conjugation * as involution and L is a quadratic splitting field of D, is Morita equivalent to a skew symmetric form $\breve{h}$ of rank 4 over L. In particular all Hermitian forms of length 2 over $\mathrm{M}(2, L)$ with a symplectic involution are equivalent – of course by the last part of Th. 3.2.14, this is true for such Hermitian forms of *any* length and matrix algebras $\mathrm{M}(n, L)$ of *any* (even) degree with a symplectic involution. □

3.3 The Hasse Principle

In this section,

K is a global field with an involution $^-$ with fixed field K_0, of characteristic $\neq 2$,
$\mathfrak{p}$ is an arbitrary prime (finite or infinite) of K_0,
$K_{0\mathfrak{p}}$ is the corresponding completion, and
$K_\mathfrak{p} = K_{0\mathfrak{p}} \otimes_{K_0} K$.

By Th. 1.2.16, p. 23, $K_\mathfrak{p} = K_\mathfrak{P}$ or $K_{0\mathfrak{p}} \oplus K_{0\mathfrak{p}}$; in the first case $\mathfrak{P}$ is the unique extension of $\mathfrak{p}$ to K and either $K = K_0$ (and $\mathfrak{P} = \mathfrak{p}$), or K/K_0 and $K_\mathfrak{P}/K_{0\mathfrak{p}}$ are both quadratic extension of fields (and $\mathfrak{P}/\mathfrak{p}$ is inert or ramified). In the second case K/K_0 is a quadratic extension, $\mathfrak{p}$ has two extensions $\mathfrak{P}_1$ and $\mathfrak{P}_2$ to K ("$\mathfrak{p}$ splits in K"), and for each of them, $K_{\mathfrak{P}_i} = K_{0\mathfrak{p}}$ under the "inclusion" $K_{0\mathfrak{p}} \xrightarrow{\iota_i} K_{\mathfrak{P}_i}$. If we write $K = K_0(\sqrt{\delta})$ with $\delta \in K_0$, these three cases can be distinguished by

$$K_\mathfrak{p} = \begin{cases} K_\mathfrak{P} & \text{if } \delta \in K_0^2 \quad (K = K_0 \text{ and } \mathfrak{P} = \mathfrak{p}), \\ K_\mathfrak{P} & \text{if } \delta \notin K_{0\mathfrak{p}}^2 \quad (K \neq K_0 \text{ and } \mathfrak{p} \text{ inert or ramified in } K/K_0), \\ K_{0\mathfrak{p}} \oplus K_{0\mathfrak{p}} & \text{if } \delta \in K_{0\mathfrak{p}}^2 \backslash K_0^2 \quad (K \neq K_0 \text{ and } \mathfrak{p} \text{ splits in } K/K_0). \end{cases}$$

These are standard facts of number theory – see e.g. Prop. 8.2, p. 163, [53].

Note that $K_\mathfrak{p}$ is a separable (commutative) $K_{0\mathfrak{p}}$-algebra.

The involution $^-$ on K has an extension to $K_\mathfrak{p} = K_{0\mathfrak{p}} \otimes_{K_0} K$ given by $\overline{\alpha \otimes \beta} = \alpha \otimes \bar{\beta}$ and is, respectively,

$$(K_\mathfrak{p}, {}^-) = \begin{cases} (K_\mathfrak{P}, \mathrm{id}), \\ (K_\mathfrak{P}, {}^-), \\ (K_{0\mathfrak{p}} \oplus K_{0\mathfrak{p}}, \leftrightarrow), \end{cases}$$

where, in the second case, $^-$ on $K_\mathfrak{P}$ is the unique nontrivial automorphism of $K_\mathfrak{P}/K_{0\mathfrak{p}}$ and in the third $\leftrightarrow$ is the "interchange" $\overleftrightarrow{(\alpha, \beta)} = (\beta, \alpha)$ (and $K_{0\mathfrak{p}} \to K_{0\mathfrak{p}} \oplus K_{0\mathfrak{p}}$ is the diagonal map). This is trivial in the first case, and clear also in the second since the involution on $K_\mathfrak{p}$ is the identity on $K_{0\mathfrak{p}}$ but not on $K_\mathfrak{p}$. The third case follows at once from the fact that the involution on $K_\mathfrak{p}$ is a nontrivial $K_{0\mathfrak{p}}$-linear homomorphism on the semisimple $K_{0\mathfrak{p}}$-algebra $K_{0\mathfrak{p}} \oplus K_{0\mathfrak{p}}$.

If V is a vector space over K, $V_\mathfrak{p}$ denotes the $K_\mathfrak{p}$-module $K_\mathfrak{p} \otimes_K V = K_{0\mathfrak{p}} \otimes_{K_0} V$ (with the obvious identification $(\alpha \otimes_{K_0} \beta) \otimes_K v = \alpha \otimes_{K_0} \beta v$ for $\alpha \in K_{0\mathfrak{p}}, \beta \in K$).

If $(A, {}^-)$ is a separable involution algebra over $(K, {}^-)$, then $A_\mathfrak{p}$ is a separable algebra over $K_{0\mathfrak{p}}$, has the "tensor product" involution $^-$

$$(A_\mathfrak{p}, {}^-) = (K_\mathfrak{p}, {}^-) \otimes_{(K, {}^-)} (A, {}^-) = (K_{0\mathfrak{p}}, \mathrm{id}) \otimes_{(K_0, \mathrm{id})} (A, {}^-),$$

and is an involution algebra over $(K_\mathfrak{p}, {}^-)$ (Th. 1.3.2, p. 29). A sesquilinear form $f : V \times V \to (A, {}^-)$ has a unique extension to a sesquilinear form

$$f_\mathfrak{p} : V_\mathfrak{p} \times V_\mathfrak{p} \to (A_\mathfrak{p}, {}^-)$$

determined by

$$\begin{aligned} f_\mathfrak{p}(\alpha \otimes_K v, \beta \otimes_K u) &= \alpha\bar{\beta} \otimes_K f(v, u) \quad (\alpha, \beta \in K_\mathfrak{p}) \\ &= \alpha\bar{\beta} f(v, u), \end{aligned}$$

if A is identified with a K-subalgebra of $A_{\mathfrak{p}}$ via $a \to 1 \otimes a$ (cf. (3.27), p. 110). Equivalently,

$$\begin{aligned} f_{\mathfrak{p}}(\alpha \otimes_{K_0} v, \beta \otimes_{K_0} u) &= \alpha\beta \otimes_{K_0} f(v,u) \quad (\alpha, \beta \in K_{0\mathfrak{p}}) \\ &= \alpha\beta f(v,u). \end{aligned}$$

If f is ε-Hermitian or nonsingular, then so is $f_{\mathfrak{p}}$. Cf. Th. 3.2.3, p. 111 and its preamble.

If $\mathfrak{p}$ splits in K,

$$A_{\mathfrak{p}} \cong (K_{0\mathfrak{p}} \oplus K_{0\mathfrak{p}}) \otimes_K A \cong (K_{0\mathfrak{p}} \otimes_K A) \oplus (K_{0\mathfrak{p}} \otimes_K A).$$

This gives us an isomorphism of involution algebras

$$(A_{\mathfrak{p}}, {}^{-}) \cong ((K_{0\mathfrak{p}} \otimes_K A) \oplus (K_{0\mathfrak{p}} \otimes_K A), {}^{-})$$

where the involution on the right-hand side is $\overline{(\alpha \otimes a, \beta \otimes b)} = (\beta \otimes \bar{b}, \alpha \otimes \bar{a})$. By Ex. 8, p. 26, $(A_{\mathfrak{p}}, {}^{-})$ is a hyperbolic involution algebra.

Therefore by Corollary 3.1.5, p. 98, we get the following lemma:

Lemma 3.3.1 *If $\mathfrak{p}$ splits in K and V is a module over the semisimple algebra A with $(K, {}^{-})$-involution ${}^{-}$, any nonsingular ε-Hermitian form $V_{\mathfrak{p}} \times V_{\mathfrak{p}} \to (A_{\mathfrak{p}}, {}^{-})$ is hyperbolic, and so is determined up to equivalence by the isomorphism class of V as an A-module.* □

As a first step in proving the Hasse Principle for isometry representations (Th. 3.3.3), we prove a form of it for central simple K-algebras:

Theorem 3.3.1 *Let $(A, {}^{-})$ be a central simple algebra with $(K, {}^{-})$-involution, and let $h : V \times V \to (A, {}^{-})$ be a nonsingular ε-Hermitian form. Then the "Hasse Principle"*

for all nonsingular ε-Hermitian forms $g : V \times V \to (A, {}^{-})$,
$g \cong h$ if and only if $g_{\mathfrak{p}} \cong h_{\mathfrak{p}}$ for all primes $\mathfrak{p}$ of K_0

holds except in the following situation (in which case it does not *hold):*

$K = K_0$, and

$A \cong$ a matrix algebra over a quaternion algebra which is nonsplit at more than 2 primes, and

g (and h) are skew Hermitian when $(A, {}^{-})$ is of symplectic type, or Hermitian when $(A, {}^{-})$ is of orthogonal type.

The last condition is equivalent, by (3.39), p. 120, to the adjoint involutions of g and h being orthogonal.

The proof of the theorem consists of applying Morita theory and the usual Hasse Principle over division algebras (cf. p. 63). By the results in §3.2.7, we may suppose that

$A = \mathrm{M}(n, D)$ for some central division algebra D,

$\bar{a} = \underline{h}_0^{-1} a^{t-} \underline{h}_0$ where

${}^{-}$ on D is an involution of the same kind as that on A, and is conjugation if D is a quaternion algebra and ${}^{-}$ is of the first kind,

$\underline{h}_0$ is an ε_0-Hermitian matrix in $D^{n \times n}$,

$\varepsilon_0 = \pm 1$, and $= 1$ if and only if $(A, {}^{-})$ and $(D, {}^{-})$ have the same type,

$V = D^{n \times m}$,

$h(u, v) = u \underline{h} v^{t-} \underline{h}_0$ for all $u, v \in D^{n \times m}$ for some nonsingular $\varepsilon\varepsilon_0$-Hermitian matrix $\underline{h} \in D^{m \times m}$.

Let $\breve{\varepsilon} = \varepsilon\varepsilon_0$. The nonsingular $\breve{\varepsilon}$-Hermitian form

$$\breve{h}(u,v) = u\underline{h}v^{t-} \quad (u,v \in D^{1\times m})$$

over $(D, {}^-)$ is Morita equivalent to h.

Of course these same considerations apply to g,

$$g(u,v) = u\underline{g}v^{t-}\underline{h}_0 \quad \text{for all } u,v \in D^{n\times m}$$

for some nonsingular $\breve{\varepsilon}$-Hermitian matrix $\underline{g} \in D^{m\times m}$, resulting in a form $\breve{g}$ over $(D, {}^-)$ which is Morita equivalent to g. Thus $g \cong h$ is equivalent to $\breve{g} \cong \breve{h}$ by "Morita theory" (Th. 3.2.14).

Assume first that $g_\mathfrak{p} \cong h_\mathfrak{p}$ for all primes $\mathfrak{p}$ of K_0. Then

$$\breve{g}_\mathfrak{p} \cong \breve{h}_\mathfrak{p} \tag{3.55}$$

where, e.g., $\breve{g}_\mathfrak{p} = (\breve{g})_\mathfrak{p}$. This is clear if $\mathfrak{p}$ splits in K, by Lem. 3.3.1 (applied with $A = D$). Otherwise $K_\mathfrak{p}$ is a field and $A_\mathfrak{p}$ and $D_\mathfrak{p}$ are (central) simple algebras, $A_\mathfrak{p} = \mathrm{M}(n, D_\mathfrak{p})$. Now $h_\mathfrak{p}(u,v) = u\underline{h}v^{t-}\underline{h}_0$ $(u,v \in D_\mathfrak{p}^{n\times m})$,[7] and so $g_\mathfrak{p} \cong h_\mathfrak{p}$ means that there is $x \in \mathbf{GL}(m, D_\mathfrak{p})$ such that $x\underline{h}x^{t-} = \underline{g}$ (Cor. 3.2.3). Since $\breve{h}(u,v) = u\underline{h}v^{t-}$ $(u,v \in D^{1\times m})$, it is apparent that (3.55) holds in this case too since

$$\breve{g}_\mathfrak{p}(u,v) = u\underline{g}v^{t-} \text{ and } \breve{h}_\mathfrak{p}(u,v) = u\underline{h}v^{t-} \ (u,v \in D_\mathfrak{p}^{1\times m}).$$

We can now apply the Hasse Principle (p. 63) to conclude that $\breve{g}$ and $\breve{h}$ – and so also g and h – are equivalent unless they are skew Hermitian forms and D is a quaternion algebra which is nonsplit at more than 2 primes. In the latter case there *are* skew Hermitian forms $\breve{g}$ which are locally equivalent but not globally equivalent to $\breve{h}$ (*ibid*); thus the Morita equivalent forms g and h over $(A, {}^-)$ are also locally equivalent but not globally equivalent. Since $\varepsilon\varepsilon_0 = -1$ in this case, h and g are either Hermitian ($\varepsilon = 1$) and $(A, {}^-)$ is orthogonal ($\varepsilon_0 = -1$) – cf. Lem. 1.3.2 (p. 28) and Th. 1.3.6 (p. 30) – or h and g are skew Hermitian and $(A, {}^-)$ is symplectic. □

Exercise. This exercise refers to the proof of (3.55). Assume $\mathfrak{p}$ is not split in K, so $K_\mathfrak{p}$ is a field. Let $D_\mathfrak{p} = \mathrm{M}(l, E)$ where E is a central division algebra over $K_\mathfrak{p}$, and let ${}^\sim$ be an involution on E of the same kind as ${}^-$ on $D_\mathfrak{p}$. Let T stand for transpose in the case of matrices over E, and choose $\underline{h}_1 \in \mathbf{GL}(l, E)$ such that

$$\bar{d} = \underline{h}_1^{-1} d^{T\sim} \underline{h}_1 \text{ for all } d \in D_\mathfrak{p} = \mathrm{M}(l, E);$$

it can be chosen so that $\underline{h}_1^{T\sim} = \varepsilon_1\underline{h}_1$, $\varepsilon_1 = 1$ if the involutions are of the same type, $= -1$ if they are not.

(a) If $b \in D_\mathfrak{p}^{r\times s} = E^{lr\times ls}$, show that

$$b^{t-} = (\underline{h}_1^{(s)})^{-1} b^{T\sim} \underline{h}_1^{(r)},$$

where for any square matrix X, $X^{(t)}$ stands for the diagonal block matrix with t copies of X on the diagonal. (On the left side, b is viewed as a matrix in $D_\mathfrak{p}^{r\times s}$ and on the right side as a matrix in $E^{lr\times ls}$).

(b) Show that

$$h_\mathfrak{p}(u,v) = u\underline{h}(\underline{h}_1^{(m)})^{-1} v^{T\sim} \underline{h}_1^{(n)} \underline{h}_0,$$

[7] The matrix $\underline{h}$ is *not* in general a discriminant matrix of $h_\mathfrak{p}$ – see the exercise below.

and that if $\underline{e} \in \mathrm{M}(ln, E)$,

$$(\underline{h}_1^{(n)}\underline{h}_0)^{-1}\underline{e}^{T\sim}(\underline{h}_1^{(n)}\underline{h}_0) = \bar{\underline{e}}$$

where $\bar{\underline{e}}$ denotes the result of applying the original involution on A, extended to $A_\mathfrak{p} = \mathrm{M}(ln, E)$, to $\underline{e}$. Deduce that

$$\underline{h}(\underline{h}_1^{(m)})^{-1} \in \mathrm{M}(lm, E),$$

is a discriminant matrix of $h_\mathfrak{p}$ and that it is $\varepsilon_1\varepsilon$-Hermitian (with respect to the involution $T\sim$). □

Theorem 3.3.2 Hasse Principle for ε-Hermitian forms over separable involution algebras. *Let $(A, ^-)$ be a separable algebra with involution over $(K, ^-)$ where K is a global field with characteristic $\neq 2$, and $h : V \times V \to (A, ^-)$ is a nonsingular ε-Hermitian form.*

Then the "Hasse Principle"

for all nonsingular ε-Hermitian forms $g : V \times V \to (A, ^-)$,
$g \cong h$ if and only if $g_\mathfrak{p} \cong h_\mathfrak{p}$ for all primes $\mathfrak{p}$ of K_0

holds if and only if $A_iV = 0$ for every simple direct summand $(A_i, ^-)$ of $(A, ^-)$ which satisfies the following:

(*) *$A_i \cong$ a matrix algebra over a quaternion algebra which is nonsplit at more than 2 primes (of its center), and is an involution algebra of orthogonal type if h is Hermitian, and of symplectic type if h is skew Hermitian.*

Note that this means that the Hasse Principle *always* holds if $K \neq K_0$, since the involution on the center of any simple component of A which is stable under the involution, is nontrivial – so there is no A_i satisfying (*).

If

$$(A, ^-) = (A_1, ^-) \oplus (A_2, ^-) \oplus \cdots \oplus (A_k, ^-)$$

is the decomposition of $(A, ^-)$ into simple involution algebras, then any ε-Hermitian form $h : V \times V \to (A, ^-)$ is the vector sum of ε-Hermitian forms $h_i : A_iV \times A_iV \to (A_i, ^-)$, and the theorem holds for $(A, ^-)$ if and only if it holds for each $(A_i, ^-)$. So it suffices to prove the theorem for each of these h_i.

If $A_iV \neq 0$ and (*) holds, then the Hasse Principle does *not* hold by Th. 3.3.1, so the condition that $A_iV = 0$ if (*) holds is necessary.

Assume therefore that (*) does *not* hold, and that $g_\mathfrak{p} \cong h_\mathfrak{p}$ for all $\mathfrak{p}$.

If $(A_i, ^-)$ is hyperbolic, the Hasse Principle follows at once from Cor. 3.1.6, p. 98 since if two K-spaces have different dimensions, so do their completions over $K_\mathfrak{p}$ for any prime $\mathfrak{p}$.

Suppose that A is simple, with center L. Then

$$(K_{0\mathfrak{p}} \otimes_{K_0} L, ^-) = (L_{\mathfrak{P}_1}, ^-) \oplus \cdots \oplus (L_{\mathfrak{P}_t}, ^-)$$

where the $(L_{\mathfrak{P}_i}, ^-)$ are simple involution algebras – that is to say $L_{\mathfrak{P}_i}$ is either a field or a direct sum of two fields which are interchanged by the involution. Then

$$\begin{aligned}(A_\mathfrak{p}, ^-) = (K_{0\mathfrak{p}} \otimes_{K_0} A, ^-) &= (K_{0\mathfrak{p}} \otimes_{K_0} L \otimes_L A, ^-)\\ &= (L_{\mathfrak{P}_1} \otimes_L A, ^-) \oplus \cdots \oplus (L_{\mathfrak{P}_t} \otimes_L A, ^-)\\ &= (A_{\mathfrak{P}_1}, ^-) \oplus \cdots \oplus (A_{\mathfrak{P}_t}, ^-)\end{aligned}$$

is the decomposition of $(A_\mathfrak{p}, {}^-)$ into simple involution algebras – $A_{\mathfrak{P}_i}$ is a simple algebra if $L_{\mathfrak{P}_i}$ is a field, otherwise a hyperbolic involution algebra.

A sesquilinear form $f_\mathfrak{p}$ over $(A_\mathfrak{p}, {}^-)$ is a vector sum of forms $f_{\mathfrak{P}_1}, \ldots, f_{\mathfrak{P}_t}$ over $(A_{\mathfrak{P}_1}, {}^-), \ldots, (A_{\mathfrak{P}_t}, {}^-)$, and so the assumption that $g_\mathfrak{p} \cong h_\mathfrak{p}$ for all primes $\mathfrak{p}$ of K_0 implies that $g_\mathfrak{P} \cong h_\mathfrak{P}$ for all primes $\mathfrak{P}$ of L lying over $\mathfrak{p}$. Thus $g \cong h$ by Th. 3.3.1. $\square$

We now come to the principal goal of this section, the Hasse Principle for isometry representations. This is a corollary of the foregoing theorem, using the category equivalence between isometry representations with respect to an ε-reflexive form over $(K, {}^-)$ and ε-Hermitian forms over $(KG, {}^-)$.

Any isometry representation $\rho : G \to \mathbf{I}(V, h)$ can be extended to an isometry representation $\rho_\mathfrak{p} : G \to \mathbf{I}(V_\mathfrak{p}, h_\mathfrak{p})$ in the obvious way,

$$\rho_\mathfrak{p}(s)(\alpha \otimes_{K_0} v) = \alpha \otimes_{K_0} \rho(s)v \quad (\alpha \in K_{0\mathfrak{p}}, v \in V). \tag{3.56}$$

If $K \neq K_0$ and $\mathfrak{p}$ splits in K, then $(K_\mathfrak{p}, {}^-)$ splits into a hyperbolic involution algebra $(K_{0\mathfrak{p}} \oplus K_{0\mathfrak{p}}, \leftrightarrow)$ in which $h_\mathfrak{p}$ takes its values. Thus (3.56) is an extension of the notion of isometry representation as defined earlier, but should pose no difficulty.

Theorem 3.3.3 Hasse Principle for isometry representations. *Let G be an arbitrary finite group, K a global field with characteristic $\nmid 2|G|$, and $\rho : G \to \mathbf{I}(V, h)$ an isometry representation. If ρ' is another isometry representation of G, then the "Hasse Principle"*

$$\rho' \cong \rho \text{ if and only if } \rho'_\mathfrak{p} \cong \rho_\mathfrak{p} \text{ for all primes } \mathfrak{p} \text{ of } K_0$$

holds if and only if[8] $A_i V = 0$ *for every simple direct summand A_i of KG which satisfies the following:*

(**) *$A_i \cong$ a matrix algebra over a quaternion algebra which is nonsplit at more than 2 primes of its center, and the restriction of the standard $(K, {}^-)$-involution on KG to A_i is an involution of orthogonal type if ρ is orthogonal, and of symplectic type if ρ is symplectic.*

An involution ${}^-$ is defined on $K_\mathfrak{p}G$, even when $K_\mathfrak{p}$ is not a field, by $\sum_s \alpha s \to \sum_s \bar{\alpha} s^{-1}$, and then the canonical isomorphism

$$(KG)_\mathfrak{p} = K_\mathfrak{p} \otimes_K KG \cong K_\mathfrak{p}G, \quad \alpha \otimes \beta s \to \alpha\beta s \;\; (\alpha \in K_\mathfrak{p}, \beta \in K),$$

is an isomorphism of involution algebras.

Similarly the ε-Hermitian form

$$\widehat{h_\mathfrak{p}} : V_\mathfrak{p} \times V_\mathfrak{p} \to (K_\mathfrak{p}G, {}^-)$$

is defined by

$$\widehat{h_\mathfrak{p}}(u, v) = \sum_s h_\mathfrak{p}(u, sv)s,$$

which of course agrees with the earlier definition ((2.5), p. 85) when $K_\mathfrak{p}$ *is* a field, $\mathrm{Tr}_{K_\mathfrak{p}G/K_\mathfrak{p}}$ is defined by

$$\mathrm{Tr}_{K_\mathfrak{p}G/K_\mathfrak{p}} \sum_s \alpha_s s = \alpha_1,$$

[8] It is understood here that G acts on V through the representation ρ.

and if $\rho' : G \to \mathbf{I}(V', h')$, it is easy to check that $\rho'_{\mathfrak{p}} \cong \rho_{\mathfrak{p}}$ is equivalent to $\widehat{h'_{\mathfrak{p}}} \cong \widehat{h_{\mathfrak{p}}}$, since $\mathrm{Tr}_{K_{\mathfrak{p}}G/K_{\mathfrak{p}}} \widehat{h_{\mathfrak{p}}} = h_{\mathfrak{p}}$ and $\mathrm{Tr}_{K_{\mathfrak{p}}G/K_{\mathfrak{p}}} \widehat{h'_{\mathfrak{p}}} = h'_{\mathfrak{p}}$.

It follows easily from Lem. 3.1.1, p. 93, that $\widehat{h_{\mathfrak{p}}}$ is nonsingular (in the obvious sense if $K_{\mathfrak{p}}$ is not a field – note that $K_{\mathfrak{p}}G$ is in any case a semisimple algebra over $K_{0\mathfrak{p}}$), and moreover $\widehat{h_{\mathfrak{p}}} = \hat{h}_{\mathfrak{p}}$.

The sufficiency now follows easily: $\rho' \cong \rho$ if and only if $\hat{h}' \cong \hat{h}$ if and only if $\hat{h}'_{\mathfrak{p}} \cong \hat{h}_{\mathfrak{p}}$ for all $\mathfrak{p}$ (Th. 3.3.2), if and only if $\widehat{h'_{\mathfrak{p}}} \cong \widehat{h_{\mathfrak{p}}}$ for all $\mathfrak{p}$, if and only if $\rho'_{\mathfrak{p}} \cong \rho_{\mathfrak{p}}$ for all $\mathfrak{p}$.

Conversely, suppose there is a simple summand A_i of KG satisfying (**), with $A_iV \neq 0$. By Th. 3.3.2, there is a nonsingular ε-Hermitian form $\hat{g} : V \times V \to (KG, ^-)$ such that

$$\hat{g}_{\mathfrak{p}} \cong \hat{h}_{\mathfrak{p}} \text{ for all } \mathfrak{p}, \text{ but } \hat{g} \not\cong \hat{h}. \tag{3.57}$$

Let $g = \mathrm{Tr}_{KG/K}\,\hat{g}$. Since KG acts on V via ρ,

$$g(\rho(s)u, \rho(s)v) = \mathrm{Tr}_{KG/K}\,\hat{g}(su, sv) = \mathrm{Tr}_{KG/K}\big(s(\hat{g}(u,v))s^{-1}\big) = g(u,v),$$

so $\rho G \subset \mathbf{I}(V, g)$. Define $\rho' = \rho$, considered as a homomorphism $\rho' : G \to \mathbf{I}(V, g)$. Then by (3.57), $\rho'_{\mathfrak{p}} \cong \rho_{\mathfrak{p}}$ for all $\mathfrak{p}$ but $\rho' \not\cong \rho$. □

Corollary 3.3.1 *The Hasse Principle for Isometry Representations always holds in the following cases:*

(a) The representations are unitary.

(b) The characteristic of K is $\neq 0$.

(c) The representations are of odd type – e.g. $|G|$ is odd or G has a normal Abelian subgroup H of odd index $[G : H]$ (cf. p. 78).

(d) G is nilpotent and K contains either a cubic or quartic root of unity.

(a) If the representation is unitary, (**) does not occur.

(b) Suppose the characteristic is $p \neq 0$. The simple components of $\mathbb{F}_pG$ are matrix algebras over finite field extensions of $\mathbb{F}_p$ since there are no finite dimensional (noncommutative) division algebras over a finite field (Ex. 3, p. 17). Thus the simple components of KG are matrix algebras over finite extensions of K. Again (**) does not apply.

(c) Any A_i which satisfies (**) has even degree since it is a matrix algebra over a nonsplit quaternion algebra, and so $A_iV = 0$ if the representation is of odd type. (Note that the fact that the degree divides $[G : H]$ holds in general only if $\mathrm{char}\, K = 0$; but if $\mathrm{char}\, K \neq 0$, (b) applies).

(d) The A_i have index 1 in this case – see [24], 14.5, p. 77. □

3.4 Lifting a local representation to a global representation

We now make a change in notation: in this section, $\mathfrak{p}$ will be a prime of K (an equivalence class of valuations, Archimedean or nonArchimedean) and $\mathfrak{p}_0 = \mathfrak{p}|_{K_0}$ will be the prime of K_0 lying below $\mathfrak{p}$. $K_{\mathfrak{p}}$ is a completion. We shall assume that the automorphism $^-$ of K can be extended to $K_{\mathfrak{p}}$; this is equivalent to $\bar{\mathfrak{p}} = \mathfrak{p}$, i.e. for any $|\ | \in \mathfrak{p}$, $|\bar{\alpha}| = |\alpha|$ for all $\alpha \in K$ (Prop. 9.6, p. 170, [53]). Of course this is trivially the case if $^-$ is the identity on K.

The fact that $\bar{\mathfrak{p}} = \mathfrak{p}$ means that $\mathfrak{p}_0$ does not split in K, and so in particular

$$K_{0\mathfrak{p}_0} \otimes_{K_0} K \cong K_{\mathfrak{p}} \tag{3.58}$$

via the map which takes $\alpha_0 \otimes \beta$ to $\alpha_0\beta$ – in fact this is an isomorphism of fields with involution. The proof of the following theorem uses the topology induced by $\mathfrak{p}_0$ on (finite dimensional) vector spaces over $K_{0\mathfrak{p}_0}$. If W is such a space and $\{w_i\}$ a basis, there is a "norm" on W defined by $||w|| = \max_i |\alpha_i|_{\mathfrak{p}_0}$ where $w = \sum_i \alpha_i w_i$, $\alpha_i \in K_{0\mathfrak{p}_0}$, and $|\ |_{\mathfrak{p}_0} \in \mathfrak{p}_0$, from which we get a topology on W by using the sets $N_\varepsilon = \{w \in W :\ ||w|| < \varepsilon\}$ ($\varepsilon > 0$ in $\mathbb{R}$) as a basis for the neighbourhood system of $0 \in W$. Then $\{w_0 + N_\varepsilon\}$ is a basis for the neighbourhood system of $w_0 \in W$. The resulting topology is easily seen to be independent of the choice of $|\ |_{\mathfrak{p}_0}$ since any other choice is a power $|\ |_{\mathfrak{p}_0}^\theta$ for a positive real number θ.

If a different basis is chosen, say $w_i' = \sum_j \beta_{ij} w_j$ for $i = 1, \ldots, n$, and if $||\ ||'$ is the associated norm, then $||w|| \le n||w||'||\beta||$ where $||\beta|| = \max_{i,j} |\beta_{ij}|_{\mathfrak{p}_0}$. Thus the topology defined by $\{w_i'\}$ is finer than that defined by $\{w_i\}$, and by symmetry the two topologies coincide.

Note that if W is spanned over $K_{0\mathfrak{p}_0}$ by a set of vectors, the K_0-vector space spanned by the set is dense in W. For example if $W_0 \subset W$ is a K_0-vector space such that $K_{0\mathfrak{p}_0} W_0 = W$, then W_0 is dense in W.

Theorem 3.4.1 *Suppose that $\mathfrak{p}$ is a prime of K such that the involution $^-$ on K can be extended to an involution $^-$ on the completion $K_{\mathfrak{p}}$. Let $(A, ^-)$ be a separable K-algebra with $(K, ^-)$-involution, define*

$$(A_{\mathfrak{p}}, ^-) = (K_{\mathfrak{p}}, ^-) \otimes_{(K,^-)} (A, ^-),$$

and let $g : W \times W \to (A_{\mathfrak{p}}, ^-)$ be a nonsingular ε-Hermitian form. Then g is equivalent to the completion $h_{\mathfrak{p}} : V_{\mathfrak{p}} \times V_{\mathfrak{p}} \to (A_{\mathfrak{p}}, ^-)$ of some ε-Hermitian form $h : V \times V \to (A, ^-)$ if and only if W is $A_{\mathfrak{p}}$-isomorphic to the completion of an A-module. If we anticipate the terminology of §4.1, $[W, g]$ is in the image of $\hat{\mathrm{W}}_\varepsilon(A, ^-) \to \hat{\mathrm{W}}_\varepsilon(A_{\mathfrak{p}}, ^-)$ if and only if $[W] \in \mathrm{K}_0(A_{\mathfrak{p}})$ is in the image of $\mathrm{K}_0(A) \to \mathrm{K}_0(A_{\mathfrak{p}})$.

The necessity is clear, so assume W is isomorphic, as an $A_{\mathfrak{p}}$-module, to the completion of an A-module V, $W \cong V_{\mathfrak{p}}$. Let

$$(A, ^-) = (A_1, ^-) \oplus (A_2, ^-) \oplus \cdots \oplus (A_l, ^-)$$

be the decomposition into simple (separable) $(K, ^-)$-involution algebras,

$$(A_{\mathfrak{p}}, ^-) = (A_{1\mathfrak{p}}, ^-) \oplus (A_{2\mathfrak{p}}, ^-) \oplus \cdots \oplus (A_{l\mathfrak{p}}, ^-)$$

its completion, and

$$(W, g) = (A_{1\mathfrak{p}} W, g_1) \perp (A_{2\mathfrak{p}} W, g_2) \perp \cdots \perp (A_{l\mathfrak{p}} W, g_l)$$

the corresponding decomposition of (W, g) (where g_i is the restriction of g to $A_{i\mathfrak{p}} W$). Also

$$V_{\mathfrak{p}} = A_{\mathfrak{p}} V_{\mathfrak{p}} = A_{1\mathfrak{p}} V \oplus A_{2\mathfrak{p}} V \oplus \cdots \oplus A_{l\mathfrak{p}} V$$

since, identifying $V \subset V_{\mathfrak{p}}$, $A_{i\mathfrak{p}} V_{\mathfrak{p}} = A_{i\mathfrak{p}} V$ for all i. Therefore

$$A_{i\mathfrak{p}} W \cong A_{i\mathfrak{p}} V = K_{\mathfrak{p}} A_i V = (A_i V)_{\mathfrak{p}} \ \text{ for each } i,$$

i.e. $A_{i\mathfrak{p}} W$ is isomorphic to the completion of the A_i-module $A_i V$.

Therefore it suffices to prove the theorem in the special case when $(A, ^-)$ is a *simple* separable involution algebra, and we assume this from now on.

Suppose first that $(A, ^-)$ is hyperbolic, say $A = B \oplus \bar{B}$ with B (and $\bar{B}$) simple K-algebras. Then $(A_\mathfrak{p}, ^-)$ is also hyperbolic, $A_\mathfrak{p} = B_\mathfrak{p} \oplus \bar{B}_\mathfrak{p}$, so g is an ε-hyperbolic form (Cor. 3.1.5, p. 98). Let V be an A-module with completion $V_\mathfrak{p} \cong W$. Then $V = BV \oplus \bar{B}V$; since $\bar{B}V$ is a right B-module, as is the B-dual $(BV)^*$, it suffices to show that the BV and $\bar{B}V$ have the same length since then V supports an ε-hyperbolic form over $(A, ^-)$, whose completion is necessarily $\cong g$. Now

$$V_\mathfrak{p} = A_\mathfrak{p} V_\mathfrak{p} = B_\mathfrak{p} V_\mathfrak{p} \oplus \bar{B}_\mathfrak{p} V_\mathfrak{p}$$

supports an ε-hyperbolic form over $(A_\mathfrak{p}, ^-)$ since W does. Since $B_\mathfrak{p}\bar{B}_\mathfrak{p} = 0$, $B_\mathfrak{p} V_\mathfrak{p}$ and $\bar{B}_\mathfrak{p} V_\mathfrak{p}$ are totally isotropic with respect to this form, so by Th. 3.1.2, p. 97, and (3.12), $B_\mathfrak{p} V_\mathfrak{p}$ and $\bar{B}_\mathfrak{p} V_\mathfrak{p}$ have the same length, and therefore BV and $\bar{B}V$ have the same length since, e.g., simple modules over B and $\bar{B} \cong B^o$ have the same length when completed at $\mathfrak{p}$.

Now suppose that A is simple, say with center L. Then

$$\begin{aligned} A_\mathfrak{p} &= K_\mathfrak{p} \otimes_K A = K_\mathfrak{p} \otimes_K L \otimes_L A \\ &= (\oplus_{\mathfrak{P}|\mathfrak{p}} L_\mathfrak{P}) \otimes_L A = \oplus_{\mathfrak{P}|\mathfrak{p}} A_\mathfrak{P} \end{aligned}$$

where the $\mathfrak{P}$ are the primes of L lying above $\mathfrak{p}$ (cf. Th. 1.2.16, p. 23) and $A_\mathfrak{P} = L_\mathfrak{P} \otimes_L A$ is a central simple $L_\mathfrak{P}$-algebra. Then

$$V_\mathfrak{p} = A_\mathfrak{p} V_\mathfrak{p} = \oplus_{\mathfrak{P}|\mathfrak{p}}(A_\mathfrak{P} V_\mathfrak{p}) = \oplus_{\mathfrak{P}|\mathfrak{p}}(A_\mathfrak{P} W). \tag{3.59}$$

Note also that if $\mathfrak{p}_0 := \mathfrak{p}|_{K_0}$ and $V_{\mathfrak{p}_0} := K_{0\mathfrak{p}_0} \otimes_{K_0} V$, then

$$V_{\mathfrak{p}_0} = K_{0\mathfrak{p}_0} \otimes_{K_0} (K \otimes_K V) = (K_{0\mathfrak{p}_0} \otimes_{K_0} K) \otimes_K V = K_\mathfrak{p} \otimes_K V = V_\mathfrak{p}$$

by (3.58), that is to say, $V_{\mathfrak{p}_0}$ can be identified with $V_\mathfrak{p}$ via $\alpha_0 \otimes_{K_0} v = \alpha_0 \otimes_K v$ $(\alpha_0 \in K_{0\mathfrak{p}_0})$. We similarly identify $A_{\mathfrak{p}_0}$ with $A_\mathfrak{p}$ – more precisely the map $\alpha_0 \otimes_{K_0} A \to \alpha_0 \otimes_K A$ provides an isomorphism

$$(A_{\mathfrak{p}_0}, ^-) = (K_{0\mathfrak{p}_0}, \mathrm{id}) \otimes_{(K_0,\mathrm{id})} (A, ^-) \;\to\; (A_\mathfrak{p}, ^-) = (K_\mathfrak{p}, ^-) \otimes_{(K,^-)} (A, ^-)$$

which we use to identify the involution algebras $(A_{\mathfrak{p}_0}, ^-) = (A_\mathfrak{p}, ^-)$. It is straightforward to check that this implies that the isomorphism of $V_{\mathfrak{p}_0}$ and $V_\mathfrak{p}$ used in identifying them is an $(A_{\mathfrak{p}_0}, ^-) = (A_\mathfrak{p}, ^-)$-isomorphism. Thus

$$\mathrm{End}_{A_{\mathfrak{p}_0}} V_{\mathfrak{p}_0} = \mathrm{End}_{A_\mathfrak{p}} V_\mathfrak{p}.$$

Moreover the identifications respect the topologies since, for example, the identifying map $K_{0\mathfrak{p}_0} \otimes_{K_0} V \to K_\mathfrak{p} \otimes_K V$ is a homeomorphism.

We next show that there is at least one nonsingular ε-Hermitian form

$$h_1 : V \times V \to (A, ^-)$$

on V. Suppose there is no such form. Then by Cor. 3.2.2, p. 115, A is quasisplit (i.e. $\cong \mathrm{M}(n, L)$ for some n), $m := \mathrm{len}_A V$ is odd, and either $\varepsilon = 1$ and $(A, ^-)$ is symplectic, or $\varepsilon = -1$ and $(A, ^-)$ is orthogonal. But then for any prime $\mathfrak{P}|\mathfrak{p}$ of L, $A_\mathfrak{P} = L_\mathfrak{P} \otimes_L A$ is split over $L_\mathfrak{P}$ and $(A_\mathfrak{P}, ^-)$ is of symplectic type if $\varepsilon = 1$ or of orthogonal type if $\varepsilon = -1$. Since $V \cong L^{n\times m}$, $V_\mathfrak{P} = L_\mathfrak{P} \otimes_L V \cong L_\mathfrak{P}^{n\times m}$ also has odd length over $A_\mathfrak{P}$ for each $\mathfrak{P}$ – namely m – and so it does not support a nonsingular ε-Hermitian form over $(A_\mathfrak{P}, ^-)$ (*ibid*). This is a contradiction since there *is* a nonsingular ε-Hermitian form f on the module

$$V_\mathfrak{p} = \oplus_{\mathfrak{P}|\mathfrak{p}} A_\mathfrak{P} V_\mathfrak{p}$$

since g is one on $W \cong V_{\mathfrak{p}}$, and

$$\oplus_{\mathfrak{P}|\mathfrak{p}} A_{\mathfrak{P}} V_{\mathfrak{p}} = \perp_{\mathfrak{P}|\mathfrak{p}} A_{\mathfrak{P}} V_{\mathfrak{p}} = \perp_{\mathfrak{P}|\mathfrak{p}} V_{\mathfrak{P}}$$

is an orthogonal decomposition with respect to f since $A_{\mathfrak{P}} A_{\mathfrak{P}'} = 0$ if $\mathfrak{P} \neq \mathfrak{P}'$; thus

$$f|_{V_{\mathfrak{P}} \times V_{\mathfrak{P}}} : V_{\mathfrak{P}} \times V_{\mathfrak{P}} \to (A_{\mathfrak{P}}, {}^-)$$

is a nonsingular ε-Hermitian form.

There are two "completions" of h_1, namely

$$h_{1\mathfrak{p}_0} : V_{\mathfrak{p}_0} \times V_{\mathfrak{p}_0} \to (A_{\mathfrak{p}_0}, {}^-) \text{ and } h_{1\mathfrak{p}} : V_{\mathfrak{p}} \times V_{\mathfrak{p}} \to (A_{\mathfrak{p}}, {}^-),$$

given by

$$h_{1\mathfrak{p}_0}(\alpha_0 \otimes u, \beta_0 \otimes v) = (\alpha_0\beta_0) \otimes h_1(u,v) \text{ and } h_{1\mathfrak{p}}(\alpha \otimes u, \beta \otimes v) = (\alpha\bar{\beta}) \otimes h_1(u,v),$$

and, as expected, they are equal under our identifications since the diagram

$$\begin{array}{ccc} V_{\mathfrak{p}_0} \times V_{\mathfrak{p}_0} & \xrightarrow{h_{1\mathfrak{p}_0}} & (A_{\mathfrak{p}_0}, {}^-) \\ \downarrow & & \downarrow \\ V_{\mathfrak{p}} \times V_{\mathfrak{p}} & \xrightarrow{h_{1\mathfrak{p}}} & (A_{\mathfrak{p}}, {}^-) \end{array}$$

is commutative.

By extending the scalars from K to $K_{\mathfrak{p}}$, we can assume that $\mathrm{End}_A V \subset \mathrm{End}_{A_{\mathfrak{p}}} V_{\mathfrak{p}} = \mathrm{End}_{A_{\mathfrak{p}_0}} V_{\mathfrak{p}_0}$. Let $\sim$ be the adjoint involution of $\mathrm{End}_{A_{\mathfrak{p}_0}} V_{\mathfrak{p}_0}$ with respect to $h_{1\mathfrak{p}_0}$; it is straightforward to check that the restriction of $\sim$ to $\mathrm{End}_K V$ is the adjoint with respect to h_1.

We first wish to show that the subset $(\mathbf{GL}_A V)_+$ of $\mathbf{GL}_A V$ of symmetric elements with respect to $\sim$ is dense in $(\mathbf{GL}_{A_{\mathfrak{p}_0}} V_{\mathfrak{p}_0})_+$ with respect to the topology stemming from the prime $\mathfrak{p}_0$.

Consider the diagram:

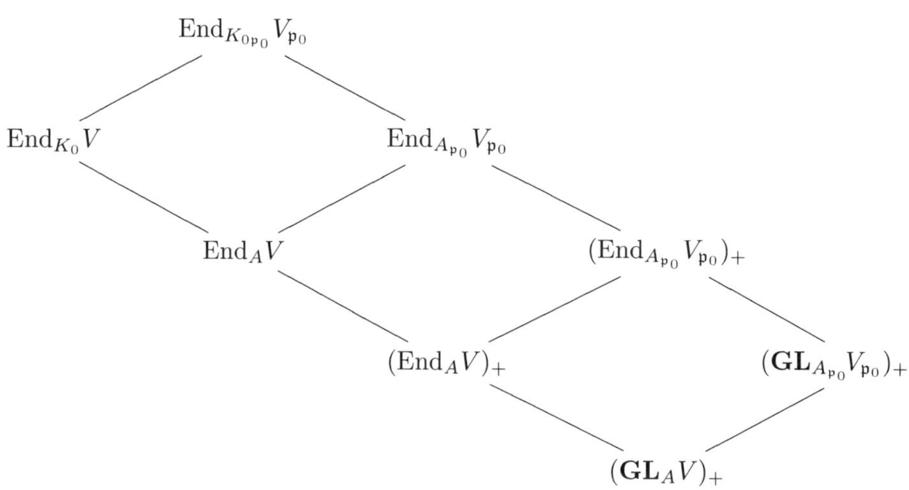

The K_0-space $\mathrm{End}_{K_0} V$ is dense in $\mathrm{End}_{K_{0\mathfrak{p}_0}} V_{\mathfrak{p}_0}$ since the "matrix units" e_{ij} corresponding to a K_0-basis $\{v_j\}$ of V,

$$e_{ij}(v_j) = v_i,$$

form a K-basis of $\mathrm{End}_{K_0} V$ and a $K_{0\mathfrak{p}_0}$-basis of $\mathrm{End}_{K_{0\mathfrak{p}_0}} V_{\mathfrak{p}_0}$.

Since $\mathrm{End}_A V$ and $\mathrm{End}_{A_{\mathfrak{p}_0}} V_{\mathfrak{p}_0}$ are the centralizers of (the images of) A and $A_{\mathfrak{p}_0}$ in $\mathrm{End}_{K_0} V$ and $\mathrm{End}_{K_{0\mathfrak{p}_0}} V_{\mathfrak{p}_0}$ respectively, the canonical homomorphism $K_{0\mathfrak{p}_0} \otimes_{K_0} \mathrm{End}_A V \to \mathrm{End}_{A_{\mathfrak{p}_0}} V_{\mathfrak{p}_0}$ is an isomorphism by Lem. 1.1, p. 11. Therefore $\mathrm{End}_A V$ spans $\mathrm{End}_{A_{\mathfrak{p}_0}} V_{\mathfrak{p}_0}$ as a vector space over $K_{0\mathfrak{p}_0}$ and so is also dense in it.

The involution $\sim$ is a K_0-linear transformation on $\mathrm{End}_A V$, with eigenvalues ± 1; let $\mathrm{End}_A V = (\mathrm{End}_A V)_+ \oplus (\mathrm{End}_A V)_-$ be its eigenspace decomposition. It is clear that

$$\mathrm{End}_{A_{\mathfrak{p}_0}} V_{\mathfrak{p}_0} = K_{0\mathfrak{p}_0} \mathrm{End}_A V = K_{0\mathfrak{p}_0} (\mathrm{End}_A V)_+ \oplus K_{0\mathfrak{p}_0} (\mathrm{End}_A V)_-$$

is also an eigenspace decomposition, with

$$(\mathrm{End}_{A_{\mathfrak{p}_0}} V_{\mathfrak{p}_0})_+ = K_{0\mathfrak{p}_0} (\mathrm{End}_A V)_+ \quad \text{and} \quad (\mathrm{End}_{A_{\mathfrak{p}_0}} V_{\mathfrak{p}_0})_- = K_{0\mathfrak{p}_0} (\mathrm{End}_A V)_-.$$

Thus $(\mathrm{End}_A V)_+$ is dense in $(\mathrm{End}_{A_{\mathfrak{p}_0}} V_{\mathfrak{p}_0})_+$.

$(\mathrm{End}_{A\mathfrak{p}_0} V_{\mathfrak{p}_0})_+$ is a $K_{0\mathfrak{p}_0}$-subspace of $\mathrm{End}_{A_{\mathfrak{p}_0}} V_{\mathfrak{p}_0}$, and $(\mathbf{GL}_{A_{\mathfrak{p}_0}} V_{\mathfrak{p}_0})_+$ is an open subset of it since it is the intersection of $(\mathrm{End}_{A\mathfrak{p}_0} V_{\mathfrak{p}_0})_+$ and the open subset of $\mathrm{End}_{K_{0\mathfrak{p}_0}} V_{\mathfrak{p}_0}$ of transformations of determinant $\neq 0$. Thus $(\mathbf{GL}_A V)_+$ is indeed dense in $(\mathbf{GL}_{A_{\mathfrak{p}_0}} V_{\mathfrak{p}_0})_+$.

Let $f : V_{\mathfrak{p}_0} \times V_{\mathfrak{p}_0} \to (A_{\mathfrak{p}_0}, {}^-)$ be the ε-Hermitian form corresponding to the form g on W via an $A_{\mathfrak{p}_0}$-isomorphism $W \to V_{\mathfrak{p}_0}$. By Prop. 3.1.1, p. 94, $f(u,v) = h_{1\mathfrak{p}_0}(u\varphi, v)$ for some (uniquely determined) $\varphi \in (\mathbf{GL}_{A_{\mathfrak{p}_0}} V_{\mathfrak{p}_0})_+$. Moreover the ε-Hermitian forms on $V_{\mathfrak{p}_0}$ which are equivalent to f are of the form $h_{1\mathfrak{p}_0}(ub\varphi\tilde{b}, v)$, $b \in \mathbf{GL}_{A_{\mathfrak{p}_0}} V_{\mathfrak{p}_0}$, i.e. they correspond to the transformations in the orbit $\{b\varphi\tilde{b} : b \in \mathbf{GL}_{A_{\mathfrak{p}_0}} V_{\mathfrak{p}_0}\}$. It suffices to show that there is an open neighborhood of φ in this orbit, since then the density of $(\mathbf{GL}_A V)_+$ in $(\mathbf{GL}_{A_{\mathfrak{p}_0}} V_{\mathfrak{p}_0})_+$ would imply the existence of $\psi \in (\mathbf{GL}_A V)_+$ such that the completion of $h_{1\psi} : V \times V \to (A, {}^-)$, $h_{1\psi}(u,v) = h_1(u\psi, v)$, is equivalent to $f(u,v)$.

With this in mind, we consider using f instead of $h_{1\mathfrak{p}_0}$ as the "base" form: If $b \in B := \mathrm{End}_{A_{\mathfrak{p}_0}} V_{\mathfrak{p}_0}$,

$$\begin{aligned} f(ub, v) &= h_{1\mathfrak{p}_0}(ub\varphi, v) = h_{1\mathfrak{p}_0}(u\varphi(\varphi^{-1}b\varphi), v) \\ &= h_{1\mathfrak{p}_0}(u\varphi, v\varphi\tilde{b}\varphi^{-1}) = f(u, v\varphi\tilde{b}\varphi^{-1}), \end{aligned}$$

so the adjoint involution of f is

$$\breve{b} = \varphi\tilde{b}\varphi^{-1}.$$

The orbit of 1 under the "$\breve{\ }$-action" $b_+ \to b b_+ \breve{b}$ of $\mathbf{GL}_{A_{\mathfrak{p}_0}} V_{\mathfrak{p}_0}$ on $(B, \breve{\ })_+$ is

$$\{b\breve{b}\} = \{b\varphi\tilde{b}\varphi^{-1}\} = \{\text{orbit } b\varphi\tilde{b} \text{ of } \varphi \text{ under the } \sim\text{-action of } \mathbf{GL}_{A_{\mathfrak{p}_0}} V_{\mathfrak{p}_0}\}\varphi^{-1}.$$

Right multiplication $\psi \to \psi\varphi^{-1}$ by φ^{-1} is a homeomorphism from the $\sim$-symmetric elements of $\mathrm{End}_{A_{\mathfrak{p}_0}} V_{\mathfrak{p}_0}$ to the $\breve{\ }$-symmetric elements, so it suffices to show that there exists an open neighborhood N of 1 in $\{b\breve{b} : b \in \mathbf{GL}_{A_{\mathfrak{p}_0}} V_{\mathfrak{p}_0}\}$ (in the topological space

$$S = (\mathrm{End}_{A_{\mathfrak{p}_0}} V_{\mathfrak{p}_0}, \breve{\ })_+$$

of $\breve{\ }$-symmetric elements) since

$$1 \in N \subset \{b\breve{b}\} \Rightarrow \varphi \in N\varphi \subset \{b\breve{b}\}\varphi = \{b\varphi\tilde{b}\}$$

so $N\varphi$ is an open neighbourhood of φ in the orbit $\{b\varphi\tilde{b}\}$ as desired.

Let $|\ |_{\mathfrak{p}_0}$ be the extension to $K_{0\mathfrak{p}_0}$ of any valuation $\in \mathfrak{p}_0$. If $X = (x_{ij})$ is a matrix with entries in $K_{0\mathfrak{p}_0}$, define $||X|| = \max_{i,j}\{|x_{ij}|_{\mathfrak{p}_0}\}$. For the moment assume the following lemma:

Lemma 3.4.1 *There is a power series*

$$\sum_{k=1}^{\infty} \alpha_k x^k \in K_0[[x]]$$

such that if n is a positive integer, $0 < r < 1$, and $X \in \mathrm{M}(n, K_{0\mathfrak{p}_0})$ satisfies $||X|| < \frac{r}{n}$, the series $\sum_{k=1}^{\infty} \alpha_k X^k$ converges to $Y \in \mathrm{M}(n, K_{0\mathfrak{p}_0})$ and satisfies $(1+Y)^2 = 1 + X$.

Now identify $\mathrm{End}_{A_{\mathfrak{p}_0}} V_{\mathfrak{p}_0} \subset \mathrm{End}_{K_{0\mathfrak{p}_0}} V_{\mathfrak{p}_0} = \mathrm{M}(n, K_{0\mathfrak{p}_0})$ $(n = \dim_{K_0} V)$, and define

$$N = \{1 + X \in S : ||X|| < \varepsilon\}.$$

Then N is an open neighbourhood of 1 in S, and by choosing ε sufficiently small, we can guarantee that the determinant is $\neq 0$ on N and so $N \subset \mathbf{GL}_{A_{\mathfrak{p}_0}} V_{\mathfrak{p}_0}$. If $1 + X \in N$ and if Y is chosen as in the lemma, then since S is a $K_{0\mathfrak{p}_0}$-subspace and $X^k \in S$ for all $k \geq 1$, the partial sums $\sum_{k=1}^{m} \alpha_k X^k$ are also in S, so $Y \in S$ since S is closed. Thus $1 + X = (1+Y)^2 = (1+Y)(1+\breve{Y})$, so $1 \in N \subset \{b\breve{b}\}$ as desired. $\square$

Proof of Lemma. If $0 < r < 1$ and $||X|| < \frac{r}{n}$,

$$||X^m|| < \frac{r^m}{n} < r^m \quad \text{for all } m \tag{3.60}$$

by induction on m.

We use the binomial series $(1+x)^{\frac{1}{2}} = \sum_{k=0}^{\infty} \alpha_k x^k$, where $\alpha_0 = 1$ and for $k \geq 1$

$$\alpha_k = \frac{\frac{1}{2}(-\frac{1}{2})(-\frac{3}{2})\cdots(-\frac{(2k-3)}{2})}{k!} = (-1)^{k-1}\frac{1\cdot 1\cdot 3\cdot\cdots\cdot(2k-3)}{k!2^k}.$$

We first show that the series is defined over any field of characteristic $\neq 2$. In other words, if the rational numbers α_k are expressed in their lowest terms (numerator and denominator relatively prime), the denominator is at most a power of 2. Now $\alpha_1 = \frac{1}{2}$, $\alpha_2 = -\frac{1}{8}$, and, for $k \geq 3$,

$$\begin{aligned}(-1)^{k-1}2^k\alpha_k = \frac{1\cdot 3\cdot 5\cdots(2k-3)}{k!} &= \frac{(2k-3)!}{2^{k-2}(k-2)!k(k-1)!} = \frac{\binom{2k-3}{k-1}}{2^{k-2}k}\\ &= \frac{(2k-3)!}{2^{k-2}(k-2)(k-3)!k!} = \frac{\binom{2k-3}{k}}{2^{k-2}(k-2)}.\end{aligned}$$

The form of the denominators $2^{k-2}k$ and $2^{k-2}(k-2)$ shows that when α_k is expressed in its lowest terms, the denominator is at most a power of 2 – in fact at most 2^{2k-1}.

The series converges absolutely for $|x|_{\mathfrak{p}} < 1$ in the Archimedean case and $|x|_{\mathfrak{p}} < |4|_{\mathfrak{p}}$ when $\mathfrak{p}$ is nonArchimedean, and since its square is $1 + x$ as a formal series over $\mathbb{Q}$, this is also true over $K_{\mathfrak{p}}$ for $|x|_{\mathfrak{p}} < 1$ respectively $|4|_{\mathfrak{p}}$.

Since $(\sum_k \alpha_k x^k)^2 = 1 + x$ is an identity in the ring of formal power series over $\mathbb{Q}$,

$$(\sum_{k=0}^{N} \alpha_k x^k)^2 = 1 + x + 0x^2 + \cdots + 0x^N + \beta_{N,N+1}x^{N+1} + \cdots + \beta_{N,2N}x^{2N} \quad (3.61)$$

for the rational numbers

$$\beta_{N,N+k} = \alpha_k\alpha_N + \alpha_{k+1}\alpha_{N-1} + \cdots + \alpha_N\alpha_k, \quad k \geq 1.$$

It follows that $4^{N+k}\beta_{N,N+k}$ is a rational integer.

Suppose first that $\mathfrak{p}_0$ is Archimedean. It is easy to see that $|\alpha_k| < 1$ for all k, and so

$$|\beta_{N,N+k}| < N - k + 1 < N + k \text{ for all } k,\ 1 \leq k \leq N.$$

We can assume that $|\ \ |_{\mathfrak{p}_0}$ is normalized so that on $\mathbb{Q}$ it is $= |\ \ |$, the ordinary absolute value. So for the matrix X, if $||X|| < \frac{r}{n}$ and $0 < r < 1$,

$$\begin{aligned} \left| \left(\sum_{k=1}^{N} \beta_{N,N+k} X^{N+k} \right)_{ij} \right|_{\mathfrak{p}_0} &= \left| \sum_{k=1}^{N} \beta_{N,N+k} (X^{N+k})_{ij} \right|_{\mathfrak{p}_0} \\ &\leq \sum_{k=1}^{N} |\beta_{N,N+k}|_{\mathfrak{p}_0} ||X^{N+k}|| \\ &< \sum_{k=1}^{N} (N+k) r^{N+k} \to 0 \text{ as } N \to \infty, \end{aligned}$$

by (3.60) since the series $\sum_{m=1}^{\infty} m r^m$ is convergent. This proves the lemma in the Archimedean case.

Now suppose $\mathfrak{p}_0$ is nonArchimedean, and assume $0 < r < |4|_{\mathfrak{p}_0}$ and $||X|| < \frac{r}{n}$. Then

$$\begin{aligned} \left| \left(\sum_{k=1}^{N} \beta_{N,N+k} X^{N+k} \right)_{ij} \right|_{\mathfrak{p}_0} &= \left| \sum_{k=1}^{N} \beta_{N,N+k} (X^{N+k})_{ij} \right|_{\mathfrak{p}_0} \\ &\leq \max_{1 \leq k \leq N} \left\{ |\beta_{N,N+k}|_{\mathfrak{p}_0} |(X^{N+k})_{ij}|_{\mathfrak{p}_0} \right\} \\ &< \frac{r^{N+k}}{|4|_{\mathfrak{p}_0}^{N+k}} \to 0 \text{ as } N \to \infty \end{aligned}$$

since $|(X^{N+k})_{ij}|_{\mathfrak{p}_0} \leq ||X^{N+k}|| < r^{N+k}$ and $|4^{N+k}\beta_{N,N+k}|_{\mathfrak{p}_0} \leq 1$ since $4^{N+k}\beta_{N,N+k}$ is a multiple of 1_{K_0}. This finishes the proof of the lemma, and so also of Th. 3.4.1. □

Corollary 3.4.1 *Suppose that $\mathfrak{p}$ is a prime of K such that the involution $^-$ on K can be extended to an involution on the completion $K_\mathfrak{p}$, also denoted by $^-$. Let $\rho : G \to \mathbf{I}(W, g)$ be an isometry representation over $(K_\mathfrak{p}, ^-)$. If the underlying linear representation $G \to \mathbf{I}(W, g) \subset \mathbf{GL}(W)$ is the completion of a linear representation of G over K, then ρ is the completion of an isometry representation $G \to \mathbf{I}(V, h)$ over $(K, ^-)$.*

This follows from the foregoing theorem, using the isometry representation $\rho : G \to \mathbf{I}(V, h)$ associated with a form $\hat{h} : V \times V \to (KG, ^-)$ and Th. 2.1.1 (p. 86). The details are omitted. □

3.5 Existence of isometry representations

In this section the theory of forms over simple algebras is applied to determine, for a given finite group G, vector space V over K and $\varepsilon = \pm 1$, the precise conditions under which V supports a nonsingular ε-reflexive form h such that there is at least one isometry representation $G \to \mathbf{I}(V, h)$. This is not much more than a restatement of the nature of indecomposable forms over a simple algebra with involution in §3.2.4. The notation in §3.2.7 (p. 118) is used freely.

Theorem 3.5.1 *Suppose that G acts on V through the linear representation $\rho : G \to \mathbf{GL}(V)$. Let $^-$ also denote the standard $(K, ^-)$-involution on KG, and let $\varepsilon = \pm 1$. Then there is a nonsingular G-invariant ε-reflexive form $h : V \times V \to (K, ^-)$ if and only if*

(a) for each simple component A of KG, $\operatorname{len}_A AV = \operatorname{len}_{\bar{A}} \bar{A}V$,

(b) for each simple quasisplit component A of KG, $\operatorname{len}_A AV$ is even if

(i) $\varepsilon = -1$ and $(A, ^-)$ is orthogonal, or

(ii) $\varepsilon = 1$ and $(A, ^-)$ is symplectic.

Suppose an invariant form h exists, and let $\hat{h} : V \times V \to (KG, ^-)$ be the associated ε-Hermitian form. The condition $\operatorname{len}_A AV = \operatorname{len}_{\bar{A}} \bar{A}V$ is trivial if $A = \bar{A}$; otherwise $A \oplus \bar{A}$ is hyperbolic, so the restriction of $\hat{h}$ to $(A \oplus \bar{A})V$ is an orthogonal direct sum of hyperbolic forms based on simple modules by Th. 3.2.5, p. 114, and so is itself hyperbolic; thus AV and $\bar{A}V$ have the same length.

Suppose that A is quasi-split and $= \bar{A}$. By Cor. 3.2.2 (p. 115), an indecomposable ε-Hermitian form over $(A, ^-)$ is hyperbolic if (i) or (ii) obtains. Since AV is a direct sum of indecomposable forms, it must have even length in these cases.

To prove the sufficiency of the conditions, it is enough to prove that there is a form of the appropriate kind on each involution isotypic component of V. Let A be a simple component of KG.

If $\bar{A} = A$, there is a nonsingular ε-Hermitian form $\hat{h} : AV \times AV \to (A, ^-)$ by Cor. 3.2.2. Then we can take $h = \operatorname{Tr}_{KG/K}\hat{h}$.

Now suppose $\bar{A} \neq A$, so $A \oplus \bar{A}$ is a simple hyperbolic involution algebra $\mathrm{H}(A)$ (p. 25). We can assume that V is an $\mathrm{H}(A)$-module such that $\operatorname{len}_A AV = \operatorname{len}_{\bar{A}}\bar{A}V$. In general the twisted dual module $\overline{U^*}$ of a (left) A-module U is a left module over $\bar{A}$, of the same length as U, and so $V = AV \oplus \bar{A}V \cong AV \oplus \overline{(AV)^*}$ (as $A \oplus \bar{A}$-modules). Thus V supports an ε-hyperbolic form $\hat{h} : V \times V \to (A \oplus \bar{A}, ^-)$ such that $\hat{h} \cong h_{\mathrm{H}(V)}$. Again we need only take $h = \operatorname{Tr}_{KG/K}\hat{h}$ to prove the theorem. □

Corollary 3.5.1 *If the involution on K is not the identity, then there is a nonsingular Hermitian form and a nonsingular skew Hermitian form on V such that ρ is an isometry representation $G \to \mathbf{I}(V, h)$ if and only if $\operatorname{len}_A AV = \operatorname{len}_{\bar{A}} \bar{A}V$ for each simple component A of KG.*

This is clear since part (b) of the theorem does not apply. □

REMARK. Th. 3.5.1 is closely related to a result of J. Ritter ((1) of the Theorem on p. 521 of [63]). Let $\varepsilon = \pm 1$, $\widehat{K}$ be an algebraic closure of K, $\mathrm{K}_0^{\varepsilon}(KG)$ the

Grothendieck group of KG-modules which support a nonsingular G-invariant ε-symmetric form over (K, id) – it is a subgroup of $\mathrm{K}_0(KG)$ – and $\varphi : \mathrm{K}_0(KG) \to \mathrm{K}_0(\widehat{K}G)$ the map given by extension of scalars. Then Ritter's result[9] (in our notation) is

$$\mathrm{K}_0^\varepsilon(KG) = \varphi^{-1}\mathrm{K}_0^\varepsilon(\widehat{K}G). \tag{3.62}$$

In other words, a KG-module V supports an orthogonal or symplectic representation $G \to \mathbf{I}(V, f)$ if and only if the $\hat{K}G$-module $V^{\widehat{K}}$ does (over $\widehat{K}$, of course).

It is clear that $\mathrm{K}_0^\varepsilon(KG) \subset \varphi^{-1}\mathrm{K}_0^\varepsilon(\widehat{K}G)$ since $\varphi(\mathrm{K}_0^\varepsilon(KG)) \subset \mathrm{K}_0^\varepsilon(\widehat{K}G)$, and we will show that Th. 3.5.1 implies the reverse containment.

Let V be a KG-module such that $V^{\widehat{K}}$ supports a nonsingular G-invariant ε-symmetric form over $(\widehat{K}, \mathrm{id})$. Let A be a simple component of KG, and suppose that $A^{\widehat{K}} = A_1 \oplus \cdots \oplus A_l$ where the A_i are simple components of $\widehat{K}G$. By the above theorem, for each i, $\mathrm{len}_{A_i}(A_i V^{\widehat{K}}) = \mathrm{len}_{\overline{A}_i}(\overline{A}_i V^{\widehat{K}})$ and is even if $\varepsilon\mathrm{type}\, A = -1$, since the type of each A_i is the same as that of A (Th. 1.3.2, p. 29). Since the $\widehat{K}$-dimension of a simple $\overline{A}_i^{\widehat{K}}$-module is the same as that of a simple $A_i^{\widehat{K}}$-module, the $\hat{K}$-spaces $A^{\widehat{K}}V^{\widehat{K}}$ and $\bar{A}^{\widehat{K}}V^{\widehat{K}}$ have the same dimension, therefore the K-spaces AV and $\bar{A}V$ also have the same dimension and so also the same length. This confirms condition (a) of the theorem.

Assume now that A is quasisplit, say $A \cong \mathrm{M}(n, L)$. Thus $l = (L : K)$. Now suppose also that $\varepsilon\mathrm{type}\, A = -1$; then $\varepsilon\mathrm{type}\, A_i = -1$ for each i, so the length of each $A_i V^{\widehat{K}}$ is even (by the theorem), and therefore its $\hat{K}$-dimension is an even multiple of n (since $A_i \cong \mathrm{M}(n, \widehat{K})$ for each i), so the $\widehat{K}$-dimension of $AV^{\widehat{K}}$ is an even multiple of $n(L : K) = \dim_K AV$. But the dimension of a simple A-module is also $n(L : K)$, so AV has even length. Thus V supports a nonsingular invariant ε-symmetric form over (K, id) by the theorem. □

3.6 Witt's Theorem for isometry representations

We give here an alternative proof, using Morita theory, of Witt's Theorem for isometry representations – cf. Cor. 3.1.8 (p. 104). In the language of invariant forms, it says that if (V_i, h_i) $(i = 1, 2, 3)$ are nonsingular G-invariant ε-reflexive spaces (over a field K of characteristic not dividing $2g$), then

$$h_1 \perp h_2 \cong_G h_1 \perp h_3 \quad \Rightarrow \quad h_2 \cong_G h_3.$$

Here $\cong_G$ stands for equivariant equivalence. We will use Morita equivalence to reduce it to the classical case of Witt's Theorem for ε-reflexive forms over division rings – see Th. 7.9.5, p. 269, [65], for example.

The category of nonsingular G-invariant ε-reflexive spaces is essentially the same as the category ${}^{KG}_{K}\mathcal{H}^*_\varepsilon$ of nonsingular $(KG, {}^-)$-invariant ε-Hermitian forms over $(K, {}^-)$, and the canonical trace $\mathrm{Tr} : (KG, {}^-) \to (K, {}^-)$ gives an isomorphism of categories

$${}^{KG}_{K}\mathcal{H}^*_\varepsilon \cong {}_{KG}\mathcal{H}^*_\varepsilon$$

by the "going down theorem", Th. 3.2.4 (p. 113). This isomorphism preserves direct sums and orthogonality, hence also orthogonal direct sums. Thus it suffices to prove "cancellation" for nonsingular ε-Hermitian forms $h : V \times V \to (KG, {}^-)$.

[9] He states it only for the case of K an algebraic number field.

Next let

$$(KG, ^-) = (A_1, ^-) \oplus \cdots \oplus (A_l, ^-)$$

be the decomposition of $(KG, ^-)$ into simple involution algebras. Then $V = A_1V \oplus \cdots \oplus A_lV$ and any $(V, h) \in {}_{KG}\mathcal{H}^*_\varepsilon$ has a decomposition

$$(V, h) = (A_1V, h_1) \times \cdots \times (A_lV, h_l)$$

into a vector sum of the nonsingular forms $h_i = h|_{V_i \times V_i} : A_iV \times A_iV \to (A_i, ^-)$. It follows easily that it suffices to prove cancellation in ${}_{A_i}\mathcal{H}^*_\varepsilon$.

If A_i is not simple, $(A_i, ^-)$ is hyperbolic and

$$(V, h) \perp (U, f) \cong (V, h) \perp (W, g) \quad \Rightarrow \quad (U, f) \cong (W, g) \tag{3.63}$$

by Cor. 3.1.6 (p. 98) since (U, f) and (W, g) are hyperbolic and have the same length.

Finally suppose that A_i is simple. If $h \perp f \cong h \perp g$ where the f, g and h are nonsingular ε-Hermitian forms over $(A_i, ^-)$, then by Morita theory, $\breve{h} \perp \breve{f} \cong \breve{h} \perp \breve{g}$ for the Morita correspondents over the division algebra similar to A_i; we then apply Witt's Theorem in this category to get $\breve{f} \cong \breve{g}$, and so also $f \cong g$, again by Morita theory. □

3.7 Notes on Chapter 3

1. The theory of parts of this chapter, especially §§3.2.1 – 3.2.3 and §3.2.9, appears in Ch. I of [35] – and in greater generality, providing a thoroughgoing background as well for approaching the "integral" theory. The basic facts in §3.2.4 were proved by McEvett [49].

2. Concerning Th. 3.2.11, p. 126. When $(A, ^-)$ is of the first kind, each $d_1 \in D$ is fixed by *some* involution of the first kind – in fact since the characteristic of K is $\neq 2$, it is fixed by an orthogonal involution (Prop. 4.17, p. 52, [36]). Thus it is not possible to strengthen the theorem in this case by restricting the domain of d_1 to some proper subset of $\dot{D}$. In principle, this may be possible when $(A, ^-)$ is of the second kind – in this case it is known (Prop. 4.18, *ibid*) that every $d \in \dot{D}$ with minimal polynomial over K of degree = the degree of D and coefficients in K_0 is fixed by some involution of the second kind on D, but to my knowledge nothing further is known about this question.

3. Morita theory for ε-Hermitian forms was introduced in [27], where it is referred to as "Hermitian Morita" theory, and is given in much greater generality. A good presentation of it (and the "ordinary" Morita theory for categories of modules) appears in [35] on pp. 51–59, where (p. 57) there is also a treatment of the "explicit" theory (as given in this book) in the case when $A = \mathrm{M}(n, (B, ^-))$ where $(B, ^-)$ is any ring with involution. D.W. Lewis and T. Unger also provide a version which is similar to ours in [44].

The more general Morita theory described in [27] or [35] provides, inter alia, a category equivalence between ${}_{(A,^-)}\mathcal{H}_\varepsilon$ and ${}_{(D,^-)}\mathcal{H}_{\varepsilon\varepsilon_0}$. This follows from Th. 3.2.14 (p. 132) (in the special case considered here – K a field), as follows.

Fix an isomorphism $\theta : (\mathrm{M}(n,D),\,^-) \to (A,\,^-)$ and let $\mathcal{M}_{\varepsilon\varepsilon_0}(D,\,^-)$ be the category of $\varepsilon\varepsilon_0$-Hermitian matrices $\underline{h} \in \mathrm{M}(m,D)$ ($\underline{h}^{t-} = \varepsilon\varepsilon_0\underline{h}$), for $m = 1,2,\ldots$ with morphisms $\underline{\varphi} \in D^{m'\times m} = \mathrm{Hom}(D^{m\times m}, D^{m'\times m'})$ acting as $\underline{h} \to \underline{\varphi}\underline{h}\underline{\varphi}^{t-}$. The map

$$\theta : \mathcal{M}_{\varepsilon\varepsilon_0}(D,\,^-) \to {}_{(A,^-)}\mathcal{H}_\varepsilon, \quad \theta(\underline{h})(u,v) = u\underline{h}v^{t-}\underline{h}_0$$

is a full and faithful functor such that each $h \in {}_{(A,^-)}\mathcal{H}_\varepsilon$ is isomorphic to a form in the image, and so is an equivalence of categories (Th. 1, p. 91, [40]). Applying this to the case $n = 1$ and $\underline{h}_0 = 1$ gives an equivalence of categories $\mathcal{M}_{\varepsilon\varepsilon_0}(D,\,^-) \to_{(D,^-)}\mathcal{H}_{\varepsilon\varepsilon_0}$, and it follows that ${}_{(A,^-)}\mathcal{H}_\varepsilon$ and ${}_{(D,^-)}\mathcal{H}_{\varepsilon\varepsilon_0}$ are equivalent.

There is however one difference between this equivalence and that in [27]: under certain circumstances, the form over D corresponding to $h \in_{(A,^-)} \mathcal{H}_\varepsilon$ in [27] has discriminant matrix $-\underline{h}$ instead of $\underline{h}$. See pp. 4698-4700, especially Lem. 1.7 and Remark 1.9, in [60]. This "anomaly" occurs if and only if $(A,\,^-)$ is quasisplit of symplectic type, but need not concern us.

4. Theorem 3.4.1 was first proved by A. Fröhlich (Th. 11.1, p. 500, [26]) when $(K,\,^-)$ is of the first kind, i.e. when $K = K_0$ – and is given here in a form which explicitly includes $(K,\,^-)$ of the second kind, but is in fact no more general than Fröhlich's theorem. □

CHAPTER 4

Equivariant Witt-Grothendieck and Witt groups

In this chapter, we begin by defining the Witt-Grothendieck and Witt groups over semisimple algebras, and then show how they are related to the equivariant Witt-Grothendieck and Witt groups (of isometry representations). The structure of the equivariant Witt-Grothendieck and Witt groups over special fields is then examined and the results applied to the equivalence of isometry representations over these fields.

4.1 The Witt-Grothendieck and Witt groups of Hermitian forms over semisimple algebras

Let $(A, ^-)$ be an involution algebra over $(K, ^-)$ which is semisimple as a K-algebra.

Let $[h]$ be the isometry class of any $h \in {}_A\mathcal{H}_\varepsilon^*$, including possibly $h = 0$. Such isometry classes form an Abelian monoid $\mathbf{M}_\varepsilon = {}_A\mathbf{M}_\varepsilon$ under the addition $[h_1] + [h_2] = [h_1 \perp h_2]$. The Grothendieck group of $\mathbf{M}_\varepsilon$ (cf. §1.6) is denoted by $\hat{\mathrm{W}}_\varepsilon(A, ^-)$ or simply by $\hat{\mathrm{W}}_\varepsilon(A)$ when $^-$ is clear from the context. We refer to it as the *Witt-Grothendieck group of nonsingular ε-Hermitian forms (over $(A, ^-)$)*, or simply as the *Witt-Grothendieck group.*

Note that $\iota : \mathbf{M}_\varepsilon \to \hat{\mathrm{W}}_\varepsilon(A)$ is injective by Witt's Theorem (Cor. 3.1.7, p. 104) and (1.47), p. 65.

Let U be an A-module (finitely generated as usual) and $h = h_{\mathrm{H}(U)}$ the ε-Hermitian hyperbolic form on $\mathrm{H}(U) = U \oplus \overline{U^*}$ (p. 97). By (3.14)

$$h_{\mathrm{H}(U_1)} \perp h_{\mathrm{H}(U_2)} \cong h_{\mathrm{H}(U_1 \oplus U_2)}. \tag{4.1}$$

Since every ε-hyperbolic module is (equivalent to one) of the form $\mathrm{H}(U)$ (Th. 3.1.2, p. 97), the set of isometry classes of hyperbolic forms is a submonoid $\mathbf{H}_\varepsilon$ of $\mathbf{M}_\varepsilon$, and we denote the subgroup of $\hat{\mathrm{W}}_\varepsilon(A)$ generated by its image by $\mathrm{H}_\varepsilon(A)$ or $\mathrm{H}_\varepsilon(A, ^-)$. (It is in fact (isomorphic to) the Grothendieck group of $\mathbf{H}_\varepsilon$).

Definition 4.1.1 *The Witt group $\mathrm{W}_\varepsilon(A, ^-) = \mathrm{W}_\varepsilon(A)$ of ε-Hermitian forms over $(A, ^-)$ is the factor group $\hat{\mathrm{W}}_\varepsilon(A)/\mathrm{H}_\varepsilon(A)$.*

If $(K, ^-)$ is of the second kind, then $\hat{\mathrm{W}}_1(A, ^-) \cong \hat{\mathrm{W}}_{-1}(A, ^-)$ and $\mathrm{W}_1(A, ^-) \cong \mathrm{W}_{-1}(A, ^-)$ (non-canonically), by scaling by $\sqrt{\delta}$ where $K = K_0(\sqrt{\delta})$.

If the involution on K is the identity, then $\mathrm{W}_1(K)$ is the classical Witt group $\mathrm{W}(K)$ (of symmetric forms), $\hat{\mathrm{W}}_{-1}(K) \cong \mathbb{Z}$ since a nonsingular skew symmetric form is determined up to equivalence by its rank (cf. p. 48), and $\mathrm{W}_{-1}(K) = 0$ since any nonsingular skew symmetric form is hyperbolic.

By Th. 3.2.2(a) (p. 110), if $\varphi : (A, ^-) \to (A', ^-)$ is a homomorphism of $(K, ^-)$-involution algebras, there are induced homomorphisms

$$\hat{\mathrm{W}}_\varepsilon(A, ^-) \to \hat{\mathrm{W}}_\varepsilon(A', ^-), \quad \mathrm{H}_\varepsilon(A, ^-) \to \mathrm{H}_\varepsilon(A', ^-), \quad \mathrm{W}_\varepsilon(A, ^-) \to \mathrm{W}_\varepsilon(A', ^-)$$

and isomorphisms

$$\hat{\mathrm{W}}_\varepsilon(A, ^-) \cong \hat{\mathrm{W}}_\varepsilon(A', ^-), \quad \mathrm{H}_\varepsilon(A, ^-) \cong \mathrm{H}_\varepsilon(A', ^-), \quad \mathrm{W}_\varepsilon(A, ^-) \cong \mathrm{W}_\varepsilon(A', ^-) \tag{4.2}$$

if φ is an isomorphism.

In fact if A and A' are simple algebras, the isomorphisms in (4.2) hold with only the assumption that $A \cong A'$ and $(A, ^-)$ and $(A', ^-)$ have the same type:

Exercises. Let A be a central simple K-algebra. Recall that by Th. 1.3.6(a) (p. 30), if $^-$ and $^\sim$ are $(K, ^-)$-involutions on A, there is $c \in A^\times$ such that $\tilde{a} = c^{-1}\bar{a}c$ and $\bar{c} = \varepsilon_0 c$, $\varepsilon_0 = 1$ if $^-$ and $^\sim$ have the same type, $= -1$ otherwise.
1. Show that $h \rightsquigarrow hc$ is a category isomorphism between ${}_{(A,^-)}\mathcal{H}^*_\varepsilon$ and ${}_{(A,^\sim)}\mathcal{H}^*_{\varepsilon\varepsilon_0}$ (where hc is the form $h(u,v)c$ and it is understood that the functor $\rightsquigarrow$ is the identity on morphisms).

2. Show that there are isomorphisms $\hat{\mathrm{W}}_\varepsilon(A, ^-) \cong \hat{\mathrm{W}}_{\varepsilon\varepsilon_0}(A, ^\sim)$ and $\mathrm{W}_\varepsilon(A, ^-) \cong \mathrm{W}_{\varepsilon\varepsilon_0}(A, ^\sim)$.

3. Now let A and A' be isomorphic simple K-algebras. Show that $\hat{\mathrm{W}}_\varepsilon(A, ^-)$ and $\mathrm{W}_\varepsilon(A, ^-)$ depend only on A and the type of $(A, ^-)$, up to isomorphism. □

This last exercise and Th. 4.1.2 (p. 154) imply that one can define the Witt-Grothendieck and Witt groups for elements of the "involutory Brauer group" $\mathrm{Br}(K, ^-)$ (cf. §4.7) – that is to say, if A and B are similar central simple K-algebras with $(K, ^-)$-involutions $^-$ and $^\sim$, respectively, of the same type, then $\hat{\mathrm{W}}_\varepsilon(A, ^-) \cong \hat{\mathrm{W}}_\varepsilon(B, ^\sim)$.

In general this isomorphism is not canonical, but we shall see later (Cor. 4.8.1, p. 191) that there *is* a canonical map $\mathrm{Sim}_\varepsilon(A, ^-) \to \mathrm{Sim}_\varepsilon(B, ^\sim)$ between *similarity* classes.

Since the category of isometry representations of G over $(K, ^-)$ with respect to ε-reflexive forms is canonically isomorphic to the category of nonsingular ε-Hermitian forms over KG with its standard $(K, ^-)$-involution (Th. 2.1.1, p. 86), the Witt-Grothendieck group of isometry representations of G over K (for given $^-$ on K and ε) – which we denote by $\hat{\mathrm{W}}_\varepsilon(K, ^-, G)$, or more often by $\hat{\mathrm{W}}_\varepsilon(K, G)$ – is canonically isomorphic to $\hat{\mathrm{W}}_\varepsilon(KG, ^-)$. Of course $\hat{\mathrm{W}}_\varepsilon(K, G)$ can also be viewed as the Witt-Grothendieck group of nonsingular G-invariant ε-reflexive forms over $(K, ^-)$.

We sometimes write ${}^{KG}_{K}\mathcal{H}_\varepsilon = {}^{G}_{K}\mathcal{H}_\varepsilon$ and ${}^{KG}_{K}\mathcal{H}^*_\varepsilon = {}^{G}_{K}\mathcal{H}^*_\varepsilon$.

For example, if $^-$ is the identity on K, $\hat{\mathrm{W}}_1(K, G)$ is the Witt-Grothendieck group of orthogonal representations of G and $\hat{\mathrm{W}}_{-1}(K, G)$ that of symplectic representations.

The Witt group $\mathrm{W}_\varepsilon(K, G)$ of isometry representations of G over K is defined as the Witt-Grothendieck group $\hat{\mathrm{W}}_\varepsilon(K, G)$ modulo the subgroup generated by hyperbolic representations – it is canonically isomorphic to $\mathrm{W}_\varepsilon(KG, ^-)$ by Cor. 3.1.2(a), p. 97.

Theorem 4.1.1 *Suppose that*

$$(A, ^-) = (A_1, ^-) \oplus (A_2, ^-) \oplus \cdots \oplus (A_l, ^-)$$

where $(A_j, ^-)$ *is a hyperbolic involution algebra for* $j > k$. *Then*

$$\hat{\mathrm{W}}_\varepsilon(A) = \hat{\mathrm{W}}_\varepsilon(A_1) \oplus \hat{\mathrm{W}}_\varepsilon(A_2) \oplus \cdots \oplus \hat{\mathrm{W}}_\varepsilon(A_l) \tag{4.3}$$

and

$$\mathrm{W}_\varepsilon(A) = \mathrm{W}_\varepsilon(A_1) \oplus \mathrm{W}_\varepsilon(A_2) \oplus \cdots \oplus \mathrm{W}_\varepsilon(A_k) \tag{4.4}$$

where $\hat{\mathrm{W}}_\varepsilon(A_i)$ *and* $\mathrm{W}_\varepsilon(A_i)$ *are identified with subgroups of* $\hat{\mathrm{W}}_\varepsilon(A)$ *and* $\mathrm{W}_\varepsilon(A)$ *respectively under canonical injections. If* $(A_j, ^-)$ *is a simple hyperbolic involution algebra, then*

$$\hat{\mathrm{W}}_\varepsilon(A_j, ^-) \cong \mathbb{Z}.$$

Suppose that $1 \le i \le l$, let $(V_i, h_i) \in {}_{A_i}\mathcal{H}_\varepsilon^*$, and let $[h_i]_i$ be its isometry class in ${}_{A_i}\mathbf{M}_\varepsilon$. If $\iota_i : A_i \to A$ is inclusion, $[h_i]_i \to [\iota_i h_i]$ is a monoid homomorphism ${}_{A_i}\mathbf{M}_\varepsilon \to {}_A\mathbf{M}_\varepsilon$, and it is easy to see that it is injective. Identify ${}_{A_i}\mathbf{M}_\varepsilon$ with its image.

If $h : V \times V \to (A, ^-)$ is an ε-Hermitian form, it is in a unique way the vector sum $h = h_1 \times h_2 \times \cdots \times h_l$ (cf. p. 95) where

$$h_i = h|_{A_iV \times A_iV} : A_iV \times A_iV \to (A_i, ^-)$$

is a (possibly 0) nonsingular ε-Hermitian form. Any isometry $(V, h) \to (V', h')$ carries A_iV to A_iV', and clearly restricts to an isometry between h_i and h_i' for all i. Thus $[h] = \sum_i [h_i]_i$ in ${}_A\mathbf{M}_\varepsilon$ for unique classes $[h_i]_i$. Conversely if $[h_i]_i \in {}_{A_i}\mathbf{M}_\varepsilon$ for $i = 1, 2, \ldots, l$, then $\sum_i [h_i]_i = [h]$ where $h := h_1 \times \cdots \times h_l$ is a form on the module $V = V_1 \oplus \cdots \oplus V_l$ (with the obvious notation), since $A_iV = V_i$ and $h|_{A_iV \times A_iV} = h_i$.

Thus ${}_A\mathbf{M}_\varepsilon = \bigoplus_i {}_{A_i}\mathbf{M}_\varepsilon$ and by Prop. 1.6.3(a), p. 66, (4.3) holds. Moreover h is hyperbolic if and only if h_i is hyperbolic for all i, and so (4.4) follows by Prop. 1.6.3(b).

The last statement follows directly from Cor. 3.1.6, p. 98. □

Now consider the Grothendieck group $\hat{\mathrm{W}}_\varepsilon(K, A) = \hat{\mathrm{W}}_\varepsilon(K, A, ^-)$ constructed from the commutative monoid of isomorphism classes of the category ${}^{(A,^-)}_{(K,^-)}\mathcal{H}_\varepsilon^*$ of nonsingular $(A, ^-)$-invariant ε-Hermitian forms. (The monoid addition is of course $[h] + [g] = [h \perp g]$). This category is isomorphic to ${}_{(A,^-)}\mathcal{H}_\varepsilon^*$ if there is[1] an involution trace $\mathrm{t} : (A, ^-) \to (K, ^-)$ (Th. 3.2.4, p. 113), and this leads to an induced isomorphism

$$\mathrm{t}_* : \hat{\mathrm{W}}_\varepsilon(A, ^-) \to \hat{\mathrm{W}}_\varepsilon(K, A, ^-) \tag{4.5}$$

determined by $[h] \to [\mathrm{t}h]$. Define $\mathrm{H}_\varepsilon(K, A)$ to be the subgroup of $\hat{\mathrm{W}}_\varepsilon(K, A)$ generated by the strongly hyperbolic invariant forms; these are the images of the hyperbolic forms in $\hat{\mathrm{W}}_\varepsilon(A, ^-)$ (Th. 3.2.4, p. 113). Put $\mathrm{W}_\varepsilon(K, A) = \hat{\mathrm{W}}_\varepsilon(K, A)/\mathrm{H}_\varepsilon(K, A)$.

Corollary 4.1.1 *With the notation of the theorem, the diagrams*

$$\begin{array}{ccc} \hat{\mathrm{W}}_\varepsilon(A_i, ^-) & \longrightarrow & \hat{\mathrm{W}}_\varepsilon(A, ^-) \\ {\scriptstyle \mathrm{t}_*}\downarrow & & \downarrow{\scriptstyle \mathrm{t}_*} \\ \hat{\mathrm{W}}_\varepsilon(K, A_i) & \longrightarrow & \hat{\mathrm{W}}_\varepsilon(K, A) \end{array}$$

[1] If A is a separable algebra over K, then $\mathrm{trd}_{A/K}$ is an involution trace – see Ex. 1, p. 35.

are commutative for $i = 1, \ldots, l$ *(where the horizontal maps are the canonical injections) and give rise to*

$$\hat{\mathrm{W}}_\varepsilon(K, A) \cong \hat{\mathrm{W}}_\varepsilon(K, A_1) \oplus \cdots \oplus \hat{\mathrm{W}}_\varepsilon(K, A_l) \tag{4.6}$$

and

$$\mathrm{W}_\varepsilon(K, A) \cong \mathrm{W}_\varepsilon(K, A_1) \oplus \cdots \oplus \mathrm{W}_\varepsilon(K, A_k). \tag{4.7}$$

The commutativity of the diagram is clear. By Example 4., p. 36, $\mathrm{t}|_{A_i} : (A_i, {}^-) \to (K, {}^-)$ is also an involution trace; thus the map $\hat{\mathrm{W}}_\varepsilon(A_i, {}^-) \to \hat{\mathrm{W}}_\varepsilon(K, A_i)$, $[h] \to [\mathrm{t}|_{A_i} h]$, is an isomorphism. (4.6) follows easily from the theorem, and the proof of (4.7) is like that of (4.4) – if V is an A-module, it is straightforward to check that an ε-hyperbolic form $h : V \times V \to (K, {}^-)$ is strongly hyperbolic if and only if all of its "component forms" $h_i : A_i V \times A_i V \to (K, {}^-)$ are strongly hyperbolic. □

The expressions (4.6) and (4.7) can also be proved directly, as were (4.3) and (4.4). Namely if $(V, h) \in {}^{(A, {}^-)}_{(K, {}^-)}\mathcal{H}^*_\varepsilon$, then $V = A_1 V \oplus \cdots \oplus A_l V$ and it is easy to see that the invariance of h implies that $V = A_1 V \perp \cdots \perp A_l V$. Clearly $h|_{A_i V \times A_i V} \in {}^{(A_i, {}^-)}_{(K, {}^-)}\mathcal{H}^*_\varepsilon$ and (4.6) follows easily. Moreover h is strongly hyperbolic if and only if each $h|_{A_i V \times A_i V}$ is and so we also get (4.7) at once.

The following important theorem is an immediate corollary of Th. 3.2.14, p. 132, and (4.2). Let A be a simple algebra with a $(K, {}^-)$-involution ${}^-$ and center L, and D a division algebra similar to A with an $(L, {}^-)$-involution ${}^\sim$. Choose an isomorphism $A \cong \mathrm{M}(n, D)$. Let ${}^-$ be the involution on $\mathrm{M}(n, D)$ so that this isomorphism is one of involution algebras:

$$(A, {}^-) \cong (\mathrm{M}(n, D), {}^-). \tag{4.8}$$

Suppose that $\bar{b} = \underline{h}_0^{-1} b^{t\sim} \underline{h}_0$ for all $b \in \mathrm{M}(n, D)$, where $\underline{h}_0^{t\sim} = \varepsilon_0 \underline{h}_0$ and $\varepsilon_0 = 1$ if $(D, {}^\sim)$ and $(A, {}^-)$ have the same type, otherwise $\varepsilon_0 = -1$ (Th. 1.3.6, p. 30).

Theorem 4.1.2 *With the above notation, there are "Morita isomorphisms"*

$$\hat{\mathrm{W}}_\varepsilon(A, {}^-) \cong \hat{\mathrm{W}}_{\varepsilon\varepsilon_0}(D, {}^\sim),\ \mathrm{H}_\varepsilon(A, {}^-) \cong \mathrm{H}_{\varepsilon\varepsilon_0}(D, {}^\sim),\ \mathrm{W}_\varepsilon(A, {}^-) \cong \mathrm{W}_{\varepsilon\varepsilon_0}(D, {}^\sim).$$

REMARKS. 1. In Th. 3.2.14, the involution on D is assumed to be conjugation (which we generally denote by *) if D is a quaternion algebra with involution of the first kind – but the isomorphisms in the above theorem still hold without that assumption.

2. The first and third of these isomorphisms are not canonical because of the factor $\alpha \in \dot{K}_0$ in Th. 3.2.10 (p. 125), as well as the choice of the involution ${}^\sim$ on D (cf. Th. 3.2.11). □

The isomorphism $\hat{\mathrm{W}}_\varepsilon(A, {}^-) \cong \hat{\mathrm{W}}_{\varepsilon\varepsilon_0}(D, {}^\sim)$ in the theorem is defined as follows: We first apply the isomorphism (4.8). As we have seen in §3.2.7, every ε-Hermitian space over $(\mathrm{M}(n, D), {}^-)$ is equivalent to one of the form $h : D^{n\times m} \times D^{n\times m} \to (\mathrm{M}(n, D), {}^-)$ where

$$h(u, v) = u\underline{h}v^{t\sim}\underline{h}_0$$

for a uniquely determined $\underline{h} \in \mathrm{M}(m, D)$. Then the first isomorphism in the theorem is determined by the map

$$[D^{n\times m}, h] \to [D^{1\times m}, \check{h}] \tag{4.9}$$

where $\breve{h}(u,v) = u\underline{h}v^{t\sim}$. Since h is an ε-Hermitian form over $(\mathrm{M}(n,D),\,^-)$, the matrix $\underline{h}$ and the form $\breve{h}$ are $\varepsilon\varepsilon_0$-Hermitian. The other isomorphisms in the theorem follow easily since the Morita correspondence (4.9) preserves hyperbolic forms. □

Let $(A,\,^-)$ be an arbitrary $(K,\,^-)$-involution algebra. There is a "forgetful" map $[V,h] \to [V]$ from the monoid ${}_A\mathbf{M}_\varepsilon$ of classes of nonsingular ε-Hermitian forms over $(A,\,^-)$ to the monoid $\mathbf{P}$ of isomorphism classes of A-modules, leading to a map

$$\mathrm{L} : \hat{\mathrm{W}}_\varepsilon(A,\,^-) \to \mathrm{K}_0(A) \tag{4.10}$$

which we refer to as "linearization". As an example, Th. 3.1.3 (p. 98) can be restated by saying that L is injective on $\mathrm{H}_\varepsilon(A,\,^-)$.

The following corollary of Th. 4.1.2 is readily apparent if the map $\mathrm{K}_0(A) \to \mathrm{K}_0(D)$ takes the class $[V]$ to the class $[V']$ with $\dim_D V' = \mathrm{len}_A V$:

Corollary 4.1.2 *Under the same assumptions as Th. 4.1.2, the diagram*

$$\begin{array}{ccc} \hat{\mathrm{W}}_\varepsilon(A,\,^-) & \xrightarrow{\ \mathrm{L}\ } & \mathrm{K}_0(A) \\ \wr\big\downarrow & & \big\downarrow\wr \\ \hat{\mathrm{W}}_{\varepsilon\varepsilon_0}(D,\,^\sim) & \xrightarrow{\ \mathrm{L}\ } & \mathrm{K}_0(D) \end{array}$$

is commutative. □

Consider the Abelian monoids $\mathbf{M}_\varepsilon = {}_{(A,\,^-)}\mathbf{M}_\varepsilon$ and $\mathbf{S} = {}_{(K,\,\mathrm{id}_K)}\mathbf{M}_1$. The "additive" monoid $\mathbf{S}$ also has a product determined by

$$(V,f)(W,g) = (V \otimes_K W, fg), \quad (fg)(v_1 \otimes w_1, v_2 \otimes w_2) = f(v_1,v_2)g(w_1,w_2)$$

(cf. p. 50 or 106, and Th. 3.2.1 p. 109,) which makes it into a commutative semiring with 1 the class of $\langle 1 \rangle$. Therefore $\hat{\mathrm{W}}_1(K) = \mathrm{K}(\mathbf{S})$ is a commutative ring by Prop. 1.6.2, p. 65. It follows from (b) of that proposition and the properties of the product (Th. 3.2.1) that $\hat{\mathrm{W}}_\varepsilon(A,\,^-)$ is a module over $\hat{\mathrm{W}}_1(K)$. Moreover since the product of a nonsingular ε-Hermitian form over $(A,\,^-)$ and a symmetric hyperbolic form over K is an ε-hyperbolic form over $(A,\,^-)$ (Th. 3.2.1(d)), $\mathrm{W}_\varepsilon(A,\,^-)$ is a $\mathrm{W}_1(K)$-module. Thus:

Theorem 4.1.3 *$\hat{\mathrm{W}}_\varepsilon(A)$ and $\mathrm{W}_\varepsilon(A)$ are modules over the commutative rings $\hat{\mathrm{W}}_1(K)$ and $\mathrm{W}_1(K)$ respectively. If $g \in {}_K\mathcal{H}_1^*$ and $h \in {}_A\mathcal{H}_\varepsilon^*$, then $[g][h] \in \hat{\mathrm{W}}_\varepsilon(A)$ is hyperbolic if g or h is hyperbolic.* □

4.2 Functorial properties of the Witt-Grothendieck and Witt groups of isometry representations

In this section $(A,\,^-)$ is a semisimple $(K,\,^-)$-involution algebra. With the exception of the first two topics "functoriality in $(K,\,^-)$" and "scaling", most of this section is unnecessary for subsequent results.

Functoriality in $(K,\,^-)$**.** If $(L,\,^-)/(K,\,^-)$ is an arbitrary extension of fields with involution (or more generally, if $(K,\,^-) \to (L,\,^-)$ is a homomorphism of fields with involution) and $\rho : G \to \mathbf{I}(V,h)$ is an isometry representation over $(K,\,^-)$, there is an isometry representation $\rho^L : G \to \mathbf{I}(V^L, h^L)$ over $(L, ^-)$ – cf. p. 89 – giving rise to homomorphisms ("restrictions"[2])

$$\text{Res} : \hat{\mathrm{W}}_\varepsilon(K, G) \to \hat{\mathrm{W}}_\varepsilon(L, G) \quad \text{and} \quad \text{Res} : \mathrm{W}_\varepsilon(K, G) \to \mathrm{W}_\varepsilon(L, G),$$

(the latter because scalar extension clearly takes a hyperbolic representation over K to one over L). As usual, these homomorphisms are also referred to as "extending the scalars".

Let $\mathcal{F}$ denote the category of fields with involution, of characteristic[3] $\neq 2$, and $\mathcal{A}$ the category of Abelian groups. Using the restriction maps, we can view $\hat{\mathrm{W}}_\varepsilon(K, G)$ and $\mathrm{W}_\varepsilon(K, G)$ as (covariant) functors $\mathcal{F} \to \mathcal{A}$ and the canonical map $\text{can} : \hat{\mathrm{W}}_\varepsilon(K, G) \to \mathrm{W}_\varepsilon(K, G)$ as a morphism of functors, that is to say,

$$\begin{array}{ccc} \hat{\mathrm{W}}_\varepsilon(K, G) & \xrightarrow{\text{Res}} & \hat{\mathrm{W}}_\varepsilon(L, G) \\ {\scriptstyle\text{can}}\downarrow & & \downarrow{\scriptstyle\text{can}} \\ \mathrm{W}_\varepsilon(K, G) & \xrightarrow{\text{Res}} & \mathrm{W}_\varepsilon(L, G) \end{array}$$

is commutative.

Recall that if $(L, ^-)/(K, ^-)$ is a finite separable extension of fields with involution, the field trace $\mathrm{Tr}_{L/K}$ is an involution trace (cf. Ex. 2, p. 35).

Let $\sigma : (K, ^-) \to (L, ^-)$ be a homomorphism of fields with involution such that $L/\sigma K$ is a finite separable extension. If V is a vector space over L, let V_K be the K-vector space which, as an Abelian group, is $= V$, and where K acts on it through $\alpha v = \sigma(\alpha) v$ if $\alpha \in K$ and $v \in V$.

Now let $h : V \times V \to (L, ^-)$ be a sesquilinear form. Then

$$h_K := \sigma^{-1} \circ \mathrm{Tr}_{L/\sigma K} \circ h : V_K \times V_K \to (K, ^-)$$

is also sesquilinear, and is nonsingular if h is. If h is ε-reflexive, then so is h_K. In more detail, if h is symmetric or skew-symmetric, then h_K is of the same type. But if h is Hermitian, then h_K is symmetric if $^-$ is the identity on K, otherwise Hermitian. Similarly when h is skew Hermitian, h_K is skew symmetric or skew Hermitian.

Suppose that $\varphi : (V, h) \to (V', h')$ is an isometry over L. Then φ is also an isometry $(V_K, h_K) \to (V'_K, h'_K)$ over K – in particular $\mathbf{I}(V, h) \subset \mathbf{I}(V_K, h_K)$. Thus an isometry representation $\rho : G \to \mathbf{I}(V, h)$ over L is also an isometry representation $\rho_K : G \to \mathbf{I}(V_K, h_K)$ over K, sometimes referred to as the *corestriction*[4] of ρ. Since $\rho \cong \rho'$ implies $\rho_K \cong \rho'_K$ and $(\rho \perp \rho')_K \cong \rho_K \perp \rho'_K$, there is a homomorphism

$$\text{Cor} : \hat{\mathrm{W}}_\varepsilon(L, G) \to \hat{\mathrm{W}}_\varepsilon(K, G).$$

If $\rho : G \to \mathbf{I}(V, h)$ is hyperbolic, say $\rho = \mathrm{H}(\sigma)$ where $\sigma : G \to \mathbf{GL}(U)$ is a K-linear representation and $V = \mathrm{H}(U) = U \oplus \overline{U^*}$ (so U and $\overline{U^*}$ are both totally isotropic), then $V_K = U_K \oplus \overline{U^*}_K$ with both U_K and $\overline{U^*}_K$ totally isotropic with respect to h_K and stable under G. Therefore ρ_K is hyperbolic (Cor. 3.1.2(b), p.

---

[2] This terminology comes from Galois cohomology.

[3] There is no need here to require that the characteristic of the field not divide $|G|$.

[4] This terminology is also related to a construction in non-commutative Galois cohomology– see [56].

97). It follows that there is a homomorphism Cor : $\mathrm{W}_\varepsilon(L,G) \to \mathrm{W}_\varepsilon(K,G)$, and a commutative diagram

$$\begin{array}{ccc} \hat{\mathrm{W}}_\varepsilon(L,G) & \xrightarrow{\mathrm{Cor}} & \hat{\mathrm{W}}_\varepsilon(K,G) \\ {\scriptstyle\mathrm{can}}\downarrow & & \downarrow{\scriptstyle\mathrm{can}} \\ \mathrm{W}_\varepsilon(L,G) & \xrightarrow{\mathrm{Cor}} & \mathrm{W}_\varepsilon(K,G) \end{array}$$

so as in the case of the restriction, can is a morphism of functors – this time from the subcategory of $\mathcal{F}$ which consists of the homomorphisms $\sigma : K_1 \to K_2$ such that $K_2/\sigma K_1$ is finite and separable, to the category $\mathcal{A}$ of Abelian groups. We leave the details to the interested reader.

There is a useful variation on these themes when $(KG, ^-)$ is replaced by the $(K, ^-)$-involution algebra $(A, ^-)$, involving the groups $\hat{\mathrm{W}}_\varepsilon(K, A, ^-)$ and $\mathrm{W}_\varepsilon(K, A, ^-)$. If $(L, ^-)/(K, ^-)$ is an arbitrary extension of fields with involution and $(A^L, ^-) = (L, ^-) \otimes_{(K,^-)} (A, ^-)$, there are again homomorphisms

$$\mathrm{Res} : \hat{\mathrm{W}}_\varepsilon(K, A, ^-) \to \hat{\mathrm{W}}_\varepsilon(L, A^L, ^-) \text{ and } \mathrm{Res} : \mathrm{W}_\varepsilon(K, A, ^-) \to \mathrm{W}_\varepsilon(L, A^L, ^-) \quad (4.11)$$

defined in the case of $\hat{\mathrm{W}}_\varepsilon$ by $[h] \to [h^L]$, where h is an $(A, ^-)$-invariant ε-reflexive form. The details are left as an exercise. By Th. 3.1.2, p. 97, h^L is hyperbolic if h is, and so the restriction is defined on W_ε as well.

Now assume again that L/K is a finite extension and let $\mathrm{t} : (L, ^-) \to (K, ^-)$ be an involution trace – for example $\mathrm{t} = \mathrm{Tr}_{L/K}$ if L/K is separable. It extends to a homomorphism $\mathrm{t}_A : A^L \to A$ of A-modules, satisfying $\mathrm{t}_A(\lambda \otimes_K a) = \mathrm{t}(\lambda)a$, and this in turn determines a functor $\mathrm{t}_* : {}_{(A^L,^-)}\mathcal{H}_\varepsilon \to {}_{(A,^-)}\mathcal{H}_\varepsilon$, $\mathrm{t}_*(h)(u,v) = \mathrm{t}_A(h(u,v))$, called the *(Scharlau) transfer* (induced by t). We wish to show that it determines homomorphisms

$$\mathrm{t}_* : \hat{\mathrm{W}}_\varepsilon(A^L, ^-) \to \hat{\mathrm{W}}_\varepsilon(A, ^-) \quad \text{and} \quad \mathrm{t}_* : \mathrm{W}_\varepsilon(A^L, ^-) \to \mathrm{W}_\varepsilon(A, ^-),$$

and to do this we must first show that it takes nonsingular forms to nonsingular forms, i.e.

$$\mathrm{t}_* : {}_{(A^L,^-)}\mathcal{H}^*_\varepsilon \to {}_{(A,^-)}\mathcal{H}^*_\varepsilon.$$

Since A is semisimple, it suffices to show that if $h : V \times V \to (A^L, ^-)$ is a nonsingular form, then $\mathrm{t}_* h : V \times V \to (A, ^-)$ is nondegenerate, i.e. if $v \neq 0$, then there is a $u \in V$ such that $\mathrm{t}_A(h(u,v)) \neq 0$. We can certainly find $u \in V$ such that $h(u,v) \neq 0$. Write

$$h(u,v) = \lambda_1 \otimes a_1 + \cdots + \lambda_m \otimes a_m$$

where all $\lambda_i \neq 0$ and $a_1, \ldots, a_m$ are elements of A which are linearly independent over K. Choose $\lambda \in L$ such that $\mathrm{t}(\lambda) \neq 0$. Then

$$\begin{aligned} \mathrm{t}_A(\lambda\lambda_1^{-1} h(u,v)) &= \mathrm{t}(\lambda)a_1 + \text{a } K\text{-linear combination of } a_2, \ldots, a_m \\ &\neq 0. \end{aligned}$$

Therefore $\mathrm{t}_A(h(\lambda\lambda_1^{-1}u, v)) \neq 0$ and $\mathrm{t}_* h$ is nondegenerate.

It follows easily that t_* induces a homomorphism $\hat{\mathrm{W}}_\varepsilon(A^L, ^-) \to \hat{\mathrm{W}}_\varepsilon(A, ^-)$. Since it takes hyperbolic forms to hyperbolic forms by Th. 3.1.2, it also defines a homomorphism between the Witt groups.

Exercises. 1. If $A = \mathrm{M}(n,K)$ and $A^L = \mathrm{M}(n,L)$, then $\mathrm{t}_A(\lambda_{ij}) = (\mathrm{t}\lambda_{ij})$ ($(\lambda_{ij}) \in \mathrm{M}(n,L)$).

2. Suppose $K = \mathbb{R}$, $L = \mathbb{C} = \mathbb{R}(i)$, and $\mathrm{t}(\alpha+\beta i) = \alpha$, where $\alpha, \beta \in \mathbb{R}$. Let $(A, {}^-) = (\mathrm{M}(2,\mathbb{R}), {}^t)$. If $h \in {}_{(A^{\mathbb{C}},\, {}^t)}\mathcal{S}$ is the form with discriminant matrix (z), $z \in \dot{\mathbb{C}}$,

$$h(u,v) = \begin{pmatrix} u_1 \\ u_2 \end{pmatrix} (z) \begin{pmatrix} v_1 \\ v_2 \end{pmatrix}^t \in \mathrm{M}(2,\mathbb{C}) = A^{\mathbb{C}}$$

on $V = \mathbb{C}^{2\times 1}$, then

$$h(u,v) = z \begin{pmatrix} h_{11} & h_{12} \\ h_{21} & h_{22} \end{pmatrix}$$

where, if $u_j = u_{j0} + u_{j1}i$ and $v_j = v_{j0} + v_{j1}i$ with $u_{jk}, v_{jk} \in \mathbb{R}$,

$$\begin{aligned} h_{11} &= (u_{10}v_{10} - u_{11}v_{11}) + (u_{10}v_{11} + u_{11}v_{10})i, \\ h_{12} &= (u_{10}v_{20} - u_{11}v_{21}) + (u_{10}v_{21} + u_{11}v_{20})i, \\ h_{21} &= (u_{20}v_{10} - u_{21}v_{11}) + (u_{20}v_{11} + u_{21}v_{10})i, \\ h_{22} &= (u_{20}v_{20} - u_{21}v_{21}) + (u_{20}v_{21} + u_{21}v_{20})i. \end{aligned}$$

Identify $\mathbb{C}^{2\times 1}$ with $\mathbb{R}^{2\times 2}$ via the $\mathrm{M}(2,\mathbb{R})$-isomorphism

$$\begin{pmatrix} u_{10} + u_{11}i \\ u_{20} + u_{21}i \end{pmatrix} \to \begin{pmatrix} u_{10} & u_{11} \\ u_{20} & u_{21} \end{pmatrix}.$$

(a) If $z = 1$, show that the discriminant matrix of the transfer

$$\mathrm{t}_* h : \mathbb{R}^{2\times 2} \times \mathbb{R}^{2\times 2} \to (\mathrm{M}(2,\mathbb{R}), {}^t)$$

is $\begin{pmatrix} 1 & 0 \\ 0 & -1 \end{pmatrix}$.

(b) If $z = i$, show that the discriminant matrix of $\mathrm{t}_* h$ is $\begin{pmatrix} 0 & -1 \\ -1 & 0 \end{pmatrix}$.

(c) If $z = a + bi$, $a, b \in \mathbb{R}$, not both 0, show that the discriminant matrix of $\mathrm{t}_* h$ is $\begin{pmatrix} a & -b \\ -b & -a \end{pmatrix}$.

(d) Show that $\mathrm{t}_* : \mathrm{W}_1(A^{\mathbb{C}}, {}^t) \to \mathrm{W}_1(A, {}^t)$ is the zero homomorphism.

(e) If $J = \begin{pmatrix} 0 & 1 \\ -1 & 0 \end{pmatrix}$, show that $a \to Ja^tJ^{-1}$ is a symplectic involution $\sim$ on A. If $h(u,v)$ is the form $u(1)v^tJ^{-1}$ on $\mathbb{C}^{2\times 1}$, show that $h \in {}_{(A^{\mathbb{C}},\, \sim)}\mathcal{H}_{-1}$ and that the discriminant matrix of its transfer $\mathrm{t}_* h \in {}_{(A,\, \sim)}\mathcal{H}_{-1}$ is $\begin{pmatrix} 1 & 0 \\ 0 & -1 \end{pmatrix}$. □

Example. Suppose that $(V,h) \in {}_{(A,{}^-)}\mathcal{H}_\varepsilon$, that $(L, {}^-)/(K, {}^-)$ is a finite extension and that $\mathrm{t} : (L, {}^-) \to (K, {}^-)$ is an involution trace. If $(U,g) \in {}_{(L,{}^-)}\mathcal{H}_\delta$, the products $\mathrm{Res}(h)g$ and $h\mathrm{t}_*(g)$ are defined and in ${}_{(A^L,{}^-)}\mathcal{H}_{\varepsilon\delta}$ and ${}_{(A,{}^-)}\mathcal{H}_{\varepsilon\delta}$ respectively. We claim that

$$\mathrm{t}_* \left(\mathrm{Res}(h)g\right) \cong h\mathrm{t}_*(g). \tag{4.12}$$

The left side is a form on $V^L \otimes_L U = (L \otimes_K V) \otimes_L U$ and the right side a form on $V \otimes_K U_K$; the two vector spaces are isomorphic, as K-spaces, via the composite $(\lambda \otimes v) \otimes u \to v \otimes \lambda u$ of the canonical maps

$$(L \otimes_K V) \otimes_L U \to (V \otimes_K L) \otimes_L U \to V \otimes_K (L \otimes_L U) \to V \otimes_K U.$$

This isomorphism provides the equivalence in (4.12):

$$\begin{aligned}
\mathrm{t}_*(\mathrm{Res}(h)g)\,((\lambda\otimes v)\otimes u,(\lambda'\otimes v')\otimes u') &= \mathrm{t}_A\big(\mathrm{Res}(h)\big((\lambda\otimes v)g(u,u'),\lambda'\otimes v'\big)\big)\\
&= \mathrm{t}_A\big(\mathrm{Res}(h)\big(\lambda g(u,u')\otimes v,\lambda'\otimes v'\big)\big)\\
&= \mathrm{t}_A\big((\lambda g(u,u')\bar{\lambda}')\otimes h(v,v')\big)\\
&= \mathrm{t}\big(\lambda g(u,u')\bar{\lambda}'\big)h(v,v'),\\
\text{and}\quad (h(\mathrm{t}_*g))(v\otimes\lambda u,v'\otimes\lambda' u') &= h(v((\mathrm{t}_*g)(\lambda u,\lambda' u')),v')\\
&= \mathrm{t}(g(\lambda u,\lambda' u'))h(v,v'),
\end{aligned}$$

so (4.12) holds. □

Theorem 4.2.1 *If $(L,^-)/(K,^-)$ has odd degree $(L:K)$ and $(A,^-)$ is a $(K,^-)$-involution algebra, then both of the homomorphisms (4.11) are injective.*

Suppose first that $^-$ is not the identity on L. Since $(L:K)$ is odd, $K\not\subset L_0$ (the subfield of L of elements fixed by $^-$) – thus $^-$ is also not the identity on K and $(L_0:K_0)=(L:K)$.

Back to the general case. Let $\lambda\in L_0\setminus K_0$. Then $K(\lambda)$ is a proper extension of odd degree of K, and is stable under $^-$. By induction, it suffices to prove the theorem for $(K(\lambda),^-)/(K,^-)$, and so we can assume that $L=K(\lambda)$.

Define the K-linear map $\mathrm{t}:L\to K$ by $\mathrm{t}(1)=1$ and $\mathrm{t}(\lambda^j)=0$ for $1\le j<(L:K)$. Note that t commutes with $^-$. We take $g=\langle 1\rangle$ in (4.12) (and $\delta=1$). We shall show that $\mathrm{t}_*\langle 1\rangle=\langle 1\rangle$ in the Witt group $\mathrm{W}_1(K)$. The theorem then follows since (4.12) becomes $\mathrm{t}_*\,(\mathrm{Res}(h))=h$ in the Witt group $\mathrm{W}_\varepsilon(A,^-)$, which implies that Res is injective on the Witt group level. If $\mathrm{Res}(h)=\mathrm{Res}(h')$ in the Witt-Grothendieck group $\hat{\mathrm{W}}_\varepsilon(A,^-)$, $h=h'$ in the Witt group, and so since the underlying A^L-modules of $\mathrm{Res}(h)=h^L$ and $\mathrm{Res}(h')={h'}^L$ must have the same length, the same is true of the A-modules of h and h', and so $h=h'$ in the Witt-Grothendieck group as well.

We now compute the form $\mathrm{t}_*\langle 1\rangle$ on the K-space L, using the basis $1,\lambda,\dots,\lambda^{n-1}$, $(n=[L:K])$. Clearly 1 is orthogonal to the subspace U spanned by the other basis elements, and $(\mathrm{t}_*\langle 1\rangle)(1,1)=1$. Moreover U has a totally isotropic subspace of half its dimension, namely the space spanned by $\lambda,\dots,\lambda^{(n-1)/2}$, and therefore must be hyperbolic. Thus $\mathrm{t}_*\langle 1\rangle=\langle 1\rangle$ in the Witt group over K. □

Corollary 4.2.1 *Let $\rho_i:G\to\mathbf{I}(V_i,h_i)$, $i=1,2$, be isometry representations over $(K,^-)$, and let $(L,^-)/(K,^-)$ be an extension of odd degree. Then if ρ_1^L and ρ_2^L are equivalent over L, ρ_1 and ρ_2 are equivalent over K.*

The $(KG,^-)$-form associated with ρ_i is

$$\widehat{h_{\rho_i}}(u,v)=\sum_s h_i(u,\rho_i(s)v)s,\quad (u,v\in V_i),$$

and the $(LG,^-)$-form associated with ρ_i^L is

$$\widehat{h_{\rho_i^L}}(u,v)=\sum_s h_i^L(u,\rho_i(s)v)s,\quad (u,v\in V_i^L).$$

If $u,v\in V_i$, $h_i^L(u,\rho_i(s)v)=h_i(u,\rho_i(s)v)$, so $\widehat{h_{\rho_i^L}}(u,v)=\widehat{h_{\rho_i}}(u,v)$ and

$$\widehat{h_{\rho_i^L}}=\widehat{h_{\rho_i}}^{\,L}$$

by (3.29), p. 111. The corollary now follows from the theorem, using Th. 2.1.1, p. 86. □

Exercise. If L/K is a Galois extension of odd degree, show that L has a normal orthonormal basis with respect to the symmetric bilinear form $\mathrm{Tr}_{L/K}(\lambda\mu)$. (Use Exercises 3 and 4 on p. 96 and the foregoing theorem).

Scaling. Suppose $(A, {}^-)$ is a $(K, {}^-)$-involution algebra as usual, and that $a_0 \in A^\times$ with $\bar{a}_0 = \varepsilon_0 a_0$, $\varepsilon_0 = \pm 1$. Then $\widetilde{a} = a_0^{-1}\bar{a}a_0 = \mathrm{inn}_{a_0^{-1}}\bar{a}$ defines an involution on A as well, and we look at the relationship between the Witt-Grothendieck and Witt groups of the two involution algebras $(A, {}^-)$ and $(A, {}^\sim)$.

The correspondence $h(u,v) \rightsquigarrow h(u,v)a_0$, called *scaling* (by a_0), sets up an isomorphism between the categories ${}_{(A,{}^-)}\mathcal{S}$ and ${}_{(A,{}^\sim)}\mathcal{S}$ of sesquilinear forms which preserves nonsingularity. For example

$$h(u, av)a_0 = h(u,v)\bar{a}a_0 = h(u,v)a_0 \cdot \widetilde{a},$$

and the other properties are also easy to verify. (Cf. exercises on p. 152). Moreover it induces an isomorphism between ${}_{(A,{}^-)}\mathcal{H}_\varepsilon$ and ${}_{(A,{}^\sim)}\mathcal{H}_{\varepsilon_0\varepsilon}$ and between their subcategories of hyperbolic forms, resulting in "scaling isomorphisms"

$$\hat{\mathrm{W}}_\varepsilon(A, {}^-) \cong \hat{\mathrm{W}}_{\varepsilon\varepsilon_0}(A, {}^\sim), \quad \mathrm{W}_\varepsilon(A, {}^-) \cong \mathrm{W}_{\varepsilon\varepsilon_0}(A, {}^\sim).$$

Functoriality in G. If $\varphi : H \to G$ is a homomorphism of finite groups, the map $\rho \to \rho\varphi$ leads to a homomorphism $\varphi^* : \hat{\mathrm{W}}_\varepsilon(K,G) \to \hat{\mathrm{W}}_\varepsilon(K,H)$ of Witt-Grothendieck groups in an obvious way. Since it takes hyperbolic representations of G to hyperbolic representations of H (Cor. 3.1.2(b), p. 97), there is also a corresponding homomorphism $\varphi^*\colon \mathrm{W}_\varepsilon(K,G) \to \mathrm{W}_\varepsilon(K,H)$ of Witt groups, and the canonical map $\hat{\mathrm{W}}_\varepsilon(K,G) \to \mathrm{W}_\varepsilon(K,G)$ is a morphism of contravariant functors.

If we take $H = 1$, we get

$$\hat{\mathrm{W}}_\varepsilon(K,G) \to \hat{\mathrm{W}}_\varepsilon(K, {}^-) \quad \text{and} \quad \mathrm{W}_\varepsilon(K,G) \to \mathrm{W}_\varepsilon(K, {}^-) \tag{4.13}$$

where the groups $\hat{\mathrm{W}}_\varepsilon(K, {}^-)$ and $\mathrm{W}_\varepsilon(K, {}^-)$ over $(K, {}^-)$ are the classical (nonequivariant) Grothendieck-Witt and Witt groups respectively.

These are both forgetful maps, forgetting the action of G.

Both restriction and corestriction for an extension $(L, {}^-)/(K, {}^-)$ (finite separable in the case of the corestriction) are natural with respect to a change of groups $\varphi : H \to G$: that is to say, the diagrams

$$\begin{array}{ccc} \hat{\mathrm{W}}_\varepsilon(K,G) & \xrightarrow{\mathrm{Res}} & \hat{\mathrm{W}}_\varepsilon(L,G) \\ {\scriptstyle\varphi^*}\downarrow & & \downarrow{\scriptstyle\varphi^*} \\ \hat{\mathrm{W}}_\varepsilon(K,H) & \xrightarrow{\mathrm{Res}} & \hat{\mathrm{W}}_\varepsilon(L,H) \end{array} \qquad \begin{array}{ccc} \mathrm{W}_\varepsilon(K,G) & \xrightarrow{\mathrm{Res}} & \mathrm{W}_\varepsilon(L,G) \\ {\scriptstyle\varphi^*}\downarrow & & \downarrow{\scriptstyle\varphi^*} \\ \mathrm{W}_\varepsilon(K,H) & \xrightarrow{\mathrm{Res}} & \mathrm{W}_\varepsilon(K,H) \end{array}$$

and

$$\begin{array}{ccc} \hat{\mathrm{W}}_\varepsilon(L,G) & \xrightarrow{\mathrm{Cor}} & \hat{\mathrm{W}}_\varepsilon(K,G) \\ {\scriptstyle\varphi^*}\downarrow & & \downarrow{\scriptstyle\varphi^*} \\ \hat{\mathrm{W}}_\varepsilon(L,H) & \xrightarrow{\mathrm{Cor}} & \hat{\mathrm{W}}_\varepsilon(K,H) \end{array} \qquad \begin{array}{ccc} \mathrm{W}_\varepsilon(L,G) & \xrightarrow{\mathrm{Cor}} & \mathrm{W}_\varepsilon(K,G) \\ {\scriptstyle\varphi^*}\downarrow & & \downarrow{\scriptstyle\varphi^*} \\ \mathrm{W}_\varepsilon(L,H) & \xrightarrow{\mathrm{Cor}} & \mathrm{W}_\varepsilon(K,H) \end{array}$$

are commutative.

Denote the K-linear extension of φ to $KH \to KG$ also by φ. It is a homomorphism

$$\varphi : (KH, ^-) \to (KG, ^-)$$

of $(K, ^-)$-involution algebras (where the involutions $^-$ on KH and KG are of course the standard $(K, ^-)$-involutions). Thus by Th. 3.2.2, p. 110, we get homomorphisms

$$\hat{\mathrm{W}}_\varepsilon(\varphi) : \hat{\mathrm{W}}_\varepsilon(KH, ^-) \to \hat{\mathrm{W}}_\varepsilon(KG, ^-) \text{ and } \mathrm{W}_\varepsilon(\varphi) : \mathrm{W}_\varepsilon(KH, ^-) \to \mathrm{W}_\varepsilon(KG, ^-), \tag{4.14}$$

and $\hat{\mathrm{W}}_\varepsilon(KG, ^-)$ and $\mathrm{W}_\varepsilon(KG, ^-)$ are covariant functors of G into the category of Abelian groups. Furthermore the canonical map $\hat{\mathrm{W}}_\varepsilon(KG, ^-) \to \mathrm{W}_\varepsilon(KG, ^-)$ is a morphism of functors.

In particular if we take $H = 1$, we get group homomorphisms

$$\hat{\mathrm{W}}_\varepsilon(K, ^-) \to \hat{\mathrm{W}}_\varepsilon(KG, ^-) \quad \text{and} \quad \mathrm{W}_\varepsilon(K, ^-) \to \mathrm{W}_\varepsilon(KG, ^-).$$

Induction. Let H and G be finite groups, and let X be a (KG, KH)-bimodule (which we also refer to as a (G, H)-bimodule). Then the map

$$V \rightsquigarrow X \otimes_{KH} V =: \mathrm{R}_H^G V$$

is a functor from left KH-modules to left KG-modules. We also assume that X has finite dimension over K, and view R_H^G as a functor from finitely generated KH-modules to finitely generated KG-modules. If the linear representation underlying the KH-module structure on V is $\rho : H \to \mathbf{GL}(V)$, we denote the linear representation of G afforded by $\mathrm{R}_H^G V$ by $\mathrm{R}_H^G \rho$, and call it the *representation of G induced by the (G, H)-bimodule X* (and the KH-module V or ρ).

Let $(s, h^{-1})|_X$ denote the K-linear transformation $x \to sxh^{-1}$ of X.

Lemma 4.2.1 *If* $\operatorname{char} K \nmid |H|$, *the character of the representation ρ induced by the (G, H)-bimodule X is given by*

$$\chi_{\mathrm{R}_H^G \rho}(s) = \frac{1}{|H|} \sum_{h \in H} \operatorname{tr}((s, h^{-1})|_X) \chi_\rho(h).$$

Consider first the $KH^o \otimes_K KH$-module $X \otimes_K V$ determined by $(h_1 \otimes h_2)(x \otimes v) = xh_1^{-1} \otimes h_2 v$ (where H^o is the opposite group of H). It is also a KG-module where $s \in G$ operates as $s(x \otimes v) = sx \otimes v$, and the actions of KG and $KH^o \otimes_K KH$ commute with each other. The element

$$e = \frac{1}{|H|} \sum_{h \in H} (h^{-1} \otimes h) \in KH^o \otimes_K KH = K(H^o \times H)$$

is an idempotent, and so $\pi := e|_{X \otimes_K V}$ is a projection and a KG-endomorphism. Thus its kernel and its image are KG-submodules of $X \otimes_K V$, and

$$X \otimes_K V = \ker \pi \oplus \operatorname{im} \pi. \tag{4.15}$$

Any element of the form $xh' \otimes v - x \otimes h'v$ is in $\ker \pi$ since

$$\sum_h xh'h^{-1} \otimes hv - \sum_h xh^{-1} \otimes hh'v = \sum_h xh'(hh')^{-1} \otimes (hh')v - \sum_h xh^{-1} \otimes hh'v = 0.$$

Conversely if $\sum_i x_i \otimes v_i \in \ker \pi$, i.e. $\sum_{i,h} x_i h^{-1} \otimes hv_i = 0$, then

$$\sum_i x_i \otimes v_i = \frac{1}{|H|}\sum_{i,h}(x_i \otimes v_i - x_i h^{-1} \otimes hv_i) = \frac{1}{|H|}\sum_{i,h}((x_i h^{-1})h \otimes v_i - x_i h^{-1} \otimes hv_i)$$

from which it follows that the $\ker \pi$ is spanned by all elements of the form $xh \otimes v - x \otimes hv$. This implies that

$$\operatorname{im}\pi \cong (X \otimes_K V)/\ker\pi \cong X \otimes_{KH} V = \mathrm{R}_H^G V$$

as KG-modules. Since $\pi^2 = \pi$, by (4.15) the trace of $\rho(s)$ is the trace of $(s|_{X\otimes_K V})\pi$ which is

$$\begin{aligned}\operatorname{tr}\Big(s|_{X\otimes_K V}\frac{1}{|H|}\sum_h (h^{-1}\otimes h)|_{X\otimes_K V}\Big) &= \frac{1}{|H|}\sum_{h\in H}\operatorname{tr}\left((s|_{X\otimes_K V})(h^{-1}\otimes h)|_{X\otimes_K V}\right)\\ &= \frac{1}{|H|}\sum_{h\in H}\operatorname{tr}\left((s,h^{-1})|_X \otimes h|_V\right)\\ &= \frac{1}{|H|}\sum_{h\in H}\big(\operatorname{tr}(s,h^{-1})|_X\big)\big(\operatorname{tr} h|_V\big). \qquad \square\end{aligned}$$

When H is a subgroup of G, the "standard" induced representation is that given by the (G,H)-bimodule $X = KG$. It is denoted by $\operatorname{Ind}_H^G \rho$ and its character is given by the well known formula (cf. p. 30, [68])

$$\chi_{\operatorname{Ind}_H^G \rho}(s) = \frac{1}{|H|}\sum_{\substack{t\in G\\ t^{-1}st\in H}} \chi(t^{-1}st)$$

which follows easily from the lemma, since the elements of G form a basis of KG, and in order for $t \in G$ to contribute to $\operatorname{tr}(s,h^{-1})|_{KG}$, we must have $sth^{-1} = t$, i.e. $t^{-1}st = h$ – and in this case, its contribution is 1.

We now consider induction for isometry representations. Assume as before that X is a (G,H)-bimodule, and that it now carries a nonsingular η-reflexive form b over $(K, ^-)$, $\eta = \pm 1$, which is "biinvariant":

$$b(sxh, syh) = b(x,y) \text{ for all } s \in G \text{ and } h \in H.$$

Let $\hat{b}(x,y) = \sum_{s\in G} b(s^{-1}x, y)s = \sum_{s\in G} b(x, sy)s$ be the associated $(KG, ^-)$-form. It is $(KH, ^-)$-invariant (on the right) since b is: $\hat{b}(xa, y) = \hat{b}(x, y\bar{a})$ for all $x, y \in X$ and all $a \in KH$. Thus

$$\hat{b} \in {}_{(KG,^-)}\mathcal{H}_\eta^{(KH,^-)}.$$

Suppose that $\rho : H \to \mathbf{I}(V,f)$ is an isometry representation where f is a nonsingular ε-reflexive form over $(K, ^-)$. Let $\hat{f}$ be its associated $(KH, ^-)$-form,

$$\hat{f} \in {}_{(KH,^-)}\mathcal{H}_\varepsilon$$

so the product

$$\hat{b}\hat{f} : \mathrm{R}_H^G V \times \mathrm{R}_H^G V = (X \otimes_{KH} V) \times (X \otimes_{KH} V) \to (KG, ^-)$$

is defined and is in ${}_{(KG,^-)}\mathcal{H}_{\eta\varepsilon}$ (cf. Th. 3.2.1, p. 109). Explicitly

$$\hat{b}\hat{f}(x \otimes u,\, y \otimes v) = \hat{b}(x\hat{f}(u,v),\, y).$$

Let $\mathrm{t} = \mathrm{Tr}_{KG/K} : (KG, ^-) \to (K, ^-)$ be the canonical involution trace and define $\mathrm{R}_H^G f$ to be the transfer

$$\mathrm{R}_H^G f = \mathrm{t}_*(\hat{b}\hat{f}) = \mathrm{t} \circ (\hat{b}\hat{f}) : \mathrm{R}_H^G V \times \mathrm{R}_H^G V \to (K, ^-).$$

It is a nonsingular G-invariant $\varepsilon\eta$-reflexive form over $(K, ^-)$ (Th. 3.2.4, p. 113), and so $\mathrm{R}_H^G \rho$ is an isometry representation

$$\mathrm{R}_H^G \rho : G \to \mathbf{I}(\mathrm{R}_H^G V, \mathrm{R}_H^G f).$$

We can use the definitions of $\hat{b}$ and $\hat{f}$ to give explicit formulas for R_H^G. First of all

$$\begin{aligned} \hat{b}\hat{f}(x \otimes u,\, y \otimes v) &= \sum_{s \in G} b(x \sum_{h \in H} f(u, hv)h,\, sy)s \\ &= \sum_{s,h} f(u,\, hv) b(xh,\, sy)s. \end{aligned}$$

Applying $\mathrm{Tr}_{KG/K}$, we get

$$\mathrm{R}_H^G f(x \otimes u,\, y \otimes v) = \sum_h b(xh,\, y) f(u,\, hv). \tag{4.16}$$

It is easy to check directly from this formula that $\mathrm{R}_H^G f$ is a G-invariant $\eta\varepsilon$-Hermitian form.

Theorem 4.2.2 *Let G and H be finite groups and suppose that $\varphi : G \times H^o \to \mathbf{I}(X, b)$ is an isometry representation with respect to an η-reflexive form b over $(K, ^-)$. If $\rho : H \to \mathbf{I}(V, f)$ is an isometry representation with f an ε-reflexive form, then $\mathrm{R}_H^G f = \mathrm{Tr}_{KG/K}(\hat{b}\hat{f})$ is a nonsingular G-invariant $\eta\varepsilon$-reflexive form giving rise to the isometry representation $\mathrm{R}_H^G \rho : G \to \mathbf{I}(\mathrm{R}_H^G V,\, \mathrm{R}_H^G f)$.* □

When H is a subgroup of G, the standard induced representation gives rise to at least one induced isometry representation, namely via the form $b(x, y) = \mathrm{Tr}_{KG/K}(x\bar{y})$ on $X = KG$, which is symmetric if $^-$ is the identity on K and Hermitian otherwise. We refer to the corresponding induced representation as the standard induced (isometry) representation of the isometry representation ρ. Since G is an orthonormal basis for b, it follows from (4.16) that, if $x, y \in G$, $\mathrm{R}_H^G f(x \otimes u,\, y \otimes v) = 0$ unless $xH = yH$, that is to say that if $s_1, s_2, \ldots, s_k$ is a system of left coset representatives of G/H, then the vector space $\mathrm{R}_H^G V$ is the orthogonal direct sum of the subspaces $s_i \otimes V$, $1 \le i \le k$. Furthermore

$$\mathrm{R}_H^G f(s_i \otimes u,\, s_i \otimes v) = f(u, v) \tag{4.17}$$

and so the map $v \to s_i \otimes v$ is an isometry $(V, f) \to (s_i \otimes V, R_H^G f)$.

Theorem 4.2.3 *Let H be a subgroup of G and $\rho : H \to \mathbf{I}(V, f)$ an isometry representation, where f is ε-reflexive. The form $\mathrm{R}_H^G f$ of the standard induced isometry representation $\mathrm{R}_H^G \rho : G \to \mathbf{I}(\mathrm{R}_H^G V,\, \mathrm{R}_H^G f)$ is equivalent to $(G : H)$ copies of f.* □

Remark. If $\rho : G \to \mathbf{I}(V, f)$ is an isometry representation of G (and H is a subgroup of G), one can also consider $\mathrm{R}_G^H \rho$ with respect to the (H, G)-bimodule KG. The induced (linear) representation $\mathrm{R}_G^H \rho$ in this case is simply the restriction $\rho|_H$ since the canonical isomorphism $KG \otimes_{KG} V \cong V$ is a KG-isomorphism and so also an H-isomorphism. If $\rho : G \to \mathbf{I}(V, f)$ is an isometry representation and we take $b(x, y) = \mathrm{Tr}_{KG/K}(x\bar{y})$, then if $x, y \in G$,

$$\mathrm{R}_G^H f(x \otimes u,\, y \otimes v) = \sum_s f(u,\, sv)b(xs,\, y) = f(u,\, x^{-1}yv) = f(xu,\, yv)$$

which implies that the canonical isomorphism $KG \otimes_{KG} V \cong V$ is also an isometry $(KG \otimes_{KG} V, \mathrm{R}_G^H f) \cong (V, f)$. □

There are in general many more (G, H)-biinvariant ε-reflexive forms on KG in addition to the form $\mathrm{Tr}_{KG/K}(x\bar{y})$:

Theorem 4.2.4 *Let $b : KG \times KG \to (K, ^-)$ be a sesquilinear form and $H < G$ a subgroup of G.*

(a) b is G-invariant and η-reflexive if and only if
(i) $b(s, t) = b(1, s^{-1}t)$ for all $s, t \in G$, and
(ii) $b(1, s) = \eta\overline{b(1, s^{-1})}$ for all $s \in G$.

(b) b is (G, H)-biinvariant and η-reflexive if and only if (i) and (ii) hold as well as
(iii) $b(1, h^{-1}sh) = b(1, s)$ for all $s \in G, h \in H$.

If b is G-invariant and η-reflexive (i) and (ii) follow easily. If it is also H-invariant, then $b(1, h^{-1}sh) = b(h, sh)$ by G-invariance, $= b(1, s)$ by H-invariance.

Conversely suppose (i) and (ii) hold. Then if $r \in G$,

$$b(rs, rt) = b(1, (rs)^{-1}rt) = b(1, s^{-1}t) = b(s, t)$$

so b is G-invariant. And it is η-reflexive since

$$b(s, t) = b(1, s^{-1}t) = \eta\overline{b(1, t^{-1}s)} = \eta\overline{b(t, s)}.$$

If (iii) also holds, then

$$b(sh, th) = b(1, h^{-1}s^{-1}th) = b(1, s^{-1}t) = b(s, t).$$ □

Exercise.
1. Let C be the set of "H-conjugacy" classes of G, i.e. the subsets $[s] = \{h^{-1}sh : h \in H\}$ of G.
(a) An η-reflexive (G, H)-biinvariant b form on KG defines a map $\beta : \mathrm{C} \to K$ satisfying $\beta([s]) = b(1, s)$ and

$$\beta([s^{-1}]) = \eta\overline{\beta([s])} \quad \text{for all } s \in G. \tag{4.18}$$

(b) Conversely show that if $\beta : \mathrm{C} \to K$ is a map satisfying (4.18), there is a unique η-reflexive (G, H)-biinvariant form b on KG satisfying $b(1, s) = \beta([s])$ for all $s \in G$.

(c) Order the elements of G in some way such that 1 is the smallest element and construct the multiplication table of G using this ordering. For each s, interchange rows s and s^{-1} (keeping the original labels of the rows, so that row s now contains the products $s^{-1}t$ ($t \in S$) instead of st). Now apply the composition of $G \to \mathrm{C}$ and β to the entries of the table to obtain a matrix $B \in \mathrm{M}(|G|, K)$. Show that B is the matrix of b with respect to the basis G of KG. □

Example. Suppose that $G = \mathrm{A}_4$, the alternating group on 4 letters, and let $H = \{1, (123), (132)\}$. Then

$$\begin{aligned} \mathrm{C} = \big\{&\{1\},\ \{(12)(34), (13)(24), (14)(23)\},\ \{(123)\}, \\ &\{(132)\},\ \{(124), (234), (143)\},\ \{(142), (243), (134)\}\big\}. \end{aligned} \tag{4.19}$$

Suppose that $\eta = 1$, that the involution on K is the identity, and that β sends the elements of C shown above to x, y, z, z, u, u respectively. (Note that β of the 3rd and 4th conjugacy classes must be equal by (4.18), and a similar remark applies to the 5th and 6th conjugacy classes). Then the matrix of the associated b is

$$\begin{pmatrix} x & y & y & y & z & z & u & u & u & u & u & u \\ y & x & y & y & u & u & u & u & z & u & z & u \\ y & y & x & y & u & u & u & z & u & z & u & u \\ y & y & y & x & u & u & z & u & u & u & u & z \\ z & u & u & u & x & z & u & u & u & y & y & y \\ z & u & u & u & z & x & y & y & y & u & u & u \\ u & u & u & z & u & y & x & y & y & u & u & z \\ u & u & z & u & u & y & y & x & y & z & u & u \\ u & z & u & u & u & y & y & y & x & u & z & u \\ u & u & z & u & y & u & u & z & u & x & y & y \\ u & z & u & u & y & u & u & u & z & y & x & y \\ u & u & u & z & y & u & z & u & u & y & y & x \end{pmatrix}$$

Its determinant (as calculated with Maple) is

$$(x + 3\,y + 6\,u + 2\,z)\,(x + 3\,y - 3\,u - z)^2\,(x + 2\,z - y - 2\,u)^3\,(x + u - y - z)^6,$$

so the set of solutions which lead to a nonsingular b consists of the complement of 4 hyperplanes in K^4, and so is "Zariski dense" in K^4 if K is infinite. Note that if we choose $u = z = 0$, the determinant $=_2 (x + 3y)(x - y)$ which is arbitrary mod $\dot{K}^2$. $\square$

There is also a map of Grothendieck-Witt and Witt groups connected with induction. Let X be a (G, H)-bimodule with a nonsingular biinvariant form b. Suppose that f and g are H-invariant forms on V and W respectively. If $\varphi : (V, f) \to (W, g)$ is H-equivariant, then

$$(\hat{b}\hat{g})(x \otimes \varphi u, y \otimes \varphi v) = \hat{b}(x\hat{g}(\varphi u, \varphi v), y) = \hat{b}(x\hat{f}(u, v), y) = (\hat{b}\hat{f})(x \otimes u, y \otimes v),$$

and $(1 \otimes \varphi)(sx \otimes u) = sx \otimes \varphi u = s(1 \otimes \varphi)(x \otimes u)$. It follows that $f \rightsquigarrow \mathrm{R}_H^G f$ is a functor from the category of H-invariant ε-reflexive forms to the category of G-invariant $\eta\varepsilon$-reflexive forms. Since $f = f_1 \perp f_2$ implies that $\mathrm{R}_H^G f = \mathrm{R}_H^G f_1 \perp R_H^G f_2$, we get a homomorphism $\mathrm{R}_H^G : {}^H\mathbf{M}_\varepsilon \to {}^G\mathbf{M}_{\eta\varepsilon}$ between the monoids of equivalence classes of H-invariant and G-invariant forms, and so a homomorphism

$$\mathrm{R}_H^G : \hat{\mathrm{W}}_\varepsilon(K, H) \to \hat{\mathrm{W}}_{\eta\varepsilon}(K, G). \tag{4.20}$$

Suppose that $\mathrm{H}_\varepsilon(\rho) : H \to \mathbf{I}(\mathrm{H}_\varepsilon(V), h)$ is the hyperbolic representation based on $\rho : H \to \mathbf{GL}(V)$. Then $\mathrm{R}_H^G(\mathrm{H}_\varepsilon(\rho))$ is also hyperbolic. In fact if $u, v \in V$ and $x, y \in X$, then $\hat{b}\hat{h}(x \otimes u, y \otimes v) = \hat{b}(x\hat{h}(u, v), y) = \hat{b}(0, y) = 0$, so $X \otimes V$ is a totally isotropic subspace of $X \otimes \mathrm{H}_\varepsilon(V)$ with respect to $\hat{b}\hat{h}$. Similarly $X \otimes \overline{V^*}$ is totally isotropic, and so by Th. 3.1.2 (p. 97), $\mathrm{R}_H^G(\hat{b}\hat{h})$ is a KG-hyperbolic form. Since

$\mathrm{R}_H^G(h) = \mathrm{Tr}_{KG/K}(\hat{b}\hat{h})$, $\hat{b}\hat{h} = \widehat{\mathrm{R}_H^G(h)}$ by Th. 2.1.1 (p. 86), and we see by Cor. 3.1.2 (p. 97) that $\mathrm{R}_H^G(\mathrm{H}_\varepsilon(\rho))$ is a hyperbolic representation.

It follows at once that there is a homomorphism $\mathrm{R}_H^G : \mathrm{W}_\varepsilon(K, H) \to \mathrm{W}_{\eta\varepsilon}(K, G)$ induced by (4.20).

Theorem 4.2.5 *Let* R_H^G *be the induction functor based on a nonsingular* (G, H)*-biinvariant* η*-reflexive space* (X, b)*. Then there are homomorphisms*

$$\mathrm{R}_H^G : \hat{\mathrm{W}}_\varepsilon(K, H) \to \hat{\mathrm{W}}_{\eta\varepsilon}(K, G) \quad \textit{and} \quad \mathrm{R}_H^G : \mathrm{W}_\varepsilon(K, H) \to \mathrm{W}_{\eta\varepsilon}(K, G).$$

taking the equivalence class respectively Witt class of an isometry representation of H *to the class of the induced representation of* G. □

Frobenius Reciprocity. We now consider Frobenius Reciprocity (cf. (1.59) and (1.60), p. 76) in the context of isometry representations. This applies only in the "classical" case of induction and restriction, with $H < G$ and $X = KG$ viewed as a (G, H)-bimodule (induction) or as an (H, G)-bimodule (restriction), and $b(x, y) = \mathrm{Tr}_{KG/K}(x\bar{y})$.

Let $(V, f) \in {}_{(K,^-)}^{\;G}\mathcal{H}_\varepsilon$ (the G-invariant ε-reflexive forms over $(K, ^-)$) and $(W, g) \in {}_{(K,^-)}^{\;H}\mathcal{H}_\delta$. The Frobenius Reciprocity Isomorphism (for linear representations) is

$$\begin{aligned} KG \otimes_{KH} (W \otimes_K \mathrm{Res}\, V) &\to (KG \otimes_{KH} W) \otimes_K V, \\ s \otimes (w \otimes v) &\to (s \otimes w) \otimes sv \end{aligned} \tag{4.21}$$

(cf. (1.61)). Denote by $\mathrm{Res}\, f$ the form f when considered as an H-invariant form on $\mathrm{Res}\, V$ $(= V)$: $(\mathrm{Res}\, V, \mathrm{Res}\, f) \in {}_{(K,^-)}^{\;H}\mathcal{H}_\varepsilon$. The form $g\mathrm{Res}\, f$ on $W \otimes_K \mathrm{Res}\, V$ is H-invariant (where H operates in the usual way in the tensor product: $h(w \otimes v) = hw \otimes hv$), and so the form

$$\mathrm{Ind}(g\mathrm{Res}\, f) = \mathrm{Tr}_{KG/K}\big(\hat{b}(\widehat{g\mathrm{Res}\, f})\big)$$

supported by the left side of (4.21) is G-invariant. Similarly the form $(\mathrm{Ind}\, g)f$ supported by the right side also is G-invariant. We now check that the isomorphism (4.21) is an equivariant isometry.

Since $KG \otimes_{KH} W$ is the orthogonal direct sum of the $s_i \otimes W$ (cf. p. 163), $(KG \otimes_{KH} W) \otimes_K V$ is the orthogonal direct sum of the $(s_i \otimes_{KH} W) \otimes_K V$ since

$$((\mathrm{Ind}\, g)f)((s_i \otimes w) \otimes v, (s_j \otimes w') \otimes v') = (\mathrm{Ind}\, g)(s_i \otimes w, s_j \otimes w')f(v, v') = 0$$

if $i \neq j$. Furthermore the orthogonal direct summand $s_i \otimes (W \otimes \mathrm{Res}\, V)$ of the left side of (4.21) is mapped into $(s_i \otimes_{KH} W) \otimes_K V$ for each i, and so must be mapped isomorphically onto it. Thus to show that the map in (4.21) is an isometry, it suffices to check it on these orthogonal direct summands. But this is clear since

$$\begin{aligned} &((\mathrm{Ind}\, g)f)((s_i \otimes w) \otimes s_i v, (s_i \otimes w') \otimes s_i v') \\ &\quad = (\mathrm{Ind}\, g)(s_i \otimes w, s_i \otimes w')f(s_i v, s_i v') \\ &\quad = g(w, w')f(v, v') = (gf)(w \otimes v, w' \otimes v') \\ &\quad = \mathrm{Ind}(g\mathrm{Res}\, f)(s_i \otimes (w \otimes v), s_i \otimes (w' \otimes v')) \end{aligned}$$

by (4.17). Thus we have proved

Theorem 4.2.6 *The Frobenius Reciprocity Isomorphism (4.21) is an isometry, KG-equivariant, of the $\varepsilon\delta$-Hermitian forms* $\mathrm{Ind}(g\mathrm{Res}\, f)$ *and* $(\mathrm{Ind}\, g)f$ *on the left and right sides respectively of (4.21). Thus*

$$\big(\mathrm{Ind}(W \otimes_K \mathrm{Res}\, V),\ \mathrm{Ind}(g\mathrm{Res}\, f)\big) \cong \big((\mathrm{Ind}\, W) \otimes_K V,\ (\mathrm{Ind}\, g)f\big). \qquad \square$$

Exercise. Prove this theorem by showing directly from the definitions that

$\mathrm{Ind}(g\mathrm{Res}\, f)(s\otimes(w\otimes v), s'\otimes(w'\otimes v'))$ and $((\mathrm{Ind}\, g)f)((s\otimes w)\otimes sv, (s'\otimes w')\otimes s'v')$

are both equal to

$$= \begin{cases} g(s'^{-1}sw, w')f(s'^{-1}sv, v') & \text{if } sH = s'H, \\ 0 & \text{if } sH \neq s'H. \end{cases} \qquad \square$$

The hyperbolic and linearization maps. Recall the linearization map

$$\mathrm{L} : \hat{\mathrm{W}}_\varepsilon(K, G) \to \mathrm{K}_0(KG) \tag{4.22}$$

(p. 155). There is also a map in the other direction, the "hyperbolic map"

$$\mathrm{H}_\varepsilon : \mathrm{K}_0(KG) \to \hat{\mathrm{W}}_\varepsilon(K, G) = \hat{\mathrm{W}}_\varepsilon(KG), \tag{4.23}$$

arising from the map ${}_{KG}\mathbf{P} \to {}_{KG}\mathbf{M}_\varepsilon$ which takes the isomorphism class of a KG-module U to the equivalence class of the ε-hyperbolic representation $\mathrm{H}_\varepsilon(\rho)$ based on the corresponding representation $\rho : G \to \mathbf{GL}(U)$, which we identify with the class of the ε-hyperbolic space $\mathrm{H}_\varepsilon(U)$ over $(KG, {}^-)$. These two maps satisfy

$$\mathrm{H}_{\varepsilon\delta}\Big((V, \rho) \otimes \mathrm{L}(x)\Big) = \mathrm{H}_\varepsilon(V, \rho) \otimes x \tag{4.24}$$

in $\hat{W}_{\varepsilon\delta}(K, G)$, where ρ is a linear representation on V and $x \in \hat{\mathrm{W}}_\delta(K, G)$. (Recall that $\otimes_K$ makes $\mathrm{K}_0(KG)$ into a ring – cf. p. 70). It suffices to verify this when x is (the equivalence class of) a δ-reflexive representation $G \to \mathbf{I}(U, h)$, and we do this by viewing (4.24) as a relationship in $\hat{W}_{\varepsilon\delta}(KG)$, invoking Cor. 3.1.2 (p. 97).

The left side is the $\varepsilon\delta$-hyperbolic space over $(KG, {}^-)$ based on $V \otimes U$. The right side is (the equivalence class of) the space $(\mathrm{H}_\varepsilon(V, \rho)\otimes U, \mathrm{H}_\varepsilon(\rho)h)$; its subspaces $V\otimes U$ and $\overline{V^*} \otimes U$ are both totally isotropic, so also with respect to the associated KG-form, hence by Th. 3.1.2 the right side is also the $\varepsilon\delta$-hyperbolic space based on the $(KG, {}^-)$-module $V \otimes U$.

Example. Suppose that $x \in \ker \mathrm{L}$, so $\mathrm{H}_\varepsilon(V, \rho) \otimes x = 0$. If $(V, \rho) = \langle 1 \rangle$ (with trivial action of G), then $\langle 1, -1 \rangle \otimes x = 0$, again with trivial action of G on $\langle 1, -1 \rangle$. Therefore if -1 is a square in K when ${}^-$ is the identity respectively a norm in K/K_0 when ${}^-$ is *not* the identity, $2x = 0$, i.e. $2 \ker \mathrm{L} = 0$.

Similarly if -1 is a sum of two squares respectively of two norms, say $-1 = a^2 + b^2$ respectively $a\bar{a} + b\bar{b}$, take $(V, \rho) = \langle 1, 1 \rangle$ (again with trivial action); then $\langle 1, 1, -1, -1 \rangle \otimes x = 0$. If we show that $\langle -1, -1 \rangle \cong \langle 1, 1 \rangle$, then $4 \ker \mathrm{L} = 0$. Suppose that $W = \langle -1, -1 \rangle$ in the basis w_1, w_2. Then $W = \langle 1, 1 \rangle$ in the basis $aw_1 + bw_2$, $\bar{b}w_1 - \bar{a}w_2$. $\square$

4.3 The graded Witt-Grothendieck and Witt rings

The *graded* (equivariant) Witt-Grothendieck and Witt groups are defined as

$$\begin{aligned} \hat{\mathrm{W}}(K,G) &= \hat{\mathrm{W}}_1(K,G) \oplus \hat{\mathrm{W}}_{-1}(K,G) \\ \mathrm{W}(K,G) &= \mathrm{W}_1(K,G) \oplus \mathrm{W}_{-1}(K,G) \end{aligned}$$

respectively. We shall see that the product of forms turns both of these groups into graded (commutative) rings.

Suppose that (V,g) and (W,h) are nonsingular reflexive spaces over $(K,\bar{\ })$, say δ-reflexive and ε-reflexive ($\delta,\varepsilon = \pm 1$ as usual) respectively. The product gh is defined,

$$(gh)(v_1 \otimes w_1, v_2 \otimes w_2) = g(v_1,v_2)h(w_1,w_2),$$

and is a $\delta\varepsilon$-reflexive form on $V \otimes_K W$ (cf. §3.2.1). If G acts on V and W, its action on $V \otimes W$ is $s(v \otimes w) = (sv) \otimes (sw)$, and if g and h are G-invariant, the form gh is clearly G-invariant as well.

If $\varphi : (V,g) \to (V',g')$ and $\psi : (W,h) \to (W',h')$ are equivariant isometries, then so is $\varphi \otimes \psi : (V \otimes_K W, gh) \to (V' \otimes_K W', g'h')$, and it follows that the product

$$[g][h] = [gh] \quad (g \in {}_{(K,\bar{\ })}\mathcal{H}_\delta^*, \;\; h \in {}_{(K,\bar{\ })}\mathcal{H}_\varepsilon^*)$$

is well-defined between "homogeneous elements" in $\hat{\mathrm{W}}(K,G)$ and the definition of product can be extended to nonhomogeneous elements by distributivity. It is commutative, associative and distributive when homogeneous elements are involved – for example commutativity follows from the obvious fact that the canonical isomorphism $V \otimes_K W \to W \otimes_K V$ of KG-modules is also an equivariant equivalence between gh and hg – and the general laws follow from those. The class of the "unit" form $\alpha\bar{\beta}$ ($= \langle 1 \rangle$ with the trivial representation of G) is a multiplicative identity.

Thus $\hat{\mathrm{W}}(K,G)$ is indeed a graded commutative ring with identity, which we refer to as the (graded) equivariant Witt-Grothendieck ring.

Now we consider the Witt group $\mathrm{W}(K,G)$. Since the product of any form and a totally isotropic form is totally isotropic, it follows from Cor. 3.1.2 (p. 97) that the product of any nonsingular ε-reflexive form with a δ-hyperbolic form is an $\varepsilon\delta$-hyperbolic form. Thus the subgroup of $\hat{\mathrm{W}}(K,G)$ generated by hyperbolic forms is a graded ideal, and since it is the kernel of the canonical homomorphism $\hat{\mathrm{W}}(K,G) \to \mathrm{W}(K,G)$, the Witt group inherits a graded commutative ring structure as well – whence the graded equivariant Witt ring.

Remark. In the case when $\bar{\ }$ is not the identity, $\hat{\mathrm{W}}_\mu(K,G)$ and $\mathrm{W}_\mu(K,G)$ can be defined in an obvious way for all $\mu \in K$ satisfying $\mu\bar{\mu} = 1$, and then $\oplus_\mu \hat{\mathrm{W}}_\mu(K,G)$ and $\oplus_\mu \mathrm{W}_\mu(K,G)$ can be made into graded commutative rings in the same manner. □

Note that there is a canonical map $\hat{\mathrm{W}}_\varepsilon(K) \to \hat{\mathrm{W}}_\varepsilon(K,G)$ which sends the class of the nonsingular ε-reflexive form h to the equivariant class of h where G operates trivially on the space of h. This yields a graded ring homomorphism

$$\hat{\mathrm{W}}(K) \to \hat{\mathrm{W}}(K,G)$$

through which $\hat{\mathrm{W}}(K,G)$ becomes a $\hat{\mathrm{W}}(K)$-module (as well as a $\hat{\mathrm{W}}_1(K)$-module, of course). Furthermore a hyperbolic form $h \in \hat{\mathrm{W}}_\varepsilon(K)$ is also hyperbolic in $\hat{\mathrm{W}}_\varepsilon(K,G)$

and so we also have a graded ring homomorphism

$$\mathrm{W}(K) \to \mathrm{W}(K, G).$$

which makes $\mathrm{W}(K, G)$ into a module over both $\mathrm{W}(K)$ and $\mathrm{W}_1(K)$, and makes $\mathrm{W}_1(K, G)$ into a $\mathrm{W}_1(K)$-module. (Of course $\mathrm{W}_{-1}(K) = 0$ when $(K, ^-) = (K, \mathrm{id})$).

Theorem 4.3.1 *There are canonical graded ring homomorphisms*

$$\hat{\mathrm{W}}(K) \to \hat{\mathrm{W}}(K, G) \quad \textit{and} \quad \mathrm{W}(K) \to \mathrm{W}(K, G)$$

sending the class of a form h to the equivariant class of h in which G acts trivially on the underlying space. In particular, $\hat{\mathrm{W}}_1(K, G)$ and $\hat{\mathrm{W}}(K, G)$ are $\hat{\mathrm{W}}_1(K)$-modules, and $\mathrm{W}_1(K, G)$ and $\mathrm{W}(K, G)$ are $\mathrm{W}_1(K)$-modules. □

Finally it is easy to check that the graded analogues

$$\mathrm{Res}_H^G : \hat{\mathrm{W}}(K, G) \to \hat{\mathrm{W}}(K, H) \quad \text{and} \quad \mathrm{Res}_H^G : \mathrm{W}(K, G) \to \mathrm{W}(K, H),$$

$$\hat{\mathrm{W}}(K, G) \to \hat{\mathrm{W}}(K) \quad \text{and} \quad \mathrm{W}(K, G) \to \mathrm{W}(K),$$

$$\mathrm{L} : \hat{\mathrm{W}}(K, G) \to \mathrm{K}_0(KG),$$

of the functorial properties (4.14), (4.13), and (4.10) of the Witt-Grothendieck and Witt groups hold and are ring homomorphisms, all but the last graded. The details are left to the reader.

4.4 K_0 algebraically closed

Let K_0 be an algebraically closed field $\mathbf{C}$, so $K = K_0 = \mathbf{C}$ with involution $\mathrm{id}_{\mathbf{C}}$.

The character referred to in (iv) in the next theorem is the ordinary character if the characteristic of K is 0, otherwise it is the Brauer character with respect to some p-modular system. To say that the character is real means that its values are contained in a totally real finite extension of $\mathbb{Q}$ – which is the same as saying that all of the conjugates over $\mathbb{Q}$ of the character values are real numbers (under any imbedding into $\mathbb{C}$).

Theorem 4.4.1 *Let $\mathbf{C}$ be an algebraically closed field of characteristic $p \neq 2$, and G a finite group whose order is not divisible by p. Let $\mathbf{C}G$ have the standard $(\mathbf{C}, \mathrm{id}_{\mathbf{C}})$-involution $^-$.*

(a) Suppose that $\rho : G \to \mathbf{GL}(V)$ is an irreducible representation over $\mathbf{C}$, and that A is the simple summand of $\mathbf{C}G$ corresponding to ρ. Then the following four conditions are equivalent:
  - *(i) there is a nonsingular invariant bilinear form $b : V \times V \to \mathbf{C}$.*
  - *(ii) V and $\overline{V^*}$ are isomorphic $\mathbf{C}G$-modules.*
  - *(iii) $\bar{A} = A$.*
  - *(iv) The character χ of ρ is real-valued.*

(b) Suppose there is a nonsingular invariant bilinear form b on V. It is either symmetric or skew symmetric and is unique up to nonzero scalars – that is to say the nonsingular invariant bilinear forms on V are exactly the scalar multiples of b. The involution $^-$ which A inherits from $\mathbf{C}G$ is the adjoint with respect to b and so b is symmetric if $(A, ^-)$ is orthogonal, skew symmetric if $(A, ^-)$ is symplectic.

(c) *Let* $(\mathbf{C}G, ^-) = (A_1, ^-) \oplus \cdots \oplus (A_l, ^-)$ *be the decomposition of* $(\mathbf{C}G, ^-)$ *into simple involution algebras and, for* $\varepsilon = \pm 1$,

$$\hat{\mathrm{W}}_\varepsilon(\mathbf{C}, G) = \oplus_i \hat{\mathrm{W}}_\varepsilon(\mathbf{C}, A_i) \text{ and } \mathrm{W}_\varepsilon(\mathbf{C}, G) = \oplus_i \mathrm{W}_\varepsilon(\mathbf{C}, A_i) \qquad (4.25)$$

the corresponding decomposition of the Witt-Grothendieck and Witt groups (Th. 4.1.1). For each i *and* $\varepsilon = \pm 1$, *linearization*

$$\mathrm{L} : \hat{\mathrm{W}}_\varepsilon(\mathbf{C}, A_i) \to \mathrm{K}_0(A_i) \cong \mathbb{Z} \qquad (4.26)$$

is injective, and its image is

$2\mathrm{K}_0(A_i)$	*if* $(A_i, ^-)$ *is orthogonal and* $\varepsilon = -1$, *or* $(A_i, ^-)$ *is symplectic and* $\varepsilon = 1$
$\mathbb{Z}([S_1] + [S_2])$	*if* $(A_i, ^-)$ *is hyperbolic, or*
$\mathrm{K}_0(A_i)$	*otherwise,*

where $[S_1]$ *and* $[S_2]$ *are the distinct isomorphism classes of simple modules of* A_i *(in the hyperbolic case).*

(d) *If* $(A_i, ^-)$ *is hyperbolic, then for* $\varepsilon = \pm 1$

$$\mathrm{W}_\varepsilon(\mathbf{C}, A_i) = 0.$$

(e) *If* $(A_i, ^-)$ *is not hyperbolic, i.e.* A_i *is simple, then*

$$\mathrm{W}_1(\mathbf{C}, A_i) \cong \mathbb{Z}/2\mathbb{Z}, \quad \mathrm{W}_{-1}(\mathbf{C}, A_i) = 0 \quad \text{if } (A_i, ^-) \text{ is orthogonal,}$$

and

$$\mathrm{W}_1(\mathbf{C}, A_i) = 0, \quad \mathrm{W}_{-1}(\mathbf{C}, A_i) \cong \mathbb{Z}/2\mathbb{Z} \quad \text{if } (A_i, ^-) \text{ is symplectic.}$$

(a) is a special case of Th. 3.2.7(a), p. 115.

(b) That there is a nonsingular invariant symmetric or skew symmetric form b on V follows from part (c) of Th. 3.2.7. Suppose that b' is another invariant form on V. Let $\hat{b}$ and $\hat{b}'$ be the associated forms over $(KG, ^-)$; since V is an A-module, $\hat{b}$ and $\hat{b}'$ take their values in A. Now $\hat{b}$ is nonsingular, and by Prop. 3.1.1 (p. 94), there is $\alpha \in \mathrm{End}_A V = \mathbf{C}$ such that $\hat{b}' = \alpha \hat{b}$, and so

$$b' = \mathrm{Tr}_{KG/K} \hat{b}' = \mathrm{Tr}_{KG/K} \alpha \hat{b} = \alpha b.$$

This proves the first statement of (b), and the second is simply a reflection of the invariance of b: $b(au, v) = b(u, \bar{a}v)$ for all $a \in A, u, v \in V$.

(c) If $(A_i, ^-)$ is hyperbolic, then by Cor. 3.1.5 (p. 98) and (3.14), every nonsingular ε-Hermitian form over $(A_i, ^-)$ is the orthogonal direct sum of a number of copies of the ε-hyperbolic form $h_{\mathrm{H}(S_1)}$, and it follows that $\hat{\mathrm{W}}_\varepsilon(\mathbf{C}, A_i) \cong \mathbb{Z}([S_1] + [S_2])$ via L.

If A_i is simple, it is a full matrix algebra over $\mathbf{C}$. Then the rest of (c) as well as (d) and (e) follow easily from Th. 3.5.1 (p. 146) (and the correspondence between invariant forms over $(K, ^-)$ and forms over $(A, ^-)$ – cf. Cor. 2.1.1, p. 87 – and Cor. 3.1.2(a), p. 97, and the equivalence of (a)(ii) and (a)(iii) in this theorem). □

Corollary 4.4.1 (Frobenius) *If* G *is a finite group of order* g *and* $\mathbf{C}$ *is an algebraically closed field whose characteristic does not divide* $2g$, *then two orthogonal representations of* G *over* $\mathbf{C}$ *are (orthogonally) equivalent if and only if they are linearly equivalent. Similarly two symplectic representations are equivalent if and only if they are linearly equivalent.*

The necessity is trivial. Suppose that $\rho_1 : G \to \mathbf{I}(V_1, b_1)$ and $\rho_2 : G \to \mathbf{I}(V_2, b_2)$ are both orthogonal or both symplectic representations which are linearly equivalent. Since V_j is the orthogonal direct sum of its involution isotypic components $A_i V_j$ (cf. p. 87) and since for all i $A_i V_1$ is isomorphic to $A_i V_2$, the corollary follows immediately from the injectivity of L. □

4.5 The field K_0 real closed

Throughout this section, $\mathbf{R}$ is a real closed field, $\mathbf{C} = \mathbf{R}(i)$ ($i^2 = -1$) its algebraic closure. $\mathbf{H}$ stands for the Hamiltonian quaternions $(-1,-1)_{\mathbf{R}}$. "Conjugation" on $\mathbf{C}$ and $\mathbf{H}$ is denoted by *, as is the identity $\mathrm{id}_{\mathbf{R}}$ on $\mathbf{R}$.

Theorem 4.5.1 *Suppose that K/K_0 is a definite extension – in particular K_0 is formally real. Then*

(a) all simple components of KG are stable under the standard $(K, ^-)$-involution $^-$ of KG, and

(b) on any KG-module V there is an invariant form $b : V \times V \to (K, ^-)$ which is symmetric (if $K = K_0$) respectively Hermitian (if $(K : K_0) = 2$), and is positive definite with respect to every ordering P of K_0 at which K/K_0 is definite.

Let $\mathrm{Tr} : KG \to K$ be the canonical trace. The trace form $\mathrm{Tr}(a, b) = \mathrm{Tr}(a\bar{b})$ is symmetric if $K = K_0$, Hermitian if $(K : K_0) = 2$. The elements of G form an orthonormal basis, so it is positive definite with respect to any ordering P of K_0 at which K/K_0 is definite.

Therefore if a is a nonzero element in a simple component A of KG, $0 < \mathrm{Tr}(a, a) = \mathrm{Tr}(a\bar{a})$ at any ordering at which K/K_0 is definite, which implies that $a\bar{a} \neq 0$, so $\bar{a} \in A$ since $^-$ permutes the simple components of KG. Thus $\bar{A} = A$. This is (a).

Let $b' : V \times V \to (K, ^-)$ be a symmetric respectively Hermitian form on the KG-module V which is positive definite with respect to all orderings P of K_0 at which K/K_0 is positive definite – for example choose a basis of V and declare it to be an orthonormal basis. Then the form

$$b(u, v) = \sum_{s \in G} b'(su, sv)$$

is invariant and is a symmetric respectively Hermitian form which is positive definite at every ordering P of K_0 at which K/K_0 is definite. □

We now deal with isometry representations – equivalently, nonsingular invariant reflexive forms – when $K_0 = \mathbf{R}$. There are two cases:

(i) $K = K_0 = \mathbf{R}$,

(ii) $K/K_0 = \mathbf{C}/\mathbf{R}$.

The second case is easier to handle, and we begin with it:

Theorem 4.5.2 *Let $(A, ^-)$ be a simple component of $(\mathbf{C}G, ^*)$ where * is the standard $(\mathbf{C}, ^*)$-involution on $\mathbf{C}G$, and let h be an invariant positive definite Hermitian form on a simple A-module S (Th. 4.5.1).*

(a) $(A, ^-) \cong \mathrm{M}(n, (\mathbf{C}, ^*))$.

(b) *Every invariant Hermitian form on S is of the form $\alpha h(u,v)$ with $\alpha \in \mathbf{R}$; those with $\alpha > 0$ are positive definite and form a single class $[h] \in \hat{\mathrm{W}}_1(\mathbf{C},G)$, while those with $\alpha < 0$ are negative definite and form a distinct class $[-h] \in \hat{\mathrm{W}}_1(\mathbf{C},G)$. In particular all nonzero invariant Hermitian forms are nonsingular and are either positive definite or negative definite.*

(c) *The nonsingular invariant skew Hermitian forms on S are the $\alpha h(iu,v)$ for $\alpha \in \dot{\mathbf{R}}$, and they form two classes $[h(iu,v)]$ and $[-h(iu,v)]$ in $\hat{\mathrm{W}}_{-1}(\mathbf{C},G)$.*

This is very different from the situation of $(\mathbf{C}G, {}^-)$ where ${}^-$ is the standard $(\mathbf{C}, \mathrm{id}_{\mathbf{C}})$-involution – see Cor. 4.5.2, p. 175.

By the previous theorem, A is simple; we may suppose that $A = \mathrm{M}(n,\mathbf{C})$.

(a) Let h be a positive definite invariant Hermitian form on S. By choosing a suitable basis we may assume that $\mathrm{mat}\, h = \mathrm{I}_n$, and so, identifying $S = \mathbf{C}^{1\times n}$, $h(u,v) = uv^{t*}$. If $a \in A = \mathrm{M}(n,\mathbf{C})$,

$$h(ua,v) = uav^{t*} = h(u,v\bar{a}) = u^t\bar{a}^{t*}v^*$$

which implies that $\bar{a} = a^{t*}$.

(b), (c) By Prop. 3.1.1(c) (p. 94), the invariant forms on S are those of the form $h(\alpha u,v)$, $\alpha \in \mathbf{C}$, and it is clear that $h(\alpha u,v)$ is Hermitian if and only if $\alpha \in \mathbf{R}$, skew Hermitian if and only if $\alpha \in \mathbf{R}i$. Moreover if it is Hermitian, it is positive definite if and only if $\alpha > 0$, negative definite if and only if $\alpha < 0$.

The statements about equivalence classes of invariant Hermitian forms follow easily from the fact that the positive elements of $\mathbf{R}$ are the squares $\dot{\mathbf{R}}^2$ – if $\alpha = \beta^2$, $\beta \in \dot{\mathbf{R}}$, then $h(\alpha u,v) = h(\beta u,\beta v) \cong h(u,v)$. Similarly if $\alpha > 0$, $h(-\alpha u,v) \cong -h(u,v)$ and it is clear that $[h] \neq [-h]$.

Finally any equivariant equivalence between $h(\alpha iu,v)$ and $h(-\alpha iu,v)$ ($\alpha > 0$) would also be an equivariant equivalence between $h(\alpha u,v)$ and $h(-\alpha u,v)$. Thus (c) follows. □

If $g : V \times V \to (A, {}^-)$ is a sesquilinear form and $\varphi \in \mathrm{End}_A V$, recall that $g_\varphi(u,v) = g(u\varphi,v)$ (Prop. 3.1.1(c), p. 94, now writing endomorphisms on the right).

Theorem 4.5.3 (Real counterpart of Theorem 4.4.1(b)) *Suppose that*

(i) *A is a simple component of $\mathbf{R}G$,*
(ii) *S is a simple A-module,*
(iii) *D is the division algebra $\mathrm{End}_A S$,*
(iv) *$b : S \times S \to (\mathbf{R}, \mathrm{id})$ is a positive definite invariant symmetric form (Th. 4.5.1).*

Then

(a) *D is either $\mathbf{R}$, $\mathbf{C}$ or $\mathbf{H}$.*

(b) ***Invariant symmetric forms.*** *Every nonsingular invariant symmetric form on S is a real scalar multiple of b, and the adjoint involution on D with respect to any of them is conjugation * ($=$ id on $\mathbf{R}$). There are exactly two equivariant classes of such forms on S, namely the class $[b]$ of positive definite forms and the class $[-b]$ of negative definite forms.*

(c) ***Invariant skew symmetric forms.*** *If $D = \mathbf{R}$, every invariant form on S is a real scalar multiple of b, and so in particular there are no nonzero invariant skew symmetric forms on S.*

If $D = \mathbf{C}$, the form $b_i(u,v) = b(ui,v)$ is skew symmetric and invariant and every invariant skew symmetric form on S is a real scalar multiple of it. There are two equivariant classes of nonzero invariant skew symmetric forms on S, namely $[b_i]$ and $[-b_i]$.

If $D = \mathbf{H}$, the invariant skew symmetric forms on S form a vector space of dimension 3 over $\mathbf{R}$ with basis b_i, b_j, b_k, and there is only one equivariant class of nonzero invariant skew symmetric forms.

(d) *The dimension over $\mathbf{R}$ of the space of invariant bilinear forms on S is $\dim_{\mathbf{R}} D$, that is to say it is*

$$\begin{array}{ll} 1 & \textit{if } D = \mathbf{R},\ \textit{basis } \{b\}, \\ 2 & \textit{if } D = \mathbf{C},\ \textit{basis } \{b, b_i\},\ \textit{and} \\ 4 & \textit{if } D = \mathbf{H},\ \textit{basis } \{b, b_i, b_j, b_k\}. \end{array}$$

(e) *The involution $^-$ induced on A by the standard $(\mathbf{R}, \mathrm{id})$-involution on $\mathbf{R}G$ is transpose conjugate, that is to say*

$$(A, ^-) \cong \mathrm{M}(n, (D, ^*)).$$

In particular the involution algebra

$$(A, ^-) \ \textit{is} \ \begin{cases} \textit{orthogonal if } D = \mathbf{R}, \\ \textit{symplectic if } D = \mathbf{H}, \\ \textit{unitary if } D = \mathbf{C}. \end{cases}$$

Before beginning the proof, we summarize the results of this theorem as well as the next corollary and theorem in a table. In the table, $\mathrm{M}(n, (\mathbf{C}, \mathrm{id}))^2$ is the direct sum $\mathrm{M}(n, \mathbf{C}) \oplus \mathrm{M}(n, \mathbf{C})$ and $^\leftrightarrow$ means $(a_1, a_2) \to (a_2^t, a_1^t)$. The involution $^t \mathrm{inn}\, J$ is the transpose followed by the inner automorphism with respect to an arbitrary nonsingular skew symmetric matrix $J \in \mathrm{M}(2n, \mathbf{C})$. The fourth, fifth and sixth columns refer to $(A, ^-)$-invariant (or G-invariant) forms and their equivalence classes on the simple A-module S; b is a positive definite invariant symmetric form on S. In the last column, def/F, E means that an irreducible representation ρ of G, corresponding to the summand of $(\mathbf{C}G, ^-)$ in the third column, is defined over F and its character values lie in E, and these are the minimal such fields over $\mathbf{R}$.

$(A, ^-)$	type	$(\mathbf{C}, \mathrm{id}) \otimes_{\mathbf{R}} (A, ^-)$	basis inv fms	classes sym fms	classes skew fms	$\rho, \chi(G)$
$\mathrm{M}(n, (\mathbf{R}, \mathrm{id}))$	orth.	$\mathrm{M}(n, (\mathbf{C}, \mathrm{id}))$	b	$[b], [-b]$	$\emptyset$	def/$\mathbf{R}$, $\mathbf{R}$
$\mathrm{M}(n, (\mathbf{C}, ^*))$	unit.	$\mathrm{M}(n, ((\mathbf{C}, \mathrm{id})^2, ^\leftrightarrow))$	b, b_i	$[b], [-b]$	$[b_i], [-b_i]$	def/$\mathbf{C}$, $\mathbf{C}$
$\mathrm{M}(n, (\mathbf{H}, ^*))$	sympl.	$(\mathrm{M}(2n, \mathbf{C}), {}^t \mathrm{inn} J)$	b, b_i, b_j, b_k	$[b], [-b]$	$[b_i]$	def/$\mathbf{C}$, $\mathbf{R}$

Part (a) of the theorem follows from Frobenius' Theorem, Th. 1.4.8.

(b) As usual the adjoint involution on $D = \mathrm{End}_{\mathbf{R}} S$ with respect to b will be denoted by $^-$. We note that if $d \in D$ and $d\bar{d} \in \mathbf{R}$, then $d\bar{d} \geq 0$. For otherwise $b(vd, vd) = b(v, vd\bar{d}) = d\bar{d}b(v,v) < 0$ (if $v \neq 0$). It is this simple fact that lies behind much of the behaviour of isometry representations over real fields.

Suppose that $D = \mathbf{C}$. Then $\bar{i} = \pm i$, and since $i^2 = -1$, we must have $\bar{i} = -i$ by the above remark, so $^- = {}^*$ on $\mathbf{C}$.

Let $D = \mathbf{H}$. The pure quaternions $\mathbf{H}_0 = \mathbf{R}i + \mathbf{R}j + \mathbf{R}k$ are characterized as the elements $d \in \mathbf{H}$, $d = 0$ or $d \notin \mathbf{R}$, with $d^2 \in \mathbf{R}$. It follows at once that $\mathbf{H}_0$ is stable under any involution. Since the involution is an $\mathbf{R}$-linear endomorphism of the $\mathbf{R}$-vector space $\mathbf{H}_0$ with square $= \mathrm{id}_V$, its eigenvalues are ± 1. Were d an eigenvector for the eigenvalue 1, $\bar{d}d = d^2 < 0$. Thus the involution is $d \to -d$ on $\mathbf{H}_0$, and of course is the identity on $\mathbf{R}1_{\mathbf{H}}$, hence is conjugation * on $\mathbf{H}$.

By Prop. 3.1.1 (p. 94), any invariant form $S \times S \to (\mathbf{R}, \mathrm{id})$ is of the form b_d for some $d \in D$, is symmetric if and only if $\bar{d} = d$ – i.e. $d \in \mathbf{R}$ – and is skew symmetric if and only if $\bar{d} = -d$, i.e. $D = \mathbf{C}$ and $d \in \mathbf{R}i$, or $D = \mathbf{H}$ and $d \in \mathbf{H}_0$. This proves the first statement of (b), and also proves the statements in (c) that the $\mathbf{R}$-vector space of invariant skew symmetric forms is 0 if $D = \mathbf{R}$ and has basis $b(ui, v)$ if $D = \mathbf{C}$ and basis $b(ui, v), b(uj, v), b(uk, v)$ if $D = \mathbf{H}$.

If $b(ud, v)$ and $b(ud', v)$ are symmetric, i.e. $d, d' \in \mathbf{R}$, it is trivial that they are in different equivariant classes if d and d' have different signs. If they have the same sign, then $d = \alpha^2 d'$ for some $\alpha \in \dot{\mathbf{R}}$, and it follows that the two forms are equivariantly equivalent via the map $u \to \alpha u$.

(c) By Prop. 3.1.1(f) the forms b_d and $b_{d'}$ are equivariantly equivalent if and only if there is $e \in \dot{D}$ such that $\bar{e}de = d'$. If $D = \mathbf{R}$ or $\mathbf{C}$, this is equivalent to $e\bar{e}d = d'$, which means that d and d' differ by a positive scalar (in $\mathbf{R}$). Conversely as we saw above, they are equivariantly equivalent if they differ by a positive scalar. In particular if $D = \mathbf{C}$, there are two equivalence classes of skew symmetric forms as stated.

So suppose that $D = \mathbf{H}$, and that b_d and $b_{d'}$ are nonzero invariant skew symmetric forms on V, i.e. d and d' are nonzero elements of $\mathbf{H}_0$. Since $b(ud\alpha, v)$ is equivalent to $b(ud, v)$ for any $\alpha > 0$ and since $d^2 < 0$, we can assume that $d^2 = -1$ – and similarly that $d'^2 = -1$. By the Skolem-Noether Theorem (p. 11), the isomorphism $\mathbf{R}(d) \xrightarrow{\sim} \mathbf{R}(d')$ which takes d to d' can be extended to an inner automorphism of $\mathbf{H}$, say $a \to e^{-1}ae$. Since $e\bar{e} > 0$, we can scale e to get $e\bar{e} = 1$. Then $\bar{e}de = d'$. This finishes the proof of (c).

(d) follows from (b) and (c), and Prop. 3.1.1(a) and (c).

(e) Let $\mathrm{trd} : D \to \mathbf{R}$ be the reduced trace. It is the identity if $D = \mathbf{R}$, and is $d + d^*$ in the other 2 cases – see p. 13 and §1.2.4. By (the right-handed version of) Th. 3.2.4 (p. 113), there is a unique nonsingular Hermitian form $h_0 : S \times S \to (D, \bar{\ })$ such that $\mathrm{trd} \circ h_0 = b$. If $D = \mathbf{C}$ or $\mathbf{H}$, $h_0(v, v) \in \mathbf{R}$, and so $h_0(v, v) = \frac{1}{2}\mathrm{trd}(h_0(v, v)) = \frac{1}{2}b(v, v) > 0$ if $v \neq 0$, so h_0 is also positive definite (in all three cases). We can therefore choose an orthonormal basis for S with respect to h_0, and then (e) follows by Th. 3.2.8(d) (p. 117). □

Corollary 4.5.1 *Let $(K, \bar{\ }) = (\mathbf{R}, \mathrm{id})$ or $(\mathbf{C}, {}^*)$. Let $\rho : G \to \mathbf{I}(V, h)$ and $\rho' : G \to \mathbf{I}(V', h')$ be isometry representations over $(K, \bar{\ })$ which are irreducible as linear representations, and where h and h' are symmetric if $K = \mathbf{R}$ respectively Hermitian if $K = \mathbf{C}$. Then any invariant symmetric respectively Hermitian form on V or V' is definite, and ρ and ρ' are equivalent as isometry representations if and only if*

(i) they are equivalent as linear representations, and

(ii) h and h' are both positive definite or are both negative definite.

The definiteness of the forms follows from part (b) of the theorem in the symmetric case and from Th. 4.5.2 in the Hermitian case. The necessity of (i) and (ii) is clear – suppose conversely that they hold. We may assume that $\rho' = \rho$ (as linear representations). By the same theorems $h' = \alpha h$ for some $\alpha > 0$. If $\alpha = \beta^2$, then $v \to \beta v$ is an equivalence between ρ' and ρ as isometry representations. □

Corollary 4.5.2 *Let A be a simple component of $\mathbf{C}G$, and $\rho : G \to \mathbf{GL}(V)$ an A-isotypic linear representation of G over $\mathbf{C}$ with character χ. Let $\mathbf{C}G$ and $\mathbf{R}G$ have the standard involutions, both denoted by $^-$, which are the identity on $\mathbf{C}$ and $\mathbf{R}$. Let A_0 be the simple component of $\mathbf{R}G$ for which A is a direct summand of $\mathbf{C} \otimes_{\mathbf{R}} A_0 \subset \mathbf{C} \otimes \mathbf{R}G = \mathbf{C}G$. Then A_0 (like any simple component of $\mathbf{R}G$ – cf. Th. 4.5.1(a)) is stable under the canonical involution of $\mathbf{R}G$, and A and A_0 fall into one of the following three categories:*

Unitary *$(A_0, {}^-) \cong \mathrm{M}(n, (\mathbf{C}, {}^*))$ is unitary, $A \cong \mathrm{M}(n, \mathbf{C})$, $\bar{A} \neq A$ and*

$$\begin{aligned}(\mathbf{C}, \mathrm{id}) \otimes_{(\mathbf{R},\mathrm{id})} (A_0, {}^-) &\cong ((A_0, {}^-) \oplus (A_0, {}^-), {}^{\leftrightarrow}) \\ &\cong (\mathrm{M}(n, (\mathbf{C}, \mathrm{id})) \oplus \mathrm{M}(n, (\mathbf{C}, \mathrm{id})), {}^{\leftrightarrow}).\end{aligned}$$

Thus $(\mathbf{C}, \mathrm{id}) \otimes_{(\mathbf{R},\mathrm{id})} (A_0, {}^-)$ is hyperbolic and $\mathbf{C} \otimes_{\mathbf{R}} A_0 = A \oplus \bar{A}$. Furthermore $\chi(G) \not\subset \mathbf{R}$ and ρ is not defined over $\mathbf{R}$.

Orthogonal *$\bar{A} = A$ and $(A, {}^-)$ is orthogonal. Then there is an isomorphism $(A_0, {}^-) \cong \mathrm{M}(n, (\mathbf{R}, \mathrm{id}))$, and $\mathbf{C} \otimes_{\mathbf{R}} A_0 = A$ and $(A, {}^-) \cong \mathrm{M}(n, (\mathbf{C}, \mathrm{id}))$. And ρ is defined over $\mathbf{R}$, in particular $\chi(G) \subset \mathbf{R}$.*

Symplectic *$\bar{A} = A$ and $(A, {}^-)$ is symplectic. Then $(A_0, {}^-) \cong \mathrm{M}(n, (\mathbf{H}, {}^*))$, $\mathbf{C} \otimes_{\mathbf{R}} A_0 = A \cong \mathrm{M}(2n, \mathbf{C})$, and $(A, {}^-) \cong (\mathrm{M}(2n, \mathbf{C}), {}^t \operatorname{inn} J)$, where J is an arbitrary nonsingular skew symmetric matrix. Moreover $\chi(G) \subset \mathbf{R}$, but ρ is defined over $\mathbf{R}$ if and only if $\mathrm{len}_A V$ is even.*

By the theorem, $(A_0, {}^-) \cong \mathrm{M}(n, (D, {}^*))$, $D = \mathbf{R}, \mathbf{C}$, or $\mathbf{H}$. Thus $\mathbf{C} \otimes_{\mathbf{R}} A_0 \cong \mathrm{M}(n, \mathbf{C} \otimes_{\mathbf{R}} D)$ which is, respectively, $\mathrm{M}(n, \mathbf{C})$, $\mathrm{M}(n, \mathbf{C}) \oplus \mathrm{M}(n, \mathbf{C})$, $\mathrm{M}(2n, \mathbf{C})$. $\mathbf{C} \otimes_{\mathbf{R}} A_0$ is stable under the involution of $(\mathbf{C}G, {}^-) = (\mathbf{C}, \mathrm{id}) \otimes (\mathbf{R}G, {}^-)$; when $D = \mathbf{R}$ or $\mathbf{H}$, $\mathbf{C} \otimes_{\mathbf{R}} A_0$ is simple and so $= A$ and A is therefore stable under the involution of $\mathbf{C}G$.

Suppose $D = \mathbf{R}$. By (e), $(A_0, {}^-) \cong \mathrm{M}(n, (\mathbf{R}, \mathrm{id}))$, and it follows that

$$(A, {}^-) \cong (\mathbf{C}, \mathrm{id}) \otimes_{\mathbf{R}} (A_0, {}^-) \cong \mathrm{M}(n, (\mathbf{C}, \mathrm{id})),$$

an orthogonal involution algebra. A simple A_0-module is isomorphic to $\mathbf{R}^n$, and $\mathbf{C} \otimes_{\mathbf{R}} \mathbf{R}^n \cong \mathbf{C}^n$ as an A-module, so ρ is defined over $\mathbf{R}$.

Similarly if $D = \mathbf{H}$, $(A_0, {}^-) \cong \mathrm{M}(n, (\mathbf{H}, {}^*))$ is symplectic (Th. 1.3.1) and so therefore is

$$(A, {}^-) = (\mathbf{C}, \mathrm{id}) \otimes_{(\mathbf{R},\mathrm{id})} (A_0, {}^-) \cong (\mathrm{M}(2n, \mathbf{C}), {}^t \operatorname{inn} J)$$

for any nonsingular skew symmetric matrix, by Th. 1.3.6, p. 30, and Exer. 2, p. 49. Morever a simple A_0-module is $\cong \mathbf{H}^n$ and, when the scalars are extended to $\mathbf{C}$, splits into a direct sum $\cong \mathbf{C}^{2n} \oplus \mathbf{C}^{2n}$ of two mutually isomorphic A-modules since $\dim_{\mathbf{R}} \mathbf{H}^n = 4n$ and $\dim_{\mathbf{C}} \mathbf{C}^{2n} = 2n$. This implies the last statement of the corollary since $2\chi = \chi_0$ where χ_0 is the character of the irreducible representation of G over $\mathbf{R}$ arising from the summand A_0 of $\mathbf{R}G$.

Now suppose $D = \mathbf{C}$, so

$$\begin{aligned}(\mathbf{C}, \mathrm{id}) \otimes_{(\mathbf{R},\mathrm{id})} (A_0, {}^-) &\cong (\mathbf{C}, \mathrm{id}) \otimes_{(\mathbf{R},\mathrm{id})} \mathrm{M}(n, (\mathbf{C}, {}^*)) \\ &\cong \mathrm{M}\big(n, (\mathbf{C}, \mathrm{id}) \otimes_{(\mathbf{R},\mathrm{id})} (\mathbf{C}, {}^*)\big)\end{aligned}$$

by Ex. 1, p. 24. Since $(\mathbf{C}, \mathrm{id}) \otimes_{(\mathbf{R},\mathrm{id})} (\mathbf{C}, {}^*)$ is the $(\mathbf{C}, \mathrm{id})$-involution algebra obtained by extension of scalars from the $(\mathbf{R}, \mathrm{id})$-involution algebra $(\mathbf{C}, {}^*)$, its involution $\mathrm{id} \otimes {}^*$ is $\mathbf{C}$-linear and nontrivial, and so the corresponding involution on

$\mathbf{C} \oplus \mathbf{C} \cong \mathbf{C} \otimes_{\mathbf{R}} \mathbf{C}$ must be the interchange, i.e.

$$(\mathbf{C}, \mathrm{id}) \otimes_{(\mathbf{R},\mathrm{id})} (A_0, ^-) \cong \mathrm{M}(n, ((\mathbf{C},\ \mathrm{id}) \oplus (\mathbf{C},\ \mathrm{id}),\ ^{\leftrightarrow}\ )) \\ \cong (\mathrm{M}(n, \mathbf{C}) \oplus \mathrm{M}(n, \mathbf{C}),\ ^{\leftrightarrow}\ ).$$

Thus $A \cong \mathrm{M}(n, \mathbf{C})$ and $\bar{A} \neq A$, so the character values of ρ are not in $\mathbf{R}$ (equivalence of (ii) and (v) in Th. 3.2.7(a), p. 115), so ρ is not defined over $\mathbf{R}$. This is the unitary case of the corollary. □

The statements in the corollary about χ also follow from the fact that its values generate the center of A_0 over $\mathbf{R}$ (cf. p. 78).

Consider again a simple component A of $\mathbf{R}G$, so $(A, ^-) \cong \mathrm{M}(n, (D,\ ^*\ ))$ (Th. 4.5.3(e)). If $h : D^{n\times m} \times D^{n\times m} \to \mathrm{M}(n, (D,\ ^*\ ))$ is a nonsingular ε-Hermitian form, its discriminant matrix is defined by $h(u, v) = u\underline{h}v^{t*}$, so the matrix $\underline{h}_0$ in the preamble of Th. 4.1.2 (p. 154) can be taken to be the identity matrix I_n, and in particular $\varepsilon_0 = 1$. This leads to the "Morita isomorphisms" $\hat{\mathrm{W}}_\varepsilon(A, ^-) \cong \hat{\mathrm{W}}_\varepsilon(D,\ ^*\ )$ and $\mathrm{W}_\varepsilon(A, ^-) \cong \mathrm{W}_\varepsilon(D,\ ^*\ )$ of that theorem, and so to the following table, using Table 1, p. 61:

type of $(A, ^-)$	$\hat{\mathrm{W}}_1(A, ^-)$	$\mathrm{W}_1(A, ^-)$	$\hat{\mathrm{W}}_{-1}(A, ^-)$	$\mathrm{W}_{-1}(A, ^-)$
orthogonal	$\mathbb{Z} \oplus \mathbb{Z}$	$\mathbb{Z}$	$\mathbb{Z}$	0
symplectic	$\mathbb{Z} \oplus \mathbb{Z}$	$\mathbb{Z}$	$\mathbb{Z}$	$\mathbb{Z}/2$
unitary	$\mathbb{Z} \oplus \mathbb{Z}$	$\mathbb{Z}$	$\mathbb{Z} \oplus \mathbb{Z}$	$\mathbb{Z}$

The details are left as an exercise.

Theorem 4.5.4 *Suppose that $\rho : G \to \mathbf{GL}_{\mathbf{C}}(V)$ is a linear representation over $\mathbf{C}$ with character χ. In order that all values of χ be in $\mathbf{R}$ (resp. that ρ be realizable in $\mathbf{R}$), it is necessary and sufficient that there is a nonsingular bilinear form (resp. a nonsingular symmetric bilinear form) $b : V \times V \to \mathbf{C}$ which is invariant under G.*

If $\chi(G) \subset \mathbf{R}$, then $\chi^* = \chi$ and therefore ρ and its contragredient ρ^* are equivalent (cf. (1.58), p. 74), so there is a nonsingular invariant bilinear form $V \times V \to (\mathbf{C}, \mathrm{id}_{\mathbf{C}})$ by Th. 1.7.4, p. 75 (applied with $(K, ^-) = (\mathbf{C}, \mathrm{id}_{\mathbf{C}})$). Conversely if there *is* such a form, then by the same theorem, its right adjoint is an isomorphism $V \to \overline{V^*}$ of $\mathbf{C}G$-modules, and so $\rho \cong \rho^*$ and χ is real-valued.

If ρ is defined over $\mathbf{R}$, say on the $\mathbf{R}$-space V_0, then by Th. 4.5.1, there is a nonsingular invariant symmetric bilinear form on V_0, which extends to such a form on $\mathbf{C} \otimes V_0 = V$.

Conversely assume that we have such a form b on V. Let $^-$ be the standard $(\mathbf{C}, \mathrm{id}_{\mathbf{C}})$-involution on $\mathbf{C}G$, $(\mathbf{C}G, ^-) = \oplus_i(B_i, ^-)$ the decomposition into simple involution algebras, and $V = \oplus_i B_i V$ the decomposition into involution isotypic components. Since $b(B_iV, B_jV) = b(V, \bar{B}_iB_jV) = 0$ if $i \neq j$, we can assume that $V = BV$ for some component $B = B_i$.

Let $(A_0, ^-)$ be a simple involution algebra component of $(\mathbf{R}G, ^-)$. The algebra A_0 is simple by Th. 4.5.1(a), $(A_0, ^-) \cong \mathrm{M}(n, (D,\ ^*\ ))$ where $D = \mathbf{R}, \mathbf{H}$, or $\mathbf{C}$ (Th.

4.5.3(e)), and by Cor. 4.5.2

$$(\mathbf{C}, \mathrm{id}) \otimes_{(\mathbf{R},\mathrm{id})} (A_0, {}^-) \cong \begin{cases} (\mathrm{M}(n, (\mathbf{C}, \mathrm{id})) & \text{if } D = \mathbf{R}, \\ (\mathrm{M}(2n, \mathbf{C}), {}^t \operatorname{inn} J) & \text{if } D = \mathbf{H}, \\ (\mathrm{M}(n, (\mathbf{C}, \mathrm{id})) \oplus \mathrm{M}(n, (\mathbf{C}, \mathrm{id})), \leftrightarrow) & \text{if } D = \mathbf{C},. \end{cases}$$

The three matrix involution algebras on the right are all simple as involution algebras, so one of them must be $\cong (B, {}^-)$.

If $D = \mathbf{R}$, a simple A_0-module ($\cong \mathbf{R}^n$) remains simple ($\cong \mathbf{C}^n$) when the scalars are extended from $\mathbf{R}$ to $\mathbf{C}$, and so any B-isotypic linear representation of G is defined over $\mathbf{R}$.

Now suppose that $D = \mathbf{H}$ or $\mathbf{C}$ and that $\hat{b} : V \times V \to (\mathbf{C}G, {}^-)$ is the Hermitian form associated to the form b. It takes its values in $(B, {}^-)$, and is therefore hyperbolic by Cor. 3.2.2 (p. 115) if $D = \mathbf{H}$ respectively Cor. 3.1.5 (p. 98) if $D = \mathbf{C}$. This implies that V is a direct sum of an *even* number of simple modules $\cong \mathbf{C}^{2n}$ when $D = \mathbf{H}$, and, when $D = \mathbf{C}$, is a direct sum $U \oplus W$ where U is a module over the first factor in $\mathrm{M}(n, (\mathbf{C}, \mathrm{id})) \oplus \mathrm{M}(n, (\mathbf{C}, \mathrm{id}))$ and W is a module of the *same* length as U over the second.

On the other hand a simple A_0-module ($\cong \mathbf{H}^n$ respectively $\mathbf{C}^n$) becomes, over $\mathbf{C}$, a direct sum $\cong \mathbf{C}^{2n} \oplus \mathbf{C}^{2n}$ respectively $\mathbf{C}^n \oplus \mathbf{C}^n$ of simple modules, as is easily checked by dimensions; here again the first summand in $\mathbf{C}^n \oplus \mathbf{C}^n$ is a module over the first factor in $\mathrm{M}(n, (\mathbf{C}, \mathrm{id})) \oplus \mathrm{M}(n, (\mathbf{C}, \mathrm{id}))$ and the second summand $\mathbf{C}^n$ is a module over the second factor.

In either case we see that V is the scalar extension of an A_0-module, and so ρ is defined over $\mathbf{R}$. □

We now consider orthogonal and symplectic representations over $\mathbf{R}$, and unitary representations over $\mathbf{C}$. We deal first with the orthogonal and unitary Hermitian cases since they are very similar.

Theorem 4.5.5 *Let* $(K, {}^-) = (\mathbf{R}, \mathrm{id})$ *or* $(\mathbf{C}, {}^*)$ *where* $\mathbf{R}$ *is a real closed field and* $\mathbf{C} = \mathbf{R}(i)$. *Let* $KG = A_1 \oplus \cdots \oplus A_h$ *be the decomposition into simple algebras (so* $\bar{A}_i = A_i$ *for all* i *by Th. 4.5.1).*

(a) For each i, *let* V_i *be a simple* A_i*-module and* $b_i : V_i \times V_i \to (K, {}^-)$ *an invariant positive definite symmetric respectively Hermitian form (unique up to equivariant equivalence by Th. 4.5.3). The Witt-Grothendieck group* $\hat{\mathrm{W}}_1(K, G)$ *is a free Abelian group of rank* $2h$ *with basis*

$$[b_1], \ldots, [b_h], [-b_1], \ldots, [-b_h].$$

It is also a free module of rank h *over the Witt-Grothendieck ring* $\hat{\mathrm{W}}_1(K)$, *with basis* $[b_1], \ldots, [b_h]$.

Furthermore if b *is a nonsingular invariant symmetric respectively Hermitian form on* V, *then*

$$[b] = \sum_i n_i^+ [b_i] + \sum_i n_i^- [-b_i],$$

and n_i^+ *and* n_i^- *are determined by*

$$\begin{array}{rcrcl} n_i^+ & + & n_i^- & = & \mathrm{len}_{KG} A_i V \\ (\dim_K V_i) n_i^+ & - & (\dim_K V_i) n_i^- & = & \mathrm{sgn}(b|_{A_i V}). \end{array}$$

(b) *The kernel of the canonical homomorphism* $\hat{\mathrm{W}}_1(K,G) \to \mathrm{W}_1(K,G)$ *is a free Abelian group with basis* $[b_1]+[-b_1],\dots,[b_h]+[-b_h]$.

(c) *The Witt group* $\mathrm{W}_1(K,G)$ *is a free Abelian group of rank* h *with basis consisting of the Witt classes* $[[b_1]],\dots,[[b_h]]$.
The subgroup $\hat{\mathrm{W}}_1^+(K,G)$ *of* $\hat{\mathrm{W}}_1(K,G)$ *generated by positive definite invariant forms is a free Abelian group with basis* $[b_1],\dots,[b_h]$, *and is mapped isomorphically onto* $\mathrm{W}_1(K,G)$ *by the canonical homomorphism.*

(d) *The linearization* $\mathrm{L}: \hat{\mathrm{W}}_1(K,G) \to \mathrm{K}_0(KG) \cong \mathbb{Z}^h$ *is surjective with kernel the free Abelian group with basis* $[b_1]-[-b_1],\dots,[b_h]-[-b_h]$, *and takes* $\hat{\mathrm{W}}_1^+(K,G)$ *isomorphically onto* $\mathrm{K}_0(KG)$.

(e) *The map* $[[b_V]] \to [V]$ *defines an isomorphism* $\mathrm{W}_1(K,G) \xrightarrow{\sim} \mathrm{K}_0(KG)$.

(f) *If* $\sum_i n_i^+[b_i] + \sum_i n_i^-[-b_i] \in \hat{\mathrm{W}}_1(K,G)$, $(n_i^+, n_i^- \in \mathbb{Z})$, *then there is a form* $b \in {}_{(K,{}^-)}^{\;G}\mathcal{H}_1^*$ *such that* $[b] = \sum_i n_i^+[b_i] + n_i^-[-b_i]$ *if and only if all* n_i^+ *and* n_i^- *are* ≥ 0.

Remarks. 1. Summary:

$\hat{\mathrm{W}}_1(K,G)$	$\mathrm{W}_1(K,G)$	$\ker \mathrm{L}$	$\mathrm{im}\,\mathrm{L}$
$\mathbb{Z}^{2h}$	$\mathbb{Z}^h$	$\mathbb{Z}^h$	$\mathbb{Z}^h$

These isomorphisms are consistent with those in the table on p. 176.

2. Since the correspondence $h \rightsquigarrow ih$ gives a category isomorphism from the category ${}_{(\mathbf{C},\,^*)}^{\;G}\mathcal{H}_1$ of invariant Hermitian forms over $(\mathbf{C},\,^*)$ to the category ${}_{(\mathbf{C},\,^*)}^{\;G}\mathcal{H}_{-1}$ of invariant skew Hermitian forms, the statements of the theorem apply, with the appropriate changes, in an obvious way to skew Hermitian forms as well. The details are omitted. □

(a) The canonical mapping from the monoid of equivariant classes to the Witt-Grothendieck group is injective by (1.47), p. 65, and the equivariant form of Witt's Theorem (cf. §3.6), and so these classes can be considered as elements of $\hat{\mathrm{W}}_1(K,G)$.

Recall that the canonical (involution) trace $\mathrm{Tr}_{KG/K} : (KG,{}^-) \to (K,{}^-)$ gives a category isomorphism between ${}_{(KG,{}^-)}\mathcal{H}_1$ and ${}_{(K,{}^-)}^{(KG,{}^-)}\mathcal{H}_1$ (Th. 3.2.4, p. 113), preserving orthogonality, and is also an isomorphism between the subcategories of nonsingular forms.

Since K/K_0 is definite, the simple components A_i are invariant under the involution ${}^-$ of KG (Th. 4.5.1), and so if $h: V \times V \to (KG,{}^-)$ is an arbitrary Hermitian form, V has an orthogonal decomposition

$$V = A_1V \perp A_2V \perp \cdots \perp A_hV. \tag{4.27}$$

(cf. p. 95). If $(K,{}^-) = (\mathbf{C},\,^*)$, each $(A_i,{}^-)$ is an involution algebra of the second kind, $\cong \mathrm{M}(n_i,(\mathbf{C},\,^*))$ for some n_i. If $(K,{}^-) = (\mathbf{R},\mathrm{id})$, then $(A_i,{}^-)$ is $\cong (\mathrm{M}(n_i,(D,\,^*))$ where D is $\mathbf{R}, \mathbf{C}$ or $\mathbf{H}$ (Th. 4.5.3(e)). Thus none of the components of $(KG,{}^-)$ is quasisplit and symplectic, and it follows at once from Cor. 3.2.2 (p. 115) that the underlying module of any orthogonally indecomposable Hermitian form over $(A_i,{}^-)$ is simple.

Applying this to $(V,\hat{b})$ (where b is a nonsingular invariant symmetric or Hermitian form), we see that (V,b) is equivalent to an orthogonal sum of spaces of the form (V_i,b_{ij}) where each b_{ij} is a nonsingular invariant form on V_i. But $[b_{ij}] = [\pm b_i]$

by Cor. 4.5.1, so

$$[b] = \sum_i n_i^+ [b_i] + \sum_i n_i^- [-b_i] \tag{4.28}$$

for certain nonnegative integers n_i^+, n_i^-. Since $n_i = n_i^+ + n_i^-$ is the number of times V_i appears (up to isomorphism) in the decomposition of V into simple modules, $n_i = \operatorname{len}_{KG} A_i V$. It is therefore uniquely determined by the KG-module V. Moreover

$$(\dim_K V_i) n_i^+ - (\dim_K V_i) n_i^- = \operatorname{sgn}(b|_{A_i V}), \tag{4.29}$$

and so n_i^+ and n_i^- are uniquely determined and the $[b_i]$ and $[-b_i]$ form a basis of the Witt-Grothendieck group $\hat{\mathrm{W}}_1(K,G)$ as stated.

By Th. 1.5.5 – or if we apply what we have already proved to $G = \{1\}$ – $\hat{\mathrm{W}}_1(K) = \mathbb{Z}[\langle 1 \rangle] \oplus \mathbb{Z}[\langle -1 \rangle]$. Since

$$[-b_i] = [\langle -1 \rangle][b_i]$$

where now $[\langle -1 \rangle]$ is the image of $[\langle -1 \rangle] \in \hat{\mathrm{W}}_1(K)$ in $\hat{\mathrm{W}}_1(K,G)$ (cf. Th. 4.3.1, p. 169), $\hat{\mathrm{W}}_1(K,G)$ is a free $\hat{\mathrm{W}}_1(K)$-module with basis $[b_1], \ldots, [b_h]$. (This is also easily seen without reference to Th. 4.3.1 by identifying $\hat{\mathrm{W}}_1(K)$ with the subgroup of $\hat{\mathrm{W}}_1(K,G)$ corresponding to the "trivial" summand K of KG, and using the definition of the ring structure of $\hat{\mathrm{W}}_1(K,G)$ in the first three paragraphs of §4.3).

(b) The subgroup of $\hat{\mathrm{W}}_1(K,G)$ generated by the classes of hyperbolic representations is also generated by the classes $[\mathrm{H}(V_i)]$ of hyperbolic representations based on simple KG-modules, by Cor. 3.1.3 (p. 97) and (3.14). Let $\hat{b}_i : V_i \times V_i \to (KG, {}^-)$ be the form associated to b_i; then in fact $\hat{b}_i : V_i \times V_i \to (A_i, {}^-)$, and $\hat{b}_i \perp -\hat{b}_i$ is the KG-form associated with the invariant form $b_i \perp -b_i$ on $V_i \oplus V_i$. By Prop. 3.1.2, $\hat{b}_i \perp -\hat{b}_i$ is equivalent to $h_{\mathrm{H}(V_i)}$, and so $[b_i] + [-b_i]$ is the class of the hyperbolic representation on $\mathrm{H}(V_i)$. This proves (b).

(c) It is clear that $[b_1], \ldots, [b_h]$ span $\hat{\mathrm{W}}_1^+(K,G)$, and the rest of (c) follows directly from (a) and (b).

(d) L is onto because every KG-module supports a nonsingular (in fact positive definite) symmetric respectively Hermitian invariant form. The kernel has the basis stated since the image of $[\pm b_i]$ is $[V_i]$, and the $[V_i]$ form a basis of $\mathrm{K}_0(KG)$. The last statement follows easily.

(e) The composite map $\mathrm{W}_1(K,G) \to \hat{\mathrm{W}}_1^+(K,G) \to \mathrm{K}_0(KG)$ is an isomorphism by (c) and (d).

(f) The sufficiency is clear, and the necessity was proved in (a). □

Under the assumptions of the theorem, a space $(V,b) \in {}_K^{KG}\mathcal{H}_1$ is the orthogonal direct sum (4.5) of its isotypic components; we define its *isotypic signatures* (over $\mathbf{R}$) as

$$\operatorname{sgn}_i(b) = \operatorname{sgn}(b|_{A_i V}).$$

Corollary 4.5.3 *Suppose that $\rho : G \to \mathbf{I}(b)$ and $\rho' : G \to \mathbf{I}(b')$ are isometry representations over $(\mathbf{R}, \mathrm{id})$ or $(\mathbf{C}, {}^*)$ with b and b' both symmetric or both Hermitian. Then ρ and ρ' are equivalent if and only if they are equivalent as linear representations and their forms have the same isotypic signatures:*

$$\operatorname{sgn}_i(b) = \operatorname{sgn}_i(b') \quad \textit{for } 1 \le i \le h.$$

The necessity follows from the fact that an equivariant equivalence provides an isometry between the isotypic components of the spaces of b and b', and the sufficiency from (a) and the fact that the isotypic signatures and the dimensions of the isotypic components determine the n_i^+ and the n_i^-. □

Thus we have a complete solution to the problems of equivalence for orthogonal representations over $\mathbf{R}$ and of unitary representations over $(\mathbf{C}, *)$.

We now come to symplectic representations over the real closed field $\mathbf{R}$. They differ in several ways from orthogonal and unitary representations. First of all the linearization map $\mathrm{L} : \hat{\mathrm{W}}_{-1}(\mathbf{R}, G) \to \mathrm{K}_0(\mathbf{R}G)$ is not necessarily onto, and secondly the number of classes of symplectic representations $G \to \mathbf{Sp}(V, b)$ which, as linear representations, are irreducible and equivalent, depends on the type of $(A, {}^-)$ where A is the simple component of $\mathbf{R}G$ such that $AV \neq 0$. Because of this nonuniformity, rather than dealing with $\hat{\mathrm{W}}_{-1}(\mathbf{R}, G)$ and $\mathrm{W}_{-1}(\mathbf{R}, G)$ directly, it is more convenient to determine $\hat{\mathrm{W}}_{-1}(\mathbf{R}, A_i)$ and $\mathrm{W}_{-1}(\mathbf{R}, A_i)$ for each i, where $\mathbf{R}G = \oplus_i A_i$ is the decomposition into simple algebras, and then use the fact (Cor. 4.1.1, p. 153) that

$$\hat{\mathrm{W}}_{-1}(\mathbf{R}, G) = \oplus_i \hat{\mathrm{W}}_{-1}(\mathbf{R}, A_i) \quad \text{and} \quad \mathrm{W}_{-1}(\mathbf{R}, \mathrm{G}) = \oplus_{\mathrm{i}} \mathrm{W}_{-1}(\mathbf{R}, \mathrm{A_i}).$$

Theorem 4.5.6 *Let* $\mathbf{R}$ *be a real closed field, and let* ${}^-$ *denote the standard* $(\mathbf{R}, \mathrm{id})$*-involution on* $\mathbf{R}G$ *as well as its restriction to the simple direct summands* A *of* $\mathbf{R}G$. *Let* V *be a simple* A*-module.*

(a) The number of (equivariant) equivalence classes of nonsingular G*-invariant (i.e.* $(A, {}^-)$*-invariant) skew symmetric forms on* V *is*

0 *if* $(A, {}^-)$ *is orthogonal,*
1 *if* $(A, {}^-)$ *is symplectic,*
2 *if* $(A, {}^-)$ *is unitary.*

(b) ***Orthogonal type.*** *In this case all* A*-isotypic symplectic representations are hyperbolic, so* $\mathrm{W}_{-1}(\mathbf{R}, A) = 0$ *and* $\hat{\mathrm{W}}_{-1}(\mathbf{R}, A)$ *is a free Abelian group of rank 1 with basis given by the class* $[h_{\mathrm{H}(V)}]$ *of a* (-1)*-hyperbolic representation based on a simple module* V.

The linearization map $\hat{\mathrm{W}}_{-1}(\mathbf{R}, A) \to \mathrm{K}_0(A) \cong \mathbb{Z}$ *is injective (with image* $2\mathbb{Z}$*).*

(c) ***Symplectic type.*** *If* b *is a nonsingular invariant skew symmetric form) on* V, *then* $\hat{\mathrm{W}}_{-1}(\mathbf{R}, A) = \mathbb{Z}[b]$ *is a free Abelian group of rank 1, and the linearization map* $\hat{\mathrm{W}}_{-1}(\mathbf{R}, A) \to \mathrm{K}_0(A)$ *is an isomorphism.*

The Witt group $\mathrm{W}_{-1}(\mathbf{R}, A) \cong \mathbb{Z}/2$.

(d) ***Unitary type.*** *If* b *is a nonsingular invariant skew symmetric form on* V, *then* $\hat{\mathrm{W}}_{-1}(\mathbf{R}, A)$ *is a free Abelian group of rank 2 with the two equivariant classes* $[b]$ *and* $[-b]$ *as basis.*

The linearization map $\hat{\mathrm{W}}_{-1}(\mathbf{R}, A) \to \mathrm{K}_0(A)$ *is onto and has kernel* $\mathbb{Z}([b] - [-b])$.

$\mathrm{W}_{-1}(\mathbf{R}, A)$ *is a free Abelian group of rank 1, with basis* $[[b]]$, *and the kernel of the canonical homomorphism* $\hat{\mathrm{W}}_{-1}(\mathbf{R}, A) \to \mathrm{W}_{-1}(\mathbf{R}, A)$ *is* $\mathbb{Z}([b] + [-b])$.

(e) In each of (b), (c) and (d), the class $[h] \in \hat{\mathrm{W}}_{-1}(\mathbf{R}, G)$ *of a form* $h \in {}^A_{\mathbf{R}}\mathcal{H}^*_{-1}$ *is a sum of the indicated bases with nonnegative coefficients. Conversely any such sum is the class of such a form* h.

Summary:

type of $(A, ^-)$	$\hat{\mathrm{W}}_{-1}(\mathbf{R}, A)$	$\mathrm{W}_{-1}(\mathbf{R}, A)$	$\ker \mathrm{L}$	$\mathrm{im}\, \mathrm{L}$
Orthogonal	$\mathbb{Z}$	0	0	$2\mathbb{Z}$
Symplectic	$\mathbb{Z}$	$\mathbb{Z}/2$	0	$\mathbb{Z}$
Unitary	$\mathbb{Z} \oplus \mathbb{Z}$	$\mathbb{Z}$	$\mathbb{Z}$	$\mathbb{Z}$

(Note that these results are consistent with those on p. 176). From this table, one can of course read off the analogous structures for KG. For example if o, s, u are the number of simple summands of $(KG, ^-)$ which are respectively orthogonal, symplectic and unitary,

$$\mathrm{W}_{-1}(\mathbf{R}, KG) = \mathrm{W}_{-1}(\mathbf{R}, G) \cong \{0\}^o \oplus (\mathbb{Z}/2)^s \oplus \mathbb{Z}^u = (\mathbb{Z}/2)^s \oplus \mathbb{Z}^u.$$

(a) This is Th. 4.5.3(c), p. 172.

(b) As in the proof of the previous theorem, the category ${}^{A}_{\mathbf{R}}\mathcal{H}_{-1}$ in which we are interested is isomorphic to the category ${}_{A}\mathcal{H}_{-1}$ via the canonical involution trace $\mathrm{Tr} : (\mathbf{R}G, ^-) \to (\mathbf{R}, \mathrm{id})$. Since $(A, ^-)$ is split (Th. 4.5.3(e), p. 172), it follows from Cor. 3.2.2 (p. 115) that a nonsingular skew Hermitian form over $(A, ^-)$ is an orthogonal direct sum of (mutually equivalent) hyperbolic forms based on a simple module. From this (b) follows at once.

(c) The proof here is similar to that of (b) – in fact even easier, using Th. 3.2.5(b)(i) and part (a) of this theorem. That $\mathrm{W}_{-1}(\mathbf{R}, A) \cong \mathbb{Z}/2$ then follows from the fact that an invariant hyperbolic form over $(A, ^-)$, based on a simple module, is of the form $\hat{b} \perp -\hat{b} \cong \hat{b} \perp \hat{b}$ (cf. Prop. 3.1.2, p. 97).

(d) If b is a nonsingular invariant skew symmetric form on a simple module V, the two equivariant classes of skew Hermitian forms on V are $[b]$ and $[-b]$ by Th. 4.5.3(c). By the by now familiar argument, every nonsingular invariant skew symmetric form is equivalent to an orthogonal direct sum of copies of b and $-b$, and so their classes span $\hat{\mathrm{W}}_{-1}(\mathbf{R}, A)$. Their linear independence follows from $\hat{W}_{-1}(\mathbf{R}, A) \cong \mathbb{Z} \oplus \mathbb{Z}$ – see the table on p. 176.

It is clear that the forgetful homomorphism $\hat{\mathrm{W}}_{-1}(\mathbf{R}, A) \to \mathrm{K}_0(A)$ has image $\mathbb{Z}[V] = \mathrm{K}_0(A)$ and kernel $\mathbb{Z}([b] - [-b])$. Since $b \perp -b$ is a hyperbolic plane (module of length 2), the rest of (d) follows.

(e) Clear. □

If $(A, ^-) \cong \mathrm{M}(n, (\mathbf{C}, ^*))$ and b is an $(A, ^-)$-invariant skew symmetric form over $\mathbf{R}$, then the form $b_i(u, v) = b(iu, v)$ is an $(A, ^-)$-invariant *symmetric* form over $\mathbf{R}$. Moreover two $(A, ^-)$-invariant skew symmetric forms b and b' are equivariantly equivalent if and only if b_i and b'_i are equivariantly equivalent.

Corollary 4.5.4 *Two symplectic representations*

$$\rho : G \to \mathbf{Sp}(V, b) \text{ and } \rho' : G \to \mathbf{Sp}(V', b') \tag{4.30}$$

over a real closed field $\mathbf{R}$ *are equivalent if and only if they are equivalent as linear representations, and for each unitary component* $(A, ^-)$ *of* $(\mathbf{R}G, ^-)$*, the symmetric forms* $b_i|_{AV \times AV}$ *and* $b'_i|_{AV' \times AV'}$ *have the same signatures.*

Since V is the orthogonal direct sum of the isotypic components A_iV, it suffices to prove the corollary in the case when V is isotypic, say $V = AV$. The case of $(A, ^-)$ orthogonal or symplectic follows immediately from the foregoing theorem – for example from the fact that L is injective in these cases. And the case $(A, ^-)$ unitary follows from the remarks preceding the corollary and Cor. 4.5.3. □

REMARK. The same idea used above for unitary involution algebras can be applied more generally. Namely, suppose that $(K, ^-)$ is a field with involution and $(A, ^-)$ is a unitary component of $(KG, ^-)$. If the center of A is $L = L_0(\sqrt{\alpha})$ where L_0 is the fixed field under $^-$, then two A-isotypic symplectic representations (4.30) are equivalent if and only if the $(A, ^-)$-invariant symmetric forms $b(\sqrt{\alpha}u, v)$ and $b'(\sqrt{\alpha}u, v)$ are equivalent. This replaces the equivalence problem of symplectic representations, in the case of a unitary involution algebra, by the equivalence problem for orthogonal representations – and conversely. □

4.6 K/K_0 definite

In this section we assume that K/K_0 is a definite extension – in particular $\operatorname{char} K = 0$ and K_0 is formally real.

Theorem 4.6.1 *Suppose that $(A, ^-)$ is a simple involution factor of $(KG, ^-)$ with the standard $(K, ^-)$-involution. Then (Th. 4.5.1(a)) A is a simple algebra, say $A \cong \mathrm{M}(n, D)$ where D is a division algebra. Let L be the center of A (and D) and $L_0 = (L, ^-)_+$ the subfield of symmetric elements. Then*

(a) L_0/K_0 is a totally real extension at each ordering P of K_0 at which K/K_0 is definite, and if $L \neq L_0$, L/L_0 is totally definite (and L is not formally real).

(b) If $(A, ^-)$ is orthogonal, then $D = L$ or is a quaternion algebra that splits over every real closure $\mathbf{R}$ of L. For each such $\mathbf{R}$, $(A^{\mathbf{R}}, ^-) \cong \mathrm{M}(n, (\mathbf{R}, \mathrm{id}))$ respectively $\mathrm{M}(2n, (\mathbf{R}, \mathrm{id}))$.

(c) If $(A, ^-)$ is symplectic, then D is a quaternion algebra that does not split over any real closure[5] $\mathbf{R}$ of L, and for each such $\mathbf{R}$, $(A^{\mathbf{R}}, ^-) \cong \mathrm{M}(n, (\mathbf{H}, ^*))$.

(d) Assume $(A, ^-)$ is unitary. Let $\mathbf{R}$ be a real closure of K_0 at an ordering P at which K/K_0 is definite, and let $\mathbf{C} = \mathbf{R}(\sqrt{-1})$. If $(K : K_0) = 1$ respectively 2, then $(\mathbf{R}, \mathrm{id}) \otimes_{(K,\mathrm{id})} (A, ^-)$ respectively $(\mathbf{C}, ^*) \otimes_{(K,^-)} (A, ^-)$ is a direct sum of involution algebras each of which is $\cong \mathrm{M}(nd, (\mathbf{C}, ^*))$ where $d^2 = \dim_L D$.

REMARK. The fact that D is either $= L$ or a quaternion algebra when $(A, ^-)$ is of the first kind – i.e. orthogonal or symplectic – holds for an arbitrary field (Th. 1.7.6, p. 78). Moreover when the characteristic of the field is $\neq 0$, *every* simple summand of KG is quasisplit (*ibid*). □

We begin by proving (b). In this part and in (c), we must have $L = L_0$ and so also $K = K_0$. If $\mathbf{R}$ is real closure of K, $(A^{\mathbf{R}}, ^-)$ is a direct sum of orthogonal involution algebras over $\mathbf{R}$ by Th. 1.3.2 (p. 29), and so is a direct sum of full matrix algebras over $\mathbf{R}$ by Th. 4.5.3 (p. 172) since they are simple components of $\mathbf{R}G$, in particular the center of $A^{\mathbf{R}}$ is

$$\mathbf{R} \otimes_K L = \mathbf{R} \oplus \mathbf{R} \oplus \cdots \oplus \mathbf{R}$$

[5] Such a quaternion algebra is often called *totally definite* – cf. p. 293, [55], for example.

and so $L/K = L_0/K_0$ is a totally real extension by Th. 1.4.7 (p. 40). Moreover by the above remark, D is either L or a quaternion algebra – which of course is split over any real closure $\mathbf{R}$ of L since A is.

The proof of (c) is almost the same (as well as the fact that L/K is totally real) and is left to the reader.

Now (d). Assume first that $K = K_0$. Then the proof is entirely similar to that of (b) and (c), leading to $(\mathbf{R}, \mathrm{id}) \otimes_{(K,\mathrm{id})} (A, {}^-) \cong$ a direct sum of involution algebras of the form $\mathrm{M}(nd, (\mathbf{C}, {}^*))$. Thus $\mathbf{R} \otimes_K L$ is a direct sum of copies of $\mathbf{C}$ and since the $\mathbf{R}$-algebra of symmetric elements $(\mathbf{R} \otimes L)_+$ of $\mathbf{R} \otimes L$ is $\mathbf{R} \oplus \cdots \oplus \mathbf{R}$, it follows that $\mathbf{R} \otimes_K L_0$ is a direct sum of copies of $\mathbf{R}$, so L_0/K is a totally real extension, L/L_0 is definite, and L is not formally real.

Now suppose $(K : K_0) = 2$. Since K/K_0 is definite at P, an imbedding of K into $\mathbf{C}$ is not contained in $\mathbf{R}$ and it is clear that $(\mathbf{C}, {}^*)/(K, {}^-)$ is an extension of involution fields and $(\mathbf{C}, {}^*) \otimes_{(K,{}^-)} (A, {}^-)$ is a direct sum of involution algebras of the form $\mathrm{M}(nd, (\mathbf{C}, {}^*))$. Thus

$$(\mathbf{C}, {}^*) \otimes_{(K,{}^-)} (L, {}^-) \cong (\mathbf{C}, {}^*) \oplus (\mathbf{C}, {}^*) \oplus \cdots \oplus (\mathbf{C}, {}^*).$$

But the canonical mapping

$$(\mathbf{C}, {}^*) \otimes_{(K_0,\mathrm{id})} (L_0, \mathrm{id}) \to (\mathbf{C}, {}^*) \otimes_{(K,{}^-)} (L, {}^-)$$

is an isomorphism by Lem. 1.3.1 (p. 27) and so

$$(\mathbf{C}, {}^*) \otimes_{(K_0,\mathrm{id})} (L_0, \mathrm{id}) \cong (\mathbf{C}, {}^*) \oplus (\mathbf{C}, {}^*) \oplus \cdots \oplus (\mathbf{C}, {}^*),$$

whence

$$\mathbf{R} \otimes_{K_0} L_0 \cong \mathbf{R} \oplus \mathbf{R} \oplus \cdots \oplus \mathbf{R},$$

so L_0/K_0 is totally real in this case too. □

Corollary 4.6.1 *Suppose K is formally real, let $(A, {}^-)$ be a symplectic factor of $(KG, {}^-)$ (where ${}^-$ is the standard (K, id)-involution of KG), and let L be the center of A. Then*

$$(A, {}^-) \cong \mathrm{M}(n, (D, {}^*))$$

where $D_P = L_P \otimes_L D \cong \mathbf{H}$ for each ordering P of L.

$D_P \cong \mathbf{H}$ for each ordering P by part (c) of the foregoing theorem. Note that D is also not split at any real prime $\mathfrak{P}$ of L since if P is the ordering of L corresponding to $\mathfrak{P}$ by Prop. 1.4.1 (p. 42),

$$D_{\mathfrak{P}} = L_{\mathfrak{P}} \otimes_{L_P} (L_P \otimes_L D) = L_{\mathfrak{P}} \otimes_{L_P} \mathbf{H} = \mathbb{H}.$$

Identify $A = \mathrm{M}(n, D)$. Assume first that $K = \mathbb{Q}$. By Th. 1.3.6 (p. 30) and Lem. 1.3.2 there is a matrix $b \in \mathbf{GL}(n, D)$ which is Hermitian ($b^{t*} = b$) such that

$$a^{t*} = b\bar{a}b^{-1} \quad \text{for all } a \in A = \mathrm{M}(n, D). \tag{4.31}$$

By Th. 4.5.3(e) (p. 172) we can find a matrix $c \in \mathbf{GL}(n, D_{\mathfrak{P}})$ such that the inner automorphism inn_c makes the diagram

$$\begin{array}{ccc} \mathrm{M}(n, D_{\mathfrak{P}}) & \xrightarrow{\ -\ } & \mathrm{M}(n, D_{\mathfrak{P}}) \\ {\scriptstyle \mathrm{inn}_c}\downarrow & & \downarrow{\scriptstyle \mathrm{inn}_c} \\ \mathrm{M}(n, D_{\mathfrak{P}}) & \xrightarrow{\ t*\ } & \mathrm{M}(n, D_{\mathfrak{P}}) \end{array}$$

commutative. (Here $^-$ is the unique $L_{\mathfrak{P}}$-linear extension of the involution $^-$ on $A = \mathrm{M}(n, D)$ to $L_{\mathfrak{P}} \otimes_L \mathrm{M}(n, D) = \mathrm{M}(n, D_{\mathfrak{P}})$). This is equivalent to

$$a^{t*} = c^{t*} c\bar{a}(c^{t*}c)^{-1} \quad \text{for all } a \in \mathrm{M}(n, D_{\mathfrak{P}}).$$

Thus $\alpha_{\mathfrak{P}} b = c^{t*}c$ for some $\alpha_{\mathfrak{P}} \in \dot{L}_{\mathfrak{P}}$. This can be interpreted to mean that the Hermitian form over $(D_{\mathfrak{P}}, {}^*)$ with matrix $\alpha_{\mathfrak{P}} b$ is equivalent to the form with matrix I_n, and so in particular b is positive definite at $\mathfrak{P}$ if $\alpha_{\mathfrak{P}} > 0$ and negative definite if $\alpha_{\mathfrak{P}} < 0$. Now use the approximation theorem (p. 18) to find $\beta \in \dot{L}$ which is close enough to each $\alpha_{\mathfrak{P}}$ so that βb is positive definite at each real prime of L. Since $\mathbb{Q}$ is a global field, it follows from Table 1 (p. 61) that βb is equivalent to I_n over $(D, {}^*)$, say $\beta b = e^{t*}e$, $e \in \mathbf{GL}(n, D)$. Then (4.31) becomes

$$e\bar{a}e^{-1} = (eae^{-1})^{t*},$$

i.e. inn_e is an isomorphism $(\mathrm{M}(n, D), {}^-) \xrightarrow{\sim} (\mathrm{M}(n, D), {}^{t*}) = \mathrm{M}(n, (D, {}^*))$.

Assume now that K is any formally real field. We may suppose that $\mathbb{Q} \subset K$, so $\mathbb{Q}G \subset KG$. Let A' be the simple summand of $\mathbb{Q}G$ such that A is a direct summand of the ideal $KA' \cong K \otimes_{\mathbb{Q}} A'$ of KG. Now A' is stable under the standard $(\mathbb{Q}, \mathrm{id})$-involution of $\mathbb{Q}G$ (Th. 4.5.1(a), p. 171) and since $(A, {}^-)$ is symplectic, $(A', {}^-)$ is also symplectic by Th. 1.3.2, p. 29, so $(A', {}^-) \cong \mathrm{M}(n, (D', {}^*))$ for some totally definite quaternion algebra D' by the first case. Thus $(A, {}^-)$ is isomorphic to a direct summand of

$$(K, \mathrm{id}) \otimes_{(\mathbb{Q}, \mathrm{id})} \mathrm{M}(n, (D', {}^*)) \cong \mathrm{M}(n, (K \otimes_{\mathbb{Q}} D', \mathrm{id} \otimes {}^*)).$$

Let L' be the center of D', and $K \otimes_{\mathbb{Q}} L' = L_1 \oplus \cdots \oplus L_m$ where each L_i is a finite extension of K. Then

$$K \otimes_{\mathbb{Q}} D' \cong K \otimes_{\mathbb{Q}} (L' \otimes_{L'} D') \cong (L_1 \otimes_{L'} D') \oplus \cdots \oplus (L_m \otimes_{L'} D').$$

Each $L_i \otimes_{L'} D'$ is a quaternion algebra with center L_i. Since the involution $\mathrm{id} \otimes {}^*$ (of $K \otimes_{\mathbb{Q}} D'$) is the identity on $K \otimes_{\mathbb{Q}} L'$, the involution on each L_i is also the identity, and so the involution on $L_i \otimes_{L'} D'$ is conjugation. This proves the corollary since $(A, {}^-)$ is isomorphic to one of the summands of

$$\begin{aligned} &\mathrm{M}(n, (K \otimes_{\mathbb{Q}} D', \mathrm{id} \otimes {}^*)) \cong \\ &\qquad \mathrm{M}(n, (L_1 \otimes_{L'} D', {}^*)) \oplus \cdots \oplus \mathrm{M}(n, (L_m \otimes_{L'} D', {}^*)). \end{aligned}$$

□

4.7 The involutory Schur subgroup

In this section we adopt a different approach – different than in the previous section – to the determination of the Brauer classes of the simple algebras with involution $(A, {}^-)$ which appear as direct summands of group algebras $(KG, {}^-)$ with their standard $(K, {}^-)$-involution. We assume that the characteristic is 0 since otherwise such a Brauer class is trivial – cf. Th. 1.7.6, p. 78.

Define $\mathrm{Sch}(K)$ to be the set of Brauer classes $\mathsf{a} \in \mathrm{Br}(K)$ with the property that there is a finite group G and a direct summand A of KG which is a central simple K-algebra whose Brauer class $[A] = \mathsf{a}$. We shall show that $\mathrm{Sch}(K)$ is a subgroup of $\mathrm{Br}(K)$, called the *Schur subgroup* of the Brauer group. It is of significant interest and importance in the theory of linear representations – see for example [74] and Ch. X, [1].

Suppose that A and B are central simple direct summands of KG and KH respectively, where G and H are finite groups. The tensor product $KG \otimes_K KH$

has basis $s \otimes t \quad (s \in G, t \in H)$. These basis elements form a group isomorphic to $G \times H$, so $KG \otimes_K KH \cong K(G \times H)$. Furthermore $A \otimes_K B$ is a direct summand of $KG \otimes_K KH$ and is central simple by Th. 1.2.9, so $\mathrm{Sch}(K)$ is closed under multiplication. Moreover the opposite algebra A^o is a direct summand of the group algebra KG^o of the opposite group, and so $\mathrm{Sch}(K)$ is indeed a subgroup of $\mathrm{Br}(K)$.

The Schur subgroup has been studied extensively. For example in [74], calculations are given for $\mathrm{Sch}(K)$ for many fields K, including local fields and many global fields, although the determination of $\mathrm{Sch}(K)$ is incomplete for global fields; the most extensive results for them can be found in [33].

Suppose now that K has an involution $^-$ (possibly the identity as usual), and endow each group algebra KG with the standard $(K, ^-)$-involution, also denoted by $^-$. We now define an analogue $\mathrm{Sch}(K, ^-)$ of the Schur subgroup which plays a role in the theory of isometry representations.

We begin by defining an analogue $\mathrm{Br}(K, ^-)$ of the Brauer group, which takes into account an involution on the simple algebra. Thus we consider central simple K-algebras $(A, ^-)$ with a $(K, ^-)$-involution; suppose first that the involution on K is the identity. In this case, the Brauer classes (in $\mathrm{Br}(K)$) whose algebras support an involution (of the first kind) are the elements $\mathrm{Br}_2(K)$ of order 2 (or 1) (Albert's Theorem, p. 34). The *involutory Brauer group* is then defined to be the product group

$$\mathrm{Br}(K, \mathrm{id}_K) = \mathrm{Br}_2(K) \times \{\pm 1\},$$

and the *involutory Brauer class* $[A, ^-] \in \mathrm{Br}(K, \mathrm{id}_K)$ of $(A, ^-)$ is defined to be $([A], \text{type of } (A, ^-))$. If $(B, ^\sim)$ is another central simple algebra with (K, id)-involution, it follows from Ex. 1, p. 28 that

$$[(A, ^-) \otimes_{(K, ^-)} (B, ^\sim)] = [A, ^-][B, ^\sim] \text{ and } [A^o, ^-] = [A, ^-]^{-1}. \tag{4.32}$$

If the involution $^-$ on K is *not* the identity, the *involutory Brauer group* $\mathrm{Br}(K, ^-)$ is defined to be simply those ordinary Brauer classes which support a $(K, ^-)$-involution. It is easily checked to be a subgroup of $\mathrm{Br}(K)$; it is also the kernel of the corestriction map $\mathrm{Cor}_{K/K_0} : \mathrm{Br}(K) \to \mathrm{Br}(K_0)$ (cf. Th. 1.3.7, p. 34) although we will not use this fact. It will be convenient to think of $\mathrm{Br}(K, ^-)$ in this case as

$$\mathrm{Br}(K, ^-) = (\ker \mathrm{Cor}_{K/K_0}) \times \{0\},$$

so that if we define the involutory Brauer class $[A, ^-] = ([A], \mathrm{type}(A, ^-)) = ([A], 0)$ in this case too, (4.32) once again holds.

Note that, for each $\mathsf{a} \in \mathrm{Br}(K, ^-)$, there is a central simple algebra $(A, ^-)$ with involution such that $[A, ^-] = \mathsf{a}$.

We say that the central simple algebras with involution $(A, ^-)$ and $(B, ^\sim)$ are *similar* if $[A, ^-] = [B, ^\sim]$. This means of course that A and B are similar as central simple algebras and that the involutions are of the same type.

Next we define the *involutory Schur subgroup* $\mathrm{Sch}(K, ^-)$. It consists of the involutory Brauer classes $[A, ^-] \in \mathrm{Br}(K, ^-)$ such that $(A, ^-)$ is a component of $(KG, ^-)$ for some finite group G (and the involution $^-$ on KG is of course the standard $(K, ^-)$-involution). It is straightforward to check that $\mathrm{Sch}(K, ^-)$ is a group using much the same proof as in the case of $\mathrm{Sch}(K)$, along with the fact that $(KG, ^-) \otimes_{(K, ^-)} (KH, ^-)$ and $(K(G \times H), ^-)$ can be identified with each other[6].

---

[6]The involution $^-$ on each of KG, KH and $K(G \times H)$ is the standard $(K, ^-)$-involution, of course.

It follows from Th. 1.7.6, p. 78, that $\mathrm{Sch}(K, \mathrm{id})$ consists of the split class and (certain) classes of quaternion algebras. We can think of it as consisting of the involutory Brauer classes $([D], \varepsilon)$, where $D = K$ or a central quaternion division algebra over K, such that there is a group G and a central simple summand $(A, ^-)$ of $(KG, ^-)$ such that A and D are in the same Brauer class and the type of $(A, ^-)$ is ε. In general not all of the quaternion Brauer classes will occur in this way – for example we know from Th. 4.6.1 that if K is formally real, then any such D must be either split at all real closures of K or nonsplit at all real closures, or if K is a subfield of a cyclotomic extension of $\mathbb{Q}$, its invariants (p. 17*ff*) must be "uniformly distributed" (cf. p. 93, [74]).

If $(K, ^-)$ is of the second kind, the *involutory Schur subgroup* $\mathrm{Sch}(K, ^-)$ is the subgroup of $\mathrm{Br}(K, ^-)$ defined in the same way as $\mathrm{Sch}(K, \mathrm{id})$, and can also be thought of as consisting of the Brauer classes $[D]$ of certain central division algebras.

A knowledge of the involutory Schur subgroup is useful in dealing with isometry representations (cf. the Hasse Principle for isometry representations, Th. 3.3.3, p. 138, for example) but it is not known in general – it depends heavily on the nature of the field, as does $\mathrm{Sch}(K)$ itself. There are however some general statements that can be made, and we shall also show that its calculation can be reduced to its calculation over simpler fields (subcyclotomic fields – cf. Th. 4.7.2).

When K/K_0 is a definite extension, Th. 4.6.1 can be applied usefully. Let A be a central simple summand of KG – it is stable under the standard $(K, ^-)$-involution (Th. 4.5.1(a), p. 171). Therefore if $K \neq K_0$, the forgetful homomorphism $\mathrm{Sch}(K, ^-) \overset{?}{\to} \mathrm{Sch}(K)$ is an isomorphism. When $K = K_0$ – i.e. K is formally real – we know by (b) and (c) of Th. 4.6.1, p. 182, that either $(A, ^-)$ is orthogonal and $A^{\mathbf{R}}$ is split for all real closures $\mathbf{R}$ of K, or $(A, ^-)$ is symplectic and $(A^{\mathbf{R}}, ^-) \cong \mathrm{M}(n, (\mathbf{H}, ^*))$ (for some n) for all $\mathbf{R}$. Thus the type

$$\mathrm{type}(\mathsf{a}) = \mathrm{type}(A) = \pm 1$$

of $\mathsf{a} \in \mathrm{Sch}(K)$ is well-defined in this case – in other words, $\mathrm{Sch}(K, \mathrm{id}) \overset{?}{\to} \mathrm{Sch}(K)$ is again an isomorphism.

Theorem 4.7.1 *If K/K_0 is a definite extension,* $\mathrm{Sch}(K, ^-) \overset{?}{\to} \mathrm{Sch}(K)$ *is an isomorphism.* □

A *cyclotomic* extension of a field is an extension obtained by adjoining roots of 1. A *subcyclotomic* extension is a subextension of a cyclotomic extension. It is well known that cyclotomic extensions are Abelian (cf. Ex. 6, p. 273, [53])– so subcyclotomic extensions are as well.

Let K be a field (of characteristic 0); we identify the prime subfield of K with the field $\mathbb{Q}$ of rational numbers. Let $\hat{K}$ be an algebraic closure of K, $\hat{\mathbb{Q}}_c$ the cyclotomic subfield of $\hat{K}$ (generated over $\mathbb{Q}$ by all roots of unity), and set

$$K_c = K \cap \hat{\mathbb{Q}}_c.$$

K_c does not depend on the choice of $\hat{K}$, and we refer to it as the *subcyclotomic closure* of $\mathbb{Q}$ in K. Since $\hat{\mathbb{Q}}_c/\mathbb{Q}$ is Abelian, so is $K_c/\mathbb{Q}$, and therefore K_c is stable under the involution $^-$ of K as well as "complex conjugation" on $\hat{\mathbb{Q}}_c$ – which we denote by *. Moreover the composition $^{-\,*}$ is also an involution on K_c, and we define $\check{K}$ to be its fixed field $(K_c, ^{-\,*})_+$. Thus $(\check{K}, ^-) = (\check{K}, ^*)$, its fixed subfield

$(\breve{K},\ ^*\ )_+$ is the maximal real subfield of $\breve{K}$, and $\breve{K}/(\breve{K},\ ^*\ )_+$ is definite. (Of course it is possible for the involution on $\breve{K}$ to be the identity, even if that on K is not.)

As usual $(K, ^-) \otimes_{(\breve{K}, ^-)} (\breve{K}G, ^-)$ is isomorphic with $(KG, ^-)$, and we identify them. If $\breve{A}$ is a central simple summand of $\breve{K}G$, it is stable under the standard $(\breve{K}, ^-)$-involution (Th. 4.5.1(a)), so $(K, ^-) \otimes_{(\breve{K}, ^-)} (\breve{A}, ^-)$ is a central simple direct summand of $(KG, ^-)$. The involutory Schur classes in $\mathrm{Sch}(K, ^-)$ obtained in this way from $\mathrm{Sch}(\breve{K}, ^-)$ for all finite groups G will be denoted by $K \otimes_{\breve{K}} \mathrm{Sch}(\breve{K}, ^-)$.

In a similar vein, we define $K \otimes_{\breve{K}} \mathrm{Sch}(\breve{K}) \subset \mathrm{Sch}(K)$.

Theorem 4.7.2 *(a) All simple components of $\breve{K}G$ are stable under the standard $(\breve{K}, ^-)$-involution.*

(b) The image of the forgetful map $\mathrm{Sch}(K, ^-) \xrightarrow{?} \mathrm{Sch}(K)$ is $K \otimes \mathrm{Sch}(\breve{K})$, and $\mathrm{Sch}(K, ^-) = K \otimes \mathrm{Sch}(\breve{K}, ^-)$.

(c) If $^-$ is not the identity on $\breve{K}$, $\mathrm{Sch}(\breve{K}, ^-) = \mathrm{Sch}(\breve{K})$ and $\mathrm{Sch}(K, ^-) = K \otimes \mathrm{Sch}(\breve{K})$.

Remark. Part (b) of the theorem can be restated as saying that first of all, the Schur classes over K which "support an involutory Schur class" (in $\mathrm{Sch}(K, ^-)$) are those which are defined over $\breve{K}$. (Note that $\breve{K}$ depends on the involution $^-$ on K). And secondly, when $(K, ^-) = (K, \mathrm{id})$, the possible types of an involutory Schur class over K are exactly the types which occur for the ("non-involutory") class over $\breve{K}$ from which it comes. If $(K, ^-) \neq (K, \mathrm{id})$, the involutory Schur classes over K come from involutory Schur classes of *both* orthogonal and symplectic types if $^-$ is the identity on $\breve{K}$, otherwise they come from those of unitary type. □

(a) All simple components of $\breve{K}G$ are fixed under the standard involution by Th. 4.5.1 (p. 171) since the involution $^- = {}^*$ on $\breve{K}$ is definite.

(b) Suppose that $[A, ^-] \in \mathrm{Sch}(K, ^-)$, say arising from the group algebra KG. Since A is central simple, the character of a representation of G on a simple A-module is of the form $d\chi$, d a positive integer[7], where χ is an absolutely irreducible character χ (cf. §1.7.5). Since the center of A is $K = K(\chi)$ (*ibid*), the values of χ lie in K. In fact, the values lie in K_c since the values of a character are sums of roots of unity (§1.7.2). By the equivalence of (ii) and (iv) of Th. 3.2.7(a) (p. 115), $\overline{(d\chi)^*} = d\chi$, so $\overline{\chi^*} = \chi$. Thus the values of χ actually lie in $\breve{K}$.

This implies that the simple summand $\breve{A}$ of $\breve{K}G$ which gives rise to χ over an algebraic closure, is central simple, and so $K \otimes_{\breve{K}} \breve{A} = A$. This proves that $\mathrm{Sch}(K, ^-) = K \otimes \mathrm{S}(\breve{K}, ^-)$.

The statement about the image of the forgetful map follows from the foregoing formula and the fact that $\mathrm{Sch}(\breve{K}, ^-) \xrightarrow{?} \mathrm{Sch}(\breve{K})$ is onto by (a).

Finally the first statement of (c) follows from (a), and the second from it and (b). □

When the involution on $\breve{K}$ is not the identity, part (c) reduces the calculation of the involutory Schur subgroup over K to that of the Schur subgroup over $\breve{K}$. This is occasionally useful since $\breve{K}/\mathbb{Q}$ is a subcyclotomic extension for which $\mathrm{Sch}(\breve{K})$ is known in many cases – see e.g. Chapters 7 and 8, [74].

[7] in fact, the index of A

The following theorem is proved on p. 480, [59].

Theorem 4.7.3 *(a) If K is a nonformally real algebraic number field (of finite degree over $\mathbb{Q}$), then*

$$\mathrm{Sch}(K, \mathrm{id}_K) = (K \otimes_{\check{K}} \mathrm{Sch}(\check{K})) \times \{\pm 1\}.$$

(b) If K is a finite extension of the p-adic field $\mathbb{Q}_p$, then
(i) $\mathrm{Sch}(K, {}^-) = 1$ *if* ${}^- \neq \mathrm{id}_K$.
(ii) $\mathrm{Sch}(K, \mathrm{id}_K) = \mathrm{Sch}_2(K) \times \{\pm 1\}$ *if K is an odd degree extension of an Abelian extension of $\mathbb{Q}_p$, otherwise* $\mathrm{Sch}(K, \mathrm{id}_K) = \{\pm 1\}$. □

Here $\mathrm{Sch}_2(K)$ stands for the subgroup of $\mathrm{Sch}(K)$ of elements of order 2 (or 1). $\mathrm{Sch}(K)$ was determined for a local field in Chs. 4 and 5, [74]; an alternative determination (as well as references to several other determinations) is given in [58]. Obviously since $\mathrm{Br}(K) = \mathbb{Q}/\mathbb{Z}$ (when K is a local field – cf. p. 17), $\mathrm{Sch}_2(K)$ is the group of order 1 or 2.

4.8 Similarity of isometry representations

In this book, we generally use the traditional definition of equivalence of isometry representations. There is another possibility, which has the virtue of making the representations $\rho : G \to \mathbf{I}(V, h)$ and $\rho : G \to \mathbf{I}(V, \alpha h)$, $(\alpha \in \dot{K}_0)$, "equivalent" even if the forms h and αh are not equivalent in the usual sense.

Let $(A, {}^-)$ be an involution algebra over $(K, {}^-)$. Two sesquilinear forms (V, h) and (W, g) over $(A, {}^-)$ are said to be *similar* if there is an A-isomorphism[8] $\varphi : V \to W$ and a scalar $\alpha \in \dot{K}_0$ such that $g(u\varphi, v\varphi) = \alpha h(u, v)$ for all $u, v \in V$. In this case we write $h \simeq g$ (over $(A, {}^-)$). In other words,

$$h \simeq g \quad \text{if and only if} \quad \alpha h \cong g \quad \text{for some } \alpha \in \dot{K}_0.$$

Exercise. Let A be a central simple K-algebra with $(K, {}^-)$-involutions ${}^-$ and ${}^\sim$. Show that the involution algebras $(A, {}^-)$ and $(A, {}^\sim)$ are isomorphic if and only if ${}^-$ and ${}^\sim$ are the adjoint involutions of similar reflexive forms over D, where D is the division algebra in the same Brauer class as A.

For isometry representations, similarity is defined in the obvious way: if (V, h) and (W, g) are ε-reflexive spaces over $(K, {}^-)$, two representations $\rho : G \to \mathbf{I}(V, h)$ and $\sigma : G \to \mathbf{I}(W, g)$ are *similar* if there is a K-isomorphism $\varphi : V \to W$ and a scalar $\alpha \in \dot{K}_0$ such that

$$g(u\varphi, v\varphi) = \alpha h(u, v) \text{ for all } u, v \in V \text{ and } \varphi\sigma(s) = \rho(s)\varphi \text{ for all } s \in G.$$

Again we write $\rho \simeq \sigma$ if ρ and σ are similar – this of course implies that $h \simeq g$ over $(K, {}^-)$.

The notion of similarity of representations can be expressed in terms of the associated ε-Hermitian forms over $(KG, {}^-)$ (where ${}^-$ on KG is the standard $(K, {}^-)$-involution, as usual) in exact analogy with the case of equivalence (cf. p. 86):

$$\rho \simeq \sigma \text{ if and only if } \hat{h} \simeq \hat{g}.$$

There is however one important difference between similarity and equivalence of isometry representations or of their associated forms: similarity is not "additive" in

[8] For the purposes of this section, it is convenient to write A-homomorphisms on the right.

the sense that, if $\rho_1 \simeq \rho_2$ and $\sigma_1 \simeq \sigma_2$, it is not necessarily true that $\rho_1 \oplus \sigma_1 \simeq \rho_2 \oplus \sigma_2$. This is clear since not even the underlying forms of the two sums are necessarily similar – for example $y^2 \simeq 2y^2$ but $x^2+y^2 \not\simeq x^2+2y^2$ if $2 \notin \dot{K}^2$ (and so, for example, not even the trivial representations $G \to \mathbf{O}(K^2, x^2+y^2)$ and $G \to \mathbf{O}(K^2, x^2+2y^2)$ are similar). Thus we cannot define a Witt-Grothendieck group of similarity classes of isometry representations as we did for equivalence classes. However it is possible to define a graded monoid of isometry representations in a completely different way – this is the principal goal of this section.

This situation can be ameliorated a little if we make the following definition: Let

$$(KG, {}^-) = (A_1, {}^-) \oplus \cdots \oplus (A_l, {}^-)$$

be the decomposition of $(KG, {}^-)$ into simple involution components. If $\rho : G \to \mathbf{I}(V, h)$ is an isometry representation, let $V = V_1 \perp \cdots \perp V_l$ $(V_i = A_i V)$ be the involution isotypic decomposition of V, and

$$h_i = h|_{V_i \times V_i}, \quad \hat{h}_i = \hat{h}|_{V_i \times V_i}, \quad \text{and} \quad \rho_i : G \to \mathbf{I}(V_i, h_i) \quad (\rho_i(s) = \rho(s)|_{V_i})$$

the *involution isotypic components* of h, $\hat{h}$ and ρ. If $\sigma : G \to \mathbf{I}(W, g)$ is another isometry representation of G, we say that ρ and σ – or, equivalently, $\hat{h}$ and $\hat{g}$ – are *isotypically similar* if their isotypic components ρ_i and σ_i respectively $\hat{h}_i$ and $\hat{g}_i$ are similar for all i. We write $\rho \sim \sigma$ and $\hat{h} \sim \hat{g}$ for isotypic similarity; thus $\rho \sim \sigma$ if and only if $\rho_i \simeq \sigma_i$ for all i. Of course $\rho_i \simeq \sigma_i$ and $\hat{h}_i \simeq \hat{g}_i$ are respectively equivalent to $\rho_i \sim \sigma_i$ and $\hat{h}_i \sim \hat{g}_i$, and

$$\rho \sim \sigma \text{ if and only if } \hat{h} \sim \hat{g} \text{ if and only if } \hat{h}_i \sim \hat{g}_i \text{ for all } i,$$

$$\rho \cong \sigma \;\Rightarrow\; \rho \simeq \sigma \;\Rightarrow\; \rho \sim \sigma,$$

but the opposite implications do not hold in general. Also, with respect to suitable K-bases of V and W,

$$\operatorname{mat}\rho(s) = \begin{pmatrix} \operatorname{mat}\rho_1(s) & 0 & \cdots & 0 \\ 0 & \operatorname{mat}\rho_2(s) & \cdots & 0 \\ & & \ddots & \\ 0 & 0 & \ldots & \operatorname{mat}\rho_l(s) \end{pmatrix}$$

$$= \operatorname{mat}\sigma(s) \text{ for all } s \in G \text{ if } \rho \sim \sigma.$$

Similarity classes over central simple algebras.

If $(A, {}^-)$ is a central simple algebra with $(K, {}^-)$-involution, $\mathrm{Sim}_\varepsilon(A, {}^-)$ denotes the set of similarity classes of ε-Hermitian spaces (or nonsingular forms) $h : V \times V \to (A, {}^-)$; the similarity class of (V, h) or h is denoted by $\{V, h\}$ respectively $\{h\}$.

Let $\mathbf{b} \in \mathrm{Br}(K, {}^-)$ and let D be the division algebra over K, unique up to isomorphism, such that $\mathbf{b} = ([D], \mathrm{t})$ where $\mathrm{t} = 0, \pm 1$. (So $\mathrm{t} = 0$ if $(K, {}^-)$ is of the second kind and $= \pm 1$ if $(K, {}^-)$ is of the first kind). Let $\mathcal{I}(\mathbf{b})$ be the set of isomorphism classes of central simple algebras $(\mathcal{A}, {}^\sim)$ with involution whose involutory Brauer class $[\mathcal{A}, {}^\sim] = \mathbf{b}$. Let $\varepsilon = \pm 1$. Define $\varepsilon\mathbf{b} = ([D], \varepsilon\mathrm{t})$ (which $= \mathbf{b}$ unless $(K, {}^-)$ is of the first kind and $\varepsilon = -1$).

Theorem 4.8.1 *Let $(A, ^-)$ be a central simple algebra over K with a $(K, ^-)$-involution, and suppose that* $\mathbf{b} = [A, ^-]$. *The association $(V, h) \rightsquigarrow (\mathrm{End}_A V, \mathrm{ad}_h)$ from ε-Hermitian spaces over $(A, ^-)$ to central simple algebras with $(K, ^-)$-involution induces a bijection*

$$\mathrm{Ad} : \mathrm{Sim}_\varepsilon(A, ^-) \to \mathcal{I}(\varepsilon\mathbf{b}). \tag{4.33}$$

Suppose that (V, h) and (W, g) are similar spaces in ${}_{(A,^-)}\mathcal{H}^*_\varepsilon$ – say $\varphi : (V, h) \to (W, g)$ with $g(u\varphi, v\varphi) = \alpha h(u, v)$ for some $\alpha \in \dot{K}_0$ and all $u, v \in V$. Note that $\mathrm{End}_A V \cong \mathrm{M}(m, D)$ where $m = \mathrm{len}_A V$ (Lem. 3.2.3(a), p. 119).

The map $\mathrm{End}_A V \to \mathrm{End}_A W$ given by $\theta \to \varphi^{-1}\theta\varphi$ is an isomorphism of K-algebras, and is an isomorphism

$$(\mathrm{End}_A V, \mathrm{ad}_h) \to (\mathrm{End}_A W, \mathrm{ad}_g)$$

of involution algebras since

$$\begin{aligned} g(u\varphi, v\varphi\, \mathrm{ad}_g(\varphi^{-1}\theta\varphi)) &= g(u\varphi(\varphi^{-1}\theta\varphi), v\varphi) \\ &= \alpha h(u\theta, v) = \alpha h(u, v(\mathrm{ad}_h\theta)) \\ &= g(u\varphi, v\varphi(\varphi^{-1}(\mathrm{ad}_h\theta)\varphi)) \end{aligned}$$

implies that

$$\mathrm{ad}_g(\varphi^{-1}\theta\varphi) = \varphi^{-1}(\mathrm{ad}_h\theta)\varphi.$$

Thus there is a map Ad as stated in the theorem.

We next show that Ad is surjective. Let $(\mathcal{A}, ^-)$ be a $(K, ^-)$-involution algebra whose isomorphism class is in $\mathcal{I}(\varepsilon\mathbf{b})$. Let * be a $(K, ^-)$-involution on D, orthogonal if $(K, ^-)$ is of the first kind – thus the involution ^{t*} on any matrix algebra $\mathrm{M}(n, D)$ is also orthogonal if it is of the first kind. Now $\mathcal{A} \cong \mathrm{M}(m, D)$ for some m. In showing that the isomorphism class of $(\mathcal{A}, ^-)$ is in $\mathrm{im\,Ad}$, we may in fact assume that $\mathcal{A} = \mathrm{M}(m, D)$, so $(\mathcal{A}, ^-) = (\mathrm{M}(m, D), {}^{t*}\mathrm{inn}_{\underline{g}})$ for some $\underline{g} \in \mathbf{GL}(m, D)$ such that

$$\underline{g}^{t*} = \begin{cases} \mathrm{type}(\mathcal{A}, ^-)\underline{g} = \varepsilon\mathrm{type}(A, ^-)\underline{g} & \text{if } (K, ^-) \text{ is of the first kind,} \\ \varepsilon\underline{g} & \text{if } (K, ^-) \text{ is of the second kind.} \end{cases}$$

Similarly $(A, ^-) \cong (\mathrm{M}(n, D), {}^{t*}\mathrm{inn}_{\underline{h}_A})$ where

$$\underline{h}_A^{t*} = \begin{cases} \mathrm{type}(A, ^-)\underline{h}_A & \text{if } (K, ^-) \text{ is of the first kind,} \\ \underline{h}_A & \text{if } (K, ^-) \text{ is of the second kind.} \end{cases} \tag{4.34}$$

Now define

$$g : D^{n\times m} \times D^{n\times m} \to (\mathrm{M}(n, D), {}^{t*}\mathrm{inn}_{\underline{h}_A}) \quad \text{by} \quad g(u, v) = u\underline{g}v^{t*}\underline{h}_A^{-1}.$$

A straightforward calculation shows that g is a nonsingular ε-Hermitian form, and that $\mathrm{ad}_g = {}^{t*}\mathrm{inn}_{\underline{g}}$ if we identify $\mathrm{End}_{\mathrm{M}(n,D)} D^{n\times m}$ with $\mathrm{M}(m, D)$. It follows that

$$\mathrm{Ad}\{D^{n\times m}, g\} = \text{the isomorphism class of } (\mathcal{A}, ^-).$$

Now injectivity. Suppose that $(V, h), (W, g) \in {}_{(A,^-)}\mathcal{H}^*_\varepsilon$ and that $(\mathrm{End}_A V, \mathrm{ad}_h) \cong (\mathrm{End}_A W, \mathrm{ad}_g)$; we must show that $(V, h) \simeq (W, g)$.

We may assume $A = \mathrm{M}(n, D)$ for some n, so

$$(A, ^-) = (\mathrm{M}(n, D), {}^{t*}\mathrm{inn}_{\underline{h}_A})$$

where $\underline{h}_A$ satisfies (4.34). Now V and W must have the same A-length, say m, and we may assume that $V = W = D^{n\times m}$; then

$$h(u,v) = u\underline{h}v^{t*}\underline{h}_A^{-1} \quad \text{and} \quad g(u,v) = u\underline{g}v^{t*}\underline{h}_A^{-1}$$

for certain $\underline{h}, \underline{g} \in \mathbf{GL}(m,D)$ such that $\underline{h}^{t*} = \varepsilon\text{type}(A,{}^-)\underline{h}$ and $\underline{g}^{t*} = \varepsilon\text{type}(A,{}^-)\underline{g}$ if the involutions are of the first kind, otherwise they are equal to $\varepsilon\underline{h}$ and $\varepsilon\underline{g}$ respectively. There is a canonical K-algebra isomorphism $\text{End}_A V \to \text{M}(m,D)$, $\varphi \to \underline{\varphi}$, satisfying $v\varphi = v\underline{\varphi}$ for all $v \in D^{n\times m}$, where $v\underline{\varphi}$ is of course matrix multiplication. Similarly $\text{End}_A W \to \text{M}(m,D)$. Now

$$\underline{\text{ad}_h\varphi} = \underline{h}\underline{\varphi}^{t*}\underline{h}^{-1} \quad \text{and} \quad \underline{\text{ad}_g\varphi} = \underline{g}\,\underline{\varphi}^{t*}\underline{g}^{-1},$$

and we define the involution $\text{ad}_{\underline{h}}$ of $\text{M}(m,D)$ by

$$\text{ad}_{\underline{h}}\,\underline{\varphi} = \underline{\text{ad}_h\varphi};$$

$\text{ad}_{\underline{g}}$ is similarly defined. The isomorphism $(\text{End}_A V, \text{ad}_h) \cong (\text{End}_A W, \text{ad}_g)$ is reflected in an isomorphism $(\text{M}(m,D), \text{ad}_{\underline{h}}) \to (\text{M}(m,D), \text{ad}_{\underline{g}})$. The latter is an inner automorphism $\text{inn}_{\underline{\psi}}$, so

$$\underline{\psi}\,\text{ad}_{\underline{g}}(\underline{\varphi})\underline{\psi}^{-1} = \text{ad}_{\underline{h}}(\underline{\psi}\,\underline{\varphi}\,\underline{\psi}^{-1}) \quad \text{for all } \underline{\varphi} \in \mathbf{GL}(m,D),$$

i.e.

$$\underline{\psi}\,\underline{g}\,\underline{\varphi}^{t*}\,\underline{g}^{-1}\underline{\psi}^{-1} = \underline{h}\underline{\psi}^{-t*}\underline{\varphi}^{t*}\underline{\psi}^{t*}\underline{h}^{-1} \quad \text{for all } \underline{\varphi}.$$

Thus

$$\alpha\underline{\psi}\,\underline{g} = \underline{h}\underline{\psi}^{-t*}$$

for some $\alpha \in \dot{K}$ and so

$$h(u,v) = u\underline{h}v^{t*}\underline{h}_A^{-1} = \alpha u\underline{\psi}\,\underline{g}\,\underline{\psi}^{t*}v^{t*}\underline{h}_A^{-1} = \alpha g(u\underline{\psi}, v\underline{\psi})$$

and it follows that $h \simeq g$. □

Corollary 4.8.1 *Suppose that $\varepsilon = \pm 1$.*

(a) If $(A,{}^-)$ and $(B,{}^-)$ are similar $(K,{}^-)$-involution algebras, there is a canonical bijection

$$\text{Sim}_\varepsilon(A,{}^-) \to \text{Sim}_\varepsilon(B,{}^-) \tag{4.35}$$

which preserves the lengths of the underlying modules. The composition

$$\text{Sim}_\varepsilon(A,{}^-) \to \text{Sim}_\varepsilon(B,{}^-) \to \text{Sim}_\varepsilon(C,{}^-)$$

of two such canonical bijections is another, and the inverse of a canonical bijection is also a canonical bijection.

(b) If A and B are similar as K-algebras and if $(A,{}^-)$ and $(B,{}^-)$ are involution algebras either of the second kind or of the first kind but of opposite types, there is a canonical bijection

$$\text{Sim}_\varepsilon(A,{}^-) \to \text{Sim}_{-\varepsilon}(B,{}^-) \tag{4.36}$$

which preserves the lengths of the underlying modules.

The canonical bijection (4.35) is simply the composition of the bijection (4.33) and the inverse of the bijection $\text{Sim}_\varepsilon(B,{}^-) \to \mathcal{I}(\varepsilon\mathbf{b})$ of the theorem. The lengths of the underlying modules are preserved since if $\{V,h\} \in \text{Sim}_\varepsilon(A,{}^-)$ corresponds to $\{W,g\} \in \text{Sim}_\varepsilon(B,{}^-)$, then $\text{End}_A V \cong \text{End}_B W$. The last statement of (a) follows easily from the definitions of the maps.

In the case of involutions of the second kind, the bijection (4.36) arises from scaling $h \rightsquigarrow \sqrt{\delta}h$ where $K = K_0(\sqrt{\delta})$. Suppose that the involutions are of the first kind. The bijection (4.33) is

$$\mathrm{Sim}_\varepsilon(A, {}^-) \to \mathcal{I}([A], \varepsilon \mathrm{type}\, A) = \mathcal{I}([B], (-\varepsilon)\mathrm{type}\, B),$$

which yields the bijection (4.36). □

Exercise. The canonical bijections $\mathrm{Sim}_\varepsilon(A, {}^-) \to \mathrm{Sim}_{\pm\varepsilon}(B, {}^-)$ have the property that if C is a third algebra in the same Brauer class, also with a $(K, {}^-)$-involution ${}^-$, the diagram

$$\begin{array}{ccc} \mathrm{Sim}_\varepsilon(A, {}^-) & \longrightarrow & \mathrm{Sim}_{\pm\varepsilon}(B, {}^-) \\ & \searrow & \downarrow \\ & & \mathrm{Sim}_{\pm\varepsilon}(C, {}^-) \end{array}$$

of canonical bijections is commutative (with the appropriate choice of signs depending on the types of the involutions). If

$$\varepsilon_0 = \begin{cases} \mathrm{type}\,(A, {}^-)\mathrm{type}\,(B, {}^-) & \text{if } (A, {}^-) \text{ and } (B, {}^-) \text{ are of the first kind,} \\ -1 & \text{if } (A, {}^-) \text{ and } (B, {}^-) \text{ are of the second kind,} \end{cases}$$

the inverse of the canonical bijection $\mathrm{Sim}_\varepsilon(A, {}^-) \to \mathrm{Sim}_{\varepsilon_0\varepsilon}(B, {}^-)$ is the canonical bijection $\mathrm{Sim}_{\varepsilon_0\varepsilon}(B, {}^-) \to \mathrm{Sim}_\varepsilon(A, {}^-)$. □

Suppose that $\varphi : (A, {}^-) \to (B, {}^-)$ is an isomorphism of central simple algebras with $(K, {}^-)$-involutions, and let $h : V \times V \to (A, {}^-)$ be a sesquilinear form. Make V into a B-module ${}^\varphi V$ via $bv = \varphi^{-1}(b)v$ and define

$${}^\varphi h : {}^\varphi V \times {}^\varphi V \to (B, {}^-), \quad {}^\varphi h(v_1, v_2) = \varphi(h(v_1, v_2)).$$

It is easy to check that ${}^\varphi h$ is nonsingular respectively ε-Hermitian if and only if h is. And if h is nonsingular and ε-Hermitian, then

$$(\mathrm{End}_A V, \mathrm{ad}_h) = (\mathrm{End}_B {}^\varphi V, \mathrm{ad}_{{}^\varphi h}).$$

This implies the following lemma:

Lemma 4.8.1 *Under the canonical bijection (4.35), the similarity class of (V, h) corresponds to the similarity class of $({}^\varphi V, {}^\varphi h)$.* □

Tensor products of forms and similarity classes. We now explain how to "multiply" sesquilinear forms over central simple algebras, as well as certain isotypic similarity classes.

Let $(A, {}^-)$ and $(B, {}^-)$ be central simple K-algebras with $(K, {}^-)$-involutions and suppose that $h : V \times V \to (A, {}^-)$ and $g : W \times W \to (B, {}^-)$ are sesquilinear forms. Recall that $\overline{V}$ is the *right* A-module, twisted by the involution ${}^-$ of A. There is a unique K-linear map

$$(V \otimes_K W) \otimes_K (\overline{V} \otimes_K \overline{W}) \to (A, {}^-) \otimes_{(K, {}^-)} (B, {}^-)$$

determined by

$$(v_1, w_1, \bar{v}_2, \bar{w}_2) \to h(v_1, v_2) \otimes g(w_1, w_2).$$

It is an $(A \otimes_K B, A \otimes_K B)$-bimodule homomorphism, so there exists a sesquilinear form

$$\begin{aligned} h \otimes g : (V \otimes_K W) \times (V \otimes_K W) &\to (A, {}^-) \otimes_{(K,{}^-)} (B, {}^-), \\ (h \otimes g)(v_1 \otimes w_1, v_2 \otimes w_2) &= h(v_1, v_2) \otimes g(w_1, w_2). \end{aligned}$$

Thus there is a tensor product $(V, h) \otimes_K (W, g) := (V \otimes_K W, h \otimes g)$ of sesquilinear modules over the central simple algebras $(A, {}^-)$ and $(B, {}^-)$ with $(K, {}^-)$-involutions, and the product is another sesquilinear module, over the central simple algebra $(A, {}^-) \otimes_{(K,{}^-)} (B, {}^-)$ with $(K, {}^-)$-involution.

It is straightforward to check that if h is ε-Hermitian and g is δ-Hermitian, then $h \otimes g$ is $\varepsilon\delta$-Hermitian.

Suppose h and g are nonsingular, so $V \xrightarrow{h_r} \overline{V^*}$ and $W \xrightarrow{g_r} \overline{W^*}$ are both isomorphisms. The diagram

$$\begin{array}{ccc} & & V^* \otimes_K W^* \\ & \nearrow^{h_r \otimes g_r} & \\ V \otimes_K W & & \Big\downarrow \\ & \searrow_{(h \otimes g)_r} & \\ & & (V \otimes_K W)^* \end{array}$$

of vector spaces over K is commutative, where the unmarked map is the natural map and is an isomorphism (Prop. 1.2.2(b), p. 10). Thus $h \otimes g$ is also nonsingular.

If we identify $(A \otimes_K B) \otimes_K C$ and $A \otimes_K (B \otimes_K C)$ in the usual way, it is easy to see that this multiplication is associative. Similarly it is commutative.

If $(V, h) \simeq (V', h')$ and $(W, g) \simeq (W', g')$, one can easily check that

$$h \simeq h' \text{ and } g \simeq g' \;\Rightarrow\; h \otimes g \simeq h' \otimes g'.$$

Thus the tensor product leads to an associative and commutative product

$$\begin{aligned} \mathrm{Sim}_\varepsilon(A, {}^-) \times \mathrm{Sim}_\delta(B, {}^-) &\xrightarrow{\otimes} \mathrm{Sim}_{\varepsilon\delta}\big((A, {}^-) \otimes_{(K,{}^-)} (B, {}^-)\big), \\ (\{h\}, \{g\}) &\to \{h \otimes g\}, \end{aligned} \tag{4.37}$$

which we usually write $\{h \otimes g\} = \{h\}\{g\}$.

Remark. In the case when $(A, {}^-)$ and $(B, {}^-)$ are factors of the group algebra $(KG, {}^-)$ with the standard $(K, {}^-)$-involution, this product corresponds to a product of isometry representations. Thus if $\hat{h} : V \times V \to (A, {}^-)$ is an ε-Hermitian form, the linear representation $\rho : G \to \mathbf{GL}(V)$ associated to the A-module V is an isometry representation with respect to the ε-reflexive form $h = \frac{\deg A}{|G|}\mathrm{trd}_{A/K}\hat{h} : V \times V \to (K, {}^-)$ where $\deg A$ is the degree of A (Ths. 2.1.1(p. 86) and 1.2.13 (p. 15)). And if $\hat{g} : W \times W \to (B, {}^-)$ is a δ-Hermitian form, corresponding in the same way to the isometry representation $\sigma : G \to \mathbf{I}(W, g)$, $g = \frac{\deg B}{|G|}\mathrm{trd}_{A/K}\hat{g}$, then $\hat{h}\hat{g}$ corresponds to the tensor product $\rho \otimes \sigma : G \to \mathbf{I}(V \otimes W, |G|hg)$ since

$$|G|hg = |G|\left(\frac{\deg A}{|G|}\mathrm{trd}_{A/K}\hat{h}\right)\left(\frac{\deg B}{|G|}\mathrm{trd}_{B/K}\hat{g}\right) = \frac{\deg(A \otimes_K B)}{|G|}\mathrm{trd}_{A \otimes_K B}(\hat{h} \otimes \hat{g})$$

(cf. Lem. 1.2.2, p. 13). And of course this representation is similar to the usual product $\rho \otimes \sigma : G \to \mathbf{I}(V \otimes W, hg)$. □

Let

$$\mathcal{I}(K, {}^{-}) = \bigcup_{\mathbf{b} \in \mathrm{Br}(K, {}^{-})} \mathcal{I}(\mathbf{b})$$

be the set of isomorphism classes of central simple K-algebras with a $(K, {}^{-})$-involution. The tensor product also induces a well-defined product $\mathbf{i}_1 \otimes \mathbf{i}_2$ of elements $\mathbf{i}_1, \mathbf{i}_2 \in \mathcal{I}(K, {}^{-})$; we will usually write $\mathbf{i}_1 \otimes \mathbf{i}_2 = \mathbf{i}_1 \mathbf{i}_2$.

Thus $\mathcal{I}(K, {}^{-})$ is an Abelian monoid, graded by $\mathrm{Br}(K, {}^{-})$.

We next define the similarity classes $\mathrm{Sim}_\varepsilon \mathbf{i}$ for each $\mathbf{i} \in \mathcal{I}(K, {}^{-})$: choose a representative $(A_{\mathbf{i}}, {}^{-}) \in \mathbf{i}$ for each $\mathbf{i}$, and define

$$\mathrm{Sim}_\varepsilon \mathbf{i} = \mathrm{Sim}_\varepsilon (A_{\mathbf{i}}, {}^{-}). \tag{4.38}$$

By Cor. 4.8.1 there is a canonical bijection

$$\mathrm{Sim}_\varepsilon (A, {}^{-}) \to \mathrm{Sim}_\varepsilon \mathbf{i}$$

for each $(A, {}^{-}) \in \mathbf{i}$, which has the property that if $(B, {}^{-})$ is also in $\mathbf{i}$,

$$\begin{array}{ccc} \mathrm{Sim}_\varepsilon(A, {}^{-}) & \longrightarrow & \mathrm{Sim}_\varepsilon(B, {}^{-}) \\ & \searrow & \downarrow \\ & & \mathrm{Sim}_\varepsilon \mathbf{i} \end{array}$$

is commutative (where all three maps are the canonical bijections, of course). And there is a well-defined multiplication

$$\mathrm{Sim}_\varepsilon \mathbf{i}_1 \times \mathrm{Sim}_\delta \mathbf{i}_2 \to \mathrm{Sim}_{\varepsilon\delta} \mathbf{i}_1 \mathbf{i}_2$$

given by

$$(\{h\}, \{g\}) \to \{h\}\{g\},$$

where $\{h\}\{g\} \in \mathrm{Sim}_{\varepsilon\delta}\big((A_{\mathbf{i}_1}, {}^{-}) \otimes (A_{\mathbf{i}_2}, {}^{-})\big)$ has been identified via a canonical map to an element of $\mathrm{Sim}_{\varepsilon\delta}(A_{\mathbf{i}_1\mathbf{i}_2}, {}^{-})$. Thus we have an associative and commutative monoid

$$\mathbf{M} = \bigcup_{\substack{\mathbf{i} \in \mathcal{I}(K, {}^{-}), \\ \varepsilon = \pm 1}} \mathrm{Sim}_\varepsilon \mathbf{i}$$

with an identity element, namely the similarity class $\{(K, \langle 1 \rangle)\} \in \mathrm{Sim}_1(K, {}^{-})$, and two natural gradings:

(i) a grading by $\{\pm 1\}$ via ε, and

(ii) a grading by the monoid $\mathcal{I}(K, {}^{-})$, via $\mathbf{i}$.

We now show that if a different choice of $(A_{\mathbf{i}}, {}^{-})$ is made in (4.38), say

$$\mathrm{Sim}_\varepsilon \mathbf{i} = \mathrm{Sim}_\varepsilon (A'_{\mathbf{i}}, {}^{-}) \text{ for each } \mathbf{i} \in \mathcal{I}(K, {}^{-}),$$

then there is a canonical graded isomorphism between the resulting monoid $\mathbf{M}'$ and $\mathbf{M}$. To do this, we let $\Phi : \mathbf{M} \to \mathbf{M}'$ be the map which restricts to the canonical bijection

$$\mathrm{Sim}_\varepsilon (A_{\mathbf{i}}, {}^{-}) \to \mathrm{Sim}_\varepsilon (A'_{\mathbf{i}}, {}^{-})$$

of Cor. 4.35 for each $\mathbf{i} \in \mathcal{I}(K, {}^{-})$. It is of course a bijection and it preserves both gradings. Thus it remains to show that

$$\Phi(\{h\}\{g\}) = \Phi(\{h\})\Phi(\{g\})$$

where, say, $\{(V,h)\} \in \mathrm{Sim}_\varepsilon(A_\mathbf{i}, {}^-)$ and $\{(W,g)\} \in \mathrm{Sim}_\delta(A_\mathbf{j}, {}^-)$.

Let

$$\varphi : (A_\mathbf{i}, {}^-) \to (A'_\mathbf{i}, {}^-) \quad \text{and} \quad \psi : (A_\mathbf{j}, {}^-) \to (A'_\mathbf{j}, {}^-)$$

be isomorphisms of involution algebras. Then (Lem. 4.8.1)

$$\Phi\{h\} = \{{}^\varphi h\} \quad \text{and} \quad \Phi\{g\} = \{{}^\psi g\}.$$

Now

$$\varphi \otimes \psi : (A_\mathbf{i}, {}^-) \otimes (A_\mathbf{j}, {}^-) \to (A'_\mathbf{i}, {}^-) \otimes (A'_\mathbf{j}, {}^-)$$

is also an isomorphism of involution algebras, and it follows that

$$\begin{aligned}\Phi(\{h\}\{g\}) &= \Phi(\{h \otimes g\}) \\ &= \{{}^{\varphi\otimes\psi}(h \otimes g)\} = \{{}^\varphi h \otimes {}^\psi g\} \\ &= \{{}^\varphi h\}\{{}^\psi g\} = \Phi(\{h\})\Phi(\{g\}).\end{aligned}$$

Theorem 4.8.2 *The monoid* $\mathbf{M}$ *of similarity classes is unique up to a canonical isomorphism, and the canonical bijection (4.33) induces an epimorphism*

$$\mathrm{Ad} : \mathbf{M} \to \mathcal{I}(K, {}^-) \tag{4.39}$$

which is graded with respect to the homomorphism

$$\{\pm 1\} \times \mathcal{I}(K, {}^-) \to \mathrm{Br}(K, {}^-), \quad (\varepsilon, \mathbf{i}) \to [\varepsilon\mathbf{i}]$$

of the grading monoids.

It remains to show that the canonical bijection induces a graded homomorphism. Suppose that $\{V,h\}, \{W,g\} \in \mathbf{M}$, say $\{V,h\} \in \mathrm{Sim}_\varepsilon \mathbf{i}$ and $\{W,h\} \in \mathrm{Sim}_\delta \mathbf{j}$. Then

$$\{V,h\}\{W,g\} = \{V \otimes W, h \otimes g\}.$$

There is a canonical isomorphism

$$\mathrm{End}_A V \otimes_K \mathrm{End}_B W \to \mathrm{End}_{A\otimes B}(V \otimes W),$$

which takes $\varphi \otimes \psi$ to $(v \otimes w \to v\varphi \otimes w\psi)$ (Prop. 1.2.2, p. 10). In fact it is an isomorphism of involution algebras

$$(\mathrm{End}_A V,\, \mathrm{ad}_h) \otimes_K (\mathrm{End}_B W,\, \mathrm{ad}_g) \to (\mathrm{End}_{A\otimes B}(V \otimes W),\, \mathrm{ad}_{h\otimes g})$$

since

$$\begin{aligned}(h \otimes g)\big((v \otimes w)(\varphi \otimes \psi), v' \otimes w'\big) &= h(v\varphi, v') \otimes g(w\psi, w') \\ &= h(v, v'\mathrm{ad}_h\varphi) \otimes g(w, w'\mathrm{ad}_g\psi) \\ &= (h \otimes g)\big(v \otimes w, v'\mathrm{ad}_h\varphi \otimes w'\mathrm{ad}_g\psi\big) \\ &= (h \otimes g)\big(v \otimes w, (v' \otimes w')(\mathrm{ad}_h \otimes \mathrm{ad}_g)(\varphi \otimes \psi)\big)\end{aligned}$$

implies that $\mathrm{ad}_{h\otimes g} = \mathrm{ad}_h \otimes \mathrm{ad}_g$. And this in turn implies that

$$\mathrm{Ad}\big((\mathrm{Sim}_\varepsilon\mathbf{i})(\mathrm{Sim}_\delta\mathbf{j})\big) = \mathrm{Ad}(\mathrm{Sim}_\varepsilon\mathbf{i})\mathrm{Ad}(\mathrm{Sim}_\delta\mathbf{j}).$$

The fact that it is a "graded homomorphism" follows from (4.33). □

4.9 Structure of $\hat{\mathrm{W}}_\varepsilon(K,G)$ when K_0 is not formally real

Recall our assumptions that the characteristic of K does not divide $2|G|$, and that $s(K)$ denotes the level of K (cf. p. 36). In the following theorem, $\mathrm{L} : \hat{\mathrm{W}}_\varepsilon(K,G) \to \mathrm{K}_0(KG)$ is the linearization map for G-invariant ε-reflexive forms over $(K, {}^{-})$.

Theorem 4.9.1 *If K_0 is not formally real, then*

$$2s(K_0) \ker \mathrm{L} = 0.$$

If K is algebraically closed with trivial involution,

$$\ker \mathrm{L} = 0.$$

If K_0 has cohomological dimension ≤ 1, then

$$2 \ker \mathrm{L} = 0.$$

If K_0 is a local field or a nonreal global field, then

$$4 \ker \mathrm{L} = 0.$$

If K_0 is a local field or a nonreal global field and -1 is a square in K, then

$$2 \ker \mathrm{L} = 0.$$

See also the example on p. 167, and Th. 4.10.1, p. 206.

We note that $s(K) = 1$ if K is algebraically closed, $s(K_0) = 1$ or 2 if $\dim K_0 \leq 1$, and $s(K_0) = 1, 2$ or 4 if K_0 is local or nonreal global. The reasons appear in the proof.

If K_0 is not formally real, $s(K_0)$ is a power of 2 by Pfister's Theorem (p. 37), and so the theorem implies that $\ker \mathrm{L}$ has exponent a power of 2.

Corollary 4.9.1 *If K_0 is not formally real, $\hat{\mathrm{W}}_\varepsilon(K,G)$ is the direct sum of an Abelian group annihilated by $2s(K_0)$ and a free Abelian group of rank l, where l is the number of simple involution algebra summands of $(KG, {}^{-})$.*

Since $\mathrm{K}_0(KG)$ is a finitely generated free Abelian group (of rank = the number of simple summands of KG), the image of L is also free and has rank l by Th. 3.2.5 (p. 114). Thus L splits and the corollary follows. □

REMARKS. 1. Suppose K_0 is not formally real. The torsion subgroup of $\hat{\mathrm{W}}_\varepsilon(K,G)$ is therefore $\ker \mathrm{L}$ and so is 2-primary. It is not in general finite. For example if $\dot{K}/\dot{K}^2$ is infinite and G is represented trivially on the symmetric space $\langle\alpha\rangle$ for each $\alpha \in \dot{K}$, then $\langle\alpha\rangle - \langle 1\rangle \in \ker \mathrm{L}$ but $\langle\alpha\rangle - \langle 1\rangle \neq \langle\beta\rangle - \langle 1\rangle$ in $\hat{\mathrm{W}}_\varepsilon(K,G)$ if $\alpha\dot{K}^2 \neq \beta\dot{K}^2$.

2. If K_0 is formally real, $\ker \mathrm{L}$ is *not* torsion – for example consider $\langle 1\rangle - \langle -1\rangle$. By contrast, the torsion subgroup of $\mathrm{W}_\varepsilon(K,G)$ *is* 2-primary for any field K_0 (Cor. 4.11.1, p. 211). □

3. The image of L is easily determined with no special assumptions about K, using the facts about indecomposable spaces in §3.2.4 and the exercises on p. 68. Namely if $(KG, {}^{-}) = \oplus_i (A_i, {}^{-})$ is the decomposition into simple involution algebras,

$$\mathrm{K}_0(KG, {}^{-}) = \oplus_i \mathrm{K}_0(A_i, {}^{-})$$

and the image of $\mathrm{L} : \hat{\mathrm{W}}_\varepsilon(A_i, {}^-) \to \mathrm{K}_0(A_i)$ is

$$\begin{cases} 2\mathrm{K}_0(A_i) & \text{if } A_i \text{ is quasisplit and the type of } (A_i, {}^-) \text{ is } -\varepsilon, \\ \mathrm{K}_0(A_i)_+ & \text{if } (A_i, {}^-) \text{ is hyperbolic,} \\ \mathrm{K}_0(A_i) & \text{otherwise.} \end{cases}$$

($\mathrm{K}_0(A_i)_+$ is the subset of symmetric elements under the canonical involution of $\mathrm{K}_0(A_i)$). The details are left to the reader.

Exercises. 1. (a) Show that there is a surjective homomorphism $\Delta : \hat{W}_1(K, \mathrm{id}_K, G) \to \dot{K}/\dot{K}^2$.

(b) Show that $\hat{W}_1(K, \mathrm{id}_K, G)$ is not necessarily finitely generated.

2. (a) If $r \in \mathbb{Q}$ is > 0 but $\notin \mathbb{Q}^2$, show that $[\langle r \rangle] - [\langle 1 \rangle]$ is a torsion element of $\hat{W}_1(\mathbb{Q}, G)$. (Show that it is annihilated by 4).

(b) Show that the torsion subgroup T of $\hat{W}_1(\mathbb{Q}, G)$ is not finitely generated. □

Proof of the theorem. If g is a 1-reflexive hyperbolic form over K, then by Th. 4.1.3 (p. 155), $\hat{\mathrm{W}}_\varepsilon(K,G)[g] \subset \mathrm{H}_\varepsilon(KG)$. Let $s(K_0) = 2^r$, and suppose first that $K = K_0$. The quadratic form $\langle \overbrace{1, 1, \ldots, 1}^{1+s(K)} \rangle$ is isotropic, and so the "Pfister form"

$$\langle \overbrace{1, 1, \ldots, 1}^{2s(K)} \rangle = \langle 1, 1 \rangle^{r+1}$$

is also isotropic and therefore hyperbolic by Cor. 1.5, p. 144, [65].

If $K \neq K_0$, say $K = K_0(\sqrt{\delta})$, then the trace form over K_0 of the Hermitian form $\langle 1 \rangle$ over $(K, {}^-)$ is the symmetric form $\langle 2, -2\delta \rangle$ and so the trace form of the Hermitian form $h' = \langle \overbrace{1, 1, \ldots, 1}^{2s(K_0)} \rangle$ is hyperbolic; since the trace form of a hyperbolic Hermitian form is also hyperbolic, it follows from Jacobson's theorem that h' is also hyperbolic.

Thus if $(V, h) \in {}_{KG}\mathcal{H}_\varepsilon^*$,

$$2s(K_0)[h] = [h\langle 1, 1 \rangle^{r+1}] = [\varepsilon\text{-hyperbolic form over } KG]$$

by Th. 3.2.1(d) (p. 109). Therefore if $[V, h] - [U, f] \in \ker \mathrm{L}$, $2s(K_0)h$ and $2s(K_0)f$ are hyperbolic forms on isomorphic KG-modules and are therefore equivalent (Theorem 3.1.3), whence $2s(K_0) \ker \mathrm{L} = 0$.

If K is algebraically closed (and ${}^- = \mathrm{id}_K$), then $\ker \mathrm{L} = 0$ by Frobenius' Theorem (Cor. 4.4.1). Or this can also be proved using the methods below.

To prove the rest of the theorem, it suffices to consider the linearization map $\mathrm{L} : \hat{\mathrm{W}}_\varepsilon(K, A) \to \mathrm{K}_0(A)$ when A is a simple separable $(K, {}^-)$-involution algebra (Cor. 4.1.1). In fact since the category ${}^A_K\mathcal{H}_\varepsilon^*$ of nonsingular $(A, {}^-)$-invariant ε-reflexive forms over $(K, {}^-)$ is equivalent to the category ${}_A\mathcal{H}_\varepsilon^*$ of nonsingular ε-Hermitian forms over $(A, {}^-)$, by Th. 3.2.4, we can consider instead the linearization map $\mathrm{L} : \hat{\mathrm{W}}_\varepsilon(A, {}^-) \to \mathrm{K}_0(A)$.

This leads to the following three theorems (on fields of cohomological dimension ≤ 1, local fields, and global fields respectively) which are sharper versions of Th. 4.9.1 since the kernel of $\mathrm{L} : \hat{\mathrm{W}}_\varepsilon(A) \to \mathrm{K}_0(A)$ is often annihilated by a smaller power of 2 than indicated in that theorem.

We write $\operatorname{ann}(\ker \mathrm{L}) = m$ if the annihilator of $\ker \mathrm{L}$ is $m\mathbb{Z}$.

We can assume in the proofs that A is simple since otherwise $(A, ^-)$ is hyperbolic and $m = 1$ by Cor. 3.1.5 (p. 98). By Cor. 4.1.2 (p. 155), the kernel of $\mathrm{L} : \hat{\mathrm{W}}_{\varepsilon\varepsilon_0}(D, ^-) \to \mathrm{K}_0(D)$ is isomorphic to the kernel of $\mathrm{L} : \hat{\mathrm{W}}_{\varepsilon}(A, ^-) \to \mathrm{K}_0(A)$ (where $\varepsilon_0 = 1$ if $(D, ^-)$ and $(A, ^-)$ have the same type, otherwise $\varepsilon_0 = -1$), and so we can use the former in the proofs if convenient.

Theorem 4.9.2 *Let K_0 be a field of cohomological dimension ≤ 1 and characteristic $\neq 2$, $(A, ^-)$ a separable simple $(K, ^-)$-involution algebra with center L, and* $\mathrm{L} : \hat{\mathrm{W}}_{\varepsilon}(A) \to \mathrm{K}_0(A)$ *linearization. Then* $\ker \mathrm{L} = 0$ *if*

(a) $\dot{L}^2 = \dot{L}$, or

(b) $(A, ^-)$ is unitary or hyperbolic, or

(c) $\varepsilon = -1$ and $(A, ^-)$ is orthogonal, or

(d) $\varepsilon = 1$ and $(A, ^-)$ is symplectic.

Otherwise $\operatorname{ann}(\ker \mathrm{L}) = 2$.

In other words, $\operatorname{ann}(\ker \mathrm{L}) = 2$ *if $[\dot{L} : \dot{L}^2] = 2$, and either $\varepsilon = 1$ and $(A, ^-)$ is orthogonal or $\varepsilon = -1$ and $(A, ^-)$ is symplectic; otherwise* $\ker \mathrm{L} = 0$.

If K_1 is any finite separable extension of K_0, it also has cohomological dimension ≤ 1, $\mathrm{Br}(K_1) = 1$, and $N_{K_1/K_0} K_1 = K_0$ (cf. Prop. 5, p. 93, [67]). In particular the algebra A is quasisplit, $A \cong \mathrm{M}(n, L)$. Note that $s(K_0) = 1$ or 2: if -1 is not a square in K_0, then it is a sum of 2 squares since the norm in the quadratic extension $K_0(\sqrt{-1})/K_0$ is onto.

Suppose that $(A, ^-)$ is unitary, so $\varepsilon_0 = 1$ and the fixed subfield L_0 of L is a quadratic subfield. Any 1-dimensional ε-Hermitian form over $(L, ^-)$ is equivalent to $\langle 1 \rangle$ if it is Hermitian since the norm N_{L/L_0} is onto, or to $\langle \sqrt{\delta} \rangle$ where $L = L_0(\sqrt{\delta})$ if it is skew Hermitian. Therefore L is injective in either case, so $\ker \mathrm{L} = 0$. This proves the theorem in case (b).

Suppose then that the involution on L is the identity. If $\varepsilon\varepsilon_0 = -1$, the theorem follows at once in cases (c) and (d) since $\mathrm{L} : \hat{\mathrm{W}}_{-1}(L) \to \mathrm{K}_0(L)$ is injective.

If $\varepsilon\varepsilon_0 = 1$, we are dealing with symmetric bilinear forms over L. Claim: the equivalence class of any nonsingular symmetric form over L is determined by its dimension and determinant. Once this is proved, the theorem follows in this case since if h is any such form, the determinant of $h \perp h$ is 1 (up to squares), and so if $[h] - [g] \in \ker \mathrm{L}$, $2([h] - [g]) = 0$. Thus $2 \ker \mathrm{L} = 0$. And it is clear that $\ker \mathrm{L} = 0$ if and only if every element of $\dot{L}$ is a square since this is true if and only if each nonzero 1-dimensional symmetric form is equivalent to $\langle 1 \rangle$.

Proof of claim: Let h be a nonsingular symmetric form of rank 2. If $\det h =_2 -1$, then of course h is hyperbolic. If $\det h \neq_2 -1$, the norm in the quadratic extension $L(\sqrt{-\det h})/L$ is onto, so its norm form $\langle 1, \det h \rangle$ is universal and it follows that h is equivalent to $\langle 1, \det h \rangle$. Thus every nonsingular binary symmetric form over L is equivalent to $\langle 1, \det h \rangle$, and by induction a nonsingular symmetric form h of rank n is equivalent to $\langle 1, \ldots, 1, \det h \rangle$, whence the claim. □

Corollary 4.9.2 *Suppose that the cohomological dimension of K_0 is ≤ 1 and that* $\mathrm{L} : \hat{\mathrm{W}}_{\varepsilon}(K, G) \to \mathrm{K}_0(KG)$ *is the linearization map. Then* $\ker \mathrm{L} = 0$ *if the center of each orthogonal respectively symplectic component of $(KG, ^-)$ is quadratically closed ($\dot{L}^2 = \dot{L}$) when $\varepsilon = 1$ respectively -1. Otherwise* $\operatorname{ann}(\ker \mathrm{L}) = 2$.

In particular two isometry representations of G are equivalent if they are linearly equivalent and, in the case of an orthogonal respectively symplectic representation, the center of each orthogonal respectively symplectic component of $(KG, ^-)$ is quadratically closed.

The first part amounts to another restatement of the theorem, and the second follows easily from it. □

Theorem 4.9.3 *Let K_0 be a local field of characteristic $\neq 2$, $(A, ^-)$ a separable simple $(K, ^-)$-involution algebra with center L, and $\mathrm{L} : \hat{\mathrm{W}}_\varepsilon(A) \to \mathrm{K}_0(A)$ linearization.*

(a) If $(A, ^-)$ is hyperbolic, $\ker \mathrm{L} = 0$.
(b) If $(A, ^-)$ is unitary, $\mathrm{ann}\ker \mathrm{L} = 2$.
(c) If $(A, ^-)$ is orthogonal and $\varepsilon = -1$, or $(A, ^-)$ is symplectic and $\varepsilon = 1$, then $\ker \mathrm{L} = 0$.
(d) If $(A, ^-)$ is orthogonal and $\varepsilon = 1$, or $(A, ^-)$ is symplectic and $\varepsilon = -1$, then $\mathrm{ann}\ker \mathrm{L} = 4$ *if A is quasisplit and $-1 \notin \dot{L}^2$, otherwise* $\mathrm{ann}\ker \mathrm{L} = 2$.

(a) is clear so assume that A is simple, say $A = \mathrm{M}(n, D)$. We deal with $\ker \mathrm{L} : \hat{\mathrm{W}}_{\varepsilon\varepsilon_0}(D) \to \mathrm{K}_0(D)$ as before. Let $m = \mathrm{ann}\ker \mathrm{L}$.

If $(A, ^-)$ is unitary, then A is quasisplit (cf. p. 34); by the table on p. 61 an Hermitian form h over $(L, ^-)$ is determined by its dimension and determinant $\in \dot{L}_0/\mathrm{N}_{L/L_0}\dot{L}$. Since $\dot{L}_0^2 \subset \mathrm{N}_{L/L_0}(\dot{L})$, the determinant of $h \perp h$ is 1, and so $2([h] - [g]) = 0$ if $[h] - [g] \in \ker \mathrm{L}$. Since the norm index $[\dot{L}_0 : \mathrm{N}_{L/L_0}\dot{L}]$ is > 1 (namely 2 – see p. 18), it follows easily that $\ker \mathrm{L} \neq 0$, so $m = 2$. Since skew Hermitian forms are scalar multiples of Hermitian forms, the same holds for them, and (b) is proved.

We may now assume that $(A, ^-)$ is orthogonal or symplectic.

Suppose that A is not quasisplit, i.e. D is a quaternion algebra (cf. p. 34). By the table on p. 61, the equivalence of skew Hermitian forms over $(D, ^*)$ is determined by the dimension and the determinant $\in \dot{L}/\dot{L}^2$, so by the same argument as above, $2\ker \mathrm{L} = 0$. Since there are forms of all determinants (in $\dot{L}/\dot{L}^2$) when the dimension is > 1 (cf. §1.5.7), $m = 2$. For Hermitian forms, only the dimension is needed so $\ker \mathrm{L} = 0$, i.e. $m = 1$. This proves (c) and (d) when A is not quasisplit. (Note that $\varepsilon_0 = -1$ if $(A, ^-)$ is orthogonal since "conjugate transpose" is symplectic).

Therefore we assume that A is quasisplit, as well as orthogonal or symplectic. Thus $\varepsilon_0 = 1$ if $(A, ^-)$ is orthogonal and $= -1$ if it is symplectic. If $\varepsilon\varepsilon_0 = -1$, $\hat{\mathrm{W}}_{\varepsilon\varepsilon_0}(L) \to \mathrm{K}_0(L)$ is injective since the forms are skew symmetric, so $m = 1$. This finishes the proof of (c).

Suppose then that $\varepsilon\varepsilon_0 = 1$ (case (d)), so the forms are symmetric. Certainly $\ker \mathrm{L} \neq 0$, so $m \geq 2$. The form $\langle\alpha, \alpha, \alpha, \alpha\rangle$ has determinant 1 and Hasse invariant 1, and so $4\ker \mathrm{L} = 0$ and $m = 2$ or 4. If $-1 \in \dot{L}^2$, the forms $\langle\alpha, \alpha\rangle$ and $\langle\beta, \beta\rangle$ are equivalent since the Hasse invariant of $\langle\alpha, \alpha\rangle$ is $(\alpha, \alpha)_L = (\alpha, -1)_L = 1$, and so in this case $2\ker \mathrm{L} = 0$, i.e. $m = 2$. If -1 is not a square in L, then there is $\alpha \in \dot{L}$ such that $(\alpha, -1)_L = -1$ (see the examples in §1.2.5) and so $2([\langle\alpha\rangle] - [\langle 1\rangle]) \neq 0$ and $m = 4$. □

Corollary 4.9.3 *Let K_0 be a local field and $\mathrm{L} : \hat{\mathrm{W}}_\varepsilon(G, K) \to \mathrm{K}_0(KG)$ linearization. Then* *(a)* $\ker \mathrm{L} = 0$ *if and only if*

(i) $\varepsilon = -1$ *and*

(ii) all simple components of $(KG, ^-)$ *are orthogonal or hyperbolic.*

When (ii) holds, two symplectic representations $G \to \mathbf{Sp}(n, K)$ *are equivalent if and only if they are linearly equivalent.*

(b) $\operatorname{ann}(\ker \mathrm{L}) = 2$ *if and only if*

(i) $K \neq K_0$, *or*

(ii) $\varepsilon = 1$ *and* $-1 \in \dot{K}^2$, *or*

(iii) $\varepsilon = -1$, $(KG, ^-)$ *has at least one symplectic or unitary component, and* -1 *is a square in the center of each quasisplit symplectic component.*

(c) $\operatorname{ann}(\ker \mathrm{L}) = 4$ *if and only if* $(KG, ^-)$ *has at least one quasisplit component* $(A, ^-)$ *of type* ε *such that* -1 *is not a square in its center* L, *that is to say such that either*

(i) $\varepsilon = 1$, $(A, ^-)$ *is orthogonal, and* $-1 \notin \dot{L}^2$, *or*

(ii) $\varepsilon = -1$, $(A, ^-)$ *is symplectic, and* $-1 \notin \dot{L}^2$.

(a) By case (b) of the theorem, no component $(A, ^-)$ of $(KG, ^-)$ can be unitary if $\ker \mathrm{L} = 0$ – in particular $(K, ^-) = (K, \mathrm{id})$ since otherwise the trivial component $(K, ^-)$ of $(KG, ^-)$ would be unitary. By (d), $\ker \mathrm{L} \neq 0$ if $\varepsilon = 1$ since the trivial component (K, id) of $(KG, ^-)$ is orthogonal. If $\varepsilon = -1$, the kernel of $\hat{\mathrm{W}}_{-1}(K, G) \to \mathrm{K}_0(KG)$ is 0 if and only if $(KG, ^-)$ has only orthogonal and hyperbolic components by all four parts of the theorem.

(b) If $K \neq K_0$, all simple components of $(KG, ^-)$ are unitary or hyperbolic, and (b) follows in this case since at least one – namely the trivial component $(K, ^-)$ – is unitary. Suppose that $K = K_0$. If $\varepsilon = 1$, then since the trivial component (K, id) is orthogonal, it is necessary and sufficient by (d) of the theorem that $-1 \in \dot{L}^2$. If $\varepsilon = -1$, it is clear that the condition stated in (b)(iii) is necessary and sufficient for $\operatorname{ann} \ker \mathrm{L} = 2$.

(c) Since $4 \ker \mathrm{L} = 0$ by the theorem, $\operatorname{ann}(\ker \mathrm{L}) = 4$ if and only if $(KG, ^-)$ has at least one component $(A, ^-)$ for which the annihilator of the kernel of linearization is 4, so (c) follows at once from the theorem. □

Remark. Note that if $(K, ^-)$ is of the second kind, then $\operatorname{ann} \ker \mathrm{L} = 2$ for $\varepsilon = 1$ and $\varepsilon = -1$.

Theorem 4.9.4 *Let* K/K_0 *be an indefinite extension*[9] *of global fields of characteristic* $\neq 2$, $(A, ^-)$ *a separable simple* $(K, ^-)$*-involution algebra with center* L, *and* $\mathrm{L} : \hat{\mathrm{W}}_\varepsilon(A, ^-) \to \mathrm{K}_0(A)$ *linearization. Then*

(a) If $(A, ^-)$ *is hyperbolic,* $\ker \mathrm{L} = 0$.

(b) If $(A, ^-)$ *is unitary,* $\operatorname{ann} \ker \mathrm{L} = 2$.

(c) If $(A, ^-)$ *is orthogonal and* $\varepsilon = -1$, *or* $(A, ^-)$ *is symplectic and* $\varepsilon = 1$, *then* $\ker \mathrm{L} = 0$.

(d) If $(A, ^-)$ *is orthogonal and* $\varepsilon = 1$, *or* $(A, ^-)$ *is symplectic and* $\varepsilon = -1$, *then*

$$\operatorname{ann} \ker \mathrm{L} = \begin{cases} 2 & \text{if } -1 \in \dot{L}^2, \\ 4 & \text{if } -1 \notin \dot{L}^2. \end{cases}$$

[9] cf. p. 41

(a) is clear (Cor. 3.1.5, p. 98) so assume A is simple – we may suppose that A is a matrix algebra $\mathrm{M}(n,D)$ over a division algebra D.

If $\mathfrak{p}$ runs over the primes of L_0 (the fixed field of the involution of A restricted to L), then[10]

$$\begin{array}{ccccc} \ker \mathrm{L} & \longrightarrow & \hat{\mathrm{W}}_\varepsilon(A) & \xrightarrow{\ \mathrm{L}\ } & \mathrm{K}_0(A) \\ \big\downarrow & & \big\downarrow{\scriptstyle\theta} & & \big\downarrow{\scriptstyle\eta} \\ \prod_{\mathfrak{p}}\ker \mathrm{L}_{\mathfrak{p}} & \longrightarrow & \prod_{\mathfrak{p}}\hat{\mathrm{W}}_\varepsilon(A_{\mathfrak{p}}) & \xrightarrow{\prod \mathrm{L}_{\mathfrak{p}}} & \prod_{\mathfrak{p}}\mathrm{K}_0(A_{\mathfrak{p}}) \end{array}$$

is commutative. Note that η is injective since in fact $\mathrm{K}_0(A) \to \mathrm{K}_0(A_{\mathfrak{p}})$ is injective for each $\mathfrak{p}$. If the Hasse Principle (Th. 3.3.1, p. 135) applies, i.e. if we are not in a situation which is "Morita equivalent" to skew Hermitian forms over a quaternion algebra (with the standard involution) which is nonsplit at more than two primes, then θ is also injective, and it follows that the annihilator m of $\ker \mathrm{L}$ is a divisor of the annihilator of $\prod_{\mathfrak{p}} \ker \mathrm{L}_{\mathfrak{p}}$ which is the maximum of the $m_{\mathfrak{p}}$ (the annihilator of $\ker \mathrm{L}_{\mathfrak{p}}$):

$$m \mid \max_{\mathfrak{p}} m_{\mathfrak{p}} \quad \text{if the Hasse Principle applies,} \tag{4.40}$$

since, as we shall presently see, $\mathfrak{m}_p$ is $1, 2$ or 4.

By Th. 4.9.3

$$m_{\mathfrak{p}} = 1, 2 \text{ or } 4 \text{ if } \mathfrak{p} \text{ is nonArchimedean.}$$

We now show that

$$m_{\mathfrak{p}} = 1 \text{ if } \mathfrak{p} \text{ is Archimedean.} \tag{4.41}$$

If $L = L_0$, then $K = K_0$ and $\mathfrak{p}$ is complex since K/K_0 is assumed to be indefinite. Thus

$$(A_{\mathfrak{p}}, {}^-) \cong (L_{0\mathfrak{p}}, \mathrm{id}) \cong (\mathbb{C}, \mathrm{id}) \otimes_L (A, {}^-) \cong (\mathrm{M}(n,\mathbb{C}), {}^-)$$

for some involution $^-$ on $\mathrm{M}(n,\mathbb{C})$. Since $(\mathrm{M}(n,\mathbb{C}), {}^-)$ is of the first kind, forms over it are Morita equivalent to forms over $(\mathbb{C}, \mathrm{id})$, so $\mathrm{L}_{\mathfrak{p}}$ is injective by Cor. 4.1.2 (p. 155), i.e. $m_{\mathfrak{p}} = 1$.

On the other hand if $L \neq L_0$, $L = L_0(\sqrt{\delta})$ say, then either $\mathfrak{p}$ is complex or it is real and $\delta > 0$ at $\mathfrak{p}$. In the first case

$$L_{0\mathfrak{p}} = \mathbb{C} \text{ and } (L_{\mathfrak{p}}, {}^-) = (\mathbb{C}, \mathrm{id}) \otimes_{(L_0, \mathrm{id})} (L, {}^-) \cong \big((\mathbb{C}, \mathrm{id}) \oplus (\mathbb{C}, \mathrm{id}), \ \leftrightarrow\ \big),$$

(Lem. 1.4.2(a)) and therefore

$$(A_{\mathfrak{p}}, {}^-) = (L_{\mathfrak{p}}, {}^-) \otimes_{(L, {}^-)} (A, {}^-) \cong ((\mathbb{C} \otimes A) \oplus (\mathbb{C} \otimes A), \ {}^\sim\ )$$

where the involution $^\sim$ switches the two components. Since such an involution algebra is hyperbolic (Ex. 8, p. 26), the linearization $\mathrm{L}_{\mathfrak{p}} : \hat{\mathrm{W}}_\varepsilon(A_{\mathfrak{p}}, {}^-) \to \mathrm{K}_0(A_{\mathfrak{p}})$ is injective. In the second case $L_{0\mathfrak{p}} = \mathbb{R}$ and $(L_{\mathfrak{p}}, {}^-) \cong \big((\mathbb{R}, \mathrm{id}) \oplus (\mathbb{R}, \mathrm{id}), \leftrightarrow\big)$ (Lem. 1.4.2(b)), so

$$(A_{\mathfrak{p}}, {}^-) \cong \big((\mathbb{R} \otimes A) \oplus (\mathbb{R} \otimes A), \ {}^\sim\ \big)$$

where again $^\sim$ switches the components, and so $\mathrm{L}_{\mathfrak{p}}$ is injective. This proves (4.41).

Suppose now that A is quasisplit, $A \cong \mathrm{M}(n, L)$ say, and that $\mathfrak{p}$ is a prime of L_0 such that $L_{\mathfrak{p}}$ is a field. Then $A_{\mathfrak{p}} \cong \mathrm{M}(n, L_{\mathfrak{p}})$ and $\mathrm{K}_0(A) \to \mathrm{K}_0(A_{\mathfrak{p}})$ is an

[10] For the notation $A_{\mathfrak{p}}, L_{\mathfrak{p}}$, see pp. 134*ff.* $\mathrm{L}_{\mathfrak{p}}$ is of course the linearization $\hat{\mathrm{W}}_\varepsilon(A_{\mathfrak{p}}) \to \mathrm{K}_0(A_{\mathfrak{p}})$.

isomorphism. It follows by Th. 3.4.1 (p. 140) that $\hat{\mathrm{W}}_\varepsilon(A, ^-) \overset{L_\mathfrak{p}\otimes_L}{\longrightarrow} \hat{\mathrm{W}}_\varepsilon(A_\mathfrak{p}, ^-)$ is surjective. Since

$$\begin{array}{ccc} \hat{\mathrm{W}}_\varepsilon(A, ^-) & \longrightarrow & \hat{\mathrm{W}}_\varepsilon(A_\mathfrak{p}, ^-) \\ {\scriptstyle \mathrm{L}}\downarrow & & \downarrow{\scriptstyle \mathrm{L}_\mathfrak{p}} \\ \mathrm{K}_0(A) & \xrightarrow{\sim} & \mathrm{K}_0(A_\mathfrak{p}) \end{array}$$

is commutative, we see that

$$m_\mathfrak{p} | m \quad \text{for all primes } \mathfrak{p} \text{ of } L_0 \text{ if } A \text{ is quasisplit} \tag{4.42}$$

since if $L_\mathfrak{p}$ is not a field, $A_\mathfrak{p}$ is hyperbolic and so $m_\mathfrak{p} = 1$.

$(A, ^-)$ unitary, say $L = L_0(\sqrt{\delta})$. There are infinitely many $\mathfrak{p}$ for which $\delta \notin L_{0\mathfrak{p}}^2$ by the Global Square Theorem. Let $\mathfrak{p}$ be such a prime. Then $L_{0\mathfrak{p}}(\sqrt{\delta}) = L_\mathfrak{p}$ is the completion of L at the unique prime of L lying over $\mathfrak{p}$, and $A_\mathfrak{p} = L_\mathfrak{p} \otimes_L A$ is a simple algebra.

If A is quasisplit, it follows from Th. 4.9.3(b), (4.40) and (4.42) that $m = 2$ (since if $\delta \in L_\mathfrak{p}^2$, $(A_\mathfrak{p}, ^-)$ is hyperbolic so $m_\mathfrak{p} = 1$).

If A is *not* quasisplit (and unitary), say $A = \mathrm{M}(n, D)$ where D is a central L-division algebra, let $^-$ be an $(L, ^-)$-involution on D. The diagram

$$\begin{array}{ccc} \hat{\mathrm{W}}_{-1}(A, ^-) & & \\ {\scriptstyle \sim}\downarrow & \searrow & \\ \hat{\mathrm{W}}_1(A, ^-) & \longrightarrow & \mathrm{K}_0(A), \end{array}$$

where the vertical map takes the class $[h]$ to the class $[\sqrt{\delta}h]$, is commutative, so we can assume that $\varepsilon = 1$. Since $\varepsilon_0 = 1$ also (because the involutions are unitary), ${}_{(A,^-)}\mathcal{H}_1$ is Morita equivalent to ${}_{(D,^-)}\mathcal{H}_1$, so by Cor. 4.1.2 (p. 155), we can take $A = D$. Since there are nonequivalent Hermitian forms over $(D, ^-)$ of the same rank (cf. §10.6, [65]), $m > 1$. But $m_\mathfrak{p} \leq 2$ for all $\mathfrak{p}$ (Th. 4.9.3(b)), so $m = 2$ by (4.40).

$(A, ^-)$ orthogonal or symplectic. (c) follows at once from $m | \max_\mathfrak{p} m_\mathfrak{p}$, which is $= 1$ by Th. 4.9.3(c) and the fact that $m_\mathfrak{p} = 1$ for Archimedean primes.

It remains to prove (d). Suppose first that A is quasisplit. Then $m = \max_\mathfrak{p} m_\mathfrak{p}$ by (4.40) and (4.42). If $-1 \in \dot{L}^2$, $\max_\mathfrak{p} m_\mathfrak{p} = 2$ since $-1 \in L_\mathfrak{p}^2$ for all $\mathfrak{p}$, so $m = 2$. If $-1 \notin \dot{L}^2$, let $\mathfrak{p}$ be a nonArchimedean prime at which $-1 \notin \dot{L}_\mathfrak{p}^2$. Then $m_\mathfrak{p} = 4$ by Th. 4.9.3(d).

Assume that A is nonsplit. Then under the assumptions of (d), the two categories in question are Morita equivalent to ${}_{(D,\,^*)}\mathcal{H}_{-1}^*$ (where $(D, ^*)$ is a nonsplit quaternion algebra with the standard involution) since $\varepsilon_0 = -1$, and the desired result follows from the following lemma:

Lemma 4.9.1 *Let K be a global field, and $(D, ^*)$ a standard quaternion involution algebra with center K. Let $g, h \in {}_{(D,\,^*)}\mathcal{H}_{-1}^*$ have the same rank. Then*

(a)

$$g \perp g \perp g \perp g \cong h \perp h \perp h \perp h.$$

(b) g and h can be chosen so that $g \perp g \not\cong h \perp h$ if and only if $-1 \notin \dot{K}^2$.

It suffices to show this for $g = \langle d \rangle$, $h = \langle e \rangle$ where d, e are nonzero skew elements of $(D,\ ^*\ )$.

We use the Bartels invariant $\mathrm{b}(g,h) \in \mathrm{Br}(K)/\langle [D] \rangle$ where g, h are nondegenerate skew Hermitian forms of the same rank n over $(D,\ ^*\ )$. It has the following properties (cf. Satz 4, p. 283, [3]):

(i) $\mathrm{b}(g,h)^2 = 1 = \mathrm{b}(g,g)$,
(ii) $\mathrm{b}(h,g) = \mathrm{b}(g,h)$,
(iii) $\mathrm{b}(g,h)\mathrm{b}(h,f) = \mathrm{b}(g,f)$,
(iv) $\mathrm{b}(g \perp f, h \perp f) = \mathrm{b}(g,h)$,
(v) $\mathrm{b}(g, \alpha g) = [(\alpha,\gamma)_K)]\langle [D] \rangle$.

Here $\gamma \dot{K}^2$ is the "discriminant" $(-1)^{n(n-1)/2}(\det g)\dot{K}^2$ of g, and $\alpha \in \dot{K}$. (These properties are stated in [3] for "quadratic forms" over $(D,\ ^*\ )$ but since the characteristic is $\neq 2$, they are valid for skew Hermitian forms as well).

Thus

$$\begin{aligned} &\mathrm{b}(\langle d,d,d,d \rangle, \langle e,e,e,e \rangle) \cdot \mathrm{b}\left(\langle d,d \rangle, \langle e,e \rangle\right) \\ &= \mathrm{b}\left(\langle d,d,d,d \rangle, \langle e,e,e,e \rangle\right) \mathrm{b}(\langle d,d,e,e \rangle, \langle e,e,e,e \rangle) \\ &= \mathrm{b}(\langle d,d,d,d \rangle, \langle d,d,e,e \rangle) \\ &= \mathrm{b}(\langle d,d \rangle, \langle e,e \rangle). \end{aligned}$$

Therefore $\mathrm{b}(\langle d,d,d,d \rangle, \langle e,e,e,e \rangle) = 1$ and (a) follows at once by Bartel's Theorem (p. 63).

Now (b). If $-1 \in \dot{K}^2$, then $\langle d,d \rangle \cong \langle d,-d \rangle$ which is hyperbolic, so $\langle d,d \rangle \cong \langle e,e \rangle$.

Suppose $-1 \notin \dot{K}^2$. By the Global Square Theorem, -1 is a nonsquare at an infinite number of primes of K, and by 71:19a, p. 204, [54], we can find $\alpha \in \dot{K}$ such that $(\alpha,-1)_K$ is nonsplit at a prime at which D is split. It follows that if d is any nonzero pure quaternion, then $\langle d,d \rangle$ and $\alpha\langle d,d \rangle = \langle \alpha d, \alpha d \rangle$ are not equivalent since both have discriminant -1 and

$$\mathrm{b}(\langle d,d \rangle, \alpha\langle d,d \rangle) = [(\alpha,-1)_K]\langle [D] \rangle \neq \langle [D] \rangle = \mathrm{b}(\langle d,d \rangle, \langle d,d \rangle). \qquad \square$$

This also finishes the proof of Th. 4.9.4. $\square$

Exercise. This exercise is an alternative way to show a slightly weaker result than Th. 4.9.4(d) when A is nonsplit – i.e. when A is similar to a quaternion algebra D. Let $M = L(\sqrt{\delta})$ be the quadratic extension, with nontrivial involution $^-$, where $\delta = -1$ if $-1 \notin \dot{L}^2$, and is otherwise an arbitrary nonsquare in L. Let t be the trace $\frac{1}{2}\mathrm{Tr}_{M/L}$ and $b : M \times M \to L$ the symmetric bilinear form $b(\mu,\nu) = \mathrm{t}(\mu\bar{\nu})$, equivalent to $\langle 1, -\delta \rangle$.

(a) Show that $b \perp b \cong \langle 1,1,1,1 \rangle$.

(b) Let $(D^M, ^-) = (D,\ ^*\ ) \otimes_{(L,\mathrm{id})} (M, ^-)$ and $\tau : D^M \to D$ the map determined by $\tau(d \otimes \mu) = \mathrm{t}(\mu)d$. Show that $(D^M, ^-)$ is an involution algebra over $(M, ^-)$ of the second kind, and that if $h \in {}_{(D^M, ^-)}\mathcal{H}_{-1}$, then $\tau \circ h = \tau_*(h) \in {}_{(D,\ ^*\ )}\mathcal{H}_{-1}$.

(c) Show that if $a \in D^M$, $a \neq 0$, then either $\tau a \neq 0$ or $\tau(\sqrt{\delta}a) \neq 0$. Deduce that h nonsingular implies that $\tau_* h$ is nonsingular.

(d) Show that the resulting homomorphism $\tau_* : \hat{\mathrm{W}}_{-1}(D^M) \to \hat{\mathrm{W}}_{-1}(D)$ makes the diagram

$$\begin{array}{ccc} \hat{\mathrm{W}}_{-1}(D) & \longrightarrow & \hat{\mathrm{W}}_{-1}(D \otimes M) \\ {\scriptstyle \times[b]} \big\downarrow & \swarrow {\scriptstyle \tau_*} & \\ \hat{\mathrm{W}}_{-1}(D) & & \end{array}$$

commutative, where $\times[b]$ is the map $[h] \to [hb]$ and the horizontal map is extension of scalars, $[h] \to [h^M]$. Show also that (a) implies that $2[hb] = 4[h]$.

(e) Use Th. 4.9.4(b) to show that the kernel of $\hat{\mathrm{W}}_{-1}(D^M, {}^-) \to \mathrm{K}_0(D^M)$ is annihilated by 2, so that $2[h^M]$ depends only on the D-space of h, and therefore the same is true of $2\tau_*[h^M] = 4[h]$. □

Corollary 4.9.4 *Let K/K_0 be an indefinite extension of global fields and $\mathrm{L}: \hat{\mathrm{W}}_\varepsilon(K,G) \to \mathrm{K}_0(KG)$ linearization. Then*

(a) $\ker \mathrm{L} = 0$ *if and only if $\varepsilon = -1$ and all simple components of $(KG, {}^-)$ are orthogonal or hyperbolic. Thus if all simple components of $(KG, {}^-)$ are orthogonal or hyperbolic, two symplectic representations $G \to \mathbf{Sp}(n, K)$ are equivalent if and only if they are linearly equivalent.*

(b) $\mathrm{ann}(\ker \mathrm{L}) = 2$ *if and only if*

(i) $K \neq K_0$, or

(ii) $\varepsilon = 1$ and $-1 \in \dot{K}^2$, or

(iii) $\varepsilon = -1$ and $(KG, {}^-)$ has at least one symplectic or unitary component, and -1 is a square in the center of each quasisplit symplectic component.

(c) $\mathrm{ann}(\ker \mathrm{L}) = 4$ *if and only if $K = K_0$ and either $\varepsilon = 1$ and $-1 \notin \dot{K}^2$ or $\varepsilon = -1$ and $(KG, {}^-)$ has at least one symplectic component such that -1 is a nonsquare in its center.*

(a) If $(K, {}^-)$ is of the second kind, the trivial component $(K, {}^-)$ of $(KG, {}^-)$ is unitary and, by (b) of the theorem, gives rise to a nontrivial kernel under linearization. If it is of the first kind, i.e. the trivial component $(K, {}^-)$ is orthogonal, the same is true by (d) when $\varepsilon = 1$. Therefore $\ker \mathrm{L} = 0$ implies that $(K, {}^-) = (K, \mathrm{id}_K)$ and $\varepsilon = -1$, and then that all simple components of $(KG, {}^-)$ must be hyperbolic or orthogonal. Since the converse obviously holds as well, this proves the first part of (a), and the second part follows at once.

(b) Certainly $\mathrm{ann}(\ker \mathrm{L}) = 2$ if $K \neq K_0$ since then all simple components of $(KG, {}^-)$ are unitary or hyperbolic, and at least one – the trivial component – is unitary.

So assume in the rest of the proof that $K = K_0$ – thus K is not formally real, so totally imaginary when $\mathrm{char} K = 0$.

If $\varepsilon = 1$, and $-1 \in \dot{K}^2$, the kernel of linearization for any symplectic component is 0 by (c) of the theorem, and for any orthogonal component – e.g. K – has annihilator 2 by (d). This proves the sufficiency of (b)(ii).

Now suppose that $\varepsilon = -1$ and that not every simple component of $(KG, {}^-)$ is hyperbolic or orthogonal (the case in (a)). Thus there must be unitary and/or symplectic components, and so by (b) and (d) of the theorem, the annihilator of $\ker \mathrm{L}$ must be 2, assuming that -1 is a square in the center of any symplectic component.

The only remaining cases are those described in (c). If $\varepsilon = 1$ and $-1 \notin \dot{K}^2$, then the annihilator of $\ker \mathrm{L}$ on the component $\hat{\mathrm{W}}_1(K)$ of $\hat{\mathrm{W}}_1(KG)$ is 4 by the theorem. Finally if $\varepsilon = -1$ and $(KG, ^-)$ has at least one symplectic component such that -1 is a nonsquare in its center, then it once more follows easily from the theorem that the annihilator of linearization is 4.

This proves the sufficiency of the conditions in each of (a), (b) and (c); since these conditions cover all possible cases, they are also necessary. □

4.10 Structure of $\hat{\mathrm{W}}_\varepsilon(K,G)$ when K/K_0 is a definite extension of algebraic number fields

When K_0 is formally real, the results of the previous section do not hold in general. For example if K is a formally real field with involution id_K and we take the trivial representations $\rho_\alpha : G \to \mathbf{O}(K, \langle\alpha\rangle)$ for $\alpha = 1$ and -1, then $[\rho_1] - [\rho_{-1}]$ is in the kernel of $\mathrm{L} : \hat{\mathrm{W}}_1(K,G) \to \mathrm{K}_0(K)$ but $n([\rho_1] - [\rho_{-1}]) \neq 0$ for any positive integer n since the underlying form of $n\rho_1$ is $\langle 1, 1, \ldots, 1\rangle$ and that of $n\rho_{-1}$ is $\langle -1, -1, \ldots, -1\rangle$.

When K/K_0 is a definite quadratic extension, it is clear that there is a similar example involving unitary representations.

Thus in order to get an analogous theory when K/K_0 is definite – K_0 formally real or K/K_0 a definite quadratic extension, it is necessary (at least) to take the orderings of K_0 into account. This is done by enlarging the "target group" $\mathrm{K}_0(KG)$ to a group which does that, namely

$$\mathrm{K}_0(KG) \times \prod_{P \in X_d} \hat{\mathrm{W}}_\varepsilon(K_P, G) \tag{4.43}$$

where X_d is the set of all orderings of K_0 for which K/K_0 is definite and $K_P = K_{0P} \otimes_{K_0} K$. This means that X_d consists of *all* orderings of K_0 if $K = K_0$, while if $K \neq K_0$ – say $K = K_0(\sqrt{\delta})$ – it consists of those orderings P of K_0 at which $\delta < 0$. Thus K_P is a field when $P \in X_d$, namely a real closure $\mathbf{R}$ in the first case and its algebraic closure $\mathbf{R}(\sqrt{\delta})$ in the second.

The essential point is that $\hat{\mathrm{W}}_\varepsilon(K_P, G)$ "detects" the equivariant classes at P – cf. Th. 4.5.5, p. 177 when $\varepsilon = 1$, and Th. 4.5.6, p. 180 when $\varepsilon = -1$.

It can happen of course that X_d is empty even if K_0 is formally real – when $\delta > 0$ for all orderings of K_0 (and $K \neq K_0$). The proof of Th. 4.10.1 is valid in this case as well, if we interpret the empty product $\prod_{P \in X_d} \hat{\mathrm{W}}_\varepsilon(K_P, G)$ as the one element group.

As usual, if P is an arbitrary ordering of K_0,

$$(K_P, ^-) = (K_{0P}, \mathrm{id}) \otimes_{(K_0, \mathrm{id})} (K, ^-),$$

so

$$(K_P, ^-) = \begin{cases} (\mathbf{R}, \mathrm{id}_{\mathbf{R}}) & \text{if } K = K_0, \\ (\mathbf{C}, {}^*) & \text{if } (K : K_0) = 2 \text{ and } \delta < 0 \text{ at } P, \\ (\mathbf{R} \oplus \mathbf{R}, \leftrightarrow) & \text{if } (K : K_0) = 2 \text{ and } \delta > 0 \text{ at } P, \end{cases}$$

where $\mathbf{R}$ is the real closure K_{0P}, $\mathbf{C} = \mathbf{R}(i)$ its algebraic closure, and $\leftrightarrow$ is the interchange involution (Ex. 9, p. 26).

We can choose a target group which is slightly smaller than (4.43), namely the *signature group* $\mathrm{S}_\varepsilon(K,G)$ which is defined to be the pullback

$$\Big\{(x,y) \in \mathrm{K}_0(KG) \times \prod_{P\in X_d} \hat{\mathrm{W}}(K_P,G) : \ \operatorname{im} x = \operatorname{im} y \ \text{ in } \prod_{P\in X_d} \mathrm{K}_0(K_PG)\Big\}$$

in the rectangle in the commutative diagram

$$\begin{array}{ccc}
\hat{\mathrm{W}}_\varepsilon(K,G) & & \\
 & \searrow^{\sigma_\varepsilon} \quad \searrow & \\
 & \mathrm{S}_\varepsilon(K,G) \longrightarrow & \prod_{P\in X_d} \hat{\mathrm{W}}_\varepsilon(K_P,G) \\
 \searrow & \downarrow & \downarrow \\
 & \mathrm{K}_0(KG) \longrightarrow & \prod_{P\in X_d} \mathrm{K}_0(K_PG)
\end{array} \tag{4.44}$$

The maps are the obvious ones, with the *signature (map)* σ_ε, which replaces the linearization L, defined by the universal property of the pullback. (If X_d is empty, then $\mathrm{S}_\varepsilon(K,G) = \mathrm{K}_0(KG)$ and $\sigma_\varepsilon = \mathrm{L}$ – thus the signature is a generalization of the linearization).

Our goal is to prove the following theorem:

Theorem 4.10.1 *Let K be an algebraic number field (of finite degree over $\mathbb{Q}$) such that K/K_0 is definite. Then*
(a) $4\ker\sigma_\varepsilon = 0$,
(b) if $-1 \in \dot{K}^2$, *then* $2\ker\sigma_\varepsilon = 0$,
(c) σ_ε *is onto.*

Since $\mathrm{S}_\varepsilon(K,G)$ is a subgroup of $\mathrm{K}_0(KG)\times\prod_{P\in X_d}\hat{\mathrm{W}}_\varepsilon(K_P,G)$ which is a finitely generated free Abelian group by Ths. 4.5.5 (p. 177) and 4.5.6, we get the corollary:

Corollary 4.10.1 $\hat{\mathrm{W}}_\varepsilon(K,G)$ *is the direct sum of a finitely generated free Abelian group and its torsion subgroup* $\hat{\mathrm{W}}_\varepsilon(K,G)_{\mathrm{tor}}$ *which is annihilated by 4 (or by 2 if* $-1 \in \dot{K}^2$*).* □

Note that (a) and (b) continue to hold if K/K_0 is indefinite by Th. 4.9.1 (p. 196). On the other hand the image of $\sigma_\varepsilon = \mathrm{L}$ in the indefinite case – in fact for an arbitrary K/K_0, whether global fields or not – depends on the nature of the simple involution components of $(KG, {}^-)$, and is described in Remark 3., p. 196.

As a first step in proving the theorem, we use Prop. 1.4.1 (p. 42) to replace $\hat{\mathrm{W}}_\varepsilon(K_P,G)$ by $\hat{\mathrm{W}}_\varepsilon(K_\mathfrak{p},G)$ and $\mathrm{K}_0(K_PG)$ by $\mathrm{K}_0(K_\mathfrak{p}G)$ in (4.44) where $\mathfrak{p}$ is the real prime of K_0 corresponding to P. That is to say, we replace the real closure K_{0P} of K_0 by the real completion $K_{0\mathfrak{p}}$ and put

$$(K_\mathfrak{p}, {}^-) = (K_{0\mathfrak{p}}, \mathrm{id}) \otimes_{(K_0,\mathrm{id})} (K, {}^-).$$

The nature of $(K_\mathfrak{p}, {}^-)$ is of course analogous to that of P:

$$(K_\mathfrak{p}, {}^-) = \begin{cases} (\mathbb{R}, \mathrm{id}_K) & \text{if } K = K_0, \\ (\mathbb{C}, {}^*) & \text{if } (K:K_0) = 2 \text{ and } \delta < 0 \text{ at } \mathfrak{p}, \\ (\mathbb{R}\oplus\mathbb{R}, \leftrightarrow) & \text{if } (K:K_0) = 2 \text{ and } \delta > 0 \text{ at } \mathfrak{p}, \end{cases}$$

where, as before, the entry $(\mathbb{C},\ ^*\ )$ corresponds to K/K_0 being a definite quadratic extension at $\mathfrak{p}$ and $(\mathbb{R}\oplus\mathbb{R},\leftrightarrow)$ to K/K_0 indefinite at $\mathfrak{p}$. We also think of the case $K = K_0$ as a definite "extension" at $\mathfrak{p}$.

Let $\mathfrak{X}_d$ be the set of real primes of K_0 corresponding to the orderings in X_d (cf. Prop. 1.4.1). To see that (4.44) remains valid with this change[11], i.e. that we can also define $\mathrm{S}_\varepsilon(K,G)$ as the pullback and σ_ε as the canonical map in

$$\begin{array}{ccc}
\hat{\mathrm{W}}_\varepsilon(K,G) & & \\
& \searrow^{\sigma_\varepsilon} & \\
& \mathrm{S}_\varepsilon(K,G) \longrightarrow & \prod_{\mathfrak{p}\in\mathfrak{X}_d}\hat{\mathrm{W}}_\varepsilon(K_\mathfrak{p},G) \\
& \downarrow & \downarrow \\
& \mathrm{K}_0(KG) \longrightarrow & \prod_{\mathfrak{p}\in\mathfrak{X}_d}\mathrm{K}_0(K_\mathfrak{p}G)
\end{array} \tag{4.45}$$

(with further arrows $\hat{\mathrm{W}}_\varepsilon(K,G)\to\prod_{\mathfrak{p}\in\mathfrak{X}_d}\hat{\mathrm{W}}_\varepsilon(K_\mathfrak{p},G)$ and $\hat{\mathrm{W}}_\varepsilon(K,G)\to\mathrm{K}_0(KG)$)

we may assume that K_0 is formally real. Let P be an ordering of K_0 and $K_{0P} = \mathbf{R}$ a real closure. We can assume that the algebraic closure $\mathbf{C} = K_{0P}(i)$ of K_0 at P is in $\mathbb{C}$. Then the fixed field $K_{0P} = \mathbf{R}$ of complex conjugation restricted to $\mathbf{C}$ is contained in $\mathbb{R}$, and since their orderings are respectively $\dot{\mathbf{R}}^2$ and $\dot{\mathbb{R}}^2$, $\mathbb{R}/\mathbf{R} = K_{0\mathfrak{p}}/K_{0P}$ is an order preserving extension of real closed fields.

Suppose first that $K = K_0$. Each simple component A of $\mathbf{R}G$ is stable under the standard $(\mathbf{R},\mathrm{id})$-involution of $\mathbf{R}G$ (Th. 4.5.1, p. 171), so $(A,\ ^-) \cong (\mathrm{M}(n,D),\ ^{t*}\ )$ where D is $\mathbf{R},\mathbf{C}$ or $\mathbf{H}$ (Th. 4.5.3(e)). Then the component $(\mathbb{R},\mathrm{id})\otimes_{(\mathbf{R},\mathrm{id})}(A,\ ^-)$ of $(\mathbb{R}G,\ ^-)$ is $\cong (\mathrm{M}(n,\mathbb{R}\otimes_\mathbf{R} D),\ ^{t*}\ )$ where $\mathbb{R}\otimes_\mathbf{R} D$ is (isomorphic to) $\mathbb{R},\mathbb{C}$ or $\mathbb{H}$ respectively.

Moreover it is apparent that the scalar extension $\mathbb{R}\otimes_\mathbf{R} V$ of a simple A-module V is a simple $(\mathbb{R}\otimes_\mathbf{R} A)$-module – for example

$$\dim_\mathbb{R}(\mathbb{R}\otimes_\mathbf{R} V) = \dim_\mathbf{R} V = n\dim_\mathbf{R} D = n\dim_\mathbb{R}(\mathbb{R}\otimes_\mathbf{R} D)$$

– and if h is an $(A,\ ^-)$-invariant reflexive form on an arbitrary A-module W, then $h^\mathbb{R}$ is an $(\mathbb{R}\otimes A,\ ^-)$-invariant reflexive form of the same type on $\mathbb{R}\otimes W$, and is definite, and symmetric respectively Hermitian, if h is.

The case $K \neq K_0$ and K/K_0 definite at P is entirely similar:

- each simple component A of $\mathbf{C}G$ is stable under the standard $(\mathbf{C},\ ^*\ )$-involution $^-$ of $\mathbf{C}G$,
- we can assume $(A,\ ^-) = (\mathrm{M}(n,\mathbf{C}),\ ^{t*}\ )$,
- the scalar extension $\mathbb{C}\otimes_\mathbf{C} A = \mathrm{M}(n,\mathbb{C})$ is a simple component of $\mathbb{C}G$,
- $(\mathbb{C},\ ^*\ )\otimes_{(\mathbf{C},\ ^*\ )}(A,\ ^-) = (\mathrm{M}(n,\mathbb{C}),\ ^{t*}\ )$,
- the scalar extension $\mathbb{C}\otimes_\mathbf{C} V$ of a simple A-module V is a simple $\mathbb{C}\otimes_\mathbf{C} A$-module, and
- the scalar extension $h^\mathbb{C}$ of an $(A,\ ^-)$-invariant Hermitian form h is a $(\mathbb{C}\otimes_\mathbf{C} A,\ ^-)$-invariant Hermitian form, and is definite if h is.

---

[11] We could also of course simply define $\mathrm{S}_\varepsilon(K,G)$ and σ_ε by (4.45) *ab initio*. The reader may assume this is done if s/he wishes.

Therefore if $P \in X_d$, it follows easily from Th. 4.5.5(a) (p. 177) and Th. 4.5.6 that the map $\hat{\mathrm{W}}_\varepsilon(K_P, G) \to \hat{\mathrm{W}}_\varepsilon(K_\mathfrak{p}, G)$ given by scalar extension is an isomorphism, and of course $\mathrm{K}_0(K_PG) \to \mathrm{K}_0(K_\mathfrak{p}G)$ is as well. Moreover the diagrams

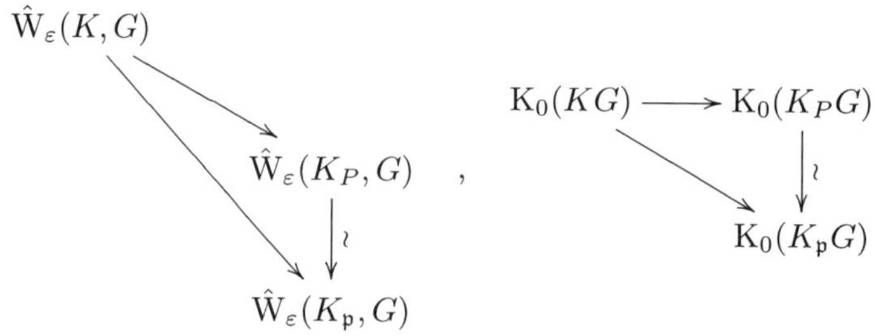

and

$$\begin{array}{ccc} \prod_{P\in X_d} \hat{\mathrm{W}}_\varepsilon(K_P, G) & \xrightarrow{\sim} & \prod_{\mathfrak{p}\in \mathfrak{X}_d} \hat{\mathrm{W}}_\varepsilon(K_\mathfrak{p}, G) \\ \downarrow & & \downarrow \\ \prod_{P\in X_d} \mathrm{K}_0(K_PG) & \xrightarrow{\sim} & \prod_{\mathfrak{p}\in \mathfrak{X}_d} \mathrm{K}_0(K_\mathfrak{p}G) \end{array}$$

are commutative, and so we can indeed replace K_P in (4.44) by $K_\mathfrak{p}$ so that $\mathrm{S}_\varepsilon(K, G)$ and σ_ε can also be defined by (4.45).

It will serve our purposes best to view the Witt-Grothendieck groups $\hat{\mathrm{W}}_\varepsilon(K, G)$ and $\hat{\mathrm{W}}_\varepsilon(K_\mathfrak{p}, G)$ as $\hat{\mathrm{W}}_\varepsilon(KG, {}^-)$ and $\hat{\mathrm{W}}_\varepsilon(K_\mathfrak{p}G, {}^-)$ instead, as explained in §4.1. Thus we will use the diagram

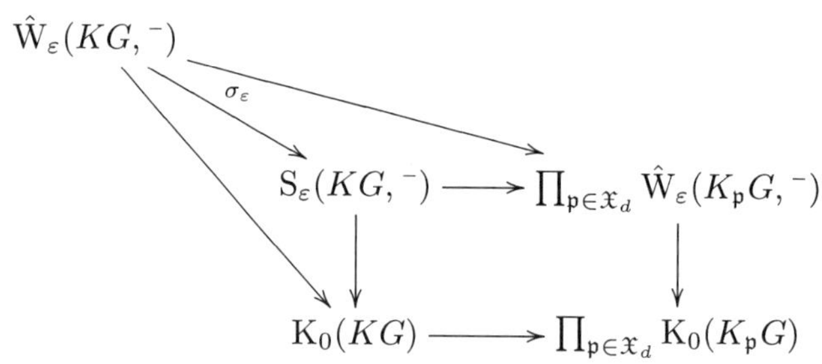

(4.46)

instead of (4.45).

If $KG = A_1 \oplus A_2 \oplus \cdots \oplus A_l$ is the decomposition into simple algebras, $\bar{A}_i = A_i$ for all i since K/K_0 is definite (Th. 4.5.1, p. 171). Thus the decomposition of $(KG, {}^-)$ into simple $(K, {}^-)$-involution algebras is

$$(KG, {}^-) = (A_1, {}^-) \oplus (A_2, {}^-) \oplus \cdots \oplus (A_l, {}^-).$$

Then

$$\hat{\mathrm{W}}_\varepsilon(KG, {}^-) = \hat{\mathrm{W}}_\varepsilon(A_1, {}^-) \oplus \hat{\mathrm{W}}_\varepsilon(A_2, {}^-) \oplus \cdots \oplus \hat{\mathrm{W}}_\varepsilon(A_l, {}^-)$$

and for each $\mathfrak{p} \in \mathfrak{X}_d$

$$\hat{\mathrm{W}}_\varepsilon(K_\mathfrak{p}G, {}^-) = \hat{\mathrm{W}}_\varepsilon(A_{1\mathfrak{p}}, {}^-) \oplus \hat{\mathrm{W}}_\varepsilon(A_{2\mathfrak{p}}, {}^-) \oplus \cdots \oplus \hat{\mathrm{W}}_\varepsilon(A_{l\mathfrak{p}}, {}^-).$$

For each i there is a diagram

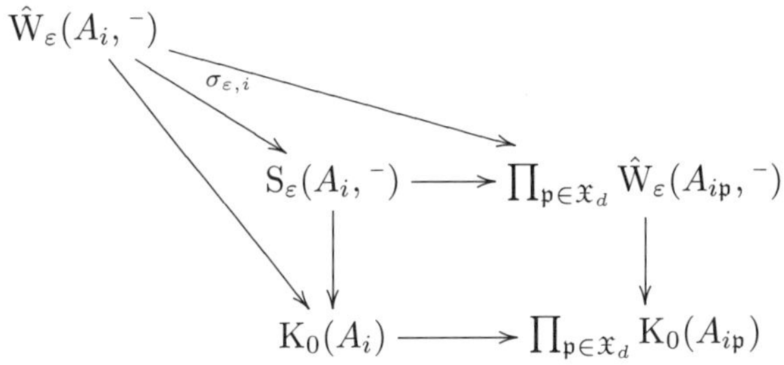

(4.47)

where $\mathrm{S}_\varepsilon(A_i, {}^-)$ is defined as the pullback in the rectangle and $\sigma_{\varepsilon,i}$ by its universal property, as before. It is straightforward to check that the diagram (4.46) is the direct sum of the diagrams (4.47) for each i, in an obvious sense.

It follows that to prove that $4\ker\sigma_\varepsilon = 0$ or $2\ker\sigma_\varepsilon = 0$ in (4.46), it suffices to prove $4\ker\sigma_{\varepsilon,i} = 0$ or $2\ker\sigma_{\varepsilon,i} = 0$ in (4.47) for each i. Similarly it suffices to show that each $\sigma_{\varepsilon,i}$ is onto.

To simplify the notation, we take an arbitrary i and write $A = A_i$ and $\sigma_{\varepsilon,i} = \sigma_\varepsilon$. We also assume that $A = \mathrm{M}(n, D)$ and $L = \mathrm{Z}(A) = \mathrm{Z}(D)$, and begin by proving (a) and (b).

Suppose that $[h] - [g] \in \ker\sigma_\varepsilon$. Thus $[h]$ and $[g]$ have the same image in $\mathrm{K}_0(A)$ – and so we can assume that h and g are forms on the same A-module V – and they also have the same image in $\hat{\mathrm{W}}_\varepsilon(A_\mathfrak{p}, {}^-)$ for each $\mathfrak{p} \in \mathfrak{X}_d$.

Now ε-Hermitian forms over $(A, {}^-)$ are Morita equivalent to $\varepsilon\varepsilon_0$-Hermitian forms over the division algebra $(D, {}^\sim)$ with involution where

$\bar{\alpha} = \widetilde{\alpha}$ for all $\alpha \in L$
${}^\sim = {}^*$ if ${}^-$ is of the 1st kind and D is a quaternion algebra,
$\bar{a} = \underline{h}_0^{-1} a^{t\sim} \underline{h}_0$ where $\underline{h}_0^{t\sim} = \varepsilon_0 \underline{h}_0$, $\varepsilon_0 = \pm 1$,
$\varepsilon_0 = 1$ if and only if ${}^-$ and ${}^\sim$ have the same type.

Thus we have a "Morita isomorphism" $\hat{\mathrm{W}}_\varepsilon(A, {}^-) \xrightarrow{\sim} \hat{\mathrm{W}}_{\varepsilon\varepsilon_0}(D, {}^\sim)$ (Th. 4.1.2, p. 154).

We first dispose of the case D a quaternion algebra with involution conjugation * and $\varepsilon\varepsilon_0 = -1$. In this case the Morita correspondents h_D and g_D in ${}_{(D,{}^*)}\mathcal{H}_{\varepsilon\varepsilon_0}$ of h and g are nonsingular skew Hermitian forms of the same rank, so by Lem. 4.9.1 (p. 202)

$$4([h_D] - [g_D]) = 0, \text{ and } 2([h_D] - [g_D]) = 0 \text{ if } -1 \in \dot{L}^2,$$

so $4([h] - [g]) = 0$, and $2([h] - [g]) = 0$ if $-1 \in \dot{K}^2$.

We now apply the Hasse Principle (Th. 3.3.2, p. 137) to the remaining cases – in them the Hasse Principle holds since none satisfies condition (*) of that theorem. Thus to show that $4([h] - [g]) = 0$, it suffices to show that

$$h_\mathfrak{p} \perp h_\mathfrak{p} \perp h_\mathfrak{p} \perp h_\mathfrak{p} \cong g_\mathfrak{p} \perp g_\mathfrak{p} \perp g_\mathfrak{p} \perp g_\mathfrak{p} \tag{4.48}$$

for all primes $\mathfrak{p}$ of K_0, where

$$h_\mathfrak{p}, g_\mathfrak{p} : V_\mathfrak{p} \times V_\mathfrak{p} \to (A_\mathfrak{p}, {}^-),$$

$$V_\mathfrak{p} = K_\mathfrak{p} \otimes_K V, \quad \text{and} \quad (A_\mathfrak{p}, {}^-) = (K_\mathfrak{p}, {}^-) \otimes_{(K,{}^-)} (A, {}^-).$$

We break the proof up into 5 cases:

1. $\mathfrak{p} \in \mathfrak{X}_d$: the fact that $[h]-[g] \in \ker\sigma_\varepsilon$ implies $[h_\mathfrak{p}] = [g_\mathfrak{p}]$ in $\hat{\mathrm{W}}_\varepsilon(A_\mathfrak{p}, {}^-)$ so (4.48) is trivial.

2. $\mathfrak{p}$ an indefinite real prime: then $(K_\mathfrak{p}, {}^-) \cong (\mathbb{R}\oplus\mathbb{R}, \leftrightarrow)$, so $(A_\mathfrak{p}, {}^-)$ is hyperbolic, and therefore $h_\mathfrak{p} \cong g_\mathfrak{p}$ (Cor. 3.1.6) since the spaces of g and h have the same image in $\mathrm{K}_0(A)$.

3. $\mathfrak{p}$ a complex prime and $(K : K_0) = 2$: then $(K_\mathfrak{p}, {}^-) \cong (\mathbb{C}\oplus\mathbb{C}, \leftrightarrow)$ and $(A_\mathfrak{p}, {}^-)$ is hyperbolic...

4. $\mathfrak{p}$ is a complex prime and $K = K_0$: then $(K_\mathfrak{p}, \mathrm{id}) \cong (\mathbb{C}, \mathrm{id})$ and $A_\mathfrak{p} \cong \mathrm{M}(n, \mathbb{C})$ with an involution of the first kind, so the Morita correspondents of $h_\mathfrak{p}$ and $g_\mathfrak{p}$ are equivalent since they are symmetric or skew symmetric forms over $\mathbb{C}$ (on isomorphic spaces). Therefore $h_\mathfrak{p} \cong g_\mathfrak{p}$.

5. $\mathfrak{p}$ a finite prime. If $\mathfrak{p}$ splits in K, then $(A_\mathfrak{p}, {}^-)$ is again hyperbolic. If it does *not* split, apply Th. 4.9.3, p. 199, (b), (c) or (d).

If $-1 \in \dot{K}^2$, the proof that $2([h]-[g]) = 0$ is identical with that for $4([h]-[g]) = 0$, taking into account the statement of Th. 4.9.1(d).

This finishes the proof of (a) and (b) of Th. 4.10.1, and we now proceed to the proof of (c).

Let $\mathfrak{X}_d = \{\mathfrak{p}_1, \ldots, \mathfrak{p}_r\}$ and let

$$\mathbf{s} = \big([V]-[W],\ [V_1, g_1]-[W_1, h_1], \ldots, [V_r, g_r]-[W_r, h_r]\big)$$

be an arbitrary element of

$$\mathrm{S}_\varepsilon(A, {}^-) \subset \mathrm{K}_0(A) \times \hat{\mathrm{W}}_\varepsilon(A_{\mathfrak{p}_1}, {}^-) \times \cdots \times \hat{\mathrm{W}}_\varepsilon(A_{\mathfrak{p}_r}, {}^-).$$

We wish to show that it is in the image of σ_ε. By definition of $\mathrm{S}_\varepsilon(A, {}^-)$,

$$[V_{\mathfrak{p}_i}] - [W_{\mathfrak{p}_i}] = [V_i] - [W_i] \text{ in } \mathrm{K}_0(A_{\mathfrak{p}_i}) \text{ for } i = 1, \ldots, r. \tag{4.49}$$

If we add the same A-module to V and W, $[V]-[W]$ does not change, so by adding the sum of a suitably large odd number of copies of V, we may assume that, for all Archimedean primes $\mathfrak{p}$ of K_0 and all simple components B of $A_\mathfrak{p}$

$$\mathrm{len}_{A_\mathfrak{p}} BV_\mathfrak{p} = \mathrm{len}_{A_\mathfrak{p}} \bar{B}V_\mathfrak{p} \text{ is even and } \mathrm{len}_{A_{\mathfrak{p}_i}} BV_{\mathfrak{p}_i} \geq \mathrm{len}_{A_{\mathfrak{p}_i}} BV_i \text{ for all } i. \tag{4.50}$$

(Note that, in any case, $\mathrm{len}_{A_\mathfrak{p}} BV_\mathfrak{p} = \mathrm{len}_{A_\mathfrak{p}} \bar{B}V_\mathfrak{p}$ by Lem. 1.3.3, p. 29.)

Next for each $i = 1, \ldots, r$, by Th. 3.2.5 (p. 114) applied for each simple *involution* component of $(A_{\mathfrak{p}_i}, {}^-)$, and by (4.50), we can add a suitable ε-Hermitian space (U_i, f_i) over $(A_{\mathfrak{p}_i}, {}^-)$ to both (V_i, g_i) and (W_i, h_i) – again without changing $\mathbf{s}$ – in order that we can assume

$$\mathrm{len}_{A_{\mathfrak{p}_i}} BV_{\mathfrak{p}_i} = \mathrm{len}_{A_{\mathfrak{p}_i}} BV_i$$

for all i and all simple components B of $A_{\mathfrak{p}_i}$, and therefore also that

$$[V_{\mathfrak{p}_i}] = [V_i] \text{ for all } i.$$

Thus by (4.49)

$$[W_{\mathfrak{p}_i}] = [W_i] \text{ for all } i.$$

As a result of these modifications, both $\mathbf{s}_V := \big([V],\ [V_1, g_1], \ldots, [V_r, g_r]\big)$ and $\mathbf{s}_W := \big([W],\ [W_1, h_1], \ldots, [W_r, h_r]\big)$ are in $\mathrm{S}_\varepsilon(A, {}^-)$, and since $\mathbf{s} = \mathbf{s}_V - \mathbf{s}_W$, it suffices to show that, say, $\mathbf{s}_V$ is in the image of σ_ε.

Let $\mathfrak{Y}$ denote the Archimedean primes of K_0 which are *not* in $\mathfrak{X}_d$ – they are the primes of K_0 which are either real and indefinite, or complex.

Let $\mathfrak{p} \in \mathfrak{Y}$ and let B be a simple algebra summand of $A_\mathfrak{p}$. If $\bar{B} \neq B$, there is an ε-hyperbolic form over $(B \oplus \bar{B}, ^-)$ on $BV_\mathfrak{p} \oplus \bar{B}V_\mathfrak{p}$ by exercises 3 and 5, p. 68, since $BV_\mathfrak{p}$ and $\bar{B}V_\mathfrak{p}$ have the same length by (4.50). If $\bar{B} = B$, it follows from the results in §3.2.4 that there is a nonsingular ε-Hermitian form over $(B, ^-)$ on $BV_\mathfrak{p}$ since it has even length by (4.50). Taking the orthogonal sum of these forms yields a nonsingular ε-Hermitian form

$$g_\mathfrak{p} : A_\mathfrak{p}V \times A_\mathfrak{p}V \to (A_\mathfrak{p}, ^-).$$

Let V_∞ be the completion of V at the prime ∞ (the absolute value) of $\mathbb{Q}$,

$$V_\infty = \mathbb{Q}_\infty \otimes_\mathbb{Q} V = (\oplus_{\mathfrak{p}|\infty} K_{0\mathfrak{p}}) \otimes_{K_0} V = \oplus_{\mathfrak{p}|\infty} V_\mathfrak{p}$$

where $\mathbb{Q}_\infty \otimes_\mathbb{Q} K_0 = \oplus_{\mathfrak{p}|\infty} K_{0\mathfrak{p}}$ (Th. 1.2.16, p. 23).

Since $[V_i] = [V_{\mathfrak{p}_i}]$ for all i, we can identify $V_i = V_{\mathfrak{p}_i}$, and define

$$g : V_\infty \times V_\infty \to (A_\infty, ^-) = \oplus_i (A_{\mathfrak{p}_i}, ^-) \oplus (\oplus_{\mathfrak{p} \in \mathfrak{Y}} (A_\mathfrak{p}, ^-))$$

by

$$(V_\infty, g) = \perp_i (V_i, g_i) \perp (\perp_{\mathfrak{p} \in \mathfrak{Y}} (V_\mathfrak{p}, g_\mathfrak{p})).$$

It follows by Th. 3.4.1 (p. 140) that there is an ε-Hermitian form $f : V \times V \to (A, ^-)$ such that $f_\infty \cong g$ and so $\sigma_\varepsilon[V, f] = \mathbf{s}_V$.

This finishes the proof of Th. 4.10.1. □

4.11 Structure of the Witt group $\mathrm{W}_\varepsilon(K,G)$

The sole purpose of this section is the proof of the following theorem:

Theorem 4.11.1 *Let $(A, ^-)$ be a simple K-algebra with a $(K, ^-)$-involution and let $\varepsilon = \pm 1$. Then the torsion subgroup of $\mathrm{W}_\varepsilon(A, ^-)$ is 2-primary.*

Since $\mathrm{W}_\varepsilon(A, ^-) = 0$ if $(A, ^-)$ is hyperbolic and $\mathrm{W}_\varepsilon(A, ^-)$ is additive in $(A, ^-)$, we immediately get

Corollary 4.11.1 *The torsion subgroup of $\mathrm{W}_\varepsilon(K, G)$ is 2-primary.* □

We begin with some facts about algebras with involution, due to E. Bayer Fluckiger and R. Parimala [6]:

Lemma 4.11.1 *(a) If L/K_0 is a (separable) field extension of 2-power degree, there exists a separable extension M/K_0 of odd degree such that LM/M can be filtered by degree 2 extensions*[12].

(b) Suppose that $(A, ^-)$ is a central simple K-algebra with a $(K, ^-)$-involution, that there is a subfield L of A such that $(L : K) = 2$ and $\bar{L} = L$, and that $^-$ is the identity on L if it is the identity on K. Then there exists $c \in A^\times$ such that $\bar{c} = -c$ and $\mathrm{inn}_c|_L$ is the nontrivial automorphism $^\sim$ of L/K.

Now suppose that $(K, ^-)$ is of the second kind with fixed field K_0 and that $(D, ^-)$ is a central K-division algebra with $(K, ^-)$-involution.

(c) There is an extension K_1/K_0 of odd degree such that $K_1 \otimes_{K_0} D$ has exponent a power of 2, i.e. the Brauer class $[K_1 \otimes_{K_0} D] \in \mathrm{Br}(K_1K)$ has order a power of 2.

---

[12]The compositum LM is well-defined by Cor. 1.2.4, p. 22, as is K_1K in (c) and MK in (d).

(d) Suppose that $D \neq K$ and that $\dim_K D$ is a power of 2. Then there is an odd degree extension M/K_0 such that $M \otimes_{K_0} D = MK \otimes_K D$ contains a field L with $[L : MK] = 2$ and $\bar{L} = L$ (if $^-$ also denotes the extension $\mathrm{id}_M \otimes_{K_0} {}^-$ of the involution to $M \otimes_{K_0} D$).

(a)

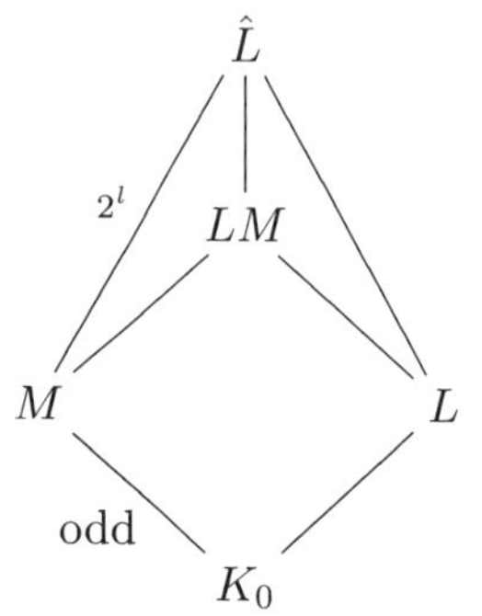

Let $\hat{L}$ be the Galois closure of L over K_0 and M the fixed field of a 2-Sylow subgroup of $\mathrm{Gal}(\hat{L}/K_0)$. Then M/K_0 certainly has odd degree. Since finite p-groups are nilpotent, $\mathrm{Gal}(\hat{L}/M)$ has a filtering by index 2 subgroups and so its composition factors are all groups of order 2. Now take a composition series of $\mathrm{Gal}(\hat{L}/M)$ through $\mathrm{Gal}(\hat{L}/LM)$. The corresponding chain of intermediate fields of $\hat{L}/M$ provides the desired filtering of LM/M.

(b) Suppose first that $^-$ is of the second kind. The involution $^\sim \circ (^-|_L)$ of L restricts to $^-$ on K, and so by the "involution extension theorem" (Th. 1.3.4, p. 30) can be extended to a $(K, ^-)$-involution $^\vee$ on A. There is a $c \in A^\times$ such that $^\vee = \mathrm{inn}_c\, ^-$ and $\bar{c} = c$ (cf. Ex. 10, p. 27). If $\lambda \in L$, $\check{\lambda} = \tilde{\bar{\lambda}} = c\bar{\lambda}c^{-1}$, and so $\mathrm{inn}_c|_L = {}^\sim$. If we now replace c by αc where $\alpha \in \dot{K}$ and $\bar{\alpha} = -\alpha$, it will have the required properties.

Suppose then that $^-$ is of the first kind, and consider first the case $\dim_K A = 4$ – i.e. A is a quaternion algebra (cf. §1.2.4). Write $L = K(l)$ where $l^2 \in K$. Then l must be a pure quaternion (*ibid*) and since $^-$ is the identity on L, it must be an orthogonal involution (Th. 1.3.1, p. 28). By Th. 1.3.6, $^- = \mathrm{inn}_c\, ^*$ (where * is conjugation) for some c such that $c^* = -c$. The fact that $\bar{l} = l$ means that $\mathrm{inn}_c l = l^* = -l$, as desired.

Finally suppose that $^-$ is of the first kind and $\dim_K A > 4$. Then the centralizer $\mathrm{Z}_A L$ is a central simple L-algebra by the Centralizer Theorem (p. 11). By the involution extension theorem, the involution $^\sim$ on L can be extended to an involution on A – and so also on $\mathrm{Z}_A L$ – which we also denote by $^\sim$. Then $^\sim = \mathrm{inn}_c\, ^-$ for some $c, \bar{c} = \pm c$ and $\mathrm{inn}_c|_L = {}^\sim$. If $\bar{c} = -c$, we are finished. Otherwise since $(\dim_K L)(\dim_K \mathrm{Z}_A L) = \dim_K A > 4$, $\mathrm{Z}_A L$ is noncommutative and so the extended involution is not the identity on $\mathrm{Z}_A L$ and we can choose $b \in \mathrm{Z}_A L$ such that $\tilde{b} = -b$. Then $\mathrm{inn}_{bc}|_L = \mathrm{inn}_c|_L$, and

$$\overline{bc} = c\bar{b} = cc^{-1}\tilde{b}c = -bc.$$

(c)

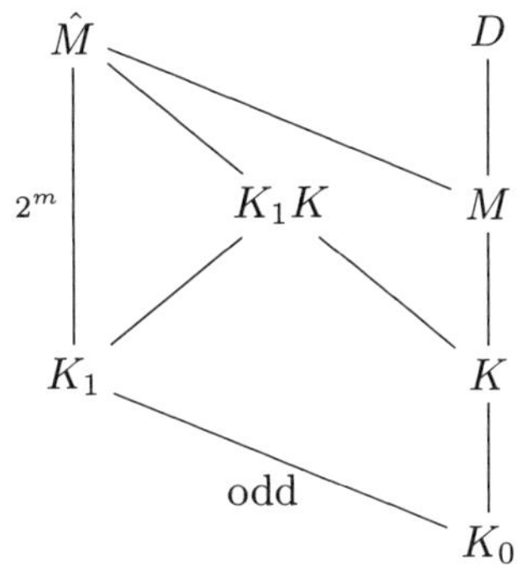

Let M be a maximal subfield of D which is separable over K_0 (cf. p. 97, [55]), $\hat{M}$ a Galois closure of M over K_0, and K_1 the fixed subfield of a Sylow 2-subgroup of $\mathrm{Gal}(\hat{M}/K_0)$. Then K_1 has odd degree over K_0, and since $\hat{M}$ is a splitting field of $K_1 \otimes_{K_0} D$ of 2-power degree over K_1K, $K_1 \otimes_{K_0} D$ has exponent a power of 2 (Ths. 28.5(i) and 29.22, [55]).

(d)

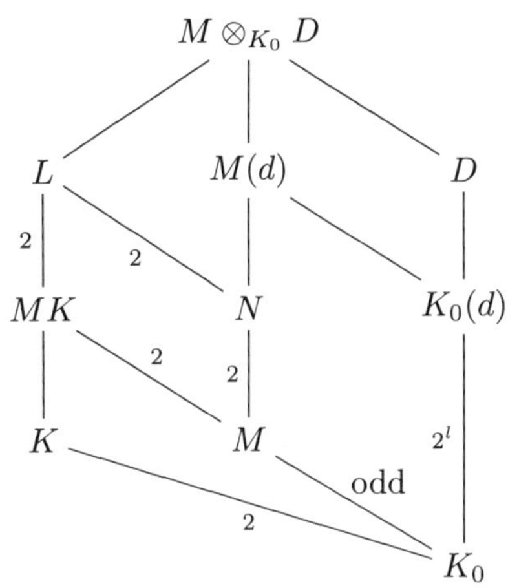

Let $d \in D \setminus K$ satisfy $\bar{d} = d$. Since $K_0(d)$ has 2-power degree over K_0, by (a) there is an extension M/K_0 of odd degree such that the extension $M(d) = MK_0(d) \subset M \otimes_{K_0} D = MK \otimes_K D$ of M contains a quadratic extension N/M. Since N and M, as subfields of $M(d)$, are element-wise fixed by $^-$ and K is not, the quadratic extensions N and MK of M are disjoint, and so $NMK = NK$ is a quadratic extension of MK, and since it is stable under $^-$, $L = NK$ satisfies the requirements. □

We now prove Th. 4.11.1 following M.G. Mahmoudi [47].

Lemma 4.11.2 *Let $(D, {}^-)$ be a central noncommutative division algebra with involution over $(K, {}^-)$ such that $\dim_K D$ is a power of 2. Then there is an extension M/K_0 of odd degree and an involution $^\sim$ on $M \otimes_{K_0} D$ with the following properties: Denote the involution $\mathrm{id} \otimes {}^-$ on $M \otimes D$ also by $^-$. There are elements i and j in $M \otimes D$ satisfying*

(a) $\tilde{i} = -i$, $\tilde{j} = -j$, $ij = -ji$, and $((KM)(i) : KM) = 2$,

(b) $^\sim = {}^-$ if $(D, {}^-)$ is an involution algebra of the second kind or a quaternion algebra with conjugation as involution, and $^\sim = \mathrm{inn}_j \circ {}^-$ otherwise.

$(D, {}^-)$ of the first kind. If $(D, {}^-)$ is a quaternion algebra with $^- = {}^*$, the theorem is obvious, so suppose that it is a quaternion algebra with an orthogonal involution (cf. Th. 1.3.1, p. 28) or that it has dimension > 4 over K. Then we can choose $d \in D \setminus K$ such that $\bar{d} = d$, and by Lem. 4.11.1(a), there is an extension M/K of odd degree such that $K(d)M/M$ has a quadratic subextension $M(i)/M$ with $i^2 \in M$. If we identify $K(d)M \subset M \otimes D$, the involution $^-$ of $M \otimes D$ is the identity on $K(d)M \supset M(i)$, and the desired element j exists by Lem. 4.11.1(b) if we define $^\sim = \mathrm{inn}_j \circ {}^-$.

$(D, {}^-)$ of the second kind. By Lem. 4.11.1(d) there is an odd degree extension M/K_0 such that the central MK-algebra $M \otimes_{K_0} D = MK \otimes_K D$ contains a quadratic

extension L/MK such that $\bar{L} = L$; let $\check{}$ be the nontrivial automorphism of L/MK. Let L_0 be the fixed field of L under $\bar{}$; L/L_0 is a quadratic extension, distinct from L/MK since $\bar{}$ is nontrivial on MK. The fact that MK is stable under $\bar{}$ implies that the conjugate of $\check{}$ by $\bar{}$ is $\check{}$ since it is of order 2 and also has fixed field MK:

$$\alpha \in MK \Rightarrow (\bar{\alpha})^{\check{}\,\bar{}} = (\bar{\alpha})^{\bar{}} = \alpha.$$

Thus the automorphisms $\bar{}\,|_L$ and $\check{}$ of L commute and generate a group $\cong \mathbb{Z}/2 \times \mathbb{Z}/2$ whose fixed field is $MK_0 = M$. Let N be the fixed field of the involution $\bar{}\,\check{}$. All

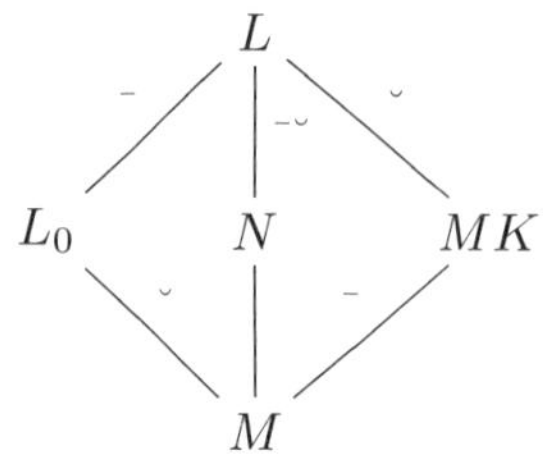

of the extensions in the diagram to the left are quadratic, with the nontrivial involutions as indicated. Let $N = M(i)$, where $i^2 \in M$. Now $\bar{}\,|_N$ and $\check{}\,|_N$ are both nontrivial, so must be equal. Therefore $\bar{i} = -i = \check{i}$ and by Lem. 4.11.1(b) there is a $j \in M \otimes_{K_0} D$ such that $\bar{j} = -j$ and $\mathrm{inn}_j|_L = \check{}$. Moreover $i \notin MK$ and so $MK(i) = L$. Thus i and j are the desired elements. □

We continue with a central division algebra $(D, \bar{})$ with a $(K, \bar{})$-involution, and elements $i, j \in \dot{D}$ satisfying

$$\bar{i} = -i, \quad \bar{j} = -j, \quad ij = -ji, \quad \text{and} \quad (K(i) : K) = 2.$$

Let $L = K(i)$ and let E be the centralizer of L in D, so E is a division algebra with center L and dimension $\frac{1}{2}\dim_K D$ over K by the Centralizer Theorem. It is clear that

$$j^2 \in \dot{E}, \quad \bar{E} = E, \quad D = E \oplus Ej, \quad jE = Ej,$$

since Ej and jE are both equal to $\{d \in D : \ d\lambda = \bar{\lambda} d \ \text{ for all } \lambda \in L\}$. Define the $(K, \bar{})$-involution $\sim$ on D by $\tilde{d} = j\bar{d}j^{-1} = \mathrm{inn}_j\, \bar{d}$ for all $d \in D$. Note that $\tilde{E} = E$.

Define $\pi_1, \pi_2 : D \to E$ by $\pi_1(e_1 + e_2 j) = e_1$ and $\pi_2(e_1 + e_2 j) = e_2$. π_1 is E-linear on both left and right, π_2 E-linear on the left and (E, inn_j)-semilinear on the right, i.e. $\pi_2(de) = \pi_2(d)(jej^{-1})$. It is also easy to check that

$$\pi_1(\bar{d}) = \overline{\pi_1(d)} \quad \text{and} \quad \pi_2(\bar{d}) = -\widetilde{\pi_2(d)},$$

and that if $h : V \times V \to (D, \bar{})$ is ε-Hermitian,

$$\pi_1(h)(u, v) = \pi_1(h(u, v)) \quad \text{and} \quad \pi_2(h)(u, v) = \pi_2(h(u, v))$$

define functors

$$\pi_1 : {}_{(D,\bar{})}\mathcal{H}_\varepsilon \to {}_{(E,\bar{})}\mathcal{H}_\varepsilon \quad \text{and} \quad \pi_2 : {}_{(D,\bar{})}\mathcal{H}_\varepsilon \to {}_{(E,\,\sim\,)}\mathcal{H}_{-\varepsilon}.$$

If $d = e_1 + e_2 j \neq 0$ and $i = 1$ or 2, either $\pi_i d$ or $\pi_i(dj)$ is $\neq 0$, so if $h(u, v) \neq 0$, either $\pi_i(h(u, v))$ or $\pi_i(h(u, jv))$ is $\neq 0$, and it follows that the functors π_i take nonsingular forms to nonsingular forms. Since they also take hyperbolic forms to hyperbolic forms, they give homomorphisms

$$\pi_1 : \hat{\mathrm{W}}_\varepsilon(D, \bar{}) \to \hat{\mathrm{W}}_\varepsilon(E, \bar{}), \quad \pi_1 : \mathrm{W}_\varepsilon(D, \bar{}) \to \mathrm{W}_\varepsilon(E, \bar{}),$$

and

$$\pi_2 : \hat{\mathrm{W}}_\varepsilon(D, \bar{}) \to \hat{\mathrm{W}}_{-\varepsilon}(E, \,\sim\,), \quad \pi_2 : \mathrm{W}_\varepsilon(D, \bar{}) \to \mathrm{W}_{-\varepsilon}(E, \,\sim\,).$$

Lemma 4.11.3 *If $\pi_1 h$ and $\pi_2 h$ are hyperbolic, then $h \perp h$ is hyperbolic.*

We may assume h is anisotropic. Suppose that $Eu + Ev$ is a hyperbolic plane with respect to $\pi_1 h$, with u and v isotropic. Then $h(u,u) = e_1 j$ for some $e_1 \in \dot{E}$. Since $h(u, ju) = -e_1 j^2 \in \dot{E}$ and $h(ju, ju) = -je_1 j^2 \in Ej$, it follows that $Du = Eu \oplus Eju$ is also a hyperbolic plane with respect to $\pi_1 h$. By Witt's theorem, the orthogonal complement $(Du)^\perp$ with respect to h is also hyperbolic with respect to $\pi_1 h$, and so by induction

$$h \cong \langle e_1 j,\, e_2 j, \ldots, e_n j\rangle, \quad e_i \in \dot{E} \text{ for all } i.$$

Since $\pi_2\langle e_i j\rangle = \langle e_i, -je_i j\rangle$,

$$\pi_2 h \cong \langle e_1, e_2, \ldots, e_n\rangle \perp \langle -je_1 j, -je_2 j, \ldots, -je_n j\rangle.$$

But

$$\pi_2 h \cong \langle e_1, e_2, \ldots, e_n\rangle \perp \langle -e_1, -e_2, \ldots, -e_n\rangle,$$

since $\pi_2 h$ is hyperbolic, and so

$$\langle e_1, e_2, \ldots, e_n\rangle \cong \langle je_1 j, je_2 j, \ldots, je_n j\rangle$$

in ${}_{(E,\,\sim)}\mathcal{H}_{-\varepsilon}$, which means there is a nonsingular matrix $T \in \mathrm{M}(n, E)$ such that $T\mathrm{diag}(e_1, e_2, \ldots, e_n)T^{t\,\sim} = \mathrm{diag}(je_1 j, je_2 j, \ldots, je_n j), = S$, say. Now if h is equivalent to $\langle e_1 j,\, e_2 j, \ldots, e_n j\rangle$ in the basis $u_1, u_2, \ldots, u_n$, then

$$h \cong \langle -je_1 j^2,\, -je_2 j^2, \ldots, -je_n j^2\rangle$$

in the basis $ju_1, ju_2, \ldots, ju_n$, i.e. $\mathrm{mat}\, h = -Sj$. This implies that h is equivalent to $\langle -e_1 j, \ldots, -e_n j\rangle$ since

$$\begin{aligned} T^{-1}(-Sj)(T^{-1})^{t\,-} &= -(T^{-1})S\left(j(T^{-1})^{t\,-}j^{-1}\right)j \\ &= -(T^{-1})S(T^{-1})^{t\,\sim}j = \mathrm{diag}(-e_1 j, \ldots, -e_n j). \end{aligned}$$

Thus $h \cong -h$ and so $h \perp h$ is hyperbolic. □

We now begin the proof proper of Th. 4.11.1. Since the definition of $\mathrm{W}_\varepsilon(A, {}^-)$ does not depend on the field K other than the fact that $\dim_K A$ is finite, we may suppose that A is central over K. By "Morita Theory" – namely Th. 4.1.2, p. 154 – we may suppose that A is a (central) division algebra D. The proof proceeds by induction on $\dim_K D$.

Suppose first that $\dim_K D = 1$, i.e. $D = K$. If ${}^-$ is the identity and $\varepsilon = -1$, we are in the case of skew symmetric forms so all forms are hyperbolic and $\mathrm{W}_{-1}(K, {}^-) = 0$. The case of ${}^-$ the identity and $\varepsilon = 1$ is that of symmetric forms and so the result follows from Pfister's Local-Global Principle (p. 52). If ${}^-$ is not the identity, we can suppose that $\varepsilon = 1$ since the correspondence $h \leftrightarrow \sqrt{\alpha}h$, where $K = K_0(\sqrt{\alpha})$, provides an isomorphism $\mathrm{W}_1(K, {}^-) \cong \mathrm{W}_{-1}(K, {}^-)$. And in this case, the theorem follows immediately from Jacobson's Theorem and Pfister's theorem since Jacobson's Theorem (p. 49) provides an injective homomorphism $\mathrm{W}_1(K, {}^-) \to \mathrm{W}_1(K_0, \mathrm{id}_{K_0})$.

Therefore we may assume that $\dim_K D > 1$ – in particular D is noncommutative.

The first step is to show that we can assume $\dim_K D$ is a power of 2.

If the involution is of the first kind, then $D \cong D^o$ and so D has order 2 in $\mathrm{Br}(K)$ and $\dim_K D$ is a power of 2 since the exponent of the Brauer class of D and the dimension of D have the same prime factors (cf. (29.24), p. 254, [55]).

Suppose that the involution is of the second kind. By Lem. 4.11.1(c) (p. 211), there is an extension K_1/K_0 of odd degree such that the Brauer class of $K_1 \otimes_{K_0} D \cong (K_1 K) \otimes_K D$ has 2-power order. The involution on D also has an extension to an involution of the second kind on $K_1 \otimes_{K_0} D$ – namely $\mathrm{id}_{K_1} \otimes_{K_0} {}^-$ – and so there is also a $(K_1K, {}^-)$-involution (also of the second kind, of course), again denoted ${}^-$, on any division algebra D_1 in the same Brauer class (Th. 1.3.7(c), p. 34). Thus

$$\mathrm{W}_1(D, {}^-) \to \mathrm{W}_1(K_1 \otimes_{K_0} D, \mathrm{id}_{K_1} \otimes_{K_0} {}^-) \to \mathrm{W}_1(D_1, {}^-)$$

where the first map is injective by Th. 4.2.1, p. 159, and the second is an isomorphism by Morita Theory (*ibid*). Thus it suffices to prove the theorem for D_1, and by replacing D by D_1, we can assume in this case also that $\dim_K D$ is a power of 2.

Now we apply Lem. 4.11.2 to this situation. This gives us first of all an extension M/K_0 of odd degree, and so the map

$$\mathrm{W}_\varepsilon(D, {}^-) \to \mathrm{W}_\varepsilon(M \otimes_{K_0} D, {}^-)$$

is injective by Th. 4.2.1. Furthermore there is an involution $\sim$ on $M \otimes_{K_0} D$ and elements i, j such that $\tilde{i} = -i, \tilde{j} = -j$, $ij = -ji$, and $((KM)(i) : KM) = 2$, and a map

$$\mathrm{W}_\varepsilon(M \otimes_{K_0} D, {}^-) \to \mathrm{W}_{\varepsilon_0\varepsilon}(M \otimes_{K_0} D, {}^\sim)$$

where either ${}^\sim = {}^-$ and $\varepsilon_0 = 1$ (in which case the map is simply the identity) or ${}^\sim = \mathrm{inn}_j \circ {}^-$ and $\varepsilon_0 = -1$ (in which case the map is obtained by scaling $h \rightsquigarrow hj^{-1}$, and is an isomorphism). Once again we can replace $(D, {}^-)$ by $(M \otimes_{K_0} D, {}^\sim)$, and so assume that D contains elements i and j such that $\bar{i} = -i, \bar{j} = -j, ij = -ji$ and $(K(i) : K) = 2$.

Suppose that (the Witt class of) h has finite order, say $2^n m$ where m is an odd integer. We wish to show that $m = 1$. Then the orthogonal sum $h' = 2^n \times h$ of 2^n copies of h has order m. We can assume that h' is anisotropic. In the notation of the foregoing lemma, the orders of $\pi_1 h'$ and $\pi_2 h'$ divide m (since π_1 and π_2 are homomorphisms) and so are odd, and therefore by the induction hypothesis (since the dimension of E is less than that of D) their orders are 1, i.e. they are both hyperbolic. It follows by the lemma that $2 \times h'$ is hyperbolic, and so $m = 1$. □

4.12 Notes on Chapter 4

1. Theorem 4.2.1 is due to E. Bayer Fluckiger and H.W. Lenstra [5] when the involution on L is the identity, and the proof given here is a simple adaptation of theirs. In fact it is pointed out in Remark 3.3 (p. 366) of [5] that, much more generally, systems of sesquilinear forms over K-algebras with involution which become isomorphic over an odd degree extension L/K are isomorphic over K.

Cor. 4.2.1 also appears in [5] (Th. 4.1) in the case of bilinear forms which are not necessarily nonsingular or ε-reflexive. This reference also contains the result in the exercise on p. 160; the proof suggested in the exercise is due to Bayer Fluckiger [4]. This exercise was first proved by P. Conner and R. Perlis in [17] in the case of an odd degree extension of $\mathbb{Q}$. Further elaborations, especially for extensions of even degree, are given in the paper [7] of Bayer Fluckiger and J.P. Serre.

2. The proof of Lemma 4.2.1 is taken from [21].

3. Exact sequences, due to D. Lewis, involving the various Witt groups defined in this chapter, can be found in his paper [42].

4. A form of Frobenius Reciprocity for orthogonal representations (cf. Th. 4.2.6, p. 167) was proved by G. Nebe [52].

5. The relationship (4.24) and the examples following are given by B. Kahn on p. 138, [34].

6. Corollary 4.4.1 is due to Frobenius [25] (in characteristic 0); furthermore he and Schur characterized linear representations which are equivalent to an orthogonal or symplectic representation (cf. pp. 106-109, [68]), that is to say, for which there is a G-invariant symmetric or skew symmetric form. See Th. 1.7.7 (p. 80) and Cor. 1.7.8 (p. 80).

A much more elementary matrix proof of Frobenius' Theorem is due to A.I. Malcev (pp. 184-186, [48]) for an arbitrary group G (not necessarily finite).

7. Theorem 4.5.4 is also due to Frobenius and Schur; see [68], p. 106 for a different proof.

8. Cor. 4.5.3 was first proved by Fröhlich and McEvett (Prop. 4.9, p. 128, [28]). They also showed that an invariant symmetric or Hermitian form on an irreducible KG-module is definite – our Cor. 4.5.1.

More recently, G. Nebe gave a method in [51] for calculating, under certain circumstances, the determinant or the Hasse invariant of an invariant symmetric form of a representation of a perfect group over a totally real number field, and in [52] she gives a recursion formula for the invariant symmetric form of certain irreducible representations of a symmetric group. This form is also determined, in complete generality using different methods, in [32] – we give an example in §5.6.

9. Th. 4.6.1 is due to W. Scharlau (Th. 13.5, p. 324, [65]) in the case when $K = K_0$. Parts (b) and (c) of that theorem are related to the phenomenon of "uniformly distributed invariants" of a Brauer class – see Def. 6.5, p. 93, [74].

10. Cor. 4.6.1 appears here for the first time.

11. Th. 4.8.1 and its use in this context were suggested to me by J.P. Tignol.

12. Th. 4.9.1 is due to Fröhlich (Th. 1.2, p. 474, [26]) in the case when $(K, ^-)$ is of the first kind, and extends easily to the case of the second kind included here.

13. Th. 4.9.4 is a sharpening of Th. 12.4, p. 502, [26]. When this paper appeared (1972), the equivalence theory of skew Hermitian forms over global quaternion algebras (developed by Bartels [3] and appearing in 1975) was unavailable, and Fröhlich gave instead a clever proof of a slightly weaker result, which appears in this book as the exercise on p. 203.

14. Th. 4.11.1 first appeared in the paper [64] of W. Scharlau. The proof given here is due to M.G. Mahmoudi [47].

15. If A is an arbitrary finite dimensional algebra with involution, and K is of any characteristic, C. Cibils [16] showed that $\hat{\mathrm{W}}_\varepsilon(A) \cong \hat{\mathrm{W}}_\varepsilon(A/\mathrm{rad}\,A)$ where the modified Witt group $\hat{\mathrm{W}}_\varepsilon(A)$ is defined via $(A, ^-)$-invariant ε-symmetric forms.

CHAPTER 5

Representations over finite, local and global fields

A precise characterization of the equivalence classes of isometry representations would necessarily have to take the nature of the field K into account since this is already true for reflexive forms themselves. Such a characterization has already been carried out over algebraically and real closed fields in §§4.4 and 4.5, and the conditions under which the Hasse Principle for representations over global fields holds have been determined in Th. 3.3.3. Some special results for symplectic representations over local fields appear in Cor. 4.9.3(a) and over global fields in Cor. 4.9.4(a). In this chapter we deal with further properties of isometry representations over finite, local and global fields. Although the classification theory of isometry representations in general is not known for these fields, there are some partial results which lead to classification in many circumstances.

5.1 Outline

Let G be a finite group and $\rho : G \to \mathbf{GL}(V)$ a linear representation over a finite, local or global field $(K, {}^{-})$ with involution. The principal goals of this chapter are to give at least partial solutions to the following problems:

I *Determine the (nonequivariant) equivalence classes of G-invariant reflexive forms on V, and if h is such a form – that is to say, if $\rho(G) \subset \mathbf{I}(V, h)$ – count the equivalence classes of representations $G \to \mathbf{I}(V, h)$ (orthogonal, symplectic or unitary) – in other words, count the* equivariant *equivalence classes in the equivalence class of h.*

II *Determine conditions on G or $(K, {}^{-})$ under which two orthogonal, symplectic or unitary representations $\rho : G \to \mathbf{I}(V, h)$ and $\rho' : G \to \mathbf{I}(V', h')$ are equivalent if and only if their underlying reflexive forms h and h' are "isotypically" equivalent.*

Isotypically equivalent means that corresponding involution isotypic components $(AV, h|_{AV \times AV})$ and $(AV', h'|_{AV' \times AV'})$ are (not necessarily equivariantly) equivalent for each simple involution component $(A, {}^{-})$ of $(KG, {}^{-})$. This of course implies that the lengths of corresponding isotypic components are the same, and therefore that the associated isometry representations are linearly equivalent.

If $(A, {}^{-})$ is a simple component of $(KG, {}^{-})$, an isometry representation $G \to \mathbf{I}(V, h)$ is $(A, {}^{-})$*-isotypic* (or simply A*-isotypic*) if the resulting KG-module structure on V satisfies $AV = V$ – thus h is $(A, {}^{-})$-invariant – and *has length* m if $\operatorname{len}_A V = m$.

As usual, we assume throughout that the characteristic of K does not divide $2|G| = 2g$.

There are reasonably complete solutions to **I** when K is a finite field, or when K is a local or global field and the representation is "split".

The results for **II** are strongest when the degrees of the absolutely irreducible representations of G are odd – for example if char $K = 0$ and G has odd order or is an Abelian group, or more generally has a normal Abelian subgroup of odd index – and, in the case that K is a local or global field, when the representations are split. This is done by using the solution to **I** to determine when there is a unique equivalence class of representations to each AV. See Ths. 5.2.1, Cor. 5.2.2, Th. 5.3.1, Cor. 5.4.1 and Ths. 5.4.6, 5.4.7, 5.5.1 and 5.5.2.

Strategy. We can restrict our attention to A-isotypic representations $\rho : G \to \mathbf{I}(V, h)$ where $A \cong \mathrm{M}(n, D)$ is a simple component of KG stable under the standard $(K, ^{-})$-involution. The general strategy of the proofs is as follows:

1. If $A \cong \mathrm{M}(n, D)$, view the succession of equivalences

$$\{\rho : G \to \mathbf{I}(V, h)\} \rightsquigarrow \{\hat{h} : V \times V \to (A, ^{-})\} \rightsquigarrow \{\breve{h} : \breve{V} \times \breve{V} \to (D, ^{-})\}$$

(cf. Ths. 2.1.1 (p. 86) and 3.2.14 (p. 132)) in the reverse direction. Thus we consider all possibilities for $\breve{h}$ (up to equivalence), and for each such $\breve{h}$, determine $\hat{h}$ by Morita theory, and then h by the formula $h = \frac{nd}{g}\mathrm{trd}_{A/K}\hat{h}$ (cf. Th. 1.2.13, p. 15) via the transfer theory using Th. 3.2.12, p. 128 (where $d^2 = \dim_{Z(D)} D$).

2. The (equivalence classes of) h's obtained in this way are those which admit an isometry representation $G \to \mathbf{I}(V, h)$ linearly equivalent to the (isotypic) representation of G on S^m where S is a simple A-module and $m = \dim_D \breve{V}$; the number of equivalence classes of $\breve{h}$'s which lead to h or a form equivalent to it, is the number of inequivalent isometry representations $G \to \mathbf{I}(V, h)$ of length m.

One *could* begin with the form h and then determine the $\breve{h}$'s which give rise to it by this procedure, but there is little to be gained in doing so. The procedure outlined does have the advantage of making it clear that all of the possibilities for $(A, ^{-})$-isotypic isometry representations have been taken into account.

We will denote the set of equivalence classes of nonsingular ε-Hermitian forms of rank m over $(D, ^{-})$ by ${}_{(D,^{-})}\mathbf{M}_m$, sometimes simply by ${}_D\mathbf{M}_m$.

Hyperbolic Representations. Recall that an isometry representation $\rho : G \to \mathbf{I}(V, h)$ is hyperbolic if and only if the ε-Hermitian form $\hat{h} : V \times V \to (KG, ^{-})$ is hyperbolic (cf. Cor. 3.1.2). Of course if ρ is A-isotypic where $(A, ^{-})$ is a simple involution component of $(KG, ^{-})$, then the form $\hat{h}$ can be viewed as a form over $(A, ^{-})$. Moreover if $(A, ^{-})$ is a hyperbolic algebra, then $\hat{h}$ *must* be hyperbolic (Cor. 3.1.5), and it follows in this case that two isometry representations of G which are A-isotypic are equivalent if and only if they are equivalent as linear representations, if and only if their spaces have the same length as A-modules.

This fact will be used (often implicitly) throughout this chapter.

On the other hand if A is simple and $\bar{A} = A$ and $AV = V$, then

Proposition 5.1.1 *There is an ε-hyperbolic representation $G \to \mathbf{I}(V, h)$ if (and only if) $m = \mathrm{len}_A V$ is even. And if it exists, it is unique up to equivalence.*

A representation $G \to \mathbf{I}(V, h)$ is hyperbolic if and only if the associated form $\breve{h} : D^m \times D^m \to (D, ^{-})$ is hyperbolic (Cor. 3.2.5, p. 133). And of course such a form $\breve{h}$ exists if and only if m is even, and it is unique up to equivalence. □

Corollary 5.1.1 *If V is an A-module of even length and there is exactly* one *equivalence class of A-isotypic isometry representations $G \to \mathbf{I}(V,h)$, then this class* is *the class of hyperbolic representations.* □

Notation. The notation will by and large be that introduced in Ch. 3, which we recall here, along with a few facts that will be used frequently.

A will stand for a simple algebra component of KG, stable under the standard $(K, ^-)$-involution $^-$ of KG, and S for a simple A-module. We will identify A with the matrix algebra $\mathrm{M}(n,D)$, where D is a division algebra of dimension d^2 over its center L which has degree l over K. L is also (isomorphic to) the center of A, and the subfield of L fixed by $^-$ is denoted by L_0. Let $\mathcal{G}$ be the Galois group $\mathrm{Gal}(L/K)$ – it is Abelian of order l – see p. 78.

The canonical trace $\mathrm{Tr}_{KG/K} : KG \to K$ will also be denoted by Tr; its restriction to A is $\frac{nd}{g}\mathrm{trd}_{A/K}$ (Th. 1.2.13(a), p. 15).

The division algebra D has an involution $^-$ of the same kind as A. If it is of the first kind and D is not commutative, it is assumed to be symplectic – in particular it is conjugation * if D is a quaternion algebra, which is always the case if K is a local or global field.

There is an $\underline{h}_0 \in \mathbf{GL}(n,D)$ such that $\bar{a} = \underline{h}_0^{-1}a^{t-}\underline{h}_0$ for all $a \in A = \mathrm{M}(n,D)$, where $\underline{h}_0^{t-} = \varepsilon_0\underline{h}_0$, $\varepsilon_0 = \pm 1$, and $= 1$ if $(A, ^-)$ is of the second kind, i.e. is unitary. If $(A, ^-)$ is of the first kind, $\varepsilon_0 = 1$ if and only if $(A, ^-)$ has the same type (orthogonal or symplectic) as $\mathrm{M}(n,(D, ^-))$ – or equivalently, as $(D, ^-)$. When the index of A is 1, this means that $\varepsilon_0 = 1$ if $(A, ^-)$ is orthogonal, $= -1$ if $(A, ^-)$ is symplectic, since the transpose ^{t} is orthogonal. When the index is > 1, it is the opposite: $\varepsilon_0 = -1$ if $(A, ^-)$ is orthogonal, $= 1$ if $(A, ^-)$ is symplectic, because of our assumption that $(D, ^-)$ is symplectic if it is of the first kind and D is noncommutative.

If V is an A-module of length m we generally identify it with $D^{n\times m}$. Then a sesquilinear form $\hat{h} : V \times V \to (A, ^-)$ is given by $\hat{h}(u,v) = u\underline{\hat{h}}v^{t-}\underline{h}_0$ for a matrix $\underline{\hat{h}} \in \mathrm{M}(m,D)$. If the form $\hat{h}$ is ε-Hermitian (i.e. $\mathrm{Tr}\,\hat{h}$ is ε-reflexive), the matrix $\underline{\hat{h}}$ is $\breve{\varepsilon}$-Hermitian where $\breve{\varepsilon} = \varepsilon\varepsilon_0$, and conversely. The $\breve{\varepsilon}$-Hermitian form $u\underline{\hat{h}}v^{t-}$ ($u, v \in D^{1\times m}$) over $(D, ^-)$ (which is Morita equivalent to $\hat{h}$) is denoted by $\check{h}$.

The "basic sign relationship"

$$\varepsilon\varepsilon_0\breve{\varepsilon} = 1 \tag{5.1}$$

is used constantly in this chapter.

We summarize:

(a) $\hat{h}$ is ε-Hermitian and $\mathrm{Tr}\,\hat{h} = \frac{nd}{g}\mathrm{trd}_A\hat{h}$ is ε-reflexive,

(b) $\bar{a} = \underline{h}_0^{-1}a^{t-}\underline{h}_0$ and $\underline{h}_0$ is ε_0-Hermitian,

(c) the matrix $\underline{\hat{h}}$ and the form $\check{h}$ are $\breve{\varepsilon}$-Hermitian.

If (V,h) is a reflexive space and we speak of A-isotypic representations $G \to \mathbf{I}(V,h)$ where A is simple, it will be understood that $\dim_K V$ is an integral multiple $m \dim_K S$ of the dimension of a simple A-module S. The integer m is called the *latent A-length* or simply the *latent length* of V.

5.2 Symplectic representations

Because of the fact that two nonsingular skew symmetric forms are equivalent if they have the same rank, symplectic representations are considerably easier to deal with than are orthogonal or unitary representations and so we begin with them.

Theorem 5.2.1 *We use the notation outlined above.*

Problem **I**. *Let V be an A-module of length m.*

$(A, ^-)$ orthogonal. *If A has index 1 (i.e. is quasisplit), there exists an $(A, ^-)$-isotypic symplectic representation $G \to \mathbf{Sp}(V, h)$ – in other words, V supports a nonsingular $(A, ^-)$-invariant skew symmetric form h – if and only m is even; it has only one equivariant equivalence class consisting of the hyperbolic representations of length m. This is the case in particular if K is finite or real closed since A is then necessarily quasisplit – cf. Ex. 3 on p. 17 and Th. 4.5.3(e), p. 172.*

If K is local or global and A has index > 1, there is an $(A, ^-)$-isotypic symplectic representation $G \to \mathbf{Sp}(V, h)$ for any m, unique up to equivalence, and consisting of hyperbolic representations if (and only if) m is even.

$(A, ^-)$ symplectic. *The equivalence classes of $(A, ^-)$-isotypic symplectic representations $G \to \mathbf{Sp}(V, h)$ are in bijective correspondence with the equivalence classes of symmetric bilinear forms of rank m over L if A has index 1 and with the classes of skew Hermitian forms of rank m over $(D, ^-)$ if the index of A is > 1.*

It follows that the number of equivalence classes of such (length m) symplectic representations $G \to \mathbf{Sp}(V, h)$ is

2	*if K is finite,*
$(2 - \delta_{1m})[\dot{L} : \dot{L}^2]$	*if K is local and A has index 1,*
$[\dot{L} : \dot{L}^2] - \delta_{1m}$	*if K is local and A has index > 1,*
1	*if K is real closed,*
∞	*if K is global.*

(δ_{ij} is the Kronecker delta.)

$(A, ^-)$ unitary. *The equivalence classes of $(A, ^-)$-isotypic symplectic representations $G \to \mathbf{Sp}(V, h)$ are in bijective correspondence with the skew Hermitian forms of rank m over $(D, ^-)$.*

Thus the number of equivalence classes of such symplectic representations $G \to \mathbf{Sp}(V, h)$ is

1	*if K is finite,*
2	*if K is local,*
$m + 1$	*if K is real closed,*
∞	*if K is global.*

Problem **II**. *Two symplectic representations $\rho : G \to \mathbf{Sp}(V, h)$ and $\rho' : G \to \mathbf{Sp}(V', h')$ of G over K are equivalent if and only if they are linearly equivalent, if at least one of the following conditions holds:*

(a) G is a symmetric group.
(b) K is finite and $\rho(A)V = 0$ if $(A, ^-)$ is a symplectic component of $(KG, ^-)$.
(c) ρ is split and $\rho(A)V = 0$ if $(A, ^-)$ is a symplectic component of $(KG, ^-)$.
(d) ρ is of odd representation type and if $(B, ^-)$ is a simple unitary involution component of $(KG, ^-)$ such that $\rho(B)V \neq 0$, then $(B, ^-)$ is hyperbolic.

Since h is skew symmetric, $\varepsilon = -1$ throughout and so $\breve{\varepsilon} = -\varepsilon_0$.

Consider first **I**.

$(A, ^-)$ orthogonal. If A has index 1, $\varepsilon_0 = 1$ so $\breve{\varepsilon} = -1$ by the basic sign relationship (5.1) and A-isotypic symplectic representations $G \to \mathbf{Sp}(V, h)$ correspond bijectively to equivalence classes of skew symmetric forms over L. Thus **I** in this case follows from Cor. 5.1.1. (It also follows from the results in §3.2.4 – see Cor. 3.2.2, p. 115.)

Now suppose that A has index > 1, i.e. $D \neq L$. Since we assume that the involution $^-$ on D is symplectic, $\varepsilon_0 = -1$. Thus $\breve{\varepsilon} = 1$ and $\hat{\underline{h}}$ must be an Hermitian matrix, so by Morita theory the equivalence classes of symplectic representations on V are in bijective correspondence with the equivalence classes of Hermitian forms of rank m over $(D, ^-)$. There is exactly one such class when K is local, by Table 1 (p. 61).

Suppose K is global. We must count the number of equivalence classes of Hermitian forms $\breve{h}$ of rank m – say on $D^{1\times m}$ – over $(D, ^*)$, a quaternion division algebra with conjugation as involution. If D splits at the prime $\mathfrak{p}$ of its center L, $\breve{h}_\mathfrak{p}$ is an Hermitian form

$$\breve{h}_\mathfrak{p} : L_\mathfrak{p}^{2\times 2m} \times L_\mathfrak{p}^{2\times 2m} \to (\mathrm{M}(2, L_\mathfrak{p}), ^-)$$

where $^-$ is a symplectic involution. Thus

$$\breve{h}_\mathfrak{p}(u, v) = u\underline{g}v^t\underline{h}_1$$

with $\underline{h}_1$ a skew symmetric matrix of degree 2 and $\underline{g}$ a skew symmetric matrix of degree $2m$ over $L_\mathfrak{p}$, and so $\underline{g}$ – and $\breve{h}_\mathfrak{p}$ – are uniquely determined up to equivalence – that is to say their equivalence classes depend only on m, not on $\breve{h}$.

Of course D *does* split at every complex prime of its center, and also at every real prime since $(A, ^-)$ is orthogonal – cf. Th. 4.6.1(b), p. 182.

If D is *not* split at the (finite) prime $\mathfrak{p}$, $\breve{h}_\mathfrak{p}$ is an Hermitian form over the local division algebra $D_\mathfrak{p}$ and so is again uniquely determined up to equivalence by its rank (Table 1).

Therefore by the Hasse Principle for Hermitian forms over $(D, ^*)$ (7, p. 63), there is, up to equivalence, only one Hermitian form of rank m over $(D, ^*)$. This, along with Cor. 5.1.1, proves **I** in this case ($(A, ^-)$ orthogonal and K global).

$(A, ^-)$ symplectic. This means that $\underline{h}_0$ is skew symmetric ($\varepsilon_0 = -1$) if A has index 1 and is Hermitian ($\varepsilon_0 = 1$) otherwise. Thus by the basic sign relationship, h is skew symmetric if and only if $\hat{\underline{h}}$ is symmetric in the first case and skew Hermitian in the second. The cases of K finite, and of K global or local and A of index 1, follow easily, again by Table 1 and the fact that $\dot{K}/\dot{K}^2$ is infinite if K is global. If K is global and A is of index > 1, we must show that there are an infinite number of skew Hermitian forms over a quaternion division algebra $D = (\alpha, \beta)_L$. By Witt's Theorem, it suffices to do this for dimension 1. The determinant of the form $\langle xi + yj + zk\rangle$ is $(-\alpha x^2 - \beta y^2 + \alpha\beta z^2)\dot{K}^2$, and so the result follows since the quadratic form $\langle -\alpha, -\beta, \alpha\beta\rangle$ represents an infinite number of square classes (Lem. 1.5.3, p. 55).

If K is real closed, A is necessarily of index > 1 (§1.3.1); according to Table 1, skew Hermitian forms over D are classified by their rank and so there is only 1 equivalence class of symplectic representations in this case. Finally the case of K local and $(A, ^{-})$ of index > 1 follows from "Tsukamoto's Theorem" (p. 63).

$(A, ^{-})$ unitary. Then $\varepsilon_0 = 1$ and the symplectic representations of length m are in bijective correspondence with the classes of skew Hermitian forms of rank m over $(D, ^{-})$. Since the involution is of the second kind, we can instead count the number of classes of Hermitian forms. If K is finite or local, the desired result follows from Table 1 since, in the local case, $\dot{K}_0/N_{K/K_0}\dot{K}$ has order 2 by local class field theory (p. 18).

If K is real closed, then by Th. 4.5.3(e) (p. 172) $(A, ^{-}) \cong (\mathrm{M}(n, L), \ ^{t*})$ where L is the algebraic closure of K. Again by Table 1, the number of Hermitian forms of rank m over $(L, \ ^{*})$ is $m+1$. If K is global, the form $\langle 1, 1, \ldots, 1, \alpha \rangle, \alpha \in \dot{L}_0$, has determinant $\alpha N_{L/L_0}\dot{L}$ and since $\dot{L}_0/N_{L/L_0}\dot{L}$ is infinite (Lem. 1.5.4, p. 56), there are an infinite number of equivalence classes of symplectic representations of length m.

Now **II**. If G is a symmetric group, it is known that all simple components of KG are split and stable under the standard (K, id)-involution, and that the induced involution in each is orthogonal (cf. §5.6). This proves **II** in this case (by **I**). If K is finite, all simple K-algebras are quasisplit (Ex. 3, p. 17) so (b) also follows at once from **I** and Cor. 3.1.5 (p. 98), as does (c) since the assumption that ρ is split rules out $(A, ^{-})$ unitary. If ρ is of odd type and $AV \neq 0$ for some simple component of KG, then A must be quasisplit if $\bar{A} = A$ since its index is a power of 2 (p. 19), and it cannot be symplectic since the degree of a symplectic involution algebra is even (Ex. 4, p. 29), and (d) follows easily from **I**. □

From this we get a conjugacy result (cf. Th. 2.1.3, p. 89):

Corollary 5.2.1 *Let G and G' be two finite subgroups of $\mathbf{Sp}(V, h)$ of orders relatively prime to* $\operatorname{char} K$ *if* $\operatorname{char} K \neq 0$, *such that either*

(a) $G \cong$ a symmetric group, or

(b) the inclusion $G \to \mathbf{GL}(V)$ is a linear representation of odd type, and K is either finite or a splitting field for G.

Then G and G' are conjugate in $\mathbf{Sp}(V, h)$ if and only if they are conjugate in $\mathbf{GL}(V)$. □

Of course **II** implies a slightly stronger result than stated. Note also that if G has odd order – or has a normal Abelian subgroup of odd index – then the conclusion of the corollary holds if K is a splitting field of characteristic 0. Cf. p. 78.

We single out **II** for the case K a local field.

Corollary 5.2.2 *Suppose K is a local field and let $\rho : G \to \mathbf{Sp}(V, h)$ and $\rho' : G \to \mathbf{Sp}(V', h')$ be symplectic representations with the property that $AV = 0 = AV'$ for any symplectic or simple unitary component A of KG such that $\bar{A} = A$. Then $\rho \cong \rho'$ if and only if they are equivalent as linear representations.*

This is clear from the theorem, and is a generalization of Cor. 4.9.3(a), p. 199. □

Remark. We note that if $\operatorname{char} K = 0$, and if order of G is odd or, more generally, if G has a normal Abelian subgroup H of odd index, then the degree of any simple component A of KG is odd, and so in particular $(A, ^{-})$ cannot be symplectic (cf.

p. 78). Thus the condition $AV = 0 = AV'$ need only be stated for simple unitary components when this is true. □

Exercise. Let G be the cyclic group $\langle s \rangle$ of order 4, V a vector space of dimension 2 over K, and $h : V \times V \to (K, \mathrm{id})$ the skew symmetric form with matrix $\begin{pmatrix} 0 & 1 \\ -1 & 0 \end{pmatrix}$ with respect to some basis v_1, v_2. Define a linear representation $\rho : G \to \mathbf{GL}(V)$ by $\rho(s)v_1 = -v_2$, $\rho(s)v_2 = v_1$. Show that ρ is a symplectic representation $G \to \mathbf{Sp}(V, h)$ and that it is hyperbolic if and only if $-1 \in \dot{K}^2$.

5.3 Unitary and orthogonal representations over finite fields

Symplectic representations over a finite field are covered in Th. 5.2.1, so we restrict our attention to unitary and orthogonal representations in this section.

Recall the notation given in detail earlier (p. 221) – in particular that A is simple. Note that an unorthodox definition of field discriminant is used in (d) (cf. p. 57).

Theorem 5.3.1 *Let K be a finite field with an involution $^-$ and V a K-space of latent A-length m.*

Problem **I**.

Unitary representations. *There is one equivalence class of Hermitian or skew Hermitian forms on V, and one equivalence class of A-isotypic unitary representations $G \to \mathbf{U}(V, h)$ in both the Hermitian case and the skew Hermitian case.*

Orthogonal representations. *The number of equivalence classes of A-isotypic representations $G \to \mathbf{O}(V, h)$ is*

(a) 1 if $(A, ^-)$ is symplectic and m is even; the equivalence class consists of all A-isotypic hyperbolic representations on V, – in particular h is a hyperbolic form,

(b) 2 if $(A, ^-)$ is orthogonal and n is odd, 1 for each of the two classes of nonsingular symmetric forms on V,

(c) 2 if $(A, ^-)$ is orthogonal, n is even and $\det h = (\det_K(^-))^m$,

(d) 1 if $(A, ^-)$ is unitary and $\operatorname{disc} h = (\det(L/L_0))^{mn|\dot{L}_0|/|\dot{K}|}$, and

(e) 0 otherwise.

Problem **II**.

Unitary representations. *Two unitary representations of G over K are equivalent if and only if they are linearly equivalent.*

Orthogonal representations. *If $\rho : G \to \mathbf{O}(V, h)$ and $\rho' : G \to \mathbf{O}(V', h')$ are linearly equivalent, and if $AV = 0$ for every orthogonal component $(A, ^-)$ of $(KG, ^-)$ of even degree, then ρ and ρ' are equivalent if and only if h and h' are isotypically equivalent – more precisely, for each orthogonal component $(A, ^-)$ of odd degree, the restrictions of h to AV and h' to AV' are equivalent.*

Remark. Note that if $\rho : G \to \mathbf{O}(V, h)$ is an orthogonal representation and $(A, ^-)$ is an orthogonal component of $(KG, ^-)$ of even degree such that $AV \neq 0$, then it is necessarily the case that $\det h|_{AV \times AV} = (\det_K(^-))^m$ by **I**.

Similarly if $(A, ^-)$ is symplectic or unitary, then the equivalence of the isotypic components of type A of two linearly equivalent orthogonal representations follows automatically from the existence of the orthogonal representations. □

Note that $A = \mathrm{M}(n, L)$ since the Brauer group of a finite field is trivial, and that $\varepsilon = 1$ since we are dealing with unitary and orthogonal representations and we can assume that the form h in a unitary representation is Hermitian.

Unitary representations. That there is exactly one equivalence class of Hermitian or skew Hermitian forms on V is stated in Table 1, p. 61. Now consider unitary representations $G \to \mathbf{U}(V, h)$. Since $(A, ^-)$ is a $(K, ^-)$-involution algebra, it is unitary, and so $\varepsilon_0 = 1$ and $\breve{\varepsilon} = \varepsilon\varepsilon_0 = 1$, i.e. $\breve{h}$ is Hermitian. Since there is only one equivalence class of Hermitian forms of a given rank, there is exactly one equivalence class of $(A, ^-)$-isotypic unitary representations of each possible rank. (This is also true for hyperbolic factors of $(KG, ^-)$ of course.) This proves I and also II since to say that two linear representations are equivalent is the same as saying that their respective isotypic components have the same length.

Orthogonal representations. Suppose first that $(A, ^-)$ is symplectic, so $\varepsilon_0 = -1$ and $\breve{\varepsilon} = \varepsilon\varepsilon_0 = -1$, i.e. $\breve{h}$ is skew symmetric. Thus m is even and there is one and only one equivalence class of orthogonal representations for a given even positive integer m, which consists of the hyperbolic representations by Cor. 5.1.1, p. 221.

Suppose next that $(A, ^-)$ is orthogonal, so $\varepsilon_0 = 1$ and $\breve{h}$ and its discriminant matrix $\underline{\hat{h}}$ must be symmetric. By Th. 3.2.12 (p. 128), $\det(\mathrm{trd}_{A/L}\hat{h}) = (\det \underline{\hat{h}})^n (\det \underline{h}_0)^m$. Thus if n is odd, by appropriate choice of $\underline{\hat{h}}$ we get both possible determinants over L for the transfer $\frac{n}{g}\mathrm{trd}_{A/L}\hat{h}$, and so also both determinants over K by Lem. 1.5.6 (p. 57) since the norm $\mathrm{N}_{L/K}$ in a finite extension of finite fields preserves nonsquares. This means that the association $\breve{h} \rightsquigarrow h$ is bijective on equivalence classes, which proves (b).

On the other hand, if n is even, this association on equivalence classes is 2 to 1, and by Lem. 1.5.6 the image is the class with determinant

$$\begin{aligned}(\det(L/K))^{mn}\mathrm{N}_{L/K}((\det \underline{\hat{h}})^n(\det \underline{h}_0)^m) &=_2 (\mathrm{N}_{L/K} \det \underline{h}_0)^m \\ &= (\det_K(^-))^m\end{aligned}$$

by p. 32. This is (c).

If $(A, ^-)$ is unitary, $\varepsilon_0 = 1 = \breve{\varepsilon}$. Since Hermitian forms over a finite field of odd characteristic are determined up to equivalence by their rank, this means that for every m, there is exactly one equivalence class of orthogonal representations supported by a vector space of K-dimension lmn. It remains to determine which of the two equivalence classes of symmetric forms on K^{lmn} does admit an orthogonal representation.

If we begin with an Hermitian matrix $\underline{\hat{h}} \in \mathbf{GL}(m, L)$, the transfer $\mathrm{trd}_{A/L}\hat{h}$ of its associated form $\hat{h}$ is an Hermitian form of dimension mn over $(L, ^-)$, hence a sum of mn copies of $\langle 1 \rangle$. The transfer

$$\mathrm{trd}_{A/L_0}\hat{h} = \mathrm{Tr}_{L/L_0}\mathrm{trd}_{A/L}\hat{h}$$

is therefore the sum of mn copies of the binary symmetric form $\mathrm{Tr}_{L/L_0}(x\bar{y})$ over L_0, which has discriminant matrix

$$\begin{pmatrix} 2 & 0 \\ 0 & -2\delta \end{pmatrix} \qquad (L = L_0(\sqrt{\delta}))$$

with determinant $-\delta$. Therefore the discriminant of the symmetric form $\mathrm{trd}_{A/L_0}\hat{h}$ of rank $2mn$ is

$$(-1)^{2mn(2mn-1)/2}(-\delta)^{mn} = \delta^{mn},$$

so by Lem. 1.5.6 (p. 57) the symmetric form

$$h := \tfrac{n}{g}\mathrm{trd}_{A/K}\hat{h} = \tfrac{n}{g}\mathrm{Tr}_{L_0/K}\mathrm{trd}_{A/L_0}\hat{h}$$

of rank lmn over K has discriminant

$$(\tfrac{n}{g})^{lmn}(\mathrm{disc}(L_0/K))^{2mn}\mathrm{N}_{L_0/K}(\delta)^{mn} =_2 \mathrm{N}_{L_0/K}(\delta)^{mn}$$

since l is even. But if $q = |K|$,

$$\mathrm{N}_{L_0/K}\delta = \delta\delta^q \cdots \delta^{q^{l_0-1}} = \delta^{1+q+\cdots+q^{l_0-1}} = \delta^{(q^{l_0}-1)/(q-1)} = \delta^{|\dot{L}_0|/|\dot{K}|},$$

which proves (d) since $\delta =_2 \det(L/L_0)$. Thus **I** is proved and **II** follows easily. □

Corollary 5.3.1 *If K is a finite field, two subgroups of* $\mathbf{U}(V,h)$ *whose orders are not divisible by* char K, *are conjugate in* $\mathbf{U}(V,h)$ *if and only if they are conjugate in* $\mathbf{GL}(V)$. □

Exercise. Let K be a finite field (of characteristic $\neq 2$), and suppose h is the symmetric form $\langle 1, \delta\rangle$ on $V = K^{1\times 2}$ where δ is a nonsquare in K. Let

$$G = \left\{\begin{pmatrix} 1 & 0 \\ 0 & 1\end{pmatrix}, \begin{pmatrix} 1 & 0 \\ 0 & -1\end{pmatrix}\right\}, \quad G' = \left\{\begin{pmatrix} 1 & 0 \\ 0 & 1\end{pmatrix}, \begin{pmatrix} -1 & 0 \\ 0 & 1\end{pmatrix}\right\}.$$

Show that G and G' are conjugate in $\mathbf{GL}(V)$ but not in $\mathbf{O}(V,h)$.

5.4 Unitary and orthogonal representations over local fields

In this section K is a local (nonArchimedean) field, *dyadic* if 2 is not a unit, *nondyadic* otherwise. Recall that ((1.22), p. 17),

$$[\dot{K} : \dot{K}^2] = \begin{cases} 4 & \text{if } K \text{ is nondyadic,} \\ 2^{2+(K:\mathbb{Q}_2)} & \text{if } K \text{ is dyadic.}\end{cases}$$

The Witt equivalence of two symmetric forms h_1, h_2 over K is denoted by $h_1 \sim_W h_2$. The quaternary anisotropic symmetric form over K, unique up to equivalence (63:17, p. 169, [54]), is denoted by g_K. It is (equivalent to) the bilinear form of norm form xx^* of the (unique, up to isomorphism) quaternion algebra $(D, {}^*)$, and has determinant 1 and Hasse invariant $-(-1,-1)_K$ – the latter since the quaternary hyperbolic form has determinant 1 and Hasse invariant $(-1,-1)_K$. Moreover if h is a nonsingular symmetric form,

$$\text{rank } h \text{ is even and } \mathrm{disc}\, h = 1 \quad\Leftrightarrow\quad h \sim_W 0 \text{ or } g_K. \tag{5.2}$$

The next four theorems deal with unitary and orthogonal representations $G \to \mathbf{I}(V,h)$ which are A-isotypic with respect to a simple component A of KG which is stable under the standard $(K, {}^-)$-involution ${}^-$ of KG. We can assume that the K-space V has latent A-length m and, if we wish, that $A = \mathrm{M}(n, D)$ and $V = D^{n\times m}$.

See pp. 221*ff* for unexplained notation.

5.4.1 Unitary representations.

Theorem 5.4.1 *The number of equivalence classes of A-isotypic unitary representations $G \to \mathbf{U}(V, h)$ of length m is*

1 *if n is odd,*
2 *if n is even and $\det h = (\det_K(^-))^m$,*
0 *if n is even and $\det h \neq (\det_K(^-))^m$,*

where $\det_K(^-)$ is the determinant of the unitary involution of $(A, ^-)$ (cf. p. 32).

Since $\varepsilon = 1 = \varepsilon_0$, $\breve{\varepsilon} = \varepsilon\varepsilon_0 = 1$ as well, so the equivalence classes of A-isotypic unitary representations $G \to \mathbf{U}(V, h)$ of length m are in bijective correspondence with the equivalence classes ${}_{(L,^-)}\mathbf{M}_m$ of nonsingular Hermitian forms of rank m – note that $D = L$ since a (noncommutative) division algebra over a local field does not support an involution of the second kind – cf. p. 34.

Recall (Table 1, p. 61) that nonsingular Hermitian forms over a local field L are classified by their rank and determinant $\in \dot{L}_0/\mathrm{N}_{L/L_0}\dot{L} \cong \mathbb{Z}/2$ (p. 18). Thus ${}_{(L,^-)}\mathbf{M}_m$ consists of two classes.

Suppose $[\breve{h}] \in {}_{(L,^-)}\mathbf{M}_m$. If $\underline{\hat{h}} \in \mathbf{GL}(m, L)$ is the discriminant matrix of $\breve{h}$, let

$$\hat{h}(u, v) = u\underline{\hat{h}}v^{t-}\underline{h}_0 \quad (u, v \in L^{n\times m}) \tag{5.3}$$

be its Morita correspondent over $(A, ^-)$. By Th. 3.2.12 (p. 128), the transfer $\mathrm{trd}_{A/L}\hat{h}$ down to L is Hermitian with determinant $(\det \underline{\hat{h}})^n(\det \underline{h}_0)^m$, with both $\det \underline{\hat{h}}$ and $\det \underline{h}_0 \in L_0$. Thus if n is even, we get but one form over $(L, ^-)$ (up to equivalence), namely that of rank mn and determinant $(\det \underline{h}_0)^m$, while if n is odd we get both forms of rank mn since $\det \underline{\hat{h}}$ is arbitrary in $\dot{L}_0/\mathrm{N}_{L/L_0}\dot{L}$.

The transfer, via $\frac{n}{g}\mathrm{Tr}_{L/K}$, of an equivalence class $[h_1] \in {}_{(L,^-)}\mathbf{M}_{mn}$ down to ${}_{(K,^-)}\mathbf{M}_{lmn}$ has determinant

$$(\tfrac{n}{g})^{lmn}\det(L/K)^{mn}\mathrm{N}_{L/K}(\det h_1),$$

where $\det(L/K) \in \dot{K}_0$ by Lem. 1.5.6 (p. 57), and $l = (L : K)$. When n is odd, the transfers to K of the two classes in ${}_{(L,^-)}\mathbf{M}_{mn}$ therefore remain distinct by Lem. 5.4.1 below, and the theorem follows in this case.

On the other hand, when n is even, there is only one class of Hermitian forms over $(K, ^-)$ which supports unitary representations; it supports *two* classes of unitary representations and has determinant

$$\begin{aligned}(\tfrac{n}{g})^{lmn}\det(L/K)^{mn}\mathrm{N}_{L/K}(\det \underline{h}_0)^m &=_2 \mathrm{N}_{L/K}(\det_L(^-))^m \\ &= (\det_K(^-))^m\end{aligned}$$

since $\det(L/K) \in \dot{K}_0$. □

This theorem provides the following partial solution to Problem **II**:

Corollary 5.4.1 *Suppose K is a local field with a nonidentity involution $^-$, and let $\rho : G \to \mathbf{U}(V, h)$ and $\rho' : G \to \mathbf{U}(V', h')$ be unitary representations over $(K, ^-)$ such that either*
(i) $AV = 0 = AV'$ for every simple component A of KG of even degree such that $\bar{A} = A$, or
(ii) ρ and ρ' are representations of odd type, or
(iii) $\mathrm{char}\, K = 0$ *and G has a normal Abelian subgroup of odd index – e.g. $|G|$ is odd.*

Then $\rho \cong \rho'$ if and only if h and h' are are isotypically equivalent.

The necessity is trivial. To prove the sufficiency, it is enough to prove (i) since (ii) implies (i) and, when $\operatorname{char} K = 0$, (iii) also implies (i) – cf. p. 78.

The isotypic equivalence of ρ and ρ' implies that $BV \cong BV'$ as KG-modules for every simple hyperbolic involution component $(B, ^-)$ of $(KG, ^-)$, and so the subrepresentations

$$G \to \mathbf{U}(BV, h|_{BV \times BV}) \quad \text{and} \quad G \to \mathbf{U}(BV', h'|_{BV' \times BV'})$$

are equivalent (pp. 97*ff*). Since the degree of any simple summand A of KG such that $\bar{A} = A$ and $AV \neq 0 \neq AV'$ is odd, the corollary now follows at once from the theorem. □

Lemma 5.4.1 *The homomorphism $\bar{\mathrm{N}} : \dot{L}_0/\mathrm{N}_{L/L_0}\dot{L} \to \dot{K}_0/\mathrm{N}_{K/K_0}\dot{K}$ induced by N_{L_0/K_0} (or $\mathrm{N}_{L/K}$), is an isomorphism.*

Let $\mathrm{Gal}^{ab}_{L_0}$ be the Galois group of the Abelian closure L_0^{ab} of L_0 in some separable closure L_0^{sep} of L_0. Define $\mathrm{Gal}^{ab}_{K_0}$ similarly, using the same separable closure L_0^{sep}. Let $\mathrm{Gal}_{L_0} = \mathrm{Gal}(L_0^{sep}/L_0)$ and $\mathrm{Gal}_{K_0} = \mathrm{Gal}(L_0^{sep}/K_0)$ be the absolute Galois groups of L_0 and K_0. Then $\mathrm{Gal}_{L_0} \subset \mathrm{Gal}_{K_0}$ and this inclusion induces a homomorphism[1]

$$\mathrm{Gal}^{ab}_{L_0} = \mathrm{Gal}_{L_0}/\mathrm{Gal}'_{L_0} \xrightarrow{i} \mathrm{Gal}_{K_0}/\mathrm{Gal}'_{K_0} = \mathrm{Gal}^{ab}_{K_0}, \quad \sigma\mathrm{Gal}'_{L_0} \to \sigma\mathrm{Gal}'_{K_0}.$$

This fits into the commutative diagram

$$\begin{array}{ccc} \dot{L}_0 & \longrightarrow & \mathrm{Gal}^{ab}_{L_0} \\ {\scriptstyle \mathrm{N}_{L_0/K_0}} \downarrow & & \downarrow {\scriptstyle i} \\ \dot{K}_0 & \longrightarrow & \mathrm{Gal}^{ab}_{K_0} \end{array}$$

where the horizontal maps are isomorphisms given by the norm residue symbol – cf. p. 141, [14]. This implies a commutative diagram

$$\begin{array}{ccc} \dot{L}_0/\mathrm{N}_{L/L_0}\dot{L} & \longrightarrow & \mathrm{Gal}_{L/L_0} \\ {\scriptstyle \bar{\mathrm{N}}} \downarrow & & \downarrow \\ \dot{K}_0/\mathrm{N}_{K/K_0}\dot{K} & \longrightarrow & \mathrm{Gal}_{K/K_0} \end{array}$$

where the right vertical map is restriction – and is an isomorphism. (This diagram also appears in Prop. 5.8, p. 296, [53]). Since the horizontal maps are isomorphisms by local class field theory, $\bar{\mathrm{N}}$ is an isomorphism as well. □

5.4.2 Orthogonal representations. We now come to the case of orthogonal representations $G \to \mathbf{O}(V, h)$ over the local field K, which are significantly more complicated than are symplectic and unitary representations. The notation introduced on pp. 221*ff* and 227 will be used freely.

---

[1] the derived subgroup of a group H is denoted by H'.

Theorem 5.4.2 *If* $(A, {}^-)$ *is unitary, with* A *simple, the number of equivalence classes of* A*-isotypic orthogonal representations* $G \to \mathbf{O}(V, h)$ *of length* m *is*

1 *if* n *is odd and* $\det h =_2 \mathrm{N}_{L_0/K}(-\det(L/L_0))^m$,

2 *if* n *is even, and either* $h \sim_W 0$ *and* $(-1)^{mn/2}(\det_L({}^-))^m \in \mathrm{N}_{L/L_0}\dot{L}$,
or $h \sim_W g_K$ *and* $(-1)^{mn/2}(\det_L({}^-))^m \notin \mathrm{N}_{L/L_0}\dot{L}$,

0 *otherwise.*

There is no (noncommutative) division algebra with an involution of the second kind over a local field, so A is quasisplit and $\varepsilon_0 = 1$. Since $\varepsilon = 1$, $\breve{\varepsilon} = \varepsilon_0\varepsilon$ is also 1. Thus the equivalence classes of A-isotypic orthogonal representations $G \to \mathbf{O}(V, h)$ of length m are in bijective correspondence with the equivalence classes ${}_{(L,{}^-)}\mathbf{M}_m$ of nonsingular Hermitian forms of rank m. (This was also the case in the previous theorem – the difference here is that the involution on K is the identity, even though that on L is not).

Again we start with a class $[\breve{h}] \in {}_{(L,{}^-)}\mathbf{M}_m$, with discriminant matrix $\underline{\hat{h}}$ and Morita correspondent $\hat{h} : V \times V \to (A, {}^-)$ given by (5.3).

Consider first n even. Since $\mathrm{trd}_{A/L}\hat{h}$ is an Hermitian form over $(L, {}^-)$ with determinant $(\det \underline{\hat{h}})^n(\det \underline{h}_0)^m$ (Th. 3.2.12, p. 128), we can assume that its discriminant matrix is the diagonal matrix $\langle 1, \ldots, 1, (\det \underline{h}_0)^m\rangle$ of rank nm. The transfer down to L_0 via Tr_{L/L_0} of a rank 1 Hermitian form $\langle\eta\rangle$ over $L = L_0(\sqrt{\lambda_0})$, is the symmetric form

$$\langle 2\eta, -2\eta\lambda_0\rangle = \langle\eta\rangle \otimes \langle 2, -2\lambda_0\rangle,$$

so the transfer of $\hat{h}$ down to L_0 via trd_{A/L_0} is the symmetric form

$$\langle 1, \ldots, 1, (\det \underline{h}_0)^m\rangle \otimes \langle 2, -2\lambda_0\rangle \tag{5.4}$$

of rank $2mn$. The determinant of this form over L_0 is

$$((\det \underline{h}_0)^m)^2(-4\lambda_0)^{mn} =_2 1.$$

By the properties of the Hasse invariant and Lem. 1.5.1 (d) on p. 53, its Hasse invariant is

$$(-1, -1)_{L_0}^{mn/2}((-1)^{n/2}\det \underline{h}_0, \lambda_0)_{L_0}^m.$$

In particular it is independent of $\breve{h}$. Since its rank is divisible by 4 and its determinant is 1, it is $\sim_W 0$ or g_{L_0} by (5.2), and by Lem. 1.5.1(e) is $\sim_W 0$ if and only if its Hasse invariant

$$(-1, -1)_{L_0}^{mn/2}((-1)^{n/2}\det \underline{h}_0, \lambda_0)_{L_0}^m = (-1, -1)_{L_0}^{mn/2}.$$

This condition is equivalent to $((-1)^{mn/2}(\det \underline{h}_0)^m, \lambda_0)_{L_0} = 1$ which, by property 1., p. 53, is in turn equivalent to

$$(-1)^{mn/2}(\det \underline{h}_0)^m \in \mathrm{N}_{L/L_0}\dot{L}.$$

Since the hyperbolic form $\langle 1, 1, -1, -1\rangle$ over L_0 and g_{L_0} have the same rank and determinant but different Hasse invariants, their transfers down to K (via $\frac{n}{g}\mathrm{Tr}_{L_0/K}$) remain inequivalent by Milnor's Theorem (Th. 1.5.11, p. 59), and so must be hyperbolic and $\sim_W g_K$ respectively (since the transfer of the form $\sim_W g_{L_0}$ has the same determinant as the transfer of the hyperbolic form – and the latter is clearly hyperbolic). Thus in either case we get a single symmetric form $h \sim_W 0$ or g_K supporting 2 nonequivalent orthogonal representations. This finishes the proof in the case of n even since $\det \underline{h}_0 = \det_L({}^-)$.

Now suppose that n is odd and that $[\breve{h}] \in {}_{(L,^-)}\mathbf{M}_m$. One shows that, as in the case n even, the transfer to L_0 by trd_{A/L_0} of the form $\hat{h}$ Morita equivalent to $\breve{h}$ is the symmetric form

$$\langle 1, \ldots, 1, (\det \underline{\hat{h}})(\det \underline{h}_0)^m \rangle \otimes \langle 2, -2\lambda_0 \rangle \tag{5.5}$$

of rank $2mn$. Its determinant is

$$((\det \underline{\hat{h}})(\det \underline{h}_0)^m)^2 (-4\lambda_0)^{mn} =_2 (-\lambda_0)^m = (-\det(L/L_0))^m,$$

and its Hasse invariant is

$$((-1)^{m(mn-1)/2}(2 \det \underline{h}_0)^m \det \underline{\hat{h}},\ \lambda_0)_{L_0} (-1,-1)_{L_0}^{m(mn-1)/2} \tag{5.6}$$

(see below). The two possible choices for $[\breve{h}]$ lead to the two possibilities for $\det \underline{\hat{h}} \in \dot{L}_0/\mathrm{N}_{L/L_0}\dot{L}$, and so to distinct Hasse invariants. Therefore the two transfers to L_0 are distinct. Their transfers to K via $\frac{n}{g}\mathrm{Tr}_{L_0/K}$ have determinant

$$(\tfrac{n}{g})^{\frac{l}{2}(2mn)} \det(L_0/K)^{2mn} \mathrm{N}_{L_0/K}(-\det(L/L_0))^m =_2 \mathrm{N}_{L_0/K}(-\det(L/L_0))^m$$

by Lem. 1.5.8 (p. 59) since $(L_0 : K) = \frac{l}{2}$, and are inequivalent by Milnor's Theorem.

It remains to compute the Hasse invariant of (5.5). The reader is advised to skip[2] this, but it is included for completeness. By Lem. 1.5.1(d), the Hasse invariant is

$$(2,-2\lambda_0)^m \Big((\det \underline{\hat{h}})(\det \underline{\hat{h}}_0)^m, -1\Big)\Big((\det \underline{\hat{h}})(\det \underline{\hat{h}}_0)^m, -\lambda_0\Big)(-\lambda_0,-1)^{mn(mn-1)/2}.$$

Now we use the properties of the Hasse invariant on p. 53:

$$\begin{aligned} (2,-2\lambda_0)^m = (2,\lambda_0)^m &= (2^m,-1)(2^m,-\lambda_0), \\ (-\lambda_0,-1)^{mn(mn-1)/2} &= (-\lambda_0,(-1)^{m(mn-1)/2}), \end{aligned}$$

and recombine to get

$$\Big((\det \underline{\hat{h}})(2 \det \underline{\hat{h}}_0)^m, -1\Big)\Big((-1)^{m(mn-1)/2}(\det \underline{\hat{h}})(2 \det \underline{h}_0)^m, -\lambda_0\Big)$$

for the Hasse invariant. Now

$$\begin{aligned} &((\det \underline{\hat{h}})(2 \det \underline{\hat{h}}_0)^m, -1) = \\ &\quad \Big((-1)^{m(mn-1)/2}(\det \underline{\hat{h}})(2 \det \underline{\hat{h}}_0)^m, -1\Big)((-1)^{m(mn-1)/2}, -1) = \\ &\quad \Big((-1)^{m(mn-1)/2}(\det \underline{\hat{h}})(2 \det \underline{\hat{h}}_0)^m, -1\Big)(-1,-1)^{m(mn-1)/2}, \end{aligned}$$

which leads directly to (5.6). □

Theorem 5.4.3 *If $(A, ^-)$ is symplectic, the number of equivalence classes of A-isotypic orthogonal representations $G \to \mathbf{O}(V,h)$ of length m is,*

if A is quasisplit, 1 *if m is even,* 0 *otherwise;*

if A is not quasisplit, 1 *if $h \sim_W g_K$ and mn is odd,* 1 *if $h \sim_W 0$ and mn is even,* 0 *otherwise.*

[2] I've checked it several times!

REMARKS. It follows from the theorem and from Cor. 5.1.1, p. 221, that if m is even, there is exactly one equivalence class of A-isotypic orthogonal representations (when $(A, ^{-})$ is symplectic) and it consists of the hyperbolic representations of length m. On the other hand in the non-quasisplit case, if m is odd and n is even, there is a unique class of A-isotypic representations $G \to \mathbf{O}(V, h)$ with h hyperbolic, but they are *not* hyperbolic representations.

It is easily seen (for example from the proof below) that the quasisplit case holds as stated for *any* field K. This also follows from Cor. 3.2.2, p. 115. □

Either A is quasisplit and $\varepsilon_0 = -1$, or $A \cong \mathrm{M}(n, D)$ where D is a quaternion algebra with conjugation as involution and $\varepsilon_0 = 1$. Since h is symmetric, $\varepsilon = 1$ and

(i) in the quasisplit case $\breve{\varepsilon} = \varepsilon\varepsilon_0 = -1$ and the equivalence classes of A-isotypic orthogonal representations $G \to \mathbf{O}(V, h)$ of length m are in bijective correspondence with the equivalence classes ${}_{(L,^{-})}\mathbf{M}_m$ of nonsingular skew symmetric forms of rank m (of which there is of course only one for each even m and none for odd m), so the theorem follows at once in this case.

(ii) in the nonquasisplit case $\breve{\varepsilon} = 1$ and the equivalence classes of A-isotypic orthogonal representations $G \to \mathbf{O}(V, h)$ of length m are in bijective correspondence with the equivalence classes ${}_{(D,\,^*\,)}\mathbf{M}_m$ of nonsingular Hermitian forms of rank m over $(D,\ ^*\ )$, the unique quaternion algebra with conjugation as involution. But there is exactly one such class for each m, no matter if m is even or odd, since the equivalence classes of nonsingular Hermitian forms over $(D,\ ^*\ )$ are characterized by their dimension (cf. the table on p. 61). Furthermore since $\underline{h}_0$ is Hermitian, by changing the identification of A with $\mathrm{M}(n, D)$, we can assume that $\underline{h}_0 = \mathrm{I}_n$, and then by replacing $\breve{h}$ by an equivalent form, that its discriminant matrix $\underline{\hat{h}} = \mathrm{I}_m$ as well (cf. Th. 3.2.13(c), p. 128). Thus $\hat{h}(u, v) = uv^{t*}$, $u, v \in D^{n\times m}$. It is readily apparent from this that if we define the "trace" $\mathrm{trd}_{A/D}(d_{ij}) = \sum_i d_{ii}$, then $\mathrm{trd}_{A/D}\hat{h} \in {}_{(D,\,^*\,)}\mathcal{H}_1^*$ and is equivalent to the nonsingular Hermitian form on D^{nm} with discriminant matrix I_{nm} since the basis $\{e_{ij}\}$ of elementary matrices of the D-space $D^{n\times m}$ is orthonormal. Furthermore $\mathrm{trd}_{A/L}\hat{h} = \mathrm{trd}_{D/L}\mathrm{trd}_{A/D}\hat{h}$ (cf. (1.13)) and it follows that $\mathrm{trd}_{A/L}\hat{h}$ is the orthogonal direct sum of nm copies of the norm form g_L of D. Since $g_L \perp g_L$ is hyperbolic, $\mathrm{trd}_{A/L}\hat{h}$ is hyperbolic of rank $4nm$ if n or m is even – in which case $\mathrm{trd}_{A/K}\hat{h}$ is also hyperbolic, of rank $4lnm$. On the other hand if nm is odd, then $\mathrm{trd}_{A/L}\hat{h} \sim_W g_L$, so by Milnor's Theorem and the fact that the transfers of $\langle 1, 1, -1, -1\rangle$ and g_L down to K have the same determinant (Lem. 1.5.8, p. 59), $\mathrm{trd}_{A/K}\hat{h} \sim_W g_K$. □

In the next theorem,

$$\begin{aligned}
\delta &= \begin{cases} 1 & \text{if } m = 2 \text{ and } h \text{ is hyperbolic,} \\ 0 & \text{otherwise,} \end{cases} \\
\delta' &= \begin{cases} 1 & \text{if } m = 2 \text{ and } h \sim_W g_K, \\ 0 & \text{otherwise,} \end{cases} \\
\delta_{im} &= \text{the Kronecker delta.}
\end{aligned} \tag{5.7}$$

Theorem 5.4.4 *Suppose that $(A, ^{-})$ is orthogonal and quasisplit.*

(a) If n is even and $\mathrm{disc}_L(^-) \neq 1$, *the number of equivalence classes of A-isotypic orthogonal representations* $G \to \mathbf{O}(V,h)$ *of length m is*

$$\begin{array}{ll} \dfrac{[\dot{L}:\dot{L}^2]}{1+\delta_{1m}} - \delta & \textit{if } \det h = (\det_K(^-))^m, \\ 0 & \textit{otherwise.} \end{array} \tag{5.8}$$

(b) If n is even and $\mathrm{disc}_L(^-) = 1$, *the number of equivalence classes of A-isotypic orthogonal representations* $G \to \mathbf{O}(V,h)$ *of length m is*

$$\frac{2[\dot{L}:\dot{L}^2]}{1+\delta_{1m}} - \delta_{2m}.$$

The underlying form h of all of them is hyperbolic if

(i) m is even, or

(ii) $\mathrm{s}(^-) = \begin{cases} -1 & \textit{if } n \equiv 4 \textit{ or } 6 \bmod 8,\ K \textit{ is dyadic and } (L:\mathbb{Q}_2) \textit{ is odd,} \\ 1 & \textit{otherwise.} \end{cases}$

In all other cases $h \sim_W g_K$.

(In particular if n is even and $\mathrm{disc}_L(^-) = 1$, *there are no A-isotypic orthogonal representations* $G \to \mathbf{O}(V,h)$ *unless* $h \sim_W 0$ *or* g_K.*)*

(c) If n is odd and > 1, the number of equivalence classes of A-isotypic orthogonal representations $G \to \mathbf{O}(V,h)$ *of length m is*

$$\begin{array}{ll} \frac{[\mathcal{G}:\mathcal{G}^2][\dot{L}:\dot{L}^2]}{[\dot{K}:\dot{K}^2]} - \delta' & \textit{if } m>1 \textit{ and } \det h \in (\det(L/K))^m(\mathrm{N}_{L/K}\dot{L})\dot{K}^2, \\ \leq \frac{[\mathcal{G}:\mathcal{G}^2][\dot{L}:\dot{L}^2]}{[\dot{K}:\dot{K}^2]} & \textit{if } m=1 \textit{ and } \det h \in (\det(L/K))(\mathrm{N}_{L/K}\dot{L})\dot{K}^2, \\ 0 & \textit{otherwise.} \end{array}$$

In the case $m=1$, let $[h]$ and $[h']$ be equivalence classes of nonsingular symmetric forms over K of rank ln with the same determinant in $(\det(L/K))(\mathrm{N}_{L/K}\dot{L})\dot{K}^2$ *but different Hasse invariants. Then the number of classes of A-isotypic orthogonal representations* $G \to \mathbf{O}(V,h)$ *plus the number for (V',h') is* $= \frac{[\mathcal{G}:\mathcal{G}^2][\dot{L}:\dot{L}^2]}{[\dot{K}:\dot{K}^2]}$.

(d) If $n=1$, then $A=K$ and every nonsingular symmetric form h admits one and only one A-isotypic orthogonal representation $G \to \mathbf{O}(V,h)$, *and two such representations are equivalent if and only if the two forms are equivalent (simply as forms over K).*

Remarks. 1. In (b), it is easy to see, *a priori*, that if $G \to \mathbf{O}(V,h)$ is an A-isotypic orthogonal representation, $h \sim_W 0$ or g_K. Indeed, since $\mathrm{disc}(^-) = \mathrm{disc}(\langle \underline{h}_0 \rangle)$, $\det \underline{h}_0 = (-1)^{n(n-1)/2} = (-1)^{n/2}$, so by (5.11) the determinant of $\mathrm{trd}_{A/L}\hat{\underline{h}} = (-1)^{nm/2}$ and so the determinant of $\mathrm{trd}_{A/K}\hat{\underline{h}} = \frac{n}{g}\mathrm{Tr}_{L/K}\mathrm{trd}_{A/L}\hat{\underline{h}}$ is

$$(\tfrac{n}{g})^{lmn}(\det(L/K))^{mn}\mathrm{N}_{L/K}(-1)^{mn/2} =_2 (-1)^{lmn/2}$$

by Lem. 1.5.6 (p. 57) and of course its rank is lmn. Thus either it has rank $lmn \equiv 0 \bmod 4$ and determinant 1, or rank $lmn \equiv 2 \bmod 4$ and determinant -1. In the latter case the form is either a hyperbolic plane or is equivalent to a form which is the sum of a hyperbolic plane and a form of rank $\equiv 0 \bmod 4$ and determinant 1. It now follows from (5.2) that $h \sim_W 0$ or g_K.

Recall too that a hyperbolic representation of length m exists if and only if m is even, and then it is unique up to equivalence (Prop. 5.1.1, p. 220). Thus if there is more than one equivalence class of A-isotypic representations $G \to \mathbf{O}(V,h)$ with h

hyperbolic (when m is even), one of the classes is hyperbolic and the others are not. When m is odd, there are *no* hyperbolic representations of length m even though it is possible that a hyperbolic form h supports an orthogonal representation. (This remark applies generally no matter the nature of the simple component $(A, {}^-)$ of $(KG, {}^-)$).

2. Note that $\frac{[\mathcal{G}:\mathcal{G}^2][\dot{L}:\dot{L}^2]}{[\dot{K}:\dot{K}^2]} = 1$ if and only if $(L:K)$ is odd and either $L = K$ or K is nondyadic (cf. (1.22), p. 17).

3. If $l = (L:K)$ is odd, then $(\mathrm{N}_{L/K}\dot{L})\dot{K}^2 = \dot{K}$ since $(L:K) = [\dot{K} : \mathrm{N}_{L/K}\dot{L}]$ by local class field theory (cf. Ex. 4, p. 17) and so if in addition n is odd and > 1 and $m > 2$, (c) implies that *every* symmetric form h of rank lnm admits $[\dot{L} : \dot{L}^2]/[\dot{K} : \dot{K}^2]$ classes of A-isotypic orthogonal representations.

If $(L:K)$ and n are odd and $m = 2$, then the same holds except that the form $h \sim_W g_K$ of rank $2ln$ admits only $([\dot{L} : \dot{L}^2]/[\dot{K} : \dot{K}^2]) - 1$ classes.

In particular if $L = K$ or K is nondyadic, and if n and l are odd and $m \geq 2$, then every symmetric form h of rank lnm admits exactly one class of A-isotypic orthogonal representations except for the form $h \sim_W g_K$ of rank $2ln$, which admits none, while if $m = 1$, of two forms of rank ln with the same determinant, one admits A-isotypic orthogonal representations and the other does not; in particular if two orthogonal representations on a simple A-module V have equivalent forms, the representations must also be equivalent.

4. In the case n odd and $m = 1$, the symmetric forms which admit orthogonal representations are the transfers $\mathrm{Tr}_{L/K}\langle\alpha\underline{h}_0\rangle$, $\alpha \in \dot{L}$, where $\underline{h}_0$ is chosen to have determinant 1; thus the determinant of $\langle\alpha\underline{h}_0\rangle$ is α, and its Hasse-Witt invariant is $\mathrm{w}_L(\langle\underline{h}_0\rangle) = \mathrm{w}_L({}^-)$. The determinant of $\mathrm{Tr}_{L/K}\langle\alpha\underline{h}_0\rangle$ is $(\det L/K)\mathrm{N}_{L/K}(\alpha \det \underline{h}_0)$ and, in principle, its Hasse invariant (or Hasse-Witt invariant) can be calculated using Lem. 1, 3. [23]. However the result unfortunately involves the unknown factor $\mathrm{s}_K(\mathrm{Tr}_{L/K}\langle\alpha \operatorname{disc} \underline{h}_0\rangle)$, the Hasse invariant of a "scaled trace form" of the extension L/K – in other words, the determination of the Hasse invariant is reduced to the case $n = 1$. □

We now begin the proof of Th. 5.4.4. Since $\varepsilon = 1 = \varepsilon_0$, $\breve{\varepsilon} = 1$ as well, so the classes of A-isotypic orthogonal representations of length m are in bijective correspondence with the equivalence classes ${}_L\mathbf{M}_m$ of (nonsingular) symmetric forms over L of rank m. These classes are of course determined by their determinants (or discriminants) and their Hasse invariants. Thus

$$|{}_L\mathbf{M}_m| = \begin{cases} 2[\dot{L} : \dot{L}^2] & \text{if } m > 2, \\ 2[\dot{L} : \dot{L}^2] - 1 & \text{if } m = 2, \\ [\dot{L} : \dot{L}^2] & \text{if } m = 1. \end{cases} \tag{5.9}$$

(63:23, p. 171, [54] – note that the Hasse invariant is defined slightly differently there). In the case of $m = 2$, there is no form of determinant -1 and Hasse invariant -1 since a binary form with determinant -1 is necessarily a hyperbolic plane (Th. 1.5.1) which has Hasse invariant 1.

Let $[\breve{h}] \in {}_L\mathbf{M}_m$, let $\hat{\underline{h}}$ be the discriminant matrix of $\breve{h}$ and

$$\hat{h} : L^{n\times m} \times L^{n\times m} \to (A, {}^-) = (\mathrm{M}(n, L), {}^-), \qquad \hat{h}(u, v) = u\hat{\underline{h}}v^t\underline{h}_0, \tag{5.10}$$

the Morita correspondent of $\breve{h}$.

The symmetric form $h_L = \mathrm{trd}_{A/L}\hat{h}$ has discriminant matrix

$$\underline{h}_0 \otimes \underline{\hat{h}},$$

(Th. 3.2.12, p. 128), determinant

$$(\det \underline{h}_0)^m (\det \underline{\hat{h}})^n =_2 \begin{cases} (\det \underline{h}_0)^m & \text{if } n \text{ is even,} \\ (\det \underline{h}_0)^m (\det \underline{\hat{h}}) & \text{if } n \text{ is odd,} \end{cases} \tag{5.11}$$

and Hasse invariant[3] $\mathrm{s}(h_L) = \mathrm{s}(\underline{h}_0 \otimes \underline{\hat{h}})$

$$\begin{array}{ll} \mathrm{s}(\underline{h}_0)^m (\det \underline{\hat{h}}, \operatorname{disc} \underline{h}_0)_L (\det \underline{h}_0, -1)_L^{m(m-1)/2} & \text{if } n \text{ is even,} \\ \mathrm{s}(\hat{h})\mathrm{s}(\underline{h}_0)^m (\det \underline{\hat{h}}, -1)_L^{(n-1)/2} & \\ \quad \cdot (\det \underline{\hat{h}}, \det \underline{h}_0)_L^{m-1} (\det \underline{h}_0, -1)_L^{m(m-1)/2} & \text{if } n \text{ is odd,} \end{array} \tag{5.12}$$

by Lem. 1.5.1(d), p. 53.

(a) Now suppose that n is even and $\operatorname{disc}_L(\bar{\ }) \notin \dot{L}^2$. Since $\operatorname{disc}_L(\bar{\ }) = \operatorname{disc} \underline{h}_0$, (5.12) implies that we get *both* equivalence classes of forms of determinant $(\det \underline{h}_0)^m$ over L for h_L since the factor $(\det \underline{\hat{h}}, \operatorname{disc} \underline{h}_0)_L$ can be chosen to be 1 or -1 by appropriate choice of $\underline{\hat{h}}$, i.e. of $\breve{h}$, while the other factors of (5.12) remain unchanged. So by Lem. 1.5.6 and Milnor's Theorem (p. 59), both equivalence classes of determinant

$$(\tfrac{n}{g})^{lmn} (\det(L/K))^{mn} \mathrm{N}_{L/K} (\det \underline{h}_0)^m =_2 (\det_K(\bar{\ }))^m$$

in ${}_K\mathbf{M}_{lmn}$ result from the transfer from L to K via $\frac{n}{g}\mathrm{Tr}_{L/K}$. Furthermore by (5.12), one of the forms arises from those $\breve{h}$ whose determinants $\det \underline{\hat{h}}$ are in the kernel of canonical homomorphism

$$\dot{L}/\dot{L}^2 \to \dot{L}/\mathrm{N}_{L(\sqrt{\operatorname{disc} \underline{h}_0})/L} \dot{L}(\sqrt{\operatorname{disc} \underline{h}_0}) \cong \mathbb{Z}/2$$

(property 1, p. 53), and the other from those whose determinants are in the other coset of $\dot{L}/\dot{L}^2$ modulo the kernel. The first statement of the theorem now follows easily by (5.9) when $m > 2$ or $m = 1$.

Now suppose that $m = 2$. In the proof above for $m > 2$, we effectively identify each class $[\breve{h}]$ in ${}_L\mathbf{M}_m$ with its invariants $(\det \breve{h}, \mathrm{s}(\breve{h}))$, and then trace the evolution of these invariants viewed as invariants of $\underline{\hat{h}}$ (and of $\hat{h}$), first through the transfer to those of a symmetric form over L via $\mathrm{trd}_{A/L}$ and then through the transfer to those of a symmetric form over K via $\frac{n}{g}\mathrm{Tr}_{L/K}$, eventually counting the number of classes in ${}_L\mathbf{M}_m$ that give rise to the same symmetric form over K. The procedure for $m = 2$ is identical except for the fact that there is no form with invariants $(-1,-1)$. It is clear by (5.12) that this set of invariants is transformed via the first transfer by $\mathrm{trd}_{A/L}$ into the same invariants that those – namely $(-1,1)$ – of the hyperbolic plane transform into, and since the hyperbolic plane is transferred by $\frac{n}{g}\mathrm{trd}_{A/K} = (\frac{n}{g}\mathrm{Tr}_{L/K})\mathrm{trd}_{A/L}$ into a hyperbolic space (K^{nl}, h) over K, this means that there is one less class of representations into the orthogonal group $\mathbf{O}(K^{nl}, h)$. This accounts for the δ in the statement of (a) and finishes the proof of (a).

(b) Now assume that n is even and $\operatorname{disc}_L(\bar{\ }) = 1$. Then $\operatorname{disc} \underline{h}_0 = 1$, so

$$\det \underline{h}_0 = (-1)^{n(n-1)/2} = (-1)^{n/2},$$

[3] In order to simplify the notation, we use the same symbol, e.g. $\underline{h}_0$, for a form and its discriminant matrix.

and so

$$\det h_L = ((-1)^{n/2})^m (\det \underline{\hat{h}})^n =_2 (-1)^{mn/2}$$

by (5.11) and

$$\mathrm{s}(h_L) = \mathrm{s}(\underline{h}_0)^m (-1,-1)_L^{mn(m-1)/4} \tag{5.13}$$

by (5.12). Thus in this case we get only *one* equivalence class $[h_L] \in {}_L\mathbf{M}_{mn}$ as a transfer from the Morita correspondents of ${}_L\mathbf{M}_m$, and so only one class of G-invariant symmetric forms $h = \frac{n}{g}\mathrm{Tr}_{L/K} h_L$ over K. They have rank lmn and determinant

$$(\tfrac{n}{g})^{lmn}(\det(L/K))^{mn}\mathrm{N}_{L/K}(-1)^{mn/2} =_2 (-1)^{lmn/2}$$

by Lem. 1.5.6, and we now determine their equivalence class by determining their Hasse invariant.

Suppose first that $mn/2$ is even. Then $\det h_L = 1$ and $\dim h_L = mn \equiv 0 \bmod 4$, so $h_L \sim_W 0$ or $h_L \sim_W g_L$ (cf. (5.2)). Moreover by Lem. 1.5.1(e) (p. 53)

$$\begin{aligned} h_L \sim_W 0 \quad &\Leftrightarrow \quad \mathrm{s}(h_L) = (-1,-1)_L^{mn/4} \\ &\Leftrightarrow \quad \mathrm{s}(\underline{h}_0)^m (-1,-1)_L^{mn(m-1)/4} = (-1,-1)_L^{mn/4} \\ &\Leftrightarrow \quad \mathrm{s}(\underline{h}_0)^m = (-1,-1)_L^{m^2 n/4}. \end{aligned}$$

This always holds if m is even. If m is odd, $n \equiv 0 \bmod 4$ and the foregoing condition is

$$\mathrm{s}(\underline{h}_0) = (-1,-1)_L^{n/4} = \begin{cases} (-1,-1)_L & \text{if } n \equiv 4 \bmod 8 \\ 1 & \text{if } n \equiv 0 \bmod 8. \end{cases}$$

On the other hand if $mn/2$ is odd – so m is odd and $n \equiv 2 \bmod 4$ – $\det h_L = -1$ and $\dim h_L \equiv 2 \bmod 4$. Thus if b is a hyperbolic plane over L, $\det(h_L \perp b) = 1$ and $\dim(h_L \perp b) \equiv 0 \bmod 4$, and so since $h_L \sim_W h_L \perp b$, we see again that $h_L \sim_W 0$ or $h_L \sim_W g_L$. Now $\mathrm{s}(h_L \perp b) = \mathrm{s}(h_L)(-1,-1)_L$ by Lem. 1.5.1(b), and so

$$h_L \sim_W 0 \Leftrightarrow h_L \perp b \sim_W 0 \Leftrightarrow \mathrm{s}(\underline{h}_0)^m (-1,-1)_L^{\frac{mn(m-1)}{4}+1} = (-1,-1)_L^{\frac{1}{2}(\frac{mn}{2}+1)\frac{mn}{2}}$$

by (5.13) and Lem. 1.5.1(e), and this holds if and only if

$$\begin{aligned} \mathrm{s}(\underline{h}_0)^m &= (-1,-1)_L^{\frac{1}{4}(mn(m-1)+4+(\frac{mn}{2}+1)mn)} \\ &= (-1,-1)_L^{\frac{n-2}{4}} \end{aligned}$$

since both m^2 and $\left(\frac{mn}{2}\right)^2$ are $\equiv 1 \bmod 8$. Thus since m is odd and $n \equiv 2 \bmod 4$, $h_L \sim_W 0$ if and only if

$$\mathrm{s}(h_0) = \begin{cases} (-1,-1)_L & \text{when } n \equiv 6 \bmod 8, \\ 1 & \text{when } n \equiv 2 \bmod 8. \end{cases}$$

Putting this together with the above condition when $n \equiv 0 \bmod 4$, we get

$$h_L \sim_W 0 \quad \Leftrightarrow \quad m \text{ is even or } \mathrm{s}(h_0) = \begin{cases} (-1,-1)_L & \text{when } n \equiv 4 \text{ or } 6 \bmod 8, \\ 1 & \text{when } n \equiv 0 \text{ or } 2 \bmod 8 \end{cases}$$

(when n is even and $\mathrm{disc}_L(^-) = 1$). Then (b) of the theorem follows from Milnor's Theorem and the fact that $(-1,-1)_L \neq 1$ if and only if its norm form $aa^* = \alpha_0^2 + \alpha_1^2 + \alpha_2^2 + \alpha_3^2$ is anisotropic (cf. p. 146, [54]), and this happens if and only if L is dyadic and $(L : \mathbb{Q}_2)$ is odd – cf. the values of the level $s(L)$ of L on p. 37 – and of course $\mathrm{s}(^-) = \mathrm{s}(h_0)$ (cf. p. 32).

(c) Now suppose n is odd. Since $\underline{h}_0$ can be replaced by $\alpha\underline{h}_0$ for any $\alpha \in \dot{L}$, we can assume that

$$\det \underline{h}_0 =_2 1.$$

If $[\check{h}] \in {}_L\mathbf{M}_m$ and the Morita correspondent of $\check{h}$ is $\hat{h}$ (cf. (5.10)), the determinant of $h_L = \mathrm{trd}_{A/L}\hat{h}$ is $(\det \underline{h}_0)^m(\det \check{h})^n =_2 \det \check{h}$ and its Hasse invariant is

$$\mathrm{s}(h_L) = \mathrm{s}(h_0)^m(-1, \det \check{h})_L \mathrm{s}(\check{h})$$

(Lem. 1.5.1(d)). It follows that the transfer

$$ {}_L\mathbf{M}_m \to {}_L\mathbf{M}_{mn} \tag{5.14}$$

induced by $\mathrm{trd}_{A/L}$ is onto if $m > 2$, and hence bijective since ${}_L\mathbf{M}_m$ and ${}_L\mathbf{M}_{mn}$ have the same finite cardinality. This is also true for any m if $n = 1$ since then (5.14) is the identity map.

Now consider the transfer $\frac{n}{g}\mathrm{Tr}_{L/K} : {}_L\mathbf{M}_{nm} \to {}_K\mathbf{M}_{lnm}$ when $m > 2$. If $[f] \in {}_L\mathbf{M}_{nm}$ the determinant of $\frac{n}{g}\mathrm{Tr}_{L/K}(f)$ is

$$(\tfrac{n}{g})^{lnm}(\det L/K)^{nm}(\mathrm{N}_{L/K}\det f) =_2 (\tfrac{n}{g})^{lm}(\det L/K)^m(\mathrm{N}_{L/K}\det f)$$

by Lem. 1.5.6, so by Milnor's Theorem (Th. 1.5.11), the image of $\frac{n}{g}\mathrm{Tr}_{L/K}$ consists of all classes in ${}_K\mathbf{M}_{lnm}$ with determinant in

$$(\tfrac{n}{g})^{lm}(\det L/K)^m(\mathrm{N}_{L/K}\dot{L})\dot{K}^2 = (\det L/K)^m(\mathrm{N}_{L/K}\dot{L})\dot{K}^2$$

since $(\frac{n}{g})^{lm} \in \dot{K}^l \subset \mathrm{N}_{L/K}\dot{L}$. Furthermore the fibers of this transfer have the same cardinality as the kernel of the map $\bar{\mathrm{N}} : \dot{L}/\dot{L}^2 \to \dot{K}/\dot{K}^2$ induced by $\mathrm{N}_{L/K}$ which, by Lem. 1.5.5 (p. 56), is $\frac{[\dot{L}:\dot{L}^2]}{[\dot{K}:\dot{K}^2]}[\mathcal{G} : \mathcal{G}^2]$.

This finishes the proof for n odd, $m > 2$.

If n is odd and > 1 and $m = 2$, (5.14) would be bijective were it not for the "missing class" of $[{}_L\mathbf{M}_2]$ with invariants $(\det, \mathrm{s}) = (-1, -1)$. The image in ${}_L\mathbf{M}_{2n}$ of the class of a hyperbolic plane, with invariants $(-1, 1)$, is a hyperbolic space, and it follows that the other class in ${}_L\mathbf{M}_{2n}$ with the same determinant as the hyperbolic space – namely the form $\sim_W g_L$ – is the only class not in the image of (5.14). By the same argument as in the case $m > 2$, the image of the transfer map ${}_L\mathbf{M}_{2n} \to {}_K\mathbf{M}_{2ln}$ consists of all classes in ${}_K\mathbf{M}_{2ln}$ with determinant in $(\det L/K)^m(\mathrm{N}_{L/K}\dot{L})\dot{K}^2$, and the fibers again have cardinality $|\ker \bar{N}|$. Since the image of a hyperbolic space under this transfer is hyperbolic, it again follows from Milnor's Theorem that the image of a space $\sim_W g_L$ is $\sim_W g_K$. This proves the theorem when n is odd and > 1 and $m = 2$.

Now suppose n is odd and > 1, and $m = 1$. If $[\check{h}] \in {}_L\mathbf{M}_1$, say $\check{h} = \langle\alpha\rangle$, then $\mathrm{trd}_{A/L}(\hat{h})$ has discriminant matrix $\alpha \otimes \underline{h}_0 = \alpha\underline{h}_0$, which has determinant $\alpha^n =_2 \alpha$, and so the transfer ${}_L\mathbf{M}_1 \to {}_L\mathbf{M}_n$ is injective, but not bijective since only one class in ${}_L\mathbf{M}_n$ of a given determinant is in the image, namely that with Hasse-Witt invariant $\mathrm{w}_L(^-) = w_L(\langle\underline{h}_0\rangle)$ (cf. Lem. 1.5.2, p. 55). And the transfer down to K is $\frac{n}{g}\mathrm{Tr}_{L/K}\langle\alpha\underline{h}_0\rangle = \mathrm{Tr}_{L/K}\langle\frac{n}{g}\alpha\underline{h}_0\rangle$. The factor $\frac{n}{g}$ can be absorbed into α, and $h = \mathrm{Tr}_{L/K}\langle\alpha\underline{h}_0\rangle$ has determinant

$$(\det L/K)^n\mathrm{N}_{L/K}\alpha =_2 (\det L/K)\mathrm{N}_{L/K}\alpha.$$

By a previous argument the number of classes of orthogonal representations whose underlying symmetric forms have this determinant is $\frac{[\mathcal{G}:\mathcal{G}^2][\dot{L}:\dot{L}^2]}{[\dot{K}:\dot{K}^2]}$. There are exactly two such symmetric forms, with different Hasse invariants. This confirms the last statement of (c).

(d) Suppose $n = 1$. Thus $A = L$, and since A is generated over K by the restriction to G of the projection $KG \to A$ and since the involution (which inverts the elements in the image of the projection of G on L) is orthogonal and therefore the identity on L, it follows that the image of G under the projection is 1 or ± 1 and $A = L = K$. Thus an orthogonal representation $G \to \mathbf{O}(V,h)$ in this case can have *any* underlying form h and furthermore two such representations are equivalent if and only if their forms are equivalent. □

Exercise. Assume that $m = 1$, n is odd and ≥ 3, and that $\alpha \in (\det L/K)\mathrm{N}_{L/K}\dot{L}$. Show that each of the two forms over K of rank nl and determinant α support at least one orthogonal representation if any of the following holds:

(a) $\alpha =_2 1$, $l \equiv 2 \bmod 4$, $-1 \notin \dot{K}^2$.

(b) $\alpha \neq_2 1$, $l \equiv 0 \bmod 4$.

(c) $\alpha \neq_2 -1$, $l \equiv 2 \bmod 4$.

Hint: Can assume $\det \underline{h}_0 = 1$. Suppose $\alpha = (\det L/K)\mathrm{N}_{L/K}\alpha_L$, and compare the forms $\frac{n}{g}\mathrm{trd}_{L/K}\langle \alpha_L \underline{h}_0\rangle$ and $\frac{n}{g}\mathrm{trd}_{L/K}\langle \beta\alpha_L \underline{h}_0\rangle$ for $\beta \in \dot{K}$. □

In the next theorem,

$$m' = \begin{cases} 2 & \text{if } m = 1 \text{ and either } n \text{ is odd and } h \sim_W 0, \\ & \qquad\qquad \text{or } n \text{ is even and } h \sim_W g_K, \\ m & \textit{otherwise}. \end{cases}$$

Theorem 5.4.5 *Suppose that* $(A, {}^-)$ *is orthogonal and not quasisplit. The number of equivalence classes of A-isotypic orthogonal representations* $G \to \mathbf{O}(V,h)$ *is*

$$\begin{array}{ll} [\dot{L}:\dot{L}^2] - \delta_{1m} & \textit{if } \mathrm{disc}_L({}^-) = 1 \textit{ and } h \sim_W \begin{cases} 0 & \textit{if } mn \textit{ is even}, \\ g_K & \textit{if } mn \textit{ is odd}, \end{cases} \\ \frac{1}{2}[\dot{L}:\dot{L}^2] - \delta_{1m'} & \textit{if } \mathrm{disc}_L({}^-) \neq 1 \textit{ and } h \sim_W 0 \textit{ or } g_K, \\ 0 & \textit{otherwise}. \end{array}$$

Here $A = \mathrm{M}(n, D)$ where D is a quaternion division algebra over L with conjugation * as its involution. Since $(\mathrm{M}(n,D), {}^-)$ is orthogonal and $(D, {}^*)$ is symplectic, $\varepsilon_0 = -1$, i.e. $\underline{h}_0$ is skew Hermitian; and since $\varepsilon = 1$, $\breve{\varepsilon} = -1$ so ${}_D\mathbf{M}_m$ consists of the equivalence classes of skew Hermitian forms over $(D, {}^*)$ of rank m.

Since skew Hermitian forms can be diagonalized, we can suppose that $\underline{h}_0 = \langle d_{01}, \ldots, d_{0n}\rangle$ where $d_{0j}^* = -d_{0j} \in \dot{D}$ for all j, and that the matrix $\hat{\underline{h}}$ of $\breve{h}$ is $\hat{\underline{h}} = \langle d_1, \ldots, d_m\rangle$, $d_i^* = -d_i$ for all i (Th. 3.2.13(c), p. 128).

Let $B = \mathrm{M}(m, D)$. By Tsukamoto's Theorem (p. 63), the equivalence classes ${}_{(D,\,{}^*)}\mathrm{M}_m$ of skew Hermitian forms of rank m correspond bijectively via the determinant $\mathrm{nrd}_{B/L}\hat{\underline{h}}$ to the square classes $\dot{L}/\dot{L}^2$ when $m > 1$, and to all of the square classes except $-\dot{L}^2$ when $m = 1$.

By Th. 3.2.13(d) and (e), $\mathrm{trd}_{A/L}\hat{h}$ is equivalent to the orthogonal direct sum of the mn transfers

$$h_{ij}(u,v) = \mathrm{trd}_{D/L}(ud_i v^* d_{0j}) \quad (u, v \in D,\ 1 \leq i \leq m,\ 1 \leq j \leq n)$$

of one dimensional Hermitian forms over D with respect to the orthogonal involution $d \to d_{0j}^{-1} d^* d_{0j}$. By Lem. 3.2.4 (p. 130), h_{ij} is a quaternary symmetric form over L of determinant 1, and is $\sim_W 0$ if and only if the "quaternion symbol" $(d_i^2, d_{0j}^2)_L = -1$. Otherwise – i.e. if $(d_i^2, d_{0j}^2)_L = 1$ – it is $\cong g_L$ by (5.2). Since $g_L \perp g_L \sim_W 0$, it follows that $\mathrm{trd}_{A/L}\hat{h} \sim_W 0$ if and only if

$$\prod_{i,j}(d_i^2, d_{0j}^2)_L = (-1)^{mn},$$

and otherwise is $\sim_W g_L$. Now

$$\prod_i d_i^2 = \prod_i(-d_i d_i^*) = \prod_i(-\mathrm{nrd}_{D/L} d_i) = (-1)^m \mathrm{nrd}_{B/L}\,\underline{\hat{h}}$$

by (1.14), and similarly $\prod_j d_{0j}^2 = (-1)^n \mathrm{nrd}_{A/L}\,\underline{\hat{h}}_0$, so $\mathrm{trd}_{A/L}\hat{h} \sim_W 0$ if and only if

$$\left((-1)^m \mathrm{nrd}_{B/L}\underline{\hat{h}},\ (-1)^n \mathrm{nrd}_{A/L}\underline{h}_0\right)_L = (-1)^{mn},$$

i.e.

$$\mathrm{trd}_{A/L}\hat{h} \sim_W 0 \text{ if and only if } \left((-1)^m \mathrm{nrd}_{B/L}\underline{\hat{h}},\ \mathrm{disc}_L(^-)\right)_L = (-1)^{mn}. \tag{5.15}$$

Suppose first that $\mathrm{disc}_L(^-) = 1$. Then if mn is even, $\mathrm{trd}_{A/L}\hat{h} \sim_W 0$ for *all* equivalence classes ${}_{(D,\,^*\,)}\mathrm{M}_m$, and if mn is odd, $\mathrm{trd}_{A/L}\hat{h} \sim_W g_L$ for all classes. By Milnor's Theorem, this statement continues to hold if L is replaced by K. Thus for all $[\breve{h}] \in {}_{(D,\,^*\,)}\mathrm{M}_m$, $h = \frac{2n}{g}\mathrm{trd}_{A/K}\hat{h} \sim_W 0$ if mn is even and $\sim_W g_K$ if mn is odd.

Now suppose that $\mathrm{disc}_L(^-) \neq 1$. Then the symbol $(\lambda, \mathrm{disc}_L(^-))_L$ is -1 for half of the square classes $\lambda \dot{L}^2 \in \dot{L}/\dot{L}^2$ and 1 for the other half. This yields the result for this case when $m > 1$ by Tsukamoto's Theorem. If $m = 1$, then -1 is not the determinant of any class in ${}_{(D,\,^*\,)}\mathrm{M}_1$, and so there is one less equivalence class of representations for the form (hyperbolic or $\sim_W g_K$) corresponding to $(-(-1), \mathrm{disc}_L(^-))_L = (-1)^n$. Since $(1, \mathrm{disc}_L(^-))_L = 1$, this form is hyperbolic if n is even, $\sim_W g_K$ if n is odd by (5.15). □

The following theorem is a partial solution to Problem II for orthogonal representations.

Theorem 5.4.6 *Let K be a local field and $\rho : G \to \mathbf{O}(V,h)$ and $\rho' : G \to \mathbf{O}(V',h')$ orthogonal representations of G over K. If A is a simple summand of KG such that*

$$\bar{A} = A \quad \textit{and} \quad AV \neq 0,$$

assume that
(a) if $(A, ^-)$ is unitary, A has odd index,
(b) if $(A, ^-)$ is orthogonal, then A has odd index and either
(i) A is split, or
(ii) K is nondyadic and $(L:K)$ is odd.
Then $\rho \cong \rho'$ if and only if h and h' are isotypically equivalent.

This follows easily from Ths. 5.4.2, 5.4.3, 5.4.4 and 5.4.5, and the facts about hyperbolic representations (§3.1.2). We note in particular that the conditions (i) and (ii) both imply that $\frac{[\mathcal{G}:\mathcal{G}^2][\dot{L}:\dot{L}^2]}{[\dot{K}:\dot{K}^2]} = 1$. □

Isometry Representations of an Abelian group. We briefly specialize the foregoing results for Problem II for local fields to representations $G \to \mathbf{I}(V, h)$ when G is Abelian (and finite, of course).

Theorem 5.4.7 *Suppose G is Abelian and $(K, {}^-)$ is a local field with involution. Let $\rho : G \to \mathbf{I}(V, h)$ and $\rho' : G \to \mathbf{I}(V', h')$ be isometry representations of G over $(K, {}^-)$.*

(a) ***Unitary representations.*** *$\rho \cong \rho'$ if and only if h and h' are isotypically equivalent.*

(b) ***Orthogonal representations.*** *$\rho \cong \rho'$ if and only if h and h' are isotypically equivalent.*

(c) ***Symplectic representations.*** *Let e be the exponent of G, i.e. the smallest positive integer such that $s^e = 1$ for all $s \in G$. Then in order for all symplectic representations ρ and ρ' of G over K to satisfy*

$$\rho \cong \rho' \text{ if and only if } h \text{ and } h' \text{ are isotypically equivalent,} \tag{5.16}$$

it is necessary and sufficient that K contain a primitive q^{th} root of 1 for all divisors q of e which are odd primes or 4.

Since KG is Abelian, it is the direct sum of finite field extensions of K. Thus (a) follows from Cor. 5.4.1, (i).

Suppose that $(K, {}^-) = (K, \text{id})$. If L is one of the fields which are direct summands of KG and $\bar{L} = L$, then $(L, {}^-)$ is unitary if $L \neq K$ since it is generated over K by the image of the projection $G \to L$, and these images are inverted by the involution of L since the elements of G are inverted by the involution of KG. Thus the nonhyperbolic simple summands of $(KG, {}^-)$ are either of the form (K, id), which is orthogonal, or $(L, {}^-)$ with $L \neq K$, which is unitary. Thus (b) follows from Th. 5.4.6.

Now (c). If there *is* a positive divisor q of e which is an odd prime or 4 for which K does *not* contain a primitive q^{th} root of 1, then the extension L of K generated by such a root of 1 is a summand of KG since there is a surjective homomorphism $KG \to L$. Moreover the standard (K, id)-involution ${}^-$ on KG inverts the root of 1, so $\bar{L} = L$, and we can find $d \in \dot{L}$ such that $\bar{d} = -d$. Let $h : L \times L \to (L, {}^-)$ be the skew symmetric form

$$h(\lambda_1, \lambda_2) = \mathrm{Tr}_{L/K}(\lambda_1 d \bar{\lambda}_2).$$

By Th. 5.2.1 there are *two* inequivalent symplectic representations $G \to \mathbf{Sp}(V, h)$.

Conversely if there is no such divisor q of e, let L be a summand of KG such that $\bar{L} = L$. Let s_L be the image of $s \in G$ in L. If $s_L \neq \pm 1$, its order is > 2 and so some power s_L^r has order an odd prime power or 4. Therefore $s_L^r \in K$, so $s_L^r = (\bar{s}_L)^r = s_L^{-r}$ which implies that $(s_L^r)^2 = 1$, a contradiction. Therefore the image of G in L is 1 or ± 1, so in fact $L = K$. Thus *all* simple summands of KG are $= K$, and therefore are orthogonal. By Th. 5.4.6(b)(i), (5.16) follows at once. □

5.5 Global fields

In this section, K is a global field with an involution ${}^-$. Our principal goal is to consider problems **I** and **II** (p. 219) for split representations over K.

In the case of split representations, it is easy, at least in principle, to determine the isometry representations of G: if A is a simple split component of KG which is stable[4] under the standard $(K,\,^-)$-involution $^-$ of KG, say $A = \mathrm{M}(n, K)$ with $\bar{a} = \underline{h}_0^{-1} a^{t-} \underline{h}_0$ where $\underline{h}_0^{t-} = \varepsilon_0 \underline{h}_0$, by Th. 3.2.12 the association $\underline{\hat{h}} \rightsquigarrow \frac{n}{g}(\underline{h}_0^t \otimes \underline{\hat{h}})$ yields a bijective correspondence between the equivalence classes of $\breve{\varepsilon}$-reflexive forms over $(K,\,^-)$ of rank m and the equivalence classes of $(A,\,^-)$-isotypic isometry representations $G \to \mathbf{I}(K^{mn}, h)$ for $\breve{\varepsilon}\varepsilon_0$-reflexive forms h over $(K,\,^-)$ of rank mn. As we already have seen, it is easy to calculate such things as the determinant, Hasse invariant and signatures of $h = \langle \frac{n}{g}(\underline{h}_0^t \otimes \underline{\hat{h}}) \rangle$. And furthermore the Hasse Principle for isometry representations holds (Th. 3.3.3).

The case of split symplectic representations has already been covered in §5.2, so we consider only unitary and orthogonal representations here. Thus we assume throughout this section that

$$\breve{\varepsilon}\varepsilon_0 = \varepsilon = 1.$$

Suppose first that K_0 has a real prime $\mathfrak{p}$ – thus K_0 (and K) have characteristic 0 and $K_{0\mathfrak{p}} \cong \mathbb{R}$ – and let $\mathfrak{P}$ be a prime of K lying above $\mathfrak{p}$. There are 3 possibilities:

1. $K = K_0$, so $\mathfrak{P} = \mathfrak{p}$, $K_{\mathfrak{P}} = K_{0\mathfrak{p}} \cong \mathbb{R}$ and $K_{0\mathfrak{p}} \otimes_{K_0} K \cong \mathbb{R}$.
2. $K \neq K_0$ and $\mathfrak{p}$ does not split in K – so $\mathfrak{P}$ is the unique prime of K above $\mathfrak{p}$ and is complex – $K_{\mathfrak{P}} \cong \mathbb{C}$ and $K_{0\mathfrak{p}} \otimes_{K_0} K \cong \mathbb{C}$.
3. $K \neq K_0$ and $\mathfrak{p}$ splits in K – thus there are two real primes lying above $\mathfrak{p}$, $\mathfrak{P}$ and its conjugate $\bar{\mathfrak{P}}$, and $K_{\mathfrak{P}} = K_{\bar{\mathfrak{P}}} = K_{0\mathfrak{p}} \cong \mathbb{R}$ and $K_{0\mathfrak{p}} \otimes_{K_0} K \cong \mathbb{R} \oplus \mathbb{R}$.

Recall that $\mathfrak{X}_d$ is the set of all real primes of K_0 satisfying 1. or 2. – in other words $\mathfrak{X}_d$ is the set of real primes of K_0 at which K/K_0 is definite. By Th. 4.5.1 (p. 171) a simple A-module supports an invariant form (symmetric respectively Hermitian) b which is positive definite at all $\mathfrak{p} \in \mathfrak{X}_d$. Since the adjoint involution of this form applied to A is the involution $^-$, this means that we can assume that $\underline{h}_0$ (which is a discriminant matrix of b) is a positive definite matrix at all orderings of K_0 which correspond to primes in $\mathfrak{X}_d$ (Prop. 1.4.1, p. 42).

If h is any symmetric bilinear form over $\mathbb{R}$ or any Hermitian form over $(\mathbb{C},\,^*\,)$, the number of -1's in any diagonalization $\langle 1, \ldots, 1, -1, \ldots, -1 \rangle$ of h is the *negative index* of h, and the number of 1's is the *positive index*; both are independent of the diagonalization. If h is a symmetric or Hermitian form over the number field $(K,\,^-)$ with involution and $\mathfrak{p} \in \mathfrak{X}_d$, the indices of $h_{\mathfrak{p}}$ are denoted by $r_{\mathfrak{p}}^-(h)$ and $r_{\mathfrak{p}}^+(h)$ respectively, or simply $r_{\mathfrak{p}}^-$ and $r_{\mathfrak{p}}^+$ if h is clear from the context. Of course $r_{\mathfrak{p}}^- + r_{\mathfrak{p}}^+$ is the rank of h, and the indices of an Hermitian matrix $\underline{h}$ are those of the form $\langle \underline{h} \rangle$.

Lemma 5.5.1 *Let $\eta \in \dot{K}_0$. Suppose that m is a positive integer and, for each $\mathfrak{p} \in \mathfrak{X}_d$, that $r_{\mathfrak{p}}$ is an integer such that $0 \le r_{\mathfrak{p}} \le m$ and $\eta(-1)^{r_{\mathfrak{p}}} >_{\mathfrak{p}} 0$, i.e. $\eta(-1)^{r_{\mathfrak{p}}} > 0$ in $K_{0\mathfrak{p}} = \mathbb{R}$. Then there is an $m \times m$ symmetric respectively Hermitian matrix $\underline{\hat{h}}$ whose determinant is η and whose negative index at each $\mathfrak{p} \in \mathfrak{X}_d$ is $r_{\mathfrak{p}}$.*

We can assume that $\mathfrak{X}_d$ is nonempty. If $m = 1$, take $\underline{\hat{h}} = (\eta)$. Suppose $m > 1$. By the approximation theorem (p. 18), there is an $\alpha \in \dot{K}_0$ such that for all $\mathfrak{p} \in \mathfrak{X}_d$,

$$\alpha >_{\mathfrak{p}} 0 \quad \text{if } r_{\mathfrak{p}} < m,$$
$$\alpha <_{\mathfrak{p}} 0 \quad \text{if } r_{\mathfrak{p}} = m.$$

[4] Of course if it *isn't* stable, then $A \oplus \bar{A}$ is hyperbolic and the representation problem is trivial.

Put $\eta' = \eta\alpha^{-1}$ and

$$\begin{aligned} r'_{\mathfrak{p}} &= r_{\mathfrak{p}} && \text{if } r_{\mathfrak{p}} < m, \\ r'_{\mathfrak{p}} &= r_{\mathfrak{p}} - 1 && \text{if } r_{\mathfrak{p}} = m. \end{aligned}$$

Then $\eta'(-1)^{r'_{\mathfrak{p}}} >_{\mathfrak{p}} 0$ for all $\mathfrak{p} \in \mathfrak{X}_d$, and so by the induction hypothesis, there is an Hermitian matrix $\underline{\hat{h}}'$ of rank $m-1$ with $\det \underline{\hat{h}}' = \eta'$ and negative indices $r'_{\mathfrak{p}}$. Then

$$\underline{\hat{h}} = \begin{pmatrix} \underline{\hat{h}}' & 0 \\ 0 & \alpha \end{pmatrix}$$

has the required properties. □

In the case when $(K, ^-)$ is of the second kind, the elements of $\mathrm{N}_{K/K_0}\dot{K}$ are positive at each $\mathfrak{p} \in \mathfrak{X}_d$ since if $K = K_0(\sqrt{\delta})$, then $\delta < 0$ in $K_{0\mathfrak{p}}$ for all $\mathfrak{p} \in \mathfrak{X}_d$ and norms are of the form $\alpha^2 - \beta^2\delta$ $(\alpha, \beta \in \dot{K}_0)$. This means that we can refer to a norm class in $\dot{K}_0/\mathrm{N}_{K/K_0}\dot{K}$ as being positive or negative at $\mathfrak{p} \in \mathfrak{X}_d$.

Theorem 5.5.1 Split unitary representations. *Let K be a global field with a nontrivial involution $^-$. Suppose that $(A, ^-)$ is a split component of $(KG, ^-)$, say $A \cong \mathrm{M}(n, K)$. Let V be an A-module of length $m > 0$ and $h : V \times V \to (K, ^-)$ a nonsingular Hermitian form.*

I *There is an $(A, ^-)$-invariant form on V equivalent to h if and only if the positive and negative indices of h at $\mathfrak{p}$ are divisible by n for each $\mathfrak{p} \in \mathfrak{X}_d$ and, when n is even, $\det h = (\det_K(^-))^m$ (where $^-$ is the (unitary) involution on A).*

Suppose h is $(A, ^-)$-invariant.

- *(i) If n is odd, h supports exactly one $(A, ^-)$-isotypic unitary representation (up to equivalence).*
- *(ii) If n is even, the equivalence classes of unitary representations supported by h are in bijective correspondence with those norm classes in $\dot{K}_0/\mathrm{N}_{K/K_0}\dot{K}$ which are positive at each $\mathfrak{p} \in \mathfrak{X}_d$.*

II *Two split unitary representations of odd type are equivalent if and only if their forms are isotypically equivalent.*

REMARK. We recall that if K is a number field and G has odd order or has a normal Abelian subgroup of odd index, then any (linear) representation has odd type. See p. 78. □

We may suppose that $A = \mathrm{M}(n, K)$. Then for some Hermitian matrix $\underline{h}_0 \in \mathbf{GL}(n, K)$, $\bar{a} = \underline{h}_0^{-1} a^{t-} \underline{h}_0$ for all $a \in A$. The matrix $\underline{h}_0$ can be assumed to be positive definite at each $\mathfrak{p} \in \mathfrak{X}_d$ (p. 241).

Most of the theorem now follows from the fact that there is a bijection between classes of Hermitian forms over $(K, ^-)$ and classes of isotypic unitary representations, arising from $\underline{\hat{h}} \rightsquigarrow \frac{n}{g}(\underline{h}_0^t \otimes \underline{\hat{h}})$ – cf. Th. 3.2.12, p. 128. The necessity of the condition that the indices of h at each $\mathfrak{p} \in \mathfrak{X}_d$ be divisible by n is a consequence of the fact that $\underline{h}_0$ is definite at each such $\mathfrak{p}$ – e.g. consider $\underline{\hat{h}}$ to be diagonalized. That $\det h =_{\mathrm{N}} (\det \underline{h}_0)^m = (\det_K(^-))^m$ when n is even follows from $\det \frac{n}{g}(\underline{h}_0^t \otimes \underline{\hat{h}}) = (\frac{n}{g})^{mn}(\det \underline{h}_0)^m(\det \underline{\hat{h}})^n$ since $\det \underline{\hat{h}} \in \dot{K}_0$ and $\dot{K}_0^2 \subset \mathrm{N}_{K/K_0}\dot{K}$.

Conversely suppose that for all $\mathfrak{p} \in \mathfrak{X}_d$, the indices of h at $\mathfrak{p}$ are $nr_\mathfrak{p}^-$ and $nr_\mathfrak{p}^+$ for nonnegative integers $r_\mathfrak{p}^-$ and $r_\mathfrak{p}^+$, so

$$(\det h)(-1)^{nr_\mathfrak{p}^-} >_\mathfrak{p} 0.$$

Since the rank of h is nm, $r_\mathfrak{p}^-$ and $r_\mathfrak{p}^+$ are $\leq m$ for all $\mathfrak{p} \in \mathfrak{X}_d$.

Assume first that n is odd. Put

$$\eta = (\tfrac{n}{g})^m (\det \underline{h}_0)^m \det h.$$

For all $\mathfrak{p} \in \mathfrak{X}_d$,

$$\eta(-1)^{r_\mathfrak{p}^-} = (\tfrac{n}{g})^m (\det \underline{h}_0)^m (\det h)(-1)^{nr_\mathfrak{p}^-} >_\mathfrak{p} 0$$

since $\det \underline{h}_0$ is positive and n is odd, so we can apply the above lemma to find an $m \times m$ Hermitian matrix $\hat{\underline{h}}$ with determinant η and negative indices $r_\mathfrak{p}^-$. Since

$$\det(\tfrac{n}{g}(\hat{\underline{h}} \otimes \underline{h}_0)) = (\tfrac{n}{g})^{nm} \eta^n (\det \underline{h}_0)^m =_{\mathrm{N}} \det h$$

and the indices of $\frac{n}{g}(\hat{\underline{h}} \otimes \underline{h}_0^t)$ are n times those of $\hat{\underline{h}}$, namely $nr_\mathfrak{p}^-$ and $nr_\mathfrak{p}^+$, the Hermitian form $\langle \frac{n}{g}(\hat{\underline{h}} \otimes \underline{h}_0^t) \rangle$ is equivalent to h (Table 1, p. 61). Furthermore since n is odd, the determinant $(\frac{n}{g})^{nm} (\det \hat{\underline{h}})^n (\det \underline{h}_0)^m$ of h determines the determinant of *any* Hermitian matrix $\hat{\underline{h}}$ which gives rise to a unitary representation $G \to \mathbf{U}(V, h)$; moreover the indices of $\hat{\underline{h}}$ at any $\mathfrak{p} \in \mathfrak{X}_d$ are determined as well since they are those of h divided by n – namely $r_\mathfrak{p}^-$ and $r_\mathfrak{p}^+$. This proves **I** when n is odd, and **II** follows easily as well.

Now suppose n is even, and that h has determinant $(\det_K(^-))^m = (\det \underline{h}_0)^m$ and, for each $\mathfrak{p} \in \mathfrak{X}_d$, indices $nr_\mathfrak{p}^-$ and $nr_\mathfrak{p}^+$. If $\hat{\underline{h}} \in \mathbf{GL}(m, K)$ is any Hermitian matrix with indices $r_\mathfrak{p}^-$ and $r_\mathfrak{p}^+$ for each $\mathfrak{p} \in \mathfrak{X}_d$, $\langle \frac{n}{g}(\hat{\underline{h}} \otimes \underline{h}_0) \rangle$ is equivalent to h since its determinant and indices are the same as those of h. (Such $\hat{\underline{h}}$ *do* exist – for example there are such diagonal matrices in $\mathbf{GL}(m, K_0)$ by the approximation theorem (p. 18).

Furthermore since two such matrices $\hat{\underline{h}}$ have the same indices, their determinants differ by an element of $\dot{K}_0$ which is positive at each $\mathfrak{p} \in \mathfrak{X}_d$. Conversely if $\alpha \in \dot{K}_0$ is positive at each $\mathfrak{p} \in \mathfrak{X}_d$ and $\hat{\underline{h}}$ is such an Hermitian matrix, then by the foregoing lemma, there is another such matrix with determinant $\alpha(\det \hat{\underline{h}})$. This proves **I** in the case n even. □

Let $A = \mathrm{M}(n, K)$ have an orthogonal involution $\bar{a} = \underline{h}_0^{-1} a^t \underline{h}_0$; then $\underline{h}_0^t = 1$. If n is even, then

$$\det(^-) = \det \underline{h}_0$$

(p. 32) and, if also $\operatorname{disc} \langle \underline{h}_0 \rangle =_2 1$,

$$\mathrm{s}(^-) = \mathrm{s}(\langle \underline{h}_0 \rangle).$$

In the next theorem, the following terminology is used: Let $\mathfrak{p}$ be a prime of K. If $\alpha \in K$, then $\alpha_\mathfrak{p}$ is $= \alpha$ but considered as an element of $K_\mathfrak{p}$. Thus for example if h is a symmetric form over K, $(\det h)_\mathfrak{p} =_2 -1$ means that $(\det h)\dot{K}_\mathfrak{p}^2 = -\dot{K}_\mathfrak{p}^2$.

The quaternion Brauer class $(\alpha, \beta)_K$ is denoted by (α, β). The local Brauer class $(\alpha, \beta)_{K_\mathfrak{p}}$ is denoted by $(\alpha, \beta)_\mathfrak{p}$ and is considered to be ± 1. This applies as well to the Hasse invariant $\mathrm{s}(h) \in \mathrm{Br}(K)_2$ (the elements of order 1 or 2 in $\mathrm{Br}(K)$);

thus $s(h)_\mathfrak{p}$ is the $\mathfrak{p}$-component of $s(h)$ in $Br(K)_2$ – usually considered to be ± 1, and alternatively denoted by $s_\mathfrak{p}(h)$.

Theorem 5.5.2 Split orthogonal representations *Suppose that K is a global field with involution $^- = \mathrm{id}_K$.*

I *Let $(A, ^-)$ be component of $(KG, ^-)$, $A \cong \mathrm{M}(n, K)$. Let V be an A-module of length m.*

$(A, ^-)$ *symplectic*. *Every nonsingular symmetric $(A, ^-)$-invariant form $V \times V \to K$ is hyperbolic, and the orthogonal representations they afford are hyperbolic and so unique up to equivalence.*

$(A, ^-)$ *orthogonal*. *Suppose that $h : V \times V \to K$ is a nonsingular symmetric form.*
If n is odd, there is an $(A, ^-)$-invariant form $\cong h$ on V if and only if
- *(i) when $m = 1$, $h \cong \langle \alpha \underline{h}_0 \rangle$ for some $\alpha \in \dot{K}$,*
- *(ii) when $m = 2$, $s(h)_\mathfrak{p} = (-1,-1)_\mathfrak{p}^{\frac{n+1}{2}}$ for all finite primes $\mathfrak{p}$ at which $(\det h)_\mathfrak{p} =_2 -1$, and*
- *(iii) $n \,|\, \mathrm{sgn}_\mathfrak{p} h$ for all real primes $\mathfrak{p}$ of K.*

If n is even, there is an $(A, ^-)$-invariant form $\cong h$ on V if and only if
- *(iv) $\det h =_2 (\det(^-))^m$, and*
- *(v) $n | r_\mathfrak{p}^-(h)$ for all real primes $\mathfrak{p}$ of K, and*
- *(vi) $s_\mathfrak{p}(h) = (s_\mathfrak{p}(^-))^m (-1,-1)_\mathfrak{p}^{\frac{nm(m-1)}{4}}$ for all (finite or infinite) primes $\mathfrak{p}$ at which $(\mathrm{disc}\,(^-))_\mathfrak{p} =_2 1$.*

If h is $(A, ^-)$-invariant, there is (up to equivalence) only one orthogonal representation $\rho : G \to \mathbf{O}(V, h)$ if n is odd or if ρ is hyperbolic, and there are an infinite number if otherwise.

II *Two split orthogonal representations of odd type are equivalent if and only if their underlying forms are isotypically equivalent. A little more generally, two split orthogonal representations on V, for which $AV = 0$ for all simple orthogonal involution components $(A, ^-)$ of even degree, are equivalent if and only if their underlying forms are isotypically equivalent.*

REMARK. The case "$(A, ^-)$ symplectic" can occur only if K has characteristic $\neq 0$ or is totally imaginary – for if K has a real prime $\mathfrak{p}$, a simple A-module has an invariant symmetric form (which is positive definite at $\mathfrak{p}$), and so $(A, ^-)$ is orthogonal (Th. 3.2.8, p. 117). □

If $(A, ^-)$ is symplectic, $\underline{\hat{h}}$ must be skew symmetric since $\breve{\varepsilon} = \varepsilon\varepsilon_0 = -1$, and this case follows from Cor. 3.2.5 (p. 133).

Suppose that $(A, ^-)$ is orthogonal. Since $h \cong \frac{n}{g} \langle \underline{\hat{h}} \otimes \underline{h}_0 \rangle$ for some symmetric matrix $\underline{\hat{h}} \in \mathbf{GL}(m, K)$ if and only if h is equivalent to an invariant form, it suffices to find necessary and sufficient conditions for h to be equivalent to $\langle \underline{\hat{h}} \otimes \underline{h}_0 \rangle$ for some symmetric matrix $\underline{\hat{h}}$.

1. ***n* odd.**

We begin with the necessity, so suppose that $\underline{\hat{h}}$ is an $m \times m$ symmetric matrix such that $h \cong \langle \underline{\hat{h}} \otimes \underline{h}_0 \rangle$. If $\dim V = n$, $m = 1$ so $\underline{\hat{h}} = \langle \alpha \rangle$ and $\underline{\hat{h}} \otimes \underline{h}_0 = \alpha \underline{h}_0$. Thus $h \cong \langle \alpha \underline{h}_0 \rangle$ as stated. Thus (i) is necessary.

We next consider (ii), so $\dim V = 2n$. This means that $m = 2$ of course. Then

$$\det h = (\det \underline{\hat{h}})^n (\det \underline{h}_0)^2 =_2 \det \underline{\hat{h}}$$

If $(\det h)_\mathfrak{p} =_2 -1$, then $(\det \underline{\hat{h}})_\mathfrak{p} =_2 -1$ and therefore $\langle \underline{\hat{h}} \rangle$ must be a hyperbolic plane over $K_\mathfrak{p}$; thus its Hasse symbol $\mathrm{s}(\underline{\hat{h}})_\mathfrak{p}$ at $\mathfrak{p}$ is $(-1,-1)_\mathfrak{p}$. According to Lem. 1.5.1(d) (p. 53)

$$\begin{aligned} \mathrm{s}(\underline{\hat{h}} \otimes \underline{h}_0) &= \qquad (5.17) \\ &\mathrm{s}(\underline{h}_0)^m \mathrm{s}(\underline{\hat{h}})^n (\det \underline{h}_0, -1)^{\frac{m(m-1)}{2}} (\det \underline{\hat{h}}, \det \underline{h}_0)^{mn-1} (\det \underline{\hat{h}}, -1)^{\frac{n(n-1)}{2}} \\ &= \mathrm{s}(\underline{\hat{h}})(\det \underline{h}_0, -1)(\det \underline{\hat{h}}, \det \underline{h}_0)(\det \underline{\hat{h}}, -1)^{\frac{n-1}{2}}, \end{aligned}$$

and so at $\mathfrak{p}$, since $(\det \underline{\hat{h}})_\mathfrak{p} =_2 -1$ and $\mathrm{s}(\underline{\hat{h}})_\mathfrak{p} = (-1,-1)_\mathfrak{p}$,

$$\mathrm{s}(h)_\mathfrak{p} = (-1,-1)_\mathfrak{p} (-1,-1)_\mathfrak{p}^{\frac{n-1}{2}} = (-1,-1)_\mathfrak{p}^{\frac{n+1}{2}}.$$

This proves the necessity of (ii).

The necessity of (iii) follows from the fact that $h \cong \langle \underline{\hat{h}} \otimes \underline{h}_0 \rangle$ and that $\underline{h}_0$ is positive definite at any real prime $\mathfrak{p}$ – cf. p. 241.

We now consider the sufficiency for n odd; suppose first that $n = 1$, i.e. $A = K$. The image of G in $A = K$ consists of roots of unity, which are inverted by $^-$. But since $^-$ is the identity on K, the image must be $\subset \{\pm 1\}$ so *every* form h (on V) is G-invariant.

If $m = 1$, then $h \cong \langle (\alpha) \otimes \underline{h}_0 \rangle$ and so is invariant.

Thus when n is odd, it remains to prove the sufficiency when

$$n > 1 \ \text{ and } \ m > 1.$$

If suffices to show that, under these conditions, there is an $m \times m$ symmetric matrix $\underline{\hat{h}}$ such that

$$h \cong \langle \underline{\hat{h}} \otimes \underline{h}_0 \rangle. \qquad (5.18)$$

We begin by solving this equivalence locally at each prime $\mathfrak{p}$ – that is to say by showing the existence for each $\mathfrak{p}$ of a symmetric matrix $\underline{\hat{h}}_\mathfrak{p}$ satisfying

$$h \cong \langle \underline{\hat{h}}_\mathfrak{p} \otimes \underline{h}_0 \rangle \ \text{ over } K_\mathfrak{p}. \qquad (5.19)$$

Now

$$\det(\underline{\hat{h}} \otimes \underline{h}_0) =_2 (\det \underline{h}_0)^m \det \underline{\hat{h}}, \qquad (5.20)$$

and

$$\mathrm{s}(\underline{\hat{h}} \otimes \underline{h}_0) = \mathrm{s}(\underline{h}_0)^m \mathrm{s}(\underline{\hat{h}})(\det \underline{h}_0, -1)^{\frac{m(m-1)}{2}} (\det \underline{\hat{h}}, \det \underline{h}_0)^{m-1} (\det \underline{\hat{h}}, -1)^{\frac{n-1}{2}}, \qquad (5.21)$$

by (5.17) and, for any real prime $\mathfrak{p}$,

$$\mathrm{sgn}_\mathfrak{p}(\underline{\hat{h}} \otimes \underline{h}_0) = n(\mathrm{sgn}_\mathfrak{p} \underline{\hat{h}}). \qquad (5.22)$$

Thus when $\mathfrak{p}$ is real, we simply choose a symmetric matrix $\underline{\hat{h}}_\mathfrak{p} \in \mathbf{GL}(m, \mathbb{R})$ with signature $\frac{\mathrm{sgn}_\mathfrak{p} h}{n}$. So we need only deal with the finite primes.

Suppose first that $\dim V = 2n$, i.e. $m = 2$. Then the above formulas become

$$\det(\hat{\underline{h}} \otimes \underline{h}_0) =_2 \det \hat{\underline{h}}$$

and

$$\mathrm{s}(\hat{\underline{h}} \otimes \underline{h}_0) = \mathrm{s}(\hat{\underline{h}})(\det \underline{h}_0, -1)(\det \hat{\underline{h}}, \det \underline{h}_0)(\det \hat{\underline{h}}, -1)^{\frac{n-1}{2}}. \tag{5.23}$$

A solution $\hat{\underline{h}}_\mathfrak{p}$ for (5.19) must satisfy $(\det h)_\mathfrak{p} = \det(\hat{\underline{h}}_\mathfrak{p} \otimes \underline{h}_0) =_2 \det \hat{\underline{h}}_\mathfrak{p}$. Therefore if $(\det h)_\mathfrak{p} =_2 -1$, we must have $\det \hat{\underline{h}}_\mathfrak{p} = -1$, i.e. $\langle \hat{\underline{h}}_\mathfrak{p} \rangle$ is a hyperbolic plane; since it then has Hasse invariant $(-1,-1)_\mathfrak{p}$,

$$\mathrm{s}(\hat{\underline{h}} \otimes \underline{h}_0)_\mathfrak{p} = (-1,-1)_\mathfrak{p}(-1,-1)_\mathfrak{p}^{\frac{n-1}{2}}$$

which is $= \mathrm{s}(h)_\mathfrak{p}$ by (ii). Thus (5.19) is solved in the case $m = 2$ and $(\det h)_\mathfrak{p} =_2 -1$. If on the other hand $(\det h)_\mathfrak{p} \neq_2 -1$ (and $m = 2$), then we can find a symmetric $\hat{\underline{h}}_\mathfrak{p}$ in $\mathbf{GL}(2, K_\mathfrak{p})$ with determinant $(\det h)_\mathfrak{p}$ and Hasse invariant satisfying (5.23) at $\mathfrak{p}$ (cf. Th. 1.5.12, p. 64), again solving (5.19) at $\mathfrak{p}$.

We can now assume that

$$n > 1 \text{ (and odd)}, \; m > 2 \text{ and } \mathfrak{p} \text{ is finite.}$$

Again by Th. 1.5.12, there is a symmetric matrix $\hat{\underline{h}}_\mathfrak{p} \in \mathbf{GL}(m, K_\mathfrak{p})$ with arbitrary determinant and arbitrary Hasse invariant, and so it is apparent from (5.20) and (5.21) at $\mathfrak{p}$ (applied with $\hat{\underline{h}}$ replaced by $\hat{\underline{h}}_\mathfrak{p}$) and condition (iii) that, for each $\mathfrak{p}$, $\hat{\underline{h}}_\mathfrak{p}$ can be chosen so that $h \cong \langle \hat{\underline{h}}_\mathfrak{p} \otimes \underline{h}_0 \rangle$ over $K_\mathfrak{p}$.

This implies that, for all $\mathfrak{p}$,

$$\det \hat{\underline{h}}_\mathfrak{p} = (\det h)(\det \underline{h}_0)^{-m} =: d \in \dot{K}$$

and

$$\begin{aligned} s(\hat{\underline{h}}_\mathfrak{p}) &= \mathrm{s}(h)_\mathfrak{p} \mathrm{s}(\underline{h}_0)_\mathfrak{p}^m (\det \underline{h}_0, -1)_\mathfrak{p}^{\frac{m(m-1)}{2}} (\det \hat{\underline{h}}_\mathfrak{p}, \det \underline{h}_0)_\mathfrak{p}^{m-1} (\det \hat{\underline{h}}_\mathfrak{p}, -1)_\mathfrak{p}^{\frac{n-1}{2}} \\ &= \mathrm{s}(h)_\mathfrak{p} \mathrm{s}(\underline{h}_0)_\mathfrak{p}^m (\det \underline{h}_0, -1)_\mathfrak{p}^{\frac{m(m-1)}{2}} (d, \det \underline{h}_0)_\mathfrak{p}^{m-1} (d, -1)_\mathfrak{p}^{\frac{n-1}{2}}. \end{aligned}$$

It follows from Hilbert's Reciprocity law that $\prod_\mathfrak{p} s(\hat{\underline{h}}_\mathfrak{p}) = 1$ and therefore by Th. 1.5.13 that there is a symmetric matrix $\hat{\underline{h}} \in \mathbf{GL}(m, K)$ such that $\langle \hat{\underline{h}} \rangle_\mathfrak{p} \cong \langle \hat{\underline{h}}_\mathfrak{p} \rangle$ for all $\mathfrak{p}$. Thus we have "solved" (5.18) when n is odd.

It is easy to see that the association $\hat{\underline{h}} \rightsquigarrow \hat{\underline{h}} \otimes \underline{h}_0$ is injective on equivalence classes of symmetric forms (by (5.20), (5.21) and (5.22), and the equivalence theory over a number field). This proves the last statement of **I** when n is odd – that if h is invariant, there is (up to equivalence) only one orthogonal representation $G \to \mathbf{O}(V, h)$.

2. ***n* even.** Suppose that h is invariant. As in the case n odd, it is equivalent to find necessary and sufficient conditions under which there is a symmetric matrix $\hat{\underline{h}} \in \mathbf{GL}(m, K)$ such that $h \cong \langle \hat{\underline{h}} \otimes \underline{h}_0 \rangle$. The necessity of (iv) and (v) is clear. From Lem. 1.5.1(d) (p. 53),

$$\begin{aligned} \mathrm{s}(h) &= \mathrm{s}(\hat{\underline{h}} \otimes \underline{h}_0) \\ &= \mathrm{s}(\underline{h}_0)^m (\det \hat{\underline{h}}, \operatorname{disc} \underline{h}_0)(\det \underline{h}_0, -1)^{\frac{m(m-1)}{2}}. \end{aligned} \tag{5.24}$$

If $\operatorname{disc} \underline{h}_0 = 1$ at $\mathfrak{p}$, $(\det \underline{h}_0)_\mathfrak{p} = (-1)^{n/2}$ and (vi) follows at once.

Conversely suppose that (iv), (v) and (vi) prevail. First we show there is $\delta \in \dot{K}$ satisfying

(a) $(\delta, \operatorname{disc} \underline{h}_0) = \mathrm{s}(h)\mathrm{s}(\underline{h}_0)^m (\det \underline{h}_0, -1)^{m(m-1)/2}$ and

(b) the sign of δ at any real prime $\mathfrak{p}$ at which $\operatorname{disc} \underline{h}_0 =_2 1$ (i.e. at which $\operatorname{disc} \underline{h}_0 > 0$) is $(-1)^{r_\mathfrak{p}^-}$,

where $r_\mathfrak{p}^-(h) = nr_\mathfrak{p}^-$ is the negative index of h.

By Lem. 1.2.4 (p. 19), it is enough to find δ satisfying (a). By the Hilbert Reciprocity Law, the right side of (a) is $\neq 1$ at an even number of primes, and by assumption (vi), $\operatorname{disc} \underline{h}_0$ is a nonsquare at these same primes, so by Ex. 5, p. 18, there is therefore a δ satisfying (a).

Now suppose $\mathfrak{p}$ is a real prime of K. Since $\underline{h}_0$ is positive definite at all real primes,

$$\mathrm{s}(\underline{h}_0)_\mathfrak{p} = 1, \quad \det \underline{h}_{0\mathfrak{p}} =_2 1, \quad \operatorname{disc} \underline{h}_{0\mathfrak{p}} =_2 (-1)^{n/2}.$$

It follows from (a) that

$$\begin{aligned} (\delta, \operatorname{disc} \underline{h}_0)_\mathfrak{p} &= (\delta, -1)_\mathfrak{p}^{\frac{n}{2}} \\ &= \mathrm{s}(h)_\mathfrak{p} = (-1)^{\frac{nr_\mathfrak{p}^-}{2}} \end{aligned}$$

since the negative index of $h_\mathfrak{p}$ is the even integer $nr_\mathfrak{p}^-$. Let $\hat{h}_\mathfrak{p} \in \mathbf{GL}(m, K_\mathfrak{p})$ be a symmetric form over $K_\mathfrak{p} \cong \mathbb{R}$ with negative index $r_\mathfrak{p}^-$. Then

$$\det \hat{h}_\mathfrak{p} =_2 \delta \tag{5.25}$$

by (b) if $\operatorname{disc} \underline{h}_0 =_2 1$ at $\mathfrak{p}$. Suppose $\operatorname{disc} \underline{h}_0 \neq_2 1$ at $\mathfrak{p}$. Since $\operatorname{disc} \underline{h}_{0\mathfrak{p}} = (-1)^{\frac{n}{2}}$, $\frac{n}{2}$ is odd and so by (a)

$$(\delta, \operatorname{disc} \underline{h}_0)_\mathfrak{p} = (\delta, -1)_\mathfrak{p}^{\frac{n}{2}} = (\delta, -1)_\mathfrak{p} = \mathrm{s}(h)_\mathfrak{p} \text{ by (a),} \tag{5.26}$$

$$= (-1)^{\frac{nr_\mathfrak{p}^-}{2}} = (-1)^{r_\mathfrak{p}^-}. \tag{5.27}$$

Thus $(\delta, -1)_\mathfrak{p} = (-1)^{r_\mathfrak{p}^-}$ which implies that $\delta =_2 (-1)^{r_\mathfrak{p}^-}$ at $\mathfrak{p}$, and so (5.25) holds for either sign of $\operatorname{disc} \underline{h}_0$ at $\mathfrak{p}$.

Now choose a nonsingular symmetric form $\hat{h}_\mathfrak{p}$ of rank m and determinant δ for each finite or complex prime $\mathfrak{p}$. We can assume that $\mathrm{s}_\mathfrak{p}\hat{h}_\mathfrak{p} = 1$ for all but a finite number of $\mathfrak{p}$ (note that $\mathrm{s}_\mathfrak{p}\hat{h}_\mathfrak{p} = 1$ for all $\mathfrak{p}$ if $m = 1$, or $m = 2$ and $\delta =_2 -1$) and that $\prod_\mathfrak{p} \mathrm{s}_\mathfrak{p}\hat{h}_\mathfrak{p} = 1$.

It follows once more from the existence theorem over global fields (Th. 1.5.13, p. 64) that there is a symmetric form $\hat{h}$ of determinant δ over K whose completion at each prime $\mathfrak{p}$ is equivalent to $\hat{h}_\mathfrak{p}$.

Then

$$\mathrm{s}(h) = \mathrm{s}(\underline{h}_0)^m (\det \hat{\underline{h}}, \operatorname{disc} \underline{h}_0)(\det \underline{h}_0, -1)^{m(m-1)/2}$$

by (a), and this is $= \mathrm{s}(\hat{\underline{h}} \otimes \underline{h}_0)$ by (5.24).

Moreover $\det(\hat{\underline{h}} \otimes \underline{h}_0) = (\det \underline{h}_0)^m = \det h$ by (iv), and $r_\mathfrak{p}^-(\hat{\underline{h}} \otimes \underline{h}_0) = nr_\mathfrak{p}^-(\hat{\underline{h}}) = nr_\mathfrak{p}^- = r_\mathfrak{p}^-(h)$ for all real primes $\mathfrak{p}$. Thus $\langle \hat{\underline{h}} \otimes \underline{h}_0 \rangle \cong h$ as desired.

We now prove the last statement of **I** in the theorem. The case of n odd has already been dealt with, and the hyperbolic case follows from Cor. 3.2.5, p.

133, and Th. 3.1.3, p. 98. We now show that in the remaining cases (n even and ρ not hyperbolic), there are an infinite number of inequivalent representations $G \to \mathbf{O}(V,h)$ for fixed (V,h). This is done by showing that there are an infinite number of inequivalent choices for the symmetric form $\hat{h}$ over K constructed above. This follows easily from the following lemma.

Let S be the set of real primes of K at which $\operatorname{disc}\underline{h}_0 =_2 1$.

Lemma 5.5.2 *Assume* $\operatorname{disc}\underline{h}_0 \notin \dot{K}^2$ *and let* $L = K(\sqrt{\operatorname{disc}\underline{h}_0})$. *Let* N *be an arbitrary integer* > 1. *Then there exist* $\lambda_1, \lambda_2, \ldots, \lambda_N \in \dot{L}$ *such that for all* i, $\mathrm{N}_{L/K}\lambda_i > 0$ *at all* $\mathfrak{p} \in S$ *and* $\mathrm{N}_{L/K}\lambda_j \notin (\mathrm{N}_{L/K}\lambda_i)\dot{K}^2$ *for all* $j \neq i$.

By Cor. 3.7, p. 378, [53], there are infinitely many primes of K which are inert (i.e. nonsplit and unramified) in L/K; let $\mathfrak{p}_1, \ldots, \mathfrak{p}_N$ be N such finite, nondyadic (and distinct) primes. Let $\mathfrak{P}_i$ denote the unique prime of L lying above $\mathfrak{p}_i$. Every unit of $K_{\mathfrak{p}_i}$ is the norm of an element (in fact, of a unit) of $L_{\mathfrak{P}_i}$ (cf. 63:16, [54]) but not every unit is a square (this follows easily from the fact that $[\dot{K}_{\mathfrak{p}_i} : \dot{K}^2_{\mathfrak{p}_i}] = 4$ (cf. (1.22), p. 17)); let η_i be a nonsquare unit in $K_{\mathfrak{p}_i}$. For each i and j between 1 and N, choose

$$\begin{aligned} &\alpha_{ij}, \beta_{ij} \in K_{\mathfrak{p}_j}, \\ &\alpha_{i\mathfrak{p}}, \beta_{i\mathfrak{p}} \in K_{\mathfrak{p}} \quad \text{for all } \mathfrak{p} \in S, \end{aligned}$$

to satisfy

$$\begin{aligned} \mathrm{N}_{L_{\mathfrak{P}_i}/K_{\mathfrak{p}_i}}(\alpha_{ii} + \beta_{ii}\sqrt{\operatorname{disc}\underline{h}_0}) &= \alpha_{ii}^2 - \beta_{ii}^2 \operatorname{disc}\underline{h}_0 && = \eta_i \\ \mathrm{N}_{L_{\mathfrak{P}_j}/K_{\mathfrak{p}_j}}(\alpha_{ij} + \beta_{ij}\sqrt{\operatorname{disc}\underline{h}_0}) &= \alpha_{ij}^2 - \beta_{ij}^2 \operatorname{disc}\underline{h}_0 && = 1 \quad \text{for all } j \neq i, \\ &\alpha_{i\mathfrak{p}}^2 - \beta_{i\mathfrak{p}}^2 \operatorname{disc}\underline{h}_0 && > 0 \quad \text{for all } \mathfrak{p} \in S. \end{aligned}$$

Now apply the approximation theorem (p. 18) to find $\alpha_i \in K$ close enough to each of $\{\alpha_{ij} : 1 \leq j \leq N\} \cup \{\alpha_{i\mathfrak{p}} : \mathfrak{p} \in S\}$ and $\beta_i \in K$ close enough to each of $\{\beta_{ij} : 1 \leq j \leq N\} \cup \{\beta_{i\mathfrak{p}} : \mathfrak{p} \in S\}$ so that, if we define $\lambda_i = \alpha_i + \beta_i\sqrt{\operatorname{disc}\underline{h}_0} \in L$ for $i = 1, 2, \ldots, N$, then

$$\begin{aligned} \mathrm{N}_{L/K}\lambda_i &= \alpha_i^2 - \beta_i^2 \operatorname{disc}\underline{h}_0 \in \eta_i \dot{K}^2_{\mathfrak{p}_i} \\ \mathrm{N}_{L/K}\lambda_i &= \alpha_i^2 - \beta_i^2 \operatorname{disc}\underline{h}_0 \in \dot{K}^2_{\mathfrak{p}_j} \quad \text{for all } j \neq i \\ \mathrm{N}_{L/K}\lambda_i &= \alpha_i^2 - \beta_i^2 \operatorname{disc}\underline{h}_0 > 0 \quad \text{at all } \mathfrak{p} \in S. \end{aligned}$$

(cf. the Local Square Theorem). The lemma follows at once since, if $i \neq j$, $\mathrm{N}_{L/K}\lambda_i \in \dot{K}^2_{\mathfrak{p}_j}$ but $\mathrm{N}_{L/K}\lambda_i \notin \dot{K}^2_{\mathfrak{p}_i}$. □

We now refer back to the choice of δ satisfying (a) and (b). Suppose first that $\operatorname{disc}\underline{h}_0 \notin \dot{K}^2$. Recall that $(\mathrm{N}_{L/K}\lambda, \operatorname{disc}\underline{h}_0) = 1$ for all $\lambda \in \dot{L}$ (p. 17). If $\lambda_1, \ldots, \lambda_N$ are the elements of L provided by the above lemma, $\delta\mathrm{N}_{L/K}\lambda_i$ has the same properties as δ for each i, and hence can be used instead of δ itself to construct a form $\hat{h}$ of determinant $\delta\mathrm{N}_{L/K}\lambda_i$. Thus there are an infinite number of equivalence classes of orthogonal representations when n is even and $\operatorname{disc}\underline{h}_0 \notin \dot{K}^2$.

Now suppose that $\operatorname{disc}\underline{h}_0 \in \dot{K}^2$. For any integer $N > 1$, one can use the approximation theorem to find $\beta_1, \ldots, \beta_N \in K$ such each β_i is a nonsquare, is not $=_2 \beta_j$ for any $j \neq i$, and is positive at each $\mathfrak{p} \in S$. Then $\beta_i\delta$ can be used instead of δ to construct a form $\hat{h}$ of determinant $\beta_i\delta$, and so there are an infinite number of equivalence classes of orthogonal representations. This finishes the proof of Th. 5.5.2. □

5.6 Isometry representations of the symmetric group

As is the case for linear and projective representations, the isometry representation theory of $\mathbf{S}_r$ is especially explicit. The essential points are that all linear representations are split – and so the foregoing results for global fields, for example, apply – and that one can calculate the discriminant matrix $\underline{h}_0$ explicitly for a (nonsingular) symmetric form invariant under an irreducible representation $\rho : \mathbf{S}_r \to \mathbf{GL}(V)$ (at least if $\operatorname{char} K = 0$ or is $> r$). This is the same matrix (up to a nonzero scalar) involved in the involution

$$\bar{a} = \underline{h}_0^{-1} a^t \underline{h}_0 \tag{5.28}$$

on the direct summand of KG corresponding to ρ, induced by the standard involution of KG (Th. 3.2.8(a) and (d), p. 117). The procedure for the calculation of $\underline{h}_0$ is described in [32]; we shall explain it by an example. This method applies if $\operatorname{char} K = 0$ or $p > r$.

Note that, since all linear representations are split, the simple summands of KG are all defined over the prime subfield $\mathbb{F}$, that is to say, if $\mathbb{F}\mathbf{S}_r = A_1 \oplus \cdots \oplus A_k$ is the decomposition of $\mathbb{F}\mathbf{S}_r$ into simple algebra summands (each of course isomorphic to a full matrix algebra $\mathrm{M}(n_i, \mathbb{F})$), then

$$K\mathbf{S}_r = K \otimes_{\mathbb{F}} \mathbb{F}\mathbf{S}_r = (K \otimes_{\mathbb{F}} A_1) \oplus \cdots \oplus (K \otimes_{\mathbb{F}} A_k)$$

is the decomposition of $K\mathbf{S}_r$ into simple summands.

We note also that every simple involution algebra component A of $K\mathbf{S}_r$ is orthogonal when $\operatorname{char} K = 0$ or is $> r$. This follows from the existence of a nonsingular invariant symmetric form h_0 on a simple A-module, leading to (5.28). In characteristic 0, it is in fact a positive definite form defined over $\mathbb{Q}$ – cf. Th. 4.5.1 (p. 171). In characteristic p, it is described explicitly (when the characteristic is $> r$) later in this section via the example.

If K has a *nontrivial* involution and $K\mathbf{S}_r$ has the corresponding standard involution, then every simple involution component $(A, {}^-)$ of $(K\mathbf{S}_r, {}^-)$ is again a simple algebra. This follows easily from the fact that when the standard $(K, {}^-)$-involution of $K\mathbf{S}_r$ is restricted to $K_0\mathbf{S}_r$, it is the standard (K_0, id)-involution of $K_0\mathbf{S}_r$.This implies that the Hermitian matrix $\underline{h}_0$ for the unitary involution on a simple summand of $K\mathbf{S}_r$ is identical with the symmetric matrix for the involution of the first kind on the corresponding summand of $K_0\mathbf{S}_r$ (or, for that matter, of $\mathbb{F}\mathbf{S}_r$).

All of the facts cited in the following about representations of $\mathbf{S}_r$ can be found in [32].

We recall that each irreducible representation of $\mathbf{S}_r$ arises from a partition $\alpha = [\alpha_1, \alpha_2, \ldots, \alpha_k]$ of r; thus the α_i are positive integers satisfying $\alpha_1 \geq \alpha_2 \geq \cdots \geq \alpha_k$ and $\sum_i \alpha_i = r$ (cf. pp. 350 and 376 [32]). We shall describe the calculation of $\underline{h}_0$ through an example, namely the partition

$$\alpha = [3, 2, 2]$$

of 7, which gives rise to an irreducible representation of $\mathbf{S}_7$. The *standard α-tableaus* are the arrays

$$\begin{array}{lll}1&2&3\\4&5&\\6&7&\end{array}\quad\begin{array}{lll}1&2&3\\4&6&\\5&7&\end{array}\quad\begin{array}{lll}1&2&4\\3&5&\\6&7&\end{array}\quad\begin{array}{lll}1&2&4\\3&6&\\5&7&\end{array}\quad\begin{array}{lll}1&2&5\\3&4&\\6&7&\end{array}\quad\begin{array}{lll}1&2&5\\3&6&\\4&7&\end{array}$$

$$\begin{array}{lll}1&2&6\\3&4&\\5&7&\end{array}\quad\begin{array}{lll}1&2&6\\3&5&\\4&7&\end{array}\quad\begin{array}{lll}1&2&7\\3&4&\\5&6&\end{array}\quad\begin{array}{lll}1&2&7\\3&5&\\4&6&\end{array}\quad\begin{array}{lll}1&3&4\\2&5&\\6&7&\end{array}\quad\begin{array}{lll}1&3&4\\2&6&\\5&7&\end{array}$$

$$\begin{array}{lll}1&3&5\\2&4&\\6&7&\end{array}\quad\begin{array}{lll}1&3&5\\2&6&\\4&7&\end{array}\quad\begin{array}{lll}1&3&6\\2&4&\\5&7&\end{array}\quad\begin{array}{lll}1&3&6\\2&5&\\4&7&\end{array}\quad\begin{array}{lll}1&3&7\\2&4&\\5&6&\end{array}\quad\begin{array}{lll}1&3&7\\2&5&\\4&6&\end{array}$$

$$\begin{array}{lll}1&4&5\\2&6&\\3&7&\end{array}\quad\begin{array}{lll}1&4&6\\2&5&\\3&7&\end{array}\quad\begin{array}{lll}1&4&7\\2&5&\\3&6&\end{array}$$

They are characterized by the fact that the row lengths constitute the given partition of 7, and the integers in each row and in each column are increasing. The number of standard α-tableaus, 21, is the degree of the corresponding representation (3.1.13, p. 107, [32]).

There is an orthogonal basis $u_1, \cdots, u_{21}$ (with respect to $h_0 = \langle \underline{h}_0 \rangle$) of the representation space for which one can calculate $h_0(u_i, u_i)$. One starts with the ith standard α-tableau; then

$$h_0(u_i, u_i) = \prod_{x=1}^{7} \prod_{k} \tfrac{h(x,k)}{h(x,k)-1}$$

where $h(x,k)$ is a "hook length" defined as follows. $k = 1, 2, 3$ is the number of a row. $h(x,k)$ is not defined (and hence does not appear in the above product) unless the integer x appears in a row lower than row k; thus $h(1,k)$ and $h(x,3)$ are not defined for any k and x. See p. 312, [32].

Suppose x *is* in a row lower than row k. Then $h(x,k)$ is the number of integers in the "hook" consisting of

(i) the integer x,

(ii) the integers above x up to row k (but not in rows $1, \ldots, k-1$) in the same column as x, and

(iii) the integers $\le x$ in row k to the right of the column of x.

For example in the 8th standard tableau in the above display

$$\begin{array}{lll}1&2&6\\3&5&\\4&7&\end{array},$$

$h(4,1) = 4$, $h(4,2) = 2$ (and of course $h(4,3)$ is not defined). It is easy to check that

$$h_0(u_8, u_8) = \prod_{x=2}^{7} \prod_{k=1}^{2} \tfrac{h(x,k)}{h(x,k)-1} = (\tfrac{3}{2})(\tfrac{4}{3} \cdot \tfrac{2}{1})(\tfrac{2}{1})(\tfrac{4}{3} \cdot \tfrac{2}{1}) = \tfrac{2^6}{3}$$

where the parenthesized factors correspond respectively to $x = 3, 4, 5, 7$.

In this way one gets for $h_0(u_i, u_i)$, $1 \le i \le 21$,

10, 40/3, 45/4, 15, 15, 16, 16, 64/3, 18, 24, 15,

$$20, \quad 20, \quad 64/3, \quad 64/3, \quad 256/9, \quad 24, \quad 32, \quad 24, \quad 32, \quad 36.$$

After adjusting by squares,

$$h_0 \cong \langle 10,\ 30,\ 5,\ 15,\ 15,\ 1,\ 1,\ 3,\ 2,\ 6,\ 15,\ 5,\ 5,\ 3,\ 3,\ 1,\ 6,\ 2,\ 6,\ 2,\ 1\rangle.$$

Thus one can compute the determinant and Hasse invariant of h_0; they are 1 and $(3,3)_K$ respectively. This means that over a local or global field,

$$h_0 \cong \langle 1,\ 1, \ldots, 1,\ 3,\ 3\rangle.$$

We also mention, by way of an application of Th. 5.5.2 (p. 244), that if $\mathbf{S}_7 \to \mathbf{GL}(V)$ is a representation which is a direct sum of $m \geq 3$ (so $\dim V \geq 63$) irreducible representations each corresponding to the partition $[3,2,2]$, then the invariant symmetric forms on V have signature at each real prime of K divisible by 21, and conversely any such form is equivalent to an invariant form. Furthermore there is precisely one equivalence class of orthogonal representations for each equivalence class $[h]$ of symmetric forms for which $21|\mathrm{sgn}_{\mathfrak{p}} h$ for all real primes.

CHAPTER 6

Fröhlich's invariant, Clifford algebras and the equivariant Brauer-Wall group

In this chapter, we apply the Clifford algebra to orthogonal representations (of an *arbitrary* group G, not necessarily finite) and develop the theory of the equivariant Brauer-Wall group, which yields important invariants for orthogonal representations. The first two sections give a summary of the "nonequivariant" theory.

References: Ch. 9 in [65] and Chs. IV and V of [39]. The latter also presents a more extensive treatment of graded algebras.

6.1 Graded algebras and the Brauer-Wall group

A *graded* algebra[1] A over K is an algebra (of finite dimension, as usual) which has a decomposition $A = A_0 \oplus A_1$ as a vector space with the properties

$$1_A \in A_0, \quad A_0A_0 \subset A_0, \quad A_1A_1 \subset A_0, \quad A_1A_0 = A_0A_1 \subset A_1.$$

Thus a graded algebra A is a direct sum $A = A_0 \oplus A_1$ with $A_iA_j \subset A_{i+j}$ (indices mod 2). In particular, A_0 is a subalgebra of A. Such an algebra A is more properly called a $\mathbb{Z}/2\mathbb{Z}$-graded algebra, but since it is the only kind of graded algebra we will encounter, we refer to it simply as a graded algebra.

The *homogeneous* elements of A are $\mathrm{h}A = A_0 \cup A_1$, and $\partial : \mathrm{h}A \to \{0, 1\}$ is the *degree function* (not well-defined at 0). A subspace (or ideal) B of A is *graded* if $B = B_0 \oplus B_1$ with $B_i = B \cap A_i$ (in general, $B_0 \oplus B_1 \subset B$). A homomorphism φ of A into another graded algebra B is *graded* if $\varphi(A_0) \subset B_0$ and $\varphi(A_1) \subset B_1$. If $I = I_0 \oplus I_1$ is a graded two-sided ideal, then A/I is naturally a graded algebra with homogeneous components (isomorphic to) A_0/I_0 and A_1/I_1; and the canonical homomorphism $A \to A/I$ is graded.

Examples. 1. If A is any algebra and ν is an automorphism of A such that $\nu^2 = \mathrm{id}_A$, define A_0 to be the eigenspace of A corresponding to the eigenvalue 1 of ν, and A_1 to be the eigenspace for -1. Then $A = A_0 \oplus A_1$ is a graded algebra (and ν is an automorphism of A as a graded algebra). Conversely, if A is any graded algebra, the map which is 1 on A_0 and -1 on A_1 is an involutory automorphism which determines the grading of A.

2. An arbitrary algebra A can be made into a graded algebra $\langle A \rangle$ with "trivial grading" $\langle A \rangle_0 = A$ and $\langle A \rangle_1 = 0$. It corresponds to the automorphism $\nu = \mathrm{id}_A$.

3. Any quadratic extension L/K of fields has a unique automorphism of order 2, hence has a unique nontrivial grading making it into a graded K-algebra. Namely

[1] Now often called a "superalgebra".

if $A = L = K(\sqrt{\alpha})$, $A_0 = K$ and $A_1 = K\sqrt{\alpha}$. More generally any two dimensional K-algebra $L = K + K\lambda$ with $\lambda^2 = \alpha \in K$ is similarly a graded algebra (with $A_1 = K\lambda$). These graded quadratic algebras are denoted by $K\langle\sqrt{\alpha}\rangle = K\langle\lambda\rangle$.

4. If $c \in A$ satisfies $c^2 \in \dot{K}$, the inner automorphism $a \to cac^{-1}$ has order 2 (or 1) and so gives a grading on A. An important example of this is the quaternion algebra $A = (\alpha, \beta)_K = K1 + Ki + Kj + Kk$ and, say, $c = k$. Then $A_0 = K1 + Kk$ and $A_1 = Ki + Kj$. One writes $\langle\alpha, \beta\rangle_K$ for the quaternion algebra with this grading.

5. Let $V = V_0 \oplus V_1$ be a graded vector space over K (of finite dimension). This simply means that V is a direct sum of two subspaces as indicated, where V_0 is decreed to be of degree 0 and V_1 of degree 1. If $A = \mathrm{End}_K V$, define, for $i \in \mathbb{Z}/2$,

$$A_i = \{\varphi \in \mathrm{End}_K V : \varphi(V_j) \subset V_{j+i} \text{ for all } j \in \mathbb{Z}/2\}.$$

Alternatively

$$A_i = \mathrm{Hom}_K(V_0, V_i) \oplus \mathrm{Hom}_K(V_1, V_{i+1}),$$

identified with a subspace of $\mathrm{End}_K V$ in the obvious way[2]. This makes $\mathrm{End}_K V$ into a graded algebra, denoted by $\mathrm{End}_K(V_0, V_1)$ or simply $\mathrm{End}(V_0, V_1)$ if the field of scalars is clear from the context. In terms of a basis of V compatible with the direct sum, the matrices of the endomorphisms A_0 of degree 0 are of the form $\begin{pmatrix} \underline{\varphi}_{00} & 0 \\ 0 & \underline{\varphi}_{11} \end{pmatrix}$ and those A_1 of degree 1 $\begin{pmatrix} 0 & \underline{\varphi}_{01} \\ \underline{\varphi}_{10} & 0 \end{pmatrix}$. The matrix algebra $\mathrm{M}(n, K)$ with this grading is denoted by $\mathrm{M}(n_0, n_1, K)$ where $n_0 = \dim_K V_0$ and $n_1 = \dim_K V_1$. (Thus $\underline{\varphi}_{ij}$ is an $n_i \times n_j$ matrix).

More generally if D is a K-division algebra and V a right D-vector space, then $\mathrm{End}_D V$ respectively $\mathrm{M}(n, D)$ can be graded in a similar way if $V = V_0 \oplus V_1$ where the V_i are D-subspaces respectively $n = n_0 + n_1$, $n_0, n_1 \geq 0$. We write $\mathrm{End}_D(V_0, V_1)$ respectively $\mathrm{M}(n_0, n_1, D)$ for the resulting graded algebras. They are isomorphic as graded algebras.

6. Let B be any K-algebra, and let $A = B \oplus B$. We make A into a graded algebra by defining

$$A_0 = \{(b, b) | \, b \in B\}, \qquad A_1 = \{(b, -b) | \, b \in B\},$$

and write $A = B \widehat{\oplus} B$. (Recall that we are assuming $\mathrm{char}\, K \neq 2$.)

7. If A is a graded algebra, then $\mathrm{M}(k, A)$ can be graded by choosing $\mathrm{M}(k, A_0)$ to be the homogeneous component of degree 0 and $\mathrm{M}(k, A_1)$ to be that of degree 1. This graded algebra is denoted by $\widetilde{\mathrm{M}}(k, A)$. If A is a CSGA (cf. p. 255), then so is $\widetilde{\mathrm{M}}(k, A)$.

8. The most important example of a graded algebra in this book is the Clifford algebra $\mathrm{C}(V, b)$ of a symmetric space (V, b), defined in the next section. □

The key to the usefulness of the notion of graded algebra (in deriving the properties of the Clifford algebra) is the graded tensor product. Let A and B be graded algebras. The *graded tensor product* $A \widehat{\otimes}_K B$ is, as a vector space, simply the

[2]Namely the maps in $\mathrm{Hom}_K(V_0, V_i)$, for example, are extended to V by defining them to be 0 on V_1.

usual tensor product $A \otimes_K B$. (We will usually write simply $A \widehat{\otimes} B$ if the field K of scalars is clear from the context). The vector space $A \widehat{\otimes}_K B$ is the direct sum of the four subspaces $A_i \otimes B_j$ $(i, j = 0, 1)$. A multiplication is determined on $A \widehat{\otimes} B$ by

$$(a \otimes b)(a' \otimes b') = (-1)^{\partial b \partial a'}(aa' \otimes bb'),$$

between homogeneous elements, making it into an associative algebra via the distributive law. And then the grading

$$\begin{aligned}(A\widehat{\otimes}B)_0 &= (A_0 \otimes B_0) \oplus (A_1 \otimes B_1),\\ (A\widehat{\otimes}B)_1 &= (A_0 \otimes B_1) \oplus (A_1 \otimes B_0)\end{aligned}$$

makes $A \widehat{\otimes} B$ into a graded algebra.

We shall identify A and B with subalgebras of $A \widehat{\otimes} B$ in the usual way – for example $a \in A$ is identified with $a \widehat{\otimes} 1_B$. Then A and B are *graded subalgebras* of $A \widehat{\otimes} B$ in an obvious sense, and they commute in the graded sense: $ab = (-1)^{\partial a \partial b} ba$ for all homogeneous elements a, b, that is to say homogeneous elements commute (in the usual sense) unless they are both of degree 1, in which case they anticommute.

An example of the graded tensor product is

$$K\langle \alpha \rangle \mathbin{\widehat{\otimes}} K\langle \beta \rangle \cong \langle \alpha, \beta \rangle_K. \tag{6.1}$$

(Isomorphism as graded algebras).

The graded tensor product has the following universal mapping property: *if $\varphi : A \to C$ and $\psi : B \to C$ are graded algebra homomorphisms such that the images of A and B commute (in the graded sense), then there is a unique graded algebra homomorphism $\varphi \mathbin{\widehat{\otimes}} \psi : A \widehat{\otimes} B \to C$ which extends φ and ψ.* It is of course given by $(\varphi \mathbin{\widehat{\otimes}} \psi)(a \widehat{\otimes} b) = \varphi(a)\psi(b)$.

A is *graded simple* if the only two-sided graded ideals are 0 and A, and is *central* if its graded center $\widehat{\mathrm{Z}}_A$ – by definition the span by all homogeneous elements c satisfying $ca = (-1)^{\partial c \partial a} ac$ for all $a \in \mathrm{h}A$ – is $= K$ $(= K1_A)$.

Example. 9. If A is a central simple K-algebra, then $\langle A \rangle$ is a CSGA (central simple graded algebra). □

If A and B are CSGAs, then so is $A \widehat{\otimes} B$. The canonical maps

$$(A\widehat{\otimes}B)\widehat{\otimes}C \to A\widehat{\otimes}(B\widehat{\otimes}C) \quad \text{and} \quad A\widehat{\otimes}B \to B\widehat{\otimes}A \tag{6.2}$$

(the latter given by $a \otimes b \to (-1)^{\partial a \partial b} b \otimes a$ on homogeneous elements) are isomorphisms of graded algebras.

The following two theorems give the determination (due to C.T.C. Wall, [70]) of all CSGAs over a field K of characteristic $\neq 2$ – they are Ths. 3.6 and 3.8, pp. 92-94, Ch. IV, [39].

The ordinary center $\mathrm{Z}(A)$ of a CSGA A is $= K \oplus Z_1$ where $Z_1 \subset A_1$, and is therefore a graded subalgebra of A. If $Z_1 \neq 0$, A is said to be of *odd type* and we set $\tau(A) = \tau_A = 1$; otherwise A is of *even type* and $\tau(A) = \tau_A = 0$. In particular $\langle A \rangle$ is an even CSGA if A is a CSA (central simple algebra) over K.

Theorem 6.1.1 *Let A be a CSGA of odd type.*

(a) A_0 is a CSA over K, $A_1 \neq 0$, and $\mathrm{Z}(A) = \mathrm{Z}_A(A_0) = K \oplus Kz$ where $z \in Z_1$ such that $z^2 \in \dot{K}$. If $\zeta = z^2$, the square class $\zeta \dot{K}^2$ does not depend on the choice of z, and $\mathrm{Z}(A) \cong K\langle \sqrt{\zeta} \rangle$ (as graded algebras).

(b) *The canonical map* $\langle A_0\rangle \times \mathrm{Z}(A) \to A$ *provides an isomorphism*

$$A \cong \langle A_0\rangle \widehat{\otimes} K\langle\sqrt{\zeta}\,\rangle = \langle A_0\rangle \otimes K\langle\sqrt{\zeta}\,\rangle$$

of graded algebras.

(c) *If* $\zeta \in \dot{K}^2$, *then* $\mathrm{Z}(A) \cong K\widehat{\oplus}K$ *and* $A \cong A_0\widehat{\oplus}A_0$ *as graded algebras. If* $\zeta \notin \dot{K}^2$, *then* A *is a CSA over the field* $\mathrm{Z}(A) \cong K(\sqrt{\zeta}\,)$. □

Theorem 6.1.2 *Let* A *be a CSGA of even type. Then, as an ungraded algebra,* A *is a CSA over* K *isomorphic to* $\mathrm{M}(k, D)$, *say, where* D *is a central division algebra. The following hold:*

(a) *If* $A_1 = 0$, *then* $A = \langle A_0\rangle \cong \langle \mathrm{M}(k, D)\rangle$. *In this case we define* $z = 1_A$ *and* $\zeta = 1_K$.

Suppose now that $A_1 \neq 0$.

(b) $\mathrm{Z}(A_0) = \mathrm{Z}_A(A_0)$, *and there is a* $z \in \mathrm{Z}(A_0)$ *such that* $\mathrm{Z}(A_0) = K \oplus Kz$ *and* $z^2 = \zeta \in \dot{K}$. *The element* z *is uniquely determined up to a non-zero scalar, the square class* $\zeta\dot{K}^2$ *is uniquely determined, and*

$$za_iz^{-1} = (-1)^i a_i \text{ for all } a_i \in A_i.$$

(c) *If* $\zeta \in \dot{K}^2$, *then* $\mathrm{Z}(A_0) \cong K \oplus K$ *and* $A \cong \mathrm{End}(V_0, V_1)$ *for some non-zero vector spaces* V_0, V_1 *over* D. *Thus* $A_0 \cong \mathrm{M}(m, D) \oplus \mathrm{M}(n, D)$ *(where* $m + n = k$*).*

(d) *Suppose that* $\zeta \notin \dot{K}^2$ *and that the field* $\mathrm{Z}(A_0) \cong K(\sqrt{\zeta})$ *can be imbedded into* D. *Then there is a grading on* D *such that* $A \cong \widetilde{\mathrm{M}}(k, D)$. *Thus* $A_0 \cong \mathrm{M}(k, D_0)$ *is a CSA over* $\mathrm{Z}(A_0)$.

(e) *Suppose that* $\zeta \notin \dot{K}^2$ *and that the field* $\mathrm{Z}(A_0) \cong K(\sqrt{\zeta})$ cannot *be imbedded into* D. *Then* $k = 2m$ *is even,* $A \cong \langle \mathrm{M}(m, D)\rangle \widehat{\otimes} \langle -\zeta, 1\rangle_K$, *and so* $A_0 \cong \mathrm{M}(m, D) \otimes K(\sqrt{\zeta})$ *is a CSA over* $\mathrm{Z}(A_0)$. □

The square-class $\zeta\dot{K}^2$ – or the element ζ itself – is an important invariant of the CSGA A called the *quadratic invariant* of A. We denote the "characteristic element" z of A by z_A and ζ by ζ_A if it is necessary to indicate their provenance. Then

$$z_A \in A_{\tau(A)} \tag{6.3}$$

and Th. 3.12, p. 98, [39] implies that (up to a nonzero scalar)

$$z_{A\widehat{\otimes}B} = z_A \widehat{\otimes} z_B. \tag{6.4}$$

The quadratic invariant $\zeta_A\dot{K}^2$ depends only on the Brauer-Wall class $[A]$ (cf. p. 257), and z_A depends, up to non-zero scalars, only on the isomorphism class of A in the sense that if $\varphi : A \to A'$ is an isomorphism (of graded algebras), then φz_A is a non-zero scalar multiple of $z_{A'}$. We note that when A is a CSGA of even type,

$$z_A a_0 z_A^{-1} = a_0 \text{ if } a_0 \in A_0, \text{ and } z_A a_1 z_A^{-1} = -a_1 \text{ if } a_1 \in A_1. \tag{6.5}$$

We will also need the following simple extension of the Skolem-Noether theorem to the graded case:

Lemma 6.1.1 *Suppose that* σ *is a graded automorphism of the CSGA* A. *Then* σ *is an inner automorphism with respect to an element* $a \in A_0^\times \cup A_1^\times$, *unless* A *is of odd type and* σ *is not the identity on* $\mathrm{Z}(A)$; *in this exceptional case* σ *is the composition of an inner automorphism with respect to a homogeneous element*

$a \in A_0^\times$ and the automorphism $\mathrm{id}_{\langle A_0\rangle}\widehat{\otimes}\,\tilde{}\, = \mathrm{id}_{A_0} \oplus -\mathrm{id}_{A_1}$ of $A \cong \langle A_0\rangle\widehat{\otimes}K\langle\sqrt{\alpha}\rangle$ (where $\tilde{}$ is the nontrivial K-automorphism of $K\langle\sqrt{\alpha}\rangle$).

When A is concentrated in degree 0, this is simply the usual Skolem-Noether Theorem. So assume that $A_1 \neq 0$.

Suppose first that σ is the identity on $\mathrm{Z}_A(A_0)$. Then it is also the identity on $\mathrm{Z}(A_0)$.

If A has odd type, A_0 is a CSA over K (Th. 6.1.1(a)) and the usual Skolem-Noether Theorem provides $a_0 \in A_0^\times$ satisfying $\sigma|_{A_0} = \mathrm{inn}_{a_0}$. But by Th. 6.1.1(b), $A = A_0\mathrm{Z}(A)$ and since σ is the identity on $\mathrm{Z}_A(A_0) = \mathrm{Z}(A)$ (Th. 6.1.1(a)), $\sigma = \mathrm{inn}_{a_0}$.

Now suppose that A has even type. By Th. 6.1.2(b), $\mathrm{Z}(A_0) = \mathrm{Z}_A(A_0)$ so σ is the identity on the center of A_0; in the cases in (d) and (e) of that theorem, A_0 is a simple algebra so $\sigma|_{A_0} = \mathrm{inn}_{a_0}$ for some $a_0 \in A_0^\times$. In case (c), $A_0 \cong \mathrm{M}(m, D) \oplus \mathrm{M}(n, D)$ and $\mathrm{Z}(A_0) \cong K \oplus K$; the fact that σ is the identity on $\mathrm{Z}(A_0)$ implies that it stabilizes the subalgebras of A_0 which correspond to $\mathrm{M}(m, D)$ and $\mathrm{M}(n, D)$ under the isomorphism, and therefore $\sigma|_{A_0}$ is again an inner automorphism with respect to some $a_0 \in A_0^\times$. But since A is a CSA, $\sigma = \mathrm{inn}_a$ for some $a \in A^\times$; this implies that $aa_0^{-1} \in \mathrm{Z}_A(A_0) = \mathrm{Z}(A_0)$, so $a \in A_0$.

Now suppose that σ is *not* the identity on $\mathrm{Z}_A(A_0)$. We define ϑ to be K-linear automorphism $\mathrm{id}_{\langle A_0\rangle}\widehat{\otimes}\,\tilde{}$ of A in the case of odd type, and to be the inner automorphism with respect to any element of $A_1^\times$ (the elements of A_1 which are invertible in A) if the type is even – note that $A_1^\times$ is not empty since in case (c) of Th. 6.1.2 the assumption on σ implies that the summands in $\mathrm{Z}_A(A_0) = K \oplus K$ are interchanged by σ, hence also the summands in $A_0 \cong \mathrm{M}(m, D) \oplus \mathrm{M}(n, D)$, hence $m = n$ and

$$A_1 \cong \begin{pmatrix} 0^{m\times m} & D^{m\times m} \\ D^{m\times m} & 0^{m\times m} \end{pmatrix}.$$

Since $\mathrm{Z}(A_0) = K \oplus Kz_A$ and z_A anticommutes with A_1 in the even case, the graded automorphism $\vartheta\sigma$ of A *is* the identity on $\mathrm{Z}_A(A_0)$, and so the lemma follows from the first part. □

The Brauer-Wall group. Graded Brauer classes are defined for graded central simple algebras over K in a manner much like the ungraded case, namely two central simple graded algebras A and B are "similar" if there are graded vector spaces $V = V_0 \oplus V_1$ and $W = W_0 \oplus W_1$ such that

$$A\widehat{\otimes}\,\mathrm{End}(V_0, V_1) \cong B\widehat{\otimes}\,\mathrm{End}(W_0, W_1). \tag{6.6}$$

The equivalence classes again form an Abelian group, the *graded Brauer group* or – as we shall refer to it – the *Brauer-Wall group* $\mathrm{BW}(K)$. The equivalence class $[A]$ of a CSGA is called the *Brauer-Wall class* of A.

Define a product on the set

$$\mathrm{Br}(K) \times \dot{K}/\dot{K}^2 \times \mathbb{Z}/2$$

by[3]

$$(b_1, \zeta_1, \tau_1)(b_2, \zeta_2, \tau_2) = (b_1 \cdot b_2 \cdot (\varepsilon_1\zeta_1, \varepsilon_2\zeta_2)_K,\ (-1)^{\tau_1\tau_2}\zeta_1\zeta_2,\ \tau_1 + \tau_2) \tag{6.7}$$

where $(\alpha, \beta)_K$ stands for the Brauer class of the quaternion algebra with parameters α and β, and where $\varepsilon_i = (-1)^{\tau_i}$ if $\tau_1 \neq \tau_2$, and $\varepsilon_1 = 1 = \varepsilon_2$ otherwise – briefly

[3] We write ζ_1, for example, instead of $\zeta_1\dot{K}^2$.

$\varepsilon_i = (-1)^{\tau_i(\tau_1+\tau_2)}$. Let b_A denote the Brauer class of the CSA A if A is an even CSGA, respectively the Brauer class of the CSA A_0 if A is odd. It depends only on the Brauer-Wall class $[A]$ of A. The quadratic invariant ζ_A and type τ_A similarly depend only on $[A]$. Then ("Wall's Theorem" the map

$$(\mathrm{b}, \zeta, \tau) : \mathrm{BW}(K) \to \mathrm{Br}(K) \times \dot{K}/\dot{K}^2 \times \mathbb{Z}/2 \tag{6.8}$$

is a well-defined isomorphism of groups. See Ch. 5, §3, pp. 113-122, [39].

We have encountered the subgroup $\dot{K}/\dot{K}^2 \times \mathbb{Z}/2$ before – see p. 51. It is the central extension $\dot{K}/\dot{K}^2 \times_{\mathrm{z}} \mathbb{Z}/2$ where $\mathrm{z} \in \mathrm{Z}^2(\mathbb{Z}/2, \dot{K})$, $\mathrm{z}(i,j) = (-1)^{ij}$. Since the right side of (6.8) is a group, it follows that $\mathrm{z}'\left((\zeta_1, \tau_1), (\zeta_2, \tau_2)\right) = (\varepsilon_1\zeta_1, \varepsilon_2\zeta_2)_K$ is a 2-cocycle in $\mathrm{Z}^2(\dot{K}/\dot{K}^2 \times_{\mathrm{z}} \mathbb{Z}/2, \mathrm{Br}(K))$, and so the right side of (6.8) is also a central extension, with respect to z'. Thus Wall's Theorem can be restated as

$$(\mathrm{b}, \zeta, \tau) : \mathrm{BW}(K) \to \mathrm{Br}(K) \times_{\mathrm{z}'} (\dot{K}/\dot{K}^2 \times_{\mathrm{z}} \mathbb{Z}/2) \tag{6.9}$$

is an isomorphism.

6.2 Clifford algebras

This section consists of a summary of the properties of Clifford algebras that will be needed later. Good references are [65], Ch. 9, and [39], Ch. V, §§1-3.

If V is any vector space over K, $\otimes^p V$ denotes the p-fold tensor product of V with itself ($p > 1$), and $\otimes^0 V = K$, $\otimes^1 V = V$. The *tensor algebra* of V is

$$\mathrm{T}(V) = K \oplus V \oplus (V \otimes V) \oplus \cdots \oplus (\otimes^p V) \oplus \cdots,$$

with multiplication determined by juxtaposition,

$$(u_1 \otimes \cdots \otimes u_p)(v_1 \otimes \cdots \otimes v_q) = u_1 \otimes \cdots \otimes u_p \otimes v_1 \otimes \cdots \otimes v_q.$$

Definition 6.2.1 *Let (V, b) be a symmetric bilinear space (of finite dimension). The* Clifford algebra *of (V, b) is the factor algebra $\mathrm{C}(b) = \mathrm{T}(V)/I$ where I is the 2-sided ideal generated by the differences $v \otimes v - b(v, v)$ for all $v \in V$.*

We sometimes also write $\mathrm{C}(V)$ or $\mathrm{C}(V, b)$ for $\mathrm{C}(b)$.

The canonical map $V \to \mathrm{C}(V, b)$ can be shown to be injective, and we identify V with its image in $\mathrm{C}(V, b)$. V generates the Clifford algebra as an algebra. The vectors in V satisfy $v^2 = b(v, v)$, $uv + vu = 2b(u, v)$, and $uv = -vu$ if u and v are orthogonal. Moreover if $v_1, \ldots, v_n$ is a basis of V, the products

$$v_{i_1} v_{i_2} \ldots v_{i_p} \text{ with } i_1 < i_2 < \cdots < i_p \tag{6.10}$$

(including the empty product 1) form a basis of $\mathrm{C}(b)$. In particular $\dim_K \mathrm{C}(b) = 2^n$.

The Clifford algebra has the following universal mapping property: *a K-linear map $\varphi : V \to A$ of V into an algebra A satisfying $\varphi(v)^2 = b(v, v)$ for all $v \in V$, has a unique extension to an algebra homomorphism* $\mathrm{C}(b) \to A$.

We can view the tensor algebra as a $\mathbb{Z}/2\mathbb{Z}$-graded algebra $\mathrm{T}(V) = \mathrm{T}_0(V) \oplus \mathrm{T}_1(V)$ where $\mathrm{T}_0(V)$ consists of the sum of the $\otimes^p V$ with p even and $\mathrm{T}_1(V)$ the sum of those with p odd. Since the generating elements $v \otimes v - b(v, v)$ of I are in $\mathrm{T}_0(V)$, I is a homogeneous ideal and $\mathrm{C}(b)$ is therefore also a graded algebra; we denote the homogeneous parts by $\mathrm{C}_0(b)$ and $\mathrm{C}_1(b)$, or simply C_0 and C_1. C_0 is called the *even Clifford algebra*, and the products (6.10) with p even form a basis of it, and those with p odd a basis of C_1.

Suppose that (V_1, b_1) and (V_2, b_2) are symmetric bilinear spaces, and consider the graded tensor product $\mathrm{C}(b_1) \hat{\otimes} \mathrm{C}(b_2)$. If $v_1 \in V_1$ and $v_2 \in V_2$, $(v_1 \hat{\otimes} 1 + 1 \hat{\otimes} v_2)^2 =$

$b_1(v_1, v_1) + b_2(v_2, v_2)$, which implies that there is a graded algebra homomorphism $\mathrm{C}(b_1 \perp b_2) \to \mathrm{C}(b_1)\hat{\otimes}\mathrm{C}(b_2)$ which is the identity on V_1 and V_2 (when the identifications are taken into account). It is a fundamental and important property of the Clifford algebra that this homomorphism is an isomorphism – this follows immediately from the fact that it is onto and that the two algebras have the same dimension. From the point of view of orthogonal representations, this is in fact the reason for considering the graded tensor product of graded algebras.

Theorem 6.2.1 *Let (V, b) be a symmetric bilinear space of dimension n and discriminant δ.*

(a) If n is even, $\mathrm{C}(b)$ is a central simple algebra over K, and is isomorphic to a tensor product of quaternion algebras. The center of C_0 is isomorphic to $K[X]/(X^2-\delta)$. Thus if $\delta \notin \dot{K}^2$, C_0 is a central simple $K(\sqrt{\delta})$-algebra, while if $\delta \in \dot{K}^2$, it is a direct sum of two isomorphic central simple K-algebras.

(b) If n is odd, C_0 is a central simple algebra over K, and is isomorphic to a tensor product of quaternion algebras. The center of $\mathrm{C}(b)$ is isomorphic to $K[X]/(X^2 - \delta)$. If $\delta \notin \dot{K}^2$, $\mathrm{C}(b)$ is a central simple $K(\sqrt{\delta})$-algebra, while if $\delta \in \dot{K}^2$, it is a direct sum of two isomorphic central simple K-algebras. □

This provides an invariant of b in the Brauer group, namely the Brauer class $[\mathrm{C}(b)]$ if n is even, $[\mathrm{C}_0(b)]$ if n is odd – in either case of order 2 or 1. It turns out to be simply the "Witt invariant" $\mathrm{c}(b)$ introduced in §1.5.

The description of the Clifford algebra from the point of view of *graded algebras* is more satisfactory: the Clifford algebra is a central simple graded algebra, no matter if n is even or odd (Th. 2.1, p. 109, [39]). The characteristic element $z_{C(b)}$ is the product $v_1 v_2 \dots v_n$ of the vectors in an orthogonal basis of V, and the image of the Brauer-Wall class of $\mathrm{C}(b)$ in $\mathrm{BW}(K)$ under the map $(\mathrm{b}, \zeta, \tau)$ in (6.9) is $(\mathrm{c}(b), \operatorname{disc} b, \dim_{/2} b)$. The multiplication in $\mathrm{Br}(K)\dot{\times}(\dot{K}/\dot{K}^2\dot{\times}\mathbb{Z}/2)$ gives a way of computing the Witt invariant $\mathrm{c}(b_1 \perp b_2)$ in terms of those for (V_1, b_1) and (V_2, b_2),

$$\mathrm{c}(b_1 \perp b_2) = \begin{cases} \mathrm{c}(b_1)\mathrm{c}(b_2)\,(\operatorname{disc} b_1, \operatorname{disc} b_2)_K & \text{if } \dim V_1 \equiv \dim V_2 \bmod 2, \\ \mathrm{c}(b_1)\mathrm{c}(b_2)\,(\operatorname{disc} b_1, -\operatorname{disc} b_2)_K & \text{if } \dim V_1 \text{ is even and } \dim V_2 \text{ is odd.} \end{cases}$$

This formula can in turn be used to derive a formula for $\mathrm{c}(b)$ in terms of the entries in a diagonalization $b = \langle \alpha_1, \dots, \alpha_n \rangle$, which is equivalent to the definition of $\mathrm{c}(b)$ on p. 54. One begins by noting that $\mathrm{C}(\langle \alpha_1 \rangle) = K\langle \sqrt{\alpha_1} \rangle$ and

$$\mathrm{C}(\langle \alpha_1, \alpha_2 \rangle) = \mathrm{C}(\langle \alpha_1 \rangle)\widehat{\otimes}\mathrm{C}(\langle \alpha_2 \rangle) = \langle \alpha_1, \alpha_2 \rangle_K,$$

so $\mathrm{c}(\langle \alpha_1 \rangle) = 1 = \mathrm{s}(\langle \alpha_1 \rangle)$ and $\mathrm{c}(\langle \alpha_1, \alpha_2 \rangle) = (\alpha_1, \alpha_2)_K = \mathrm{s}(\langle \alpha_1, \alpha_2 \rangle)$. One then continues, using the above formula for $\mathrm{c}(b_1 \perp b_2)$ with $b_1 = \langle \alpha_1, \dots, \alpha_{n-1} \rangle$ and $b_2 = \langle \alpha_n \rangle$, first to determine $\mathrm{c}(b)$ for $n = 3, \dots, 8$, and then finally for $n = 8m + r$, assuming $m \geq 1$, for each of $r = 1, 2 \dots 8$ to obtain the definition of the Witt invariant given on p. 54. Obviously a tedious but straightforward exercise...

6.2.1 The Clifford group. Suppose that b is nonsingular. The grading on $\mathrm{C} = \mathrm{C}(b)$ defines a graded algebra automorphism $\nu = \mathrm{id}_{\mathrm{C}_0} \oplus -\mathrm{id}_{\mathrm{C}_1}$. The *Clifford group* $\Gamma = \Gamma(b)$ is defined as

$$\Gamma = \{a \in \mathrm{C}^\times : \nu(a) V a^{-1} = V\}.$$

The map

$$\pi : \Gamma \to \mathbf{GL}(V), \quad \pi(a)(v) = \nu(a)va^{-1}$$

is a homomorphism – thus it is a linear representation of Γ, called the *vector representation*. In fact the image is in the orthogonal group of b, so we have an orthogonal representation

$$\pi : \Gamma \to \mathbf{O}(b).$$

It is onto and the kernel is $\dot{K}$, giving an exact sequence

$$1 \to \dot{K} \to \Gamma(b) \xrightarrow{\pi} \mathbf{O}(b) \to 1.$$

One can show that Γ is the subgroup of the group of units of C generated by the anisotropic vectors in V,

$$\Gamma = \{u_1 u_2 \dots u_k : \ k \geq 1 \text{ and, for all } i,\ u_i \in V \text{ and } b(u_i, u_i) \neq 0\} \ \cup \ \dot{K}.$$

It follows that Γ is graded,

$$\Gamma = (\Gamma \cap \mathrm{C}_0) \ \cup \ (\Gamma \cap \mathrm{C}_1).$$

The image in $\mathbf{O}(b)$ of an anisotropic vector $u \in V \subset \mathrm{C}$ is

$$\pi(u) = -\mathrm{id}_{Ku} \perp \mathrm{id}_{(Ku)^\perp} = \sigma_u,$$

the *symmetry* of (V, b) with respect to the hyperplane $(Ku)^\perp$. Thus the image of $\Gamma \cap \mathrm{C}_0$ is the special orthogonal group $\mathbf{SO}(b)$, and the image of $\Gamma \cap \mathrm{C}_1$ is the coset of $\mathbf{O}(b)$ of orthogonal transformations of determinant -1.

A "norm" $\mathrm{N} : \Gamma \to \dot{K}$ is defined by $\mathrm{N}(a) = aa^t$, where t is the involution of C which is the identity on V – in other words $(u_1 u_2 \dots u_k)^t = u_k \dots u_2 u_1$. N is a homomorphism and

$$\mathrm{N}(u_1 u_2 \dots u_k) = b(u_1, u_1) b(u_2, u_2) \dots b(u_k, u_k).$$

Define $\mathrm{Pin}(b) = \ker \mathrm{N}$. Then we get a commutative and exact diagram

$$\begin{array}{ccccccccc}
 & & 1 & & 1 & & 1 & & \\
 & & \downarrow & & \downarrow & & \downarrow & & \\
1 & \to & \{\pm 1\} & \to & \mathrm{Pin}(b) & \xrightarrow{\pi} & \pi\mathrm{Pin}(b) & \to & 1 \\
 & & \downarrow & & \downarrow & & \downarrow & & \\
1 & \to & \dot{K} & \to & \Gamma(b) & \xrightarrow{\pi} & \mathbf{O}(b) & \to & 1 \\
 & & \downarrow{\scriptstyle 2} & & \downarrow{\scriptstyle \mathrm{N}} & & \downarrow{\scriptstyle \theta} & & \\
1 & \to & \dot{K}^2 & \to & \dot{K} & \to & \dot{K}/\dot{K}^2 & \to & 1 \\
 & & \downarrow & & & & & & \\
 & & 1 & & & & & &
\end{array}$$

θ is the *spinor norm*, and is determined, as a homomorphism, by $\theta(\sigma_u) = b(u, u)\dot{K}^2$. Replacing Γ in this diagram by $\Gamma_0 = \Gamma \cap \mathrm{C}_0$ leads to the commutative and exact diagram

$$\begin{array}{ccccccccc}
 & & 1 & & 1 & & 1 & & \\
 & & \downarrow & & \downarrow & & \downarrow & & \\
1 & \longrightarrow & \{\pm 1\} & \longrightarrow & \mathrm{Spin}(b) & \xrightarrow{\pi} & \mathbf{O}'(b) & \longrightarrow & 1 \\
 & & \downarrow & & \downarrow & & \downarrow & & \\
1 & \longrightarrow & \dot{K} & \longrightarrow & \Gamma_0(b) & \xrightarrow{\pi} & \mathbf{SO}(b) & \longrightarrow & 1 \\
 & & \downarrow{\scriptstyle 2} & & \downarrow{\scriptstyle \mathrm{N}} & & \downarrow{\scriptstyle \theta} & & \\
1 & \longrightarrow & \dot{K}^2 & \longrightarrow & \dot{K} & \longrightarrow & \dot{K}/\dot{K}^2 & \longrightarrow & 1 \\
 & & \downarrow & & & & & & \\
 & & 1 & & & & & &
\end{array}$$

where $\mathbf{O}'(b) = \pi(\mathrm{Spin})$ is the *spinorial kernel*, as it is often called, and $\mathrm{Spin}(b)$ is the *spin group*.

6.3 Equivariant graded algebras

We shall use some elementary facts about cohomology of groups in the rest of this chapter. Most books on the subject cover the facts we use – for example [69], [37], [12].

Let G be a group; it is arbitrary in this chapter, so not necessarily finite. We say that the graded algebra $A = A_0 \oplus A_1$ is G-*equivariant*, or simply *equivariant*, if there is a homomorphism

$$\sigma : G \to \mathrm{Aut}_{\mathrm{gr}} A \tag{6.11}$$

to the group of graded algebra automorphisms of A. We often write $\sigma(s)(a) = sa$ if σ is understood from the context.

Let B be another G-equivariant graded algebra, with G-action $\sigma' : G \to \mathrm{Aut}_{\mathrm{gr}} B$. The graded tensor product $A \widehat{\otimes} B$ is G-equivariant via the action $s(a \widehat{\otimes} b) = (sa) \widehat{\otimes} (sb)$, $s \in G$, and the isomorphisms (6.2) (p. 255) are clearly also G-equivariant in this case. The map associated to the G-action on $A \widehat{\otimes} B$ is denoted by $\sigma \widehat{\otimes} \sigma' : G \to \mathrm{Aut}_{\mathrm{gr}}\, A \widehat{\otimes} B$.

To define an equivalence relation among ECSGAs (equivariant central simple graded algebras) for a given group G, we consider direct sums $(V, \rho) = (V_0, \rho_0) \oplus (V_1, \rho_1)$ of (finite dimensional) K-linear representations of G. Then $s \in G$ acts on $\mathrm{End}_K(V_0, V_1)$ through conjugation by $\rho(s)$,

$$s\varphi = \rho(s)\varphi\rho(s)^{-1},$$

making it into a G-equivariant CSGA which we call a "trivial" ECSGA. In terms of matrices, this is

$$\begin{pmatrix} \underline{\rho}_0(s) & 0 \\ 0 & \underline{\rho}_1(s) \end{pmatrix} \left(\begin{pmatrix} \underline{\varphi}_{00} & 0 \\ 0 & \underline{\varphi}_{11} \end{pmatrix} + \begin{pmatrix} 0 & \underline{\varphi}_{01} \\ \underline{\varphi}_{10} & 0 \end{pmatrix} \right) \begin{pmatrix} \underline{\rho}_0(s)^{-1} & 0 \\ 0 & \underline{\rho}_1(s)^{-1} \end{pmatrix} =$$

$$\begin{pmatrix} \underline{\rho}_0(s)\underline{\varphi}_{00}\underline{\rho}_0(s)^{-1} & 0 \\ 0 & \underline{\rho}_1(s)\underline{\varphi}_{11}\underline{\rho}_1(s)^{-1} \end{pmatrix} + \begin{pmatrix} 0 & \underline{\rho}_0(s)\underline{\varphi}_{01}\underline{\rho}_1(s)^{-1} \\ \underline{\rho}_1(s)\underline{\varphi}_{10}\underline{\rho}_0(s)^{-1} & 0 \end{pmatrix}$$

where $\operatorname{mat}\rho_i(s) = \underline{\rho}_i(s)$ and $\operatorname{mat}\varphi = \begin{pmatrix} \underline{\varphi}_{00} & \underline{\varphi}_{01} \\ \underline{\varphi}_{10} & \underline{\varphi}_{11} \end{pmatrix}$.

Equivalence among ECSGAs is defined analogously to that for CSGAs, by requiring that there is an isomorphism (6.6) (p. 257) which commutes with the action of G.

The only nonobvious fact that must be verified in showing that this *is* an equivalence relation – namely that it is transitive – is that the graded tensor product

$$\operatorname{End}(V_0, V_1) \mathbin{\widehat{\otimes}} \operatorname{End}(W_0, W_1) \tag{6.12}$$

of two trivial ECSGAs is another trivial ECSGA.

First of all, by a standard fact of linear algebra (Prop. 2.5, p. 610, [41]), there is an (ungraded) algebra isomorphism

$$\operatorname{End}(V_0, V_1) \otimes \operatorname{End}(W_0, W_1) = \operatorname{End} V \otimes \operatorname{End} W \xrightarrow{\sim} \operatorname{End}(V \otimes W) \tag{6.13}$$

where $\varphi \otimes \psi$ is mapped to the endomorphism of $V \otimes W$ determined by $v \otimes w \to \varphi(v) \otimes \psi(w)$. We adopt the obvious grading

$$(V \otimes W)_0 = (V_0 \otimes W_0) \oplus (V_1 \otimes W_1), \quad (V \otimes W)_1 = (V_0 \otimes W_1) \oplus (V_1 \otimes W_0)$$

on $V \otimes W$, and the grading on $\operatorname{End}(V_0, V_1) \otimes \operatorname{End}(W_0, W_1)$ which consists of the same homogeneous components[4] as $\operatorname{End}(V_0, V_1) \mathbin{\widehat{\otimes}} \operatorname{End}(W_0, W_1)$. This done, (6.13) is an isomorphism of *graded* algebras since if $\varphi \in \operatorname{End}(V_0, V_1)_i$ and $\psi \in \operatorname{End}(W_0, W_1)_j$, then

$$\varphi \otimes \psi \in (\operatorname{End}(V_0, V_1) \otimes \operatorname{End}(W_0, W_1))_{i+j}$$

and the image of $\varphi \otimes \psi$ is in $\operatorname{End}((V \otimes W)_0, (V \otimes W)_1)_{i+j}$ since

$$(\varphi \otimes \psi)(V_k \otimes W_l) = (\varphi V_k) \otimes (\psi W_l) \subset V_{k+i} \otimes W_{l+j} \subset (V \otimes W)_{(k+l)+(i+j)}.$$

Now $\operatorname{End}(V_0, V_1) \otimes \operatorname{End}(W_0, W_1)$ is also an equivariant graded algebra (same G-action as on $\operatorname{End}(V_0, V_1) \mathbin{\widehat{\otimes}} \operatorname{End}(W_0, W_1)$, of course), and it is an easy exercise to show that (6.13) is also an *equivariant* isomorphism.

The fact that (6.12) is a trivial ECSGA is now a consequence of the following lemma (adapted from Th. 2.7, p. 89, [39]), applied to $A = \operatorname{End}(V_0, V_1)$, $B = \operatorname{End}(W_0, W_1)$ and $z = \mathrm{id}_{V_0} \oplus -\mathrm{id}_{V_1} \in A$:

Lemma 6.3.1 *Let A and B be equivariant graded algebras. Suppose there is an element $z \in \mathrm{Z}(A_0)$ such that $z^2 = 1$, $za_1 = -a_1 z$ for all $a_1 \in A_1$, and $sz = z$ for all $s \in G$. Then there is an equivariant graded isomorphism $A \widehat{\otimes} B \cong A \otimes B$.*

Let B' be the subalgebra $1 \mathbin{\widehat{\otimes}} B_0 + z \mathbin{\widehat{\otimes}} B_1$ of $A \mathbin{\widehat{\otimes}} B$. Then $A \mathbin{\widehat{\otimes}} 1$ and B' commute elementwise and they generate $A \mathbin{\widehat{\otimes}} B$ as an algebra. Therefore $A \otimes B' \cong A \mathbin{\widehat{\otimes}} B$ as algebras – in fact as EGAs if G acts on B' as a subset of $A \widehat{\otimes} B$ and the grading of B' is $B'_0 = 1 \widehat{\otimes} B_0$ and $B'_1 = z \widehat{\otimes} B_1$. The lemma now follows since $b_0 + b_1 \to 1 \mathbin{\widehat{\otimes}} b_0 + z \mathbin{\widehat{\otimes}} b_1$ is an isomorphism $B \to B'$ of EGAs. □

The resulting set of equivalence classes is the *equivariant Brauer-Wall group* $\mathrm{BW}(K, G)$ (although it remains to show that it *is* a group). We denote the equivalence class of the ECSGA A with action map $\sigma : G \to \operatorname{Aut}_{\mathrm{gr}} A$ by $[A]$ or by $[A, \sigma]$ if it is necessary to make σ explicit. Occasionally it is also convenient to write $[A] = \mathrm{BW}(A)$ or $[A, \sigma] = \mathrm{BW}(\sigma)$.

---

[4] Note that, more generally, the ordinary tensor product $A \otimes B$ of graded algebras is also a graded algebra with the same homogeneous components as $A \mathbin{\widehat{\otimes}} B$.

To show that $\widehat{\otimes}$ makes $\mathrm{BW}(K, G)$ into a group via

$$[A, \sigma][A', \sigma'] = [A \widehat{\otimes} A', \sigma \widehat{\otimes} \sigma'],$$

we note first that multiplication is well-defined, associative and commutative by the equivariant versions of (6.2) (p. 255). The class of the ECSGA $\langle K \rangle$ on which G acts trivially (or of a trivial ECSGA $\mathrm{End}(V_0, V_1)$), is the identity element.

We now show that the inverse of the class $[A, \sigma]$ of an ECSGA A is the class $[A^*, \sigma]$ of the graded opposite algebra A^* which $= A$ as a graded vector space with multiplication defined by $a \cdot b = (-1)^{\partial a \partial b} ba$ on homogeneous elements. It is shown on p. 99, [39], that (A^*, σ) is a CSGA, and it is easy to see that it is in fact an ECSGA. By Prop. 4.1 *ibid* the map

$$\theta : A \widehat{\otimes} A^* \to \mathrm{End}(A_0, A_1),$$

determined by $\theta(b \widehat{\otimes} c)(a) = (-1)^{\partial c \partial a} bac$ where $a, b, c \in \mathrm{h}(A)$, is an isomorphism of graded algebras. (Here, $\mathrm{End}(A_0, A_1)$ is the trivial ECSGA arising from the graded vector space $A = A_0 \oplus A_1$ with the linear representations $\sigma(s)|_{A_0} : G \to \mathbf{GL}(A_0)$ and $\sigma(s)|_{A_1} : G \to \mathbf{GL}(A_1)$, with direct sum $= \sigma$ of course). The isomorphism θ is equivariant since if $a, b, c \in \mathrm{h}(A) = \mathrm{h}(A^*)$,

$$\begin{aligned}
\theta(s(b \widehat{\otimes} c))(a) &= \theta(sb \widehat{\otimes} sc)(a) \\
&= (-1)^{\partial c \partial a}(sb)a(sc) = (-1)^{\partial c \partial a} s\big(b(s^{-1}a)c\big) \\
&= (s\theta(b \widehat{\otimes} c)s^{-1})(a) = (s(\theta(b \widehat{\otimes} c)))(a).
\end{aligned}$$

Thus $\mathrm{BW}(K, G)$ is an Abelian group.

Let $\varphi : \mathrm{BW}(K, G) \to \mathrm{BW}(K)$ be the forgetful map, which takes $[A, \sigma]$ to $[A]$. This is a well-defined homomorphism, and is a split epimorphism since the map $\mathrm{BW}(K) \to \mathrm{BW}(K, G)$ which takes $[A]$ to the equivariant class $[A, 1]$ of A with trivial action of G, is a splitting homomorphism. Let

$$1 \to \mathrm{BW}_0(K, G) \to \mathrm{BW}(K, G) \xrightarrow{\varphi} \mathrm{BW}(K) \to 1$$

be the associated split exact sequence, where $\mathrm{BW}_0(K, G)$ is the kernel – it is the subgroup of $\mathrm{BW}(K, G)$ of equivariant classes which are trivial in $\mathrm{BW}(K)$.

An ECSGA A determines a useful grading on G itself. Let $\psi \in \mathrm{Aut}_{gr} A$. The facts that z_A is defined up to an element of $\dot{K}$, that $z_A^2 \in \dot{K}$ and that $\psi|_K = \mathrm{id}_K$, imply that $\psi(z_A) = \pm z_A$ for any graded automorphism ψ of A – and the sign does not depend on the choice of z_A. Therefore the degree $\partial_A \psi \in \mathbb{Z}/2$ can be defined by $\psi(z_A) = (-1)^{\partial_A \psi} z_A$, and then we set $\partial_\rho s = \partial_A s = \partial_A \rho(s)$. Thus

$$s z_A = (-1)^{\partial_A s} z_A, \quad \partial_A s \in \mathbb{Z}/2. \tag{6.14}$$

It is clear that

$$\partial_A st = \partial_A s + \partial_A t \quad \text{in } \mathbb{Z}/2.$$

The action of G on A is said to be *proper* if $\partial_A s = 0$ for all $s \in G$, otherwise *improper*. (This terminology stems from the fact that, as we shall see, when A is a Clifford algebra $\mathrm{C}(V, b)$ and ρ arises from an orthogonal representation $G \to \mathbf{O}(V, b)$, $\partial_A = 0$ if and only if the orthogonal representation is proper, i.e. the image of G lies in $\mathbf{SO}(V, b)$). Clearly $\partial_A = 0$ if G has odd order, so the action of G is always proper when $|G|$ is odd.

Since $z_{A \widehat{\otimes} B} = z_A \widehat{\otimes} z_B$ (up to a nonzero scalar),

$$\partial_{A \widehat{\otimes} B} = \partial_A + \partial_B. \tag{6.15}$$

Suppose A is a trivial ECSGA $\mathrm{End}(V_0, V_1)$ (so of even type), with G-actions $\rho_i : G \to \mathbf{GL}(V_i)$. Then $A_0 = \mathrm{End}V_0 \oplus \mathrm{End}V_1$ or $\mathrm{End}V_0$ which has center $K \oplus K$ respectively K, and so it follows from the definition of z_A (Th. 6.1.2,(b), p. 256) that z_A is a nonzero scalar multiple of $(1, -1)$ respectively 1. It is fixed by any inner automorphism $\mathrm{inn}_{\rho_0(s) \oplus \rho_1(s)}$ respectively $\mathrm{inn}_{\rho_0(s)}$, and so it follows that the action of G on A is proper and $\partial_A = 0$. Thus ∂ is well defined on $\mathrm{BW}(K, G)$ and we get a homomorphism

$$\partial : \mathrm{BW}(K, G) \to \mathrm{Hom}(G, \mathbb{Z}/2). \tag{6.16}$$

The map $\tau : \mathrm{BW}(K, G) \to \mathbb{Z}/2$ given by the type of an ECSGA is also a well-defined homomorphism (since it is for $\mathrm{BW}(K)$).

We note that $\mathrm{H}^1(G, \mathbb{Z}/2) = \mathrm{Z}^1(G, \mathbb{Z}/2) = \mathrm{Hom}(G, \mathbb{Z}/2)$. The cup product

$$\smile : \mathrm{Hom}(G, \mathbb{Z}/2) \times \mathrm{Hom}(G, \mathbb{Z}/2) \to \mathrm{H}^2(G, \mathbb{Z}/2)$$

takes (∂_1, ∂_2) to the cohomology class of the cocycle $\partial_1(s)\partial_2(t) = (\partial_1 \smile \partial_2)(s, t)$ (cf. p. 110, [12] or p. 117, [37]). Denote by $(-1)^{\partial_1 \smile \partial_2}$ the cocycle $(-1)^{\partial_1(s)\partial_2(t)} \in \mathrm{Z}^2(G, \dot{K})$ whose cohomology class is the image of the cohomology class of $\partial_1 \smile \partial_2$ under the composite of the "coefficient maps" $\mathrm{H}^2(G, \mathbb{Z}/2) \xrightarrow{\sim} \mathrm{H}^2(G, \pm 1) \to \mathrm{H}^2(G, \dot{K})$.

Suppose that $\mathbb{Z}/2$ operates trivially on $\dot{K}$ and that $\alpha \in \dot{K}$. Define

$$\widetilde{\alpha}(u, v) = \alpha^{uv} \in \dot{K}, \quad u, v \in \mathbb{Z}/2, \tag{6.17}$$

where uv is the product in the ring $\mathbb{Z}/2$ (and $\alpha^0 = 1_K$, $\alpha^1 = \alpha$). By checking the two cases $u = v = w = 1$ and at least one of $u, v, w = 0$, it is easy to see that the cocycle relation $\widetilde{\alpha}(v, w)\widetilde{\alpha}(u, v + w) = \widetilde{\alpha}(u, v)\widetilde{\alpha}(u + v, w)$ holds, so $\widetilde{\alpha} \in \mathrm{Z}^2(\mathbb{Z}/2, \dot{K})$ and is normalized, i.e. $\widetilde{\alpha}(u, v) = 1$ if u or v is 0.

Lemma 6.3.2 *The cocycle $\widetilde{\alpha}$ is a coboundary if and only if $\alpha \in \dot{K}^2$.*

If $\widetilde{\alpha}$ is a coboundary, say $\widetilde{\alpha}(u, v) = c(u)c(v)c(u+v)^{-1}$, then $c(0)c(v)c(0+v)^{-1} = 1$ implies that $c(0) = 1$, and (setting $u = v = 1$) $c(1)^2 = \alpha$. Conversely if $\alpha = \beta^2$ and we set $c(u) = \beta^u$ for $u \in \mathbb{Z}/2$, i.e. $c(0) = 1$ and $c(1) = \beta$, then it is easy to see that $\widetilde{\alpha}(u, v) = c(u)c(v)c(u + v)^{-1}$ by checking cases. □

Thus if $\partial_1 \in \mathrm{Hom}(G, \mathbb{Z}/2)$, the map $\partial_1^* : \mathrm{H}^2(\mathbb{Z}/2, \dot{K}) \to \mathrm{H}^2(G, \dot{K})$ provides a cohomology class in $\mathrm{H}^2(G, \dot{K})$ for each $\alpha \in \dot{K}$, namely that of the cocycle $(\partial_1^*\widetilde{\alpha})(s, t) = \alpha^{\partial_1(s)\partial_1(t)}$ which is a coboundary if $\alpha \in \dot{K}^2$.

Let $\zeta : \mathrm{BW}(K, G) \to \dot{K}/\dot{K}^2$ be the quadratic invariant, $\zeta[A] = \zeta_A \dot{K}^2$. This is well-defined since it factors through the forgetful map, $\mathrm{BW}(K, G) \to \mathrm{BW}(K) \to \dot{K}/\dot{K}^2$ (cf. (6.9), p. 258).

Thus there are four maps

$$\begin{aligned} &\mathrm{b} : \mathrm{BW}(K, G) \to \mathrm{B}(K), && \zeta : \mathrm{BW}(K, G) \to \dot{K}/\dot{K}^2, \\ &\tau : \mathrm{BW}(K, G) \to \mathbb{Z}/2, && \partial : \mathrm{BW}(K, G) \to \mathrm{Hom}(G, \mathbb{Z}/2), \end{aligned}$$

where b is of course the composition of the forgetful map $\mathrm{BW}(K, G) \to \mathrm{BW}(K)$ and $\mathrm{b} : \mathrm{BW}(K) \to \mathrm{B}(K)$. There is one more,

$$\mathrm{f} : \mathrm{BW}(K, G) \to \mathrm{H}^2(G, \dot{K}),$$

which we now define. Let A be an even ECSGA. By Lem. 6.1.1 (p. 256), there is an exact sequence

$$1 \to \dot{K} \to \mathrm{h}(A^\times) \to \mathrm{Aut}_{\mathrm{gr}} A \to 1,$$

where of course $\mathrm{h}(A^\times) = A_0^\times \cup A_1^\times$. Thus the action $\rho : G \to \mathrm{Aut}_{\mathrm{gr}} A$ can be lifted to a "graded" projective representation $\hat{\rho} : G \to \mathrm{h}(A^\times)$ (where we assume as one usually does that $\hat{\rho}(1) = 1_A$),

$$\rho(s)a = \hat{\rho}(s)a\hat{\rho}(s)^{-1} \text{ for all } a \in A.$$

"Projective" here means that $\hat{\rho}(s)\hat{\rho}(t) = f(s,t)\hat{\rho}(st)$ for some $f(s,t) \in \dot{K}$ – then it is easy to check that $f \in \mathrm{Z}^2(G, \dot{K})$. Note also that $\hat{\rho}(s)z_A\hat{\rho}(s)^{-1} = \rho(s)(z_A) = (-1)^{\partial_A(s)} z_A$ implies that

$$\hat{\rho}(s) \in A_{\partial_A(s)} \quad \text{when } A \text{ is even,} \tag{6.18}$$

since $z_A^{-1} a_i z_A = (-1)^i a_i$ if $a_i \in A_i$ by (6.5).

If A is an odd ECSGA, there is an exact sequence

$$1 \to \dot{K} \to A_0^\times \to \mathrm{Aut} A_0 \to 1$$

by the Skolem-Noether Theorem, and we lift the map $s \to \rho(s)|_{A_0}$ to a projective representation $\hat{\rho} : G \to A_0^\times$ with associated cocycle again denoted by $f(s,t) \in \dot{K}$,

$$\rho(s)a_0 = \hat{\rho}(s)a_0\hat{\rho}(s)^{-1} \text{ for all } a_0 \in A_0.$$

Thus

$$\hat{\rho}(s) \in A_0 \quad \text{when } A \text{ is odd.} \tag{6.19}$$

A different lifting $\check{\rho}$ in *either* case (A even or odd) would satisfy $\check{\rho}(s) = c(s)\hat{\rho}(s)$ for some cochain $c(s) \in \dot{K}$, and the associated cocycle f' would be

$$f'(s,t) = \check{\rho}(s)\check{\rho}(t)\check{\rho}(st)^{-1} = \big(c(s)c(t)c(st)^{-1}\big) f(s,t)$$

which is cohomologous to f. Thus the cohomology class $\mathrm{f}_A = [f] \in \mathrm{H}^2(G, \dot{K})$ associated with ρ and A is well-defined. We shall see later (Cor. 6.3.2) that in fact f_A depends only on the class of (A, ρ) in $\mathrm{BW}(G, K)$.

REMARKS. 1. We usually denote the cocycle f of A by f_A. We will also treat the quantities z_A, ζ_A and f_A as if they are well-defined quantities of A, $\dot{K}$ and $\mathrm{Z}^2(G, \dot{K})$ respectively, even though they are in reality only well-defined in $\dot{K}z_A$, $\dot{K}/\dot{K}^2$ and $\mathrm{H}^2(G, \dot{K})$ respectively. Thus for example $\zeta_A = \alpha$ means that $\zeta_A = \alpha\dot{K}^2$, and $f_{A\hat{\otimes}B} = f_A f_B$ that the two sides are congruent modulo the coboundaries $\mathrm{B}^2(G, \dot{K})$.
2. A cocycle (or the cohomology class) associated with an odd ECSGA A is also associated with the even ECSGA $\langle A_0 \rangle$.
3. Let (A, ρ) be the trivial ECSGA based on $\sigma = \sigma_0 \oplus \sigma_1 : G \to \mathbf{GL}(V_0) \oplus \mathbf{GL}(V_1)$. Thus $\rho : G \to \mathrm{Aut}_{gr}(\mathrm{End}(V_0, V_1))$ is given by $\rho(s) = \mathrm{inn}_{\sigma(s)}$. This means that $\sigma(s)$ is a lift of $\rho(s)$, and it follows at once that $f_A(s,t) = 1$ for all $s, t \in G$, i.e. $f_A = 1$, since σ is a homomorphism. □

Lemma 6.3.3 *If A and B are even ECSGAs, then*

$$f_{A\hat{\otimes}B} = (-1)^{\partial_A \smile \partial_B} (\partial_B^* \widetilde{\zeta_A})(\partial_A^* \widetilde{\zeta_B}) f_A f_B.$$

If $B = \langle B_0 \rangle$ where B_0 is a CSA, then

$$f_{A\hat{\otimes}B} = f_A f_B.$$

REMARK. The cocycle $\partial_B^*\widetilde{\zeta_A} \in \mathrm{Z}^2(G, \dot{K})$ (for example) depends of course on the choice of a representative of the square class $\zeta_A\dot{K}^2$ (cf. (6.17)), but by Lem. 6.3.2 the cohomology class $[\widetilde{\zeta_A}] \in \mathrm{H}^2(\mathbb{Z}/2, \dot{K})$ does *not*, and so neither does $\partial_B^*[\widetilde{\zeta_A}] = [\partial_B^*\widetilde{\zeta_A}] \in \mathrm{H}^2(G, \dot{K})$ – and we are ultimately interested only in it. $\square$

We now prove the lemma. Let $\rho : G \to \mathrm{Aut_{gr}}A$ and $\sigma : G \to \mathrm{Aut_{gr}}B$ be the G-action maps and $\hat{\rho} : G \to \mathrm{h}A^\times$ and $\hat{\sigma} : G \to \mathrm{h}B^\times$ liftings of ρ and σ to projective representations with cocycles f_A and f_B, say. Note first that

$$\left(z_A^{\partial_B(s)}\hat{\rho}(s)\,\widehat{\otimes}\,z_B^{\partial_A(s)}\hat{\sigma}(s)\right)\left(\hat{\rho}(s^{-1})\,z_A^{-\partial_B(s)}\widehat{\otimes}\,\hat{\sigma}(s^{-1})\,z_B^{-\partial_A(s)}\right) = (-1)^{\partial_A(s)\partial_B(s)} f_A(s, s^{-1})\, f_B(s, s^{-1}).$$

The sign $(-1)^{\partial_A(s)\partial_B(s)}$ appears because of the definition of multiplication in a graded tensor product, and because z_A and z_B have degree 0 (since A and B are even CSGAs – cf. Th. 6.1.2) and by (6.18) $\hat{\sigma}(s)$ and $\hat{\rho}(s^{-1})$ have degree $\partial_B(s)$ and $\partial_A(s)$ respectively. Also $\widehat{\rho}(s)\widehat{\rho}(s^{-1}) = f_A(s, s^{-1})\rho(1) = f_A(s, s^{-1})$ by the definition of f_A. Thus if $\varphi(s) = z_A^{\partial_B(s)}\hat{\rho}(s)\,\widehat{\otimes}\,z_B^{\partial_A(s)}\hat{\sigma}(s)$,

$$\varphi(s)^{-1} = (-1)^{\partial_A(s)\partial_B(s)} f_A(s, s^{-1})^{-1} f_B(s, s^{-1})^{-1}\hat{\rho}(s^{-1}) z_A^{-\partial_B(s)}\,\widehat{\otimes}\,\hat{\sigma}(s^{-1}) z_B^{-\partial_A(s)}.$$

A straightforward calculation then shows that

$$\varphi(s)(a\widehat{\otimes}b)\varphi(s)^{-1} = \rho(s)(a)\widehat{\otimes}\sigma(s)(b),$$

i.e. that φ lifts $\rho\widehat{\otimes}\sigma$. We give the calculation here, although the reader might prefer to skip it at a first reading or work it out her/himself. We may suppose a and b to be homogeneous.

$$\begin{aligned}
\varphi(s)(a\widehat{\otimes}b)\varphi(s)^{-1} &= \left(z_A^{\partial_B(s)}\hat{\rho}(s)\widehat{\otimes}z_B^{\partial_A(s)}\hat{\sigma}(s)\right)\left(a\widehat{\otimes}b\right)\varphi(s)^{-1} \\
&= (-1)^{\partial_A(s)\partial_B(s)+\partial a\partial_B(s)} f_A(s, s^{-1})^{-1} f_B(s, s^{-1})^{-1} \\
&\quad \left(z_A^{\partial_B(s)}\hat{\rho}(s)a\widehat{\otimes}z_B^{\partial_A(s)}\hat{\sigma}(s)b\right)\left(\hat{\rho}(s^{-1})z_A^{-\partial_B(s)}\widehat{\otimes}\hat{\sigma}(s^{-1})z_B^{-\partial_A(s)}\right) \\
&= (-1)^{\partial_A(s)\partial_B(s)+\partial a\partial_B(s)+\partial_A(s)(\partial_B(s)+\partial b)} f_A(s, s^{-1})^{-1} f_B(s, s^{-1})^{-1} \\
&\quad z_A^{\partial_B(s)}\hat{\rho}(s)a\hat{\rho}(s^{-1})z_A^{-\partial_B(s)}\widehat{\otimes}z_B^{\partial_A(s)}\hat{\sigma}(s)b\hat{\sigma}(s^{-1})z_B^{-\partial_A(s)} \\
&= (-1)^{\partial a\partial_B(s)+\partial a\partial_B(s)+\partial_A(s)\partial b+\partial b\partial_A(s)}\rho(s)(a)\widehat{\otimes}\sigma(s)(b) \\
&= \rho(s)(a)\widehat{\otimes}\sigma(s)(b),
\end{aligned}$$

by (6.5) and since, for example, $\hat{\rho}(s)^{-1} = f_A(s, s^{-1})^{-1}\hat{\rho}(s^{-1})$, and $\partial(\rho(s)(a)) = \partial a$.

Now we use φ to find a cocycle belonging to $\rho\widehat{\otimes}\sigma$:

$$\begin{aligned}
&\varphi(s)\varphi(t)\varphi(st)^{-1}\\
&= \left(z_A^{\partial_B(s)}\hat{\rho}(s)\widehat{\otimes}z_B^{\partial_A(s)}\hat{\sigma}(s)\right)\left(z_A^{\partial_B(t)}\hat{\rho}(t)\widehat{\otimes}z_B^{\partial_A(t)}\hat{\sigma}(t)\right)\varphi(st)^{-1}\\
&= (-1)^{\partial_B(t)\partial_A(s)+\partial_B(s)\partial_A(t)+\partial_A(t)\partial_B(s)}\\
&\quad \left(z_A^{\partial_B(s)+\partial_B(t)}f_A(s,t)\hat{\rho}(st)\widehat{\otimes}z_B^{\partial_A(s)+\partial_A(t)}f_B(s,t)\hat{\sigma}(st)\right)\\
&\quad (-1)^{\partial_A(st)\partial_B(st)}f_A(st,(st)^{-1})^{-1}f_B(st,(st)^{-1})^{-1}\hat{\rho}((st)^{-1})z_A^{-\partial_B(st)}\widehat{\otimes}\\
&\qquad \hat{\sigma}((st)^{-1})z_B^{-\partial_A(st)}\\
&= (-1)^{\partial_A(s)\partial_B(t)+\partial_A(st)\partial_B(st)+\partial_A(st)\partial_B(st)}\\
&\quad f_A(s,t)f_B(s,t)f_A(st,(st)^{-1})^{-1}f_B(st,(st)^{-1})^{-1}f_A(st,(st)^{-1})f_B(st,(st)^{-1})\\
&\quad z_A^{\partial_B(s)+\partial_B(t)-\partial_B(st)}\widehat{\otimes}z_B^{\partial_A(s)+\partial_A(t)-\partial_A(st)}\\
&= (-1)^{\partial_A(s)\partial_B(t)}\zeta_A^{\partial_B(s)\partial_B(t)}\zeta_B^{\partial_A(s)\partial_A(t)}f_A(s,t)f_B(s,t)
\end{aligned}$$

Note that in the exponent of z_A in the second-last line, the terms $\partial_B(s), \partial_B(t)$ and $\partial_B(st)$ must be treated as integers, not as elements of $\mathbb{Z}/2$, since we use the law of exponents; thus for example $z^1z^1 = z^{1+1} = z^2$, not z^0. Then it is easy to show that $\partial_B(s) + \partial_B(t) - \partial_B(st) = 2\partial_B(s)\partial_B(t)$ by checking the 4 cases arising from $\partial_B(s) = 0, 1$, $\partial_B(t) = 0, 1$. Thus $z_A^{\partial_B(s)+\partial_B(t)-\partial_B(st)} = z_A^{2\partial_B(s)\partial_B(t)} = \zeta_A^{\partial_B(s)\partial_B(t)}$. Of course a similar remark applies to the power of z_B. Finally the cup product

$$\smile : \mathrm{H}^0(G,\mathbb{Z}/2)\times\mathrm{H}^0(G,\mathbb{Z}/2)\to\mathrm{H}^0(G,\mathbb{Z}/2),$$

when we identify $\mathrm{H}^0(G,\mathbb{Z}/2)$ with $\mathrm{Hom}(G,\mathbb{Z}/2)$, is the map

$$\mathrm{Hom}(G,\mathbb{Z}/2)\times\mathrm{Hom}(G,\mathbb{Z}/2)\to\mathrm{Hom}(G,\mathbb{Z}/2)$$

which takes (∂_A, ∂_B) to $(s,t)\to\partial_A(s)\partial_B(t)$.

Now for the last statement: in this case $z_B = 1_B$ (Th. 6.1.2) and it follows that $\partial_B = 0$ and $\zeta_B = 1$ – and therefore that $f_{A\widehat{\otimes}B} = f_Af_B$. □

Lemma 6.3.4 *Let A be an even ECSGA and $B = \langle B_0\rangle\widehat{\otimes}K\langle z_B\rangle$ an odd ECSGA.*

(a) The invariants of the equivariant graded subalgebra $A\widehat{\otimes}\langle B_0\rangle$ of $A\widehat{\otimes}B$ are

$$\zeta_{A\widehat{\otimes}\langle B_0\rangle} = \zeta_A,\quad \partial_{A\widehat{\otimes}\langle B_0\rangle} = \partial_A,\quad \textit{and}\quad f_{A\widehat{\otimes}\langle B_0\rangle} = f_Af_B.$$

(b) The invariants of the equivariant graded subalgebra $K\langle z_B\rangle$ of B are

$$\zeta_{K\langle z_B\rangle} = \zeta_B,\quad \partial_{K\langle z_B\rangle} = \partial_B,\quad \textit{and}\ f_{K\langle z_B\rangle} = 1.$$

By Th. 6.1.2(a) (p. 256), $z_{\langle B_0\rangle} = 1_{B_0}$, so $\zeta_{B_0} = 1$, $\partial_{B_0} = 0$, and

$$\begin{aligned}
z_{A\widehat{\otimes}\langle B_0\rangle} &= z_A\widehat{\otimes}z_{\langle B_0\rangle} \quad \text{by (6.4)},\\
&= z_A\widehat{\otimes}1_{B_0}.
\end{aligned}$$

Therefore $\zeta_{A\widehat{\otimes}\langle B_0\rangle} = \zeta_A$, $\partial_{A\widehat{\otimes}\langle B_0\rangle} = \partial_A$, and, by the previous lemma, $f_{A\widehat{\otimes}\langle B_0\rangle} = f_Af_{\langle B_0\rangle} = f_Af_B$. This is (a), and (b) follows easily from the fact that $z_{K\langle z_B\rangle} = z_B$. □

Lemma 6.3.5 *Let A and B be ECSGAs, A even and B odd. Then*

$$f_{A\widehat{\otimes}B} = \partial_A^*(\widetilde{-1})(-1)^{\partial_A\smile\partial_B}(\partial_B^*\widetilde{\zeta_A})(\partial_A^*\widetilde{\zeta_B})f_Af_B.$$

Suppose first that $B = K\langle z_B\rangle$. Since

$$(A\widehat{\otimes}K\langle z_B\rangle)_0 = (A_0\widehat{\otimes}K) \oplus (A_1\widehat{\otimes}Kz_B),$$

every element of degree 0 is a sum of elements of the form $a_0\widehat{\otimes}1$ and $a_1\widehat{\otimes}z_B$.

With the same notation as Lem. 6.3.3 and using Lem. 6.3.4, (6.3) and (6.18),

$$\left(z_A^{\partial_B(s)}\hat\rho(s)\widehat{\otimes}z_B^{\partial_A(s)}\right)\left(\hat\rho(s)^{-1}z_A^{-\partial_B(s)}\widehat{\otimes}z_B^{-\partial_A(s)}\right) = (-1)^{\partial_A(s)\partial_A(s)} = (-1)^{\partial_A(s)},$$

and so

$$\left(z_A^{\partial_B(s)}\hat\rho(s)\widehat{\otimes}z_B^{\partial_A(s)}\right)^{-1} = (-1)^{\partial_A(s)}\hat\rho(s)^{-1}z_A^{-\partial_B(s)}\widehat{\otimes}z_B^{-\partial_A(s)}.$$

Let $i = 0$ or 1. Then

$$\begin{aligned}
&(z_A^{\partial_B(s)}\hat\rho(s)\widehat{\otimes}z_B^{\partial_A(s)})(a_i\widehat{\otimes}z_B^i)(z_A^{\partial_B(s)}\hat\rho(s)\widehat{\otimes}z_B^{\partial_A(s)})^{-1}\\
&\quad= (-1)^{\partial_A(s)+i\partial_A(s)+\partial_A(s)(\partial_A(s)+i)}z_A^{\partial_B(s)}\rho(s)(a_i)z_A^{-\partial_B(s)}\widehat{\otimes}z_B^i\\
&\quad= (-1)^{i\partial_B(s)}\rho(s)(a_i)\widehat{\otimes}z_B^i = \rho(s)(a_i)\widehat{\otimes}(-1)^{i\partial_B(s)}z_B^i,
\end{aligned}$$

which is the transform of $a_i\widehat{\otimes}z_B^i$ by s (cf. (6.14)) – so $z_A^{\partial_B(s)}\hat\rho(s)\widehat{\otimes}z_B^{\partial_A(s)}$ lifts the action of s on $(A\widehat{\otimes}K\langle z_B\rangle)_0$.

Note that $z_A \in A_0$ since A is even, $\hat\rho(s) \in A_{\partial_A(s)}$ by (6.18), $z_B \in K\langle z_B\rangle_1$ – so $z_B^{\partial_A(s)} \in K\langle z_B\rangle_{\partial_A(s)}$ and

$$z_A^{\partial_B(s)}\hat\rho(s)\widehat{\otimes}z_B^{\partial_A(s)} \in A_{\partial_A(s)}\widehat{\otimes}K\langle z_B\rangle_{\partial_A(s)} \subset (A\widehat{\otimes}K\langle z_B\rangle)_0,$$

so we can use it to calculate $f_{A\widehat{\otimes}K\langle z_B\rangle}$:

$$\begin{aligned}
f_{A\widehat{\otimes}K\langle z_B\rangle}(s,t) &= \left(z_A^{\partial_B(s)}\hat\rho(s)\widehat{\otimes}z_B^{\partial_A(s)}\right)\left(z_A^{\partial_B(t)}\hat\rho(t)\widehat{\otimes}z_B^{\partial_A(t)}\right)\times\\
&\qquad\left((-1)^{\partial_A(st)}\hat\rho(st)^{-1}z_A^{-\partial_B(st)}\widehat{\otimes}z_B^{-\partial_A(st)}\right)\\
&= (-1)^{\varepsilon}z_A^{\partial_B(s)}\hat\rho(s)z_A^{\partial_B(t)}\hat\rho(t)\hat\rho(st)^{-1}z_A^{-\partial_B(st)}\widehat{\otimes}z_B^{\partial_A(s)+\partial_A(t)-\partial_A(st)} \qquad (6.20)
\end{aligned}$$

where $\varepsilon = \partial_A(st) + \partial_A(s)\partial_A(t) + \partial_A(st)\left(\partial_A(s) + \partial_A(t)\right) = \partial_A(s)\partial_A(t)$ (in $\mathbb{Z}/2$ – legitimate this time because the base of the power is -1). We next calculate the factor

$$z_A^{\partial_B(s)}\hat\rho(s)z_A^{\partial_B(t)}\hat\rho(t)\hat\rho(st)^{-1}z_A^{-\partial_B(st)} \qquad (6.21)$$

of (6.20). If $\partial_B(s) = 0$, it is

$$\hat\rho(s)z_A^{\partial_B(t)}\hat\rho(t)\hat\rho(st)^{-1}z_A^{-\partial_B(t)} = (-1)^{\partial_A(s)\partial_B(t)}f_A(s,t)$$

by (6.5) (p. 256) – note that $\hat\rho(t)\hat\rho(st)^{-1} \in A_{\partial_A(s)}$ by (6.18). A similar calculation shows that this same formula holds if $\partial_B(s) = 1$ and $\partial_B(t) = 0$. In the remaining case of $\partial_B(s) = 1 = \partial_B(t)$, the expression is equal to $z_A\hat\rho(s)z_A\hat\rho(t)\hat\rho(st)^{-1}$, which becomes $(-1)^{\partial_A(s)}\zeta_A f_A(s,t)$ if one replaces the second z_A by $z_A^{-1}\zeta_A$. Thus (6.21) is

$$(-1)^{\partial_A(s)\partial_B(t)}\zeta_A^{\partial_B(s)\partial_B(t)}f_A(s,t)$$

for arbitrary $s,t \in G$. We have already noted in the proof of Lemma 6.3.3 that

$$z_B^{\partial_A(s)+\partial_A(t)-\partial_A(st)} = \zeta_B^{\partial_A(s)\partial_A(t)}.$$

Putting this all together in (6.20) gives

$$f_{A\widehat{\otimes}K\langle z_B\rangle} = \partial_A^*(\widetilde{-1})(-1)^{\partial_A\smile\partial_B}\partial_B^*(\widetilde{\zeta_A})\partial_A^*(\widetilde{\zeta_B})f_A, \qquad (6.22)$$

which is Lem. 6.3.5 in the case $B = K\langle z_B\rangle$ since $f_{K\langle z_B\rangle} = 1$ by Lem. 6.3.4(b).

Now let B be an arbitrary odd ECSGA, $B = \langle B_0 \rangle \widehat{\otimes} K\langle z_B \rangle$ (cf. Th. 6.1.1, p. 255). The cocycle f_B associated with B is also an associated cocycle $f_{\langle B_0 \rangle}$ of the even ECSGA $\langle B_0 \rangle$ (Rem. 2, p. 265). Now

$$A\widehat{\otimes}B = \big(A\widehat{\otimes}\, \langle B_0\rangle\big)\widehat{\otimes}\, K\langle z_B\rangle,$$

and since $f_{A\widehat{\otimes}\,\langle B_0\rangle} = f_A f_B$ by Lem. 6.3.4, that same lemma and the special case (6.22) imply that

$$\begin{aligned} f_{A\widehat{\otimes}B} &= \partial^*_{A\widehat{\otimes}\langle B_0\rangle}(\widetilde{-1})(-1)^{\partial_{A\widehat{\otimes}\langle B_0\rangle}\smile\partial_B}\partial^*_B(\widetilde{\zeta_{A\widehat{\otimes}\langle B_0\rangle}})\partial^*_{A\widehat{\otimes}\langle B_0\rangle}(\widetilde{\zeta_B})f_{A\widehat{\otimes}\langle B_0\rangle} \\ &= \partial^*_A(\widetilde{-1})(-1)^{\partial_A\smile\partial_B}\partial^*_B(\widetilde{\zeta_A})\partial^*_A(\widetilde{\zeta_B})f_Af_B. \end{aligned}$$ □

Lemma 6.3.6 *If A and B are ECSGAs of odd type, then*

$$f_{A\widehat{\otimes}B} = (-1)^{\partial_A\smile\partial_B}(\partial^*_B\widetilde{\zeta_A})(\partial^*_A\widetilde{\zeta_B})f_Af_B.$$

We first consider the special case $A = K\langle z_A\rangle$ and $B = K\langle z_B\rangle$. The calculations here are similar to the previous case so we omit some of the details. Using the standard basis $z_A^i\widehat{\otimes}z_B^j$, $i, j = 0, 1$, of $K\langle z_A\rangle\widehat{\otimes}K\langle z_B\rangle = \langle\zeta_A, \zeta_B\rangle_K$ (cf. (6.1), p. 255), one first shows that $z_A^{\partial_B(s)}\widehat{\otimes}z_B^{\partial_A(s)}$ lifts the action of s on $K\langle z_A\rangle\widehat{\otimes}K\langle z_B\rangle$. One then uses this lifting to calculate $f_{K\langle z_A\rangle\widehat{\otimes}K\langle z_B\rangle}$ which turns out to be $(-1)^{\partial_A\smile\partial_B}(\partial^*_B\widetilde{\zeta_A})(\partial^*_A\widetilde{\zeta_B})$. This proves the formula in this special case since $f_{K\langle z_A\rangle} = 1 = f_{K\langle z_B\rangle}$ by Lem. 6.3.4.

In the general case $A \cong \langle A_0\rangle\widehat{\otimes}K\langle z_A\rangle$ and $B \cong \langle B_0\rangle\widehat{\otimes}K\langle z_B\rangle$, so since $\langle A_0\rangle\widehat{\otimes}\langle B_0\rangle \cong \langle A_0\otimes B_0\rangle$,

$$A\widehat{\otimes}B \cong \langle A_0\otimes B_0\rangle\widehat{\otimes}\left(K\langle z_A\rangle\widehat{\otimes}K\langle z_B\rangle\right).$$

By Lem. 6.3.3, $f_{\langle A_0\rangle\widehat{\otimes}\langle B_0\rangle} = f_{\langle A_0\rangle}f_{\langle B_0\rangle} = f_Af_B$ by Rem. 2, p. 265. Then another application of Lem. 6.3.3, this time to $\langle A_0\otimes B_0\rangle = \langle A_0\rangle\widehat{\otimes}\langle B_0\rangle$ and $K\langle z_A\rangle\widehat{\otimes}K\langle z_B\rangle$, using the special case, yields the required formula. □

We now combine the formulas of Lemmas 6.3.3, 6.3.5 and 6.3.6:

Theorem 6.3.1 *Let A and B be ECSGAs with cocycles $f_A, f_B \in \mathrm{Z}^2(G, \dot{K})$ respectively. Then*

$$f_{A\widehat{\otimes}B} = (\partial^*_A(\widetilde{-1}))^{\tau_B(\tau_A+\tau_B)}(\partial^*_B(\widetilde{-1}))^{\tau_A(\tau_A+\tau_B)}(-1)^{\partial_A\smile\partial_B}(\partial^*_B\widetilde{\zeta_A})(\partial^*_A\widetilde{\zeta_B})f_Af_B.$$ □

The exponents $\tau_B(\tau_A+\tau_B)$ and $\tau_A(\tau_A+\tau_B)$ can be computed in either $\mathbb{Z}$ or in $\mathbb{Z}/2$ since the cohomology classes of $\partial^*_A(\widetilde{-1}))$ and $\partial^*_B(\widetilde{-1}))$ are in a subgroup of $\mathrm{H}^2(G, \dot{K})$ of exponent 2, namely the image of $\mathrm{H}^2(\mathbb{Z}/2, \dot{K}) = \dot{K}/\dot{K}^2 \to \mathrm{H}^2(G, \dot{K})$.

Corollary 6.3.1 *If G has odd order, $f_{A\widehat{\otimes}B} = f_Af_B$.*

This is clear since $\partial_A = 0$ for any ECSGA A, so also $\partial^*_A = 0$. □

Corollary 6.3.2 *The cohomology class $\mathrm{f}_A = [f_A] \in \mathrm{H}^2(G, \dot{K})$ depends only on the equivalence class $[A]$ of A (and accordingly we put $\mathrm{f}_{[A]} = \mathrm{f}_A$).*

If B is a trivial ECSGA, the invariants τ_B, ∂_B, ζ_B and $[f_B]$ are all trivial. This follows for τ_B and ζ_B by Wall's Theorem (p. 258) since the Brauer-Wall class of a trivial algebra is $0 \in \mathrm{BW}(K)$, for ∂_B by p. 264, and for $[f_B]$ by Rem. 3, p. 265. Therefore $A\widehat{\otimes}B$ also has cohomology class $[f_A]$ by the theorem. □

Lemma 6.3.7 *Let* $\partial_1 \in \operatorname{Hom}(G, \mathbb{Z}/2)$. *There is an ECSGA* C *with*

$$\partial_C = \partial_1, \ \ \mathrm{f}_C = 1, \ \ \mathrm{b}_C = 1, \ \ \zeta_C = 1, \ \ \textit{and} \ \ \tau_C = 0.$$

Take C to be the CSGA $\mathrm{M}(1,1,K)$, and let $J \in C_1$ be the matrix $\begin{pmatrix} 0 & 1 \\ 1 & 0 \end{pmatrix}$. Make C into an ECSGA by defining the action of $s \in G$ to be the identity if $\partial_1(s) = 0$ and inn_J if $\partial_1(s) = 1$. Then C has the desired properties. □

In order to prove the main theorem of this chapter (Th. 6.3.2), we need to briefly consider projective representations. Recall that such a representation is a map $\widetilde{\rho} : G \to \mathbf{GL}(W)$, where W is a (finite dimensional) vector space over K, satisfying $\widetilde{\rho}(s)\widetilde{\rho}(t) = f(s,t)\widetilde{\rho}(st)$ for scalars $f(s,t) \in \dot{K}$, for all $s, t \in G$; then f is a cocycle, $f \in \mathrm{Z}^2(G, \dot{K})$.

We shall need to know that, given an arbitrary $f \in \mathrm{Z}^2(G, \dot{K})$, there is at least one projective representation with f as cocycle. Consider the "twisted" group algebra $W = K_f G$ – this is the algebra with basis $\{\widetilde{s},\ s \in G\}$ and multiplication $\widetilde{s}\widetilde{t} = f(s,t)\widetilde{st}$. It is easy to check that this makes $K_f G$ into an associative algebra over K, and that the "left regular representation" $\widetilde{\rho} : G \to \mathbf{GL}(W)$, defined by $\widetilde{\rho}(s)(w) = \widetilde{s}w$, has the desired property.

Let $\mathrm{z} \in \mathrm{Z}^2(\mathbb{Z}, \dot{K})$ be the 2-cocycle and $\dot{K}/\dot{K}^2 \times_{\mathrm{z}} \mathbb{Z}/2$ the central extension defined on p. 51, and H the group $\operatorname{Hom}(G, \mathbb{Z}/2) \times (\dot{K}/\dot{K}^2 \times_{\mathrm{z}} \mathbb{Z}/2)$. If A is an ECSGA, let[5]

$$\theta_{[A]} = (\partial_A, \zeta_A, \tau_A) \in H.$$

It is a well-defined homomorphism

$$\theta = (\partial, \zeta, \tau) : \mathrm{BW}(K, G) \to H$$

since $[A] \to (\zeta_A, \tau_A)$ is by Wall's Theorem (p. 258) and ∂ is by (6.16). Since $(\zeta, \tau) : \mathrm{BW}(K) \to \dot{K}/\dot{K}^2 \times_{\mathrm{z}} \mathbb{Z}/2$ is also surjective by Wall's Theorem, and since for any ECSGA A,

$$(\zeta, \tau)([A \widehat{\otimes} C]) = (\zeta, \tau)([A]) \quad \text{and} \quad \partial_{[A \widehat{\otimes} C]} = \partial_{[A]} + \partial_1$$

where C is the ECSGA of Lem. 6.3.7, it follows that

$$\theta : \mathrm{BW}(K, G) \to H \text{ is an epimorphism of groups} \tag{6.23}$$

since the forgetful homomorphism $\varphi : \mathrm{BW}(K, G) \to \mathrm{BW}(K)$ is onto.

Suppose $h_i = (\partial_i, \zeta_i, \tau_i) \in H$ for $i = 1, 2$. Define

$$F(h_1, h_2) = \\ \partial_1^*(\widetilde{-1})^{\tau_2(\tau_1+\tau_2)} \partial_2^*(\widetilde{-1})^{\tau_1(\tau_1+\tau_2)} (-1)^{\partial_1 \smile \partial_2} \big(\partial_2^* \widetilde{\zeta}_1\big)\big(\partial_1^* \widetilde{\zeta}_2\big) \ \in \ \mathrm{Z}^2(G, \dot{K}) \tag{6.24}$$

(G acting trivially on $\dot{K}$). Denote its cohomology class by $\Phi(h_1, h_2) \in \mathrm{H}^2(G, \dot{K})$. Thus

$$\Phi : H \times H \to \mathrm{H}^2(G, \dot{K}),$$

and $\Phi(1_H, h) = \Phi(h, 1_H) = 1$ for all $h \in H$.

Consider the multiplication on the set product[6] $\mathrm{H}^2(G, \dot{K}) \times H$ defined by

$$(\phi_1, h_1)(\phi_2, h_2) = (\Phi(h_1, h_2)\phi_1\phi_2, h_1 h_2), \tag{6.25}$$

[5] As is done earlier (p. 257), we use ζ_A, for example, instead of $\zeta_A \dot{K}^2$.

[6] It will of course be the central extension $\mathrm{H}^2(G, \dot{K}) \times_\Phi H$ once we show that Φ is a cocycle.

and the map

$$\psi = (\mathrm{f}, \theta) : \mathrm{BW}(K, G) \to \mathrm{H}^2(G, \dot{K}) \times H.$$

Then

$$\begin{aligned}\psi([A][B]) &= \psi([A \widehat{\otimes} B]) \\ &= (\mathrm{f}_{[A\widehat{\otimes}B]}, \theta_{[A\widehat{\otimes}B]}) \\ &= (\Phi(\theta_A, \theta_B)\mathrm{f}_A\mathrm{f}_B, \theta_{[A]}\theta_{[B]}) \text{ by Th. 6.3.1 and (6.23),} \\ &= \psi([A])\psi([B]) \text{ by (6.25).}\end{aligned}$$

We now show that ψ is an epimorphism – from which it follows that $\mathrm{H}^2(G, \dot{K}) \times H$ is a group (under the multiplication given in (6.25)), thence that $\Phi \in \mathrm{Z}^2(H, \mathrm{H}^2(G, \dot{K}))$ where H acts trivially on $\mathrm{H}^2(G, \dot{K})$, and therefore that $\mathrm{H}^2(G, \dot{K}) \times H$ is the central extension $\mathrm{H}^2(G, \dot{K}) \times_\Phi H$.

Let f be an arbitrary cocycle in $\mathrm{Z}^2(G, \dot{K})$ and $\widetilde{\rho} : G \to \mathbf{GL}(W)$ a projective representation with cocycle f (cf. p. 270). Then $\rho(s)(a) := \widetilde{\rho}(s)a\widetilde{\rho}(s)^{-1}$ makes $A = \langle \mathrm{End} W \rangle$ into an ECSGA with invariants $f_A = f$, $\partial_A = 0$, $\zeta_A = 1$ and $\tau_A = 0$. Thus $\mathrm{f}_{[A]} = [f]$ and $\theta[A] = 1$, and since $\Phi(1, h) = 1$ for any $h \in H$, $\psi([A \widehat{\otimes} B]) = ([f]\mathrm{f}_{[B]}, \theta_{[B]})$. Since θ is an epimorphism, we can conclude that ψ is also. Thus $\mathrm{H}^2(G, \dot{K}) \times H$ is indeed a group, namely the central extension $\mathrm{H}^2(G, \dot{K}) \times_\Phi H$, and $\Phi \in \mathrm{Z}^2(H, \mathrm{H}^2(G, \dot{K}))$.

Consider the central extension

$$\begin{aligned}1 \to \mathrm{H}^2(G, \dot{K}) \to \mathrm{H}^2(G, \dot{K}) \times_{\pi^*\Phi} (\mathrm{Hom}(G, \mathbb{Z}/2) \times \mathrm{BW}(K)) \\ \to \mathrm{Hom}(G, \mathbb{Z}/2) \times \mathrm{BW}(K) \to 1\end{aligned}$$

with cocycle $\pi^*\Phi \in \mathrm{Z}^2\big(\mathrm{Hom}(G, \mathbb{Z}/2) \times \mathrm{BW}(K),\ \mathrm{H}^2(G, \dot{K})\big)$ where

$$\begin{aligned}\pi = (\mathrm{id}, (\zeta, \tau)) : \mathrm{Hom}(G, \mathbb{Z}/2) \times \mathrm{BW}(K) \\ \to \mathrm{Hom}(G, \mathbb{Z}/2) \times (\dot{K}/\dot{K}^2 \times_{\mathrm{z}} \mathbb{Z}/2) = H\end{aligned}$$

is a homomorphism (by Wall's Theorem).

If we apply the Wall isomorphism $\mathrm{BW}(K) \xrightarrow{\sim} \mathrm{Br}(K) \times_{\mathrm{z}'} (\dot{K}/\dot{K}^2 \times_{\mathrm{z}} \mathbb{Z}/2)$ (cf. (6.9)), this means we have a group structure on

$$\mathrm{H}^2(G, \dot{K}) \times_{\pi^*\Phi} \Big(\mathrm{Hom}(G, \mathbb{Z}/2) \times \Big(\mathrm{Br}(K) \times_{\mathrm{z}'} (\dot{K}/\dot{K}^2 \times_{\mathrm{z}} \mathbb{Z}/2)\Big)\Big)$$

with multiplication given by

$$\begin{aligned}&(\mathrm{f}_1, \partial_1, \mathrm{b}_1, \zeta_1, \tau_1) \cdot (\mathrm{f}_2, \partial_2, \mathrm{b}_2, \zeta_2, \tau_2) \\ &\quad = \big(\mathrm{f}_1\mathrm{f}_2\Phi(\partial_1, \zeta_1, \tau_1; \partial_2, \zeta_2, \tau_2),\ \partial_1 + \partial_2, \\ &\qquad \mathrm{b}_1\mathrm{b}_2\big((-1)^{\tau_1(\tau_1+\tau_2)}\zeta_1,\ (-1)^{\tau_2(\tau_1+\tau_2)}\zeta_2\big)_K,\ (-1)^{\tau_1\tau_2}\zeta_1\zeta_2,\ \tau_1 + \tau_2\big).\end{aligned}$$

We now come to one of the main theorems of this chapter.

Theorem 6.3.2 *The map*

$$\begin{aligned}&(\mathrm{f}, \partial, \mathrm{b}, \zeta, \tau) : \mathrm{BW}(K, G) \to \\ &\qquad \mathrm{H}^2(G, \dot{K}) \times_{\pi^*\Phi} \Big(\mathrm{Hom}(G, \mathbb{Z}/2) \times \Big(\mathrm{Br}(K) \times_{\mathrm{z}'} (\dot{K}/\dot{K}^2 \times_{\mathrm{z}} \mathbb{Z}/2)\Big)\Big)\end{aligned}$$

is an isomorphism.

We will identify $\mathrm{BW}(K)$ with $\mathrm{Br}(K)\times_{z'}(\dot{K}/\dot{K}^2\times_z\mathbb{Z}/2)$ via Wall's isomorphism. Since the forgetful map $\varphi:\mathrm{BW}(K,G)\to\mathrm{BW}(K)$ is onto, the homomorphism

$$(\partial,\varphi):\mathrm{BW}(K,G)\to\mathrm{Hom}(G,\mathbb{Z}/2)\times\mathrm{BW}(K)$$

is also onto, using the same argument used in showing that θ is onto. To see that $(\mathrm{f},\partial,\varphi)$ is onto, note that if $f\in\mathrm{Z}^2(G,\dot{K})$ and $\hat{\rho}:G\to\mathbf{GL}(W)$ is any projective representation with cocycle f and $A=\langle\mathrm{End}\,W\rangle$ is the ECSGA considered earlier, the invariants $\partial_A,\mathrm{b}_A,\zeta_A$ and τ_A of A are trivial, while $\mathrm{f}_A=[f]$. If B is an arbitrary ECSGA, by Wall's Theorem and (6.7) (p. 257),

$$\begin{aligned}
\tau_{[A\hat{\otimes}B]} &= \tau_{[A]}+\tau_{[B]}=\tau_{[B]},\\
\zeta_{[A\hat{\otimes}B]} &= (-1)^{\tau_{[A]}\tau_{[B]}}\zeta_{[A]}\zeta_{[B]}=\zeta_{[B]},\\
\mathrm{b}_{[A\hat{\otimes}B]} &= \mathrm{b}_{[A]}\mathrm{b}_{[B]}\big(\varepsilon_1\zeta_{[A]},\varepsilon_2\zeta_{[B]}\big)_K=\mathrm{b}_{[B]}\quad(\text{since }\varepsilon_1=1=\zeta_{[A]}),\\
\partial_{[A\hat{\otimes}B]} &= \partial_{[A]}+\partial_{[B]}=\partial_{[B]}.
\end{aligned}$$

Moreover it follows easily from Th. 6.3.1 that

$$f_{[A\hat{\otimes}B]}=f_{[A]}f_{[B]}=f f_{[B]},$$

and so $(\mathrm{f},\partial,\varphi)$ is onto.

Now we show that $(\mathrm{f},\partial,\varphi)$ is also injective. Suppose $[A,\rho]\in\ker(\mathrm{f},\partial,\varphi)$. Since $[A]=1$ in $\mathrm{BW}(K)$, we can assume $A=\mathrm{End}(V_0,V_1)$ for some graded vector space $V=V_0\oplus V_1$. Then $\mathrm{Z}(A_0)=\{\alpha(\mathrm{id}_{V_0})\oplus\beta(\mathrm{id}_{V_1}):\ \alpha,\beta\in K\}$; since $z_A\in\mathrm{Z}(A_0)$ and $z_A^2\in\dot{K}$ (cf. Th. 6.1.2, p. 256), we can take $z_A=\mathrm{id}_{V_0}\oplus-\mathrm{id}_{V_1}$ or $=\mathrm{id}_V$ if V_1 or $V_0=0$. Since $\partial_A=0$ means that $\rho(s)(z_A)=(-1)^{\partial_A(\rho(s))}z_A=z_A$ for all $s\in G$ and since $\rho(s)$ fixes both id_{V_0} and id_{V_1} or interchanges them, it must fix them and it follows that $\rho(s)(\mathrm{End}\,V_0)=\mathrm{End}\,V_0$ and $\rho(s)(\mathrm{End}\,V_1)=\mathrm{End}\,V_1$ for all s.

Therefore we can choose a lifting $\hat{\rho}:G\to A_0^\times=\mathbf{GL}(V_0)\oplus\mathbf{GL}(V_1)$ of $\rho:G\to\mathrm{Aut}_{\mathrm{gr}}A$ with cocycle $f_A(s,t)=\hat{\rho}(s)\hat{\rho}(t)\hat{\rho}(st)^{-1}$ (p. 265). Since $\mathrm{f}_A=[f_A]=1$, there is a cochain $c:G\to\dot{K}$ such that $f_A(s,t)=c(st)c(s)^{-1}c(t)^{-1}$ for all $s,t\in G$. The map $\sigma:G\to A_0^\times$ defined by $\sigma(s)=c(s)\hat{\rho}(s)$ for all $s\in G$ is also a lifting of ρ and is a homomorphism $\sigma:G\to\mathbf{GL}(V_0)\oplus\mathbf{GL}(V_1)$. Thus σ is a direct sum $\sigma_0\oplus\sigma_1$ of two linear representations $\sigma_0:G\to\mathbf{GL}(V_0)$ and $\sigma_1:G\to\mathbf{GL}(V_1)$. It follows that A is a trivial ECSGA, and so $\ker(\mathrm{f},\partial,\varphi)=0$ and $(\mathrm{f},\partial,\varphi)$ is an isomorphism. □

Corollary 6.3.3 *The forgetful map* $\mathrm{BW}(K,G)\to\mathrm{BW}(K)$ *is a split epimorphism,*

$$1\to\mathrm{BW}_0(K,G)\to\mathrm{BW}(K,G)\to\mathrm{BW}(K)\to 1,$$

where $\mathrm{BW}_0(K,G)$ *is the subgroup of ECSGAs which are trivial as CSGAs. This subgroup is in turn a central extension*

$$0\to\mathrm{H}^2(G,\dot{K})\to\mathrm{BW}_0(K,G)\to\mathrm{Hom}(G,\mathbb{Z}/2)\to 0$$

belonging to the "cup product" $\smile$, *more precisely to the cocycle*

$$(\partial_1,\partial_2)\to\Big([(s,t)\to(-1)^{\partial_1(s)\partial_2(t)}]\in\mathrm{H}^2(G,\dot{K})\Big)$$

in $\mathrm{Z}^2(\mathrm{Hom}(G,\mathbb{Z}/2),\mathrm{H}^2(G,\dot{K}))$.

The forgetful map φ is split by the map which takes $[A] \in \mathrm{BW}(K)$ to the equivariant equivalence class of A with trivial G-action, and so is a split epimorphism. If we identify

$$\mathrm{BW}(K,G) = \mathrm{H}^2(G,\dot{K})\dot{\times}\big(\mathrm{Hom}(G,\mathbb{Z}/2) \times \mathrm{BW}(K)\big)$$

via the isomorphism of the theorem, then

$$\mathrm{BW}_0(K,G) = \mathrm{H}^2(G,\dot{K})\dot{\times}\,(\mathrm{Hom}(G,\mathbb{Z}/2) \times \{1\})\,.$$

This is the central extension of $\mathrm{Hom}(G,\mathbb{Z}/2)$ by $\mathrm{H}^2(G,\dot{K})$ with respect to the cocycle

$$\eta(\partial_1,1,0;\partial_2,1,0) = [(-1)^{\partial_1 \smile \partial_2}] \in \mathrm{H}^2(G,\dot{K})$$

by (6.24). □

6.4 The equivariant Clifford algebra and Witt group

Let $b : V \times V \to K$ be a nonsingular symmetric bilinear form. The Clifford algebra $\mathrm{C}(V,b)$ is a CSGA, with $\mathrm{C}(V,b)_0$ the subalgebra spanned by even products of vectors of V, and $\mathrm{C}(V,b)_1$ the subspace spanned by odd products. (See §6.2 for its basic properties). These homogeneous components are usually written $\mathrm{C}_0(V,b)$ and $\mathrm{C}_1(V,b)$ (or more simply $\mathrm{C}_i(b)$ or even C_i). The Clifford algebra satisfies (p. 259) the fundamental isomorphism

$$\mathrm{C}(V_1,b_1)\widehat{\otimes}\mathrm{C}(V_2,b_2) \cong \mathrm{C}(V_1 \perp V_2, b_1 \perp b_2).$$

The group G is again arbitrary in this section.

If $\rho : G \to \mathbf{O}(V,b)$ is an orthogonal representation, the universal mapping property of the Clifford algebra shows that the action of G on V can be extended to an action $\rho_C : G \to \mathrm{Aut}_{\mathrm{gr}}\mathrm{C}(V,b)$, obviously satisfying

$$\rho_C(s)(v_1v_2\cdots v_k) = \rho(s)(v_1)\rho(s)(v_2)\cdots\rho(s)(v_k).$$

In this way $\mathrm{C}(V,b)$ becomes an ECSGA, the *Clifford algebra* $\mathrm{C}(\rho)$ *of the orthogonal representation* ρ. The class $[\mathrm{C}(\rho)] \in \mathrm{BW}(K,G)$ is the "Brauer-Wall class $\mathrm{BW}(\rho)$ of ρ", which by Th. 6.3.2 can be viewed as a quintuple

$$\begin{aligned}\mathrm{BW}(\rho) &= (\mathrm{f}_\rho, \partial_\rho, \mathrm{b}_\rho, \zeta_\rho, \tau_\rho) \\ &\in \mathrm{H}^2(G,\dot{K}) \times_{\pi^*\Phi} \Big(\mathrm{Hom}(G,\mathbb{Z}/2) \times \Big(\mathrm{Br}(K) \times_{\mathrm{z}'} (\dot{K}/\dot{K}^2 \times_{\mathrm{z}} \mathbb{Z}/2)\Big)\Big)\,. \end{aligned} \tag{6.26}$$

We now show how these invariants are related directly to those of (V,b) and ρ:

$$\begin{aligned} \tau_\rho &= \tau_C = \dim_{/2} V, \\ z_\rho &= z_C = v_1v_2\ldots v_n, \\ \zeta_\rho &= \zeta_C = \operatorname{disc} b, \\ \mathrm{b}_\rho &= \mathrm{b}_\mathrm{C} = \mathrm{c}(b), \\ (-1)^{\partial_\rho(s)} &= \det \rho(s) \ \text{ for all } s \in G, \\ \mathrm{f}_\rho &= \text{Fröhlich's cohomology class of } \rho, \end{aligned}$$

where $\{v_1, v_2, \ldots, v_n\}$ is any orthogonal basis of (V,b), and $\mathrm{c}(b)$ is the Witt invariant (cf. p. 54). Fröhlich's cohomology class has not been defined earlier. (The invariants $\tau_\rho, z_\rho, \zeta_\rho$ and b_ρ depend only on the Clifford algebra as a CSGA – hence we also write $\tau_\rho = \tau_C$, for example. Recall also that z_ρ and ζ_ρ are defined only up to elements of $\dot{K}$ and $\dot{K}^2$ respectively, in spite of the presence of the equality signs!)

Write $\mathrm{Z}(C) = K \oplus Z_1$ where $Z_1 \subset C_1$ (cf. p. 255). Then $\tau_C = 0$ if $Z_1 = 0$, $= 1$ if $Z_1 \neq 0$. Since $v_i v_j = -v_j v_i$ if $i \neq j$, it is clear that $v_1 v_2 \ldots v_n$ commutes with all v_i if n is odd, and so in this case the nonzero element $v_1 v_2 \ldots v_n \in Z_1$ and C has odd type: $\tau_C = 1$. And of course we can take $z_C = v_1 v_2 \ldots v_n$, and $\zeta_C = z_C^2 = (-1)^{n(n-1)/2} v_1^2 v_2^2 \cdots v_n^2, = \operatorname{disc} b$, as an easy calculation shows. If n is even, $v_1 v_2 \ldots v_n$ commutes with all $v_i v_j$ $(i \neq j)$. Thus it is in $\mathrm{Z}(C_0)$ and C_0 has a center larger than K, so C must have even type by Th. 6.1.1(a) (p. 255), i.e. $\tau_C = 0$. Thus in general $\tau_C = \dim_{/2} V$. Moreover by Th. 6.1.2(b), $z_C = v_1 v_2 \ldots v_n$ and again $\zeta_C = z_C^2 = \operatorname{disc} b$.

A proof that $\mathrm{b}_C = \mathrm{c}(b)$ is given on p. 259.

By the "Cartan-Dieudonné Theorem" ([39], p. 18 – or it can also be deduced from the fact that the vector representation $\pi : \Gamma \to \mathbf{O}(V, b)$ is onto), every $\varphi \in \mathbf{O}(V, b)$ is a product of symmetries (cf. p. 260), $\varphi = \sigma_{u_1} \sigma_{u_2} \cdots \sigma_{u_m}$. Now $\sigma z_C = -z_C$ for any symmetry σ. Indeed we can choose an orthogonal basis of (V, b) which contains the vector u for which $\sigma = \sigma_u$. Then all vectors in the orthogonal basis are fixed by σ except that $\sigma u = -u$, and it follows at once that $\sigma z_C = -z_C$ and $\partial_\rho \sigma = 1$. Thus if $\varphi = \rho(s)$, $\partial_\rho(\rho(s)) \equiv m \bmod 2$ and since the determinant of any symmetry is -1,

$$\det \rho(s) = (-1)^m = (-1)^{\partial_\rho(\rho(s))}.$$

"Fröhlich's cohomology class" is defined in (4.3) and (5.2), [26], exactly as we have defined f_A on p. 265*ff*.

Now suppose that $\sigma : G \to \mathbf{O}(V', b')$ is another orthogonal representation. We know that $C(V, b) \widehat{\otimes} C(V', b') \cong C(V \perp V', b \perp b')$ as CSGAs, and it is easy to see that this is as well an isomorphism of ECSGAs. Thus

$$\mathrm{BW}(\rho \perp \sigma) = \mathrm{BW}(\rho)\mathrm{BW}(\sigma),$$

and we get a homomorphism

$$\hat{\mathrm{W}}_1(K, G) \xrightarrow{\mathrm{BW}} \mathrm{BW}(K, G) \tag{6.27}$$

where $\hat{\mathrm{W}}_1(K, G)$ is the Witt-Grothendieck group of orthogonal representations of G.

The group law on

$$\mathrm{H}^2(G, \dot{K}) \times_{\pi^*\Phi} \Big(\mathrm{Hom}(G, \mathbb{Z}/2) \times \Big(\mathrm{Br}(K) \times_{\mathrm{z}'} (\dot{K}/\dot{K}^2 \times_{\mathrm{z}} \mathbb{Z}/2)\Big)\Big)$$

given on p. 271, gives the laws

$$\begin{aligned}
\tau_{\rho\perp\sigma} &= \tau_\rho + \tau_\sigma \\
\zeta_{\rho\perp\sigma} &= (-1)^{\tau_\rho\tau_\sigma}\zeta_\rho\zeta_\sigma \\
\mathrm{b}_{\rho\perp\sigma} &= \mathrm{b}_\rho\mathrm{b}_\sigma\big((-1)^{\tau_\rho(\tau_\rho+\tau_\sigma)}\zeta_\rho,\, (-1)^{\tau_\sigma(\tau_\rho+\tau_\sigma)}\zeta_\sigma\big)_K \\
\partial_{\rho\perp\sigma} &= \partial_\rho + \partial_\sigma \\
\mathrm{f}_{\rho\perp\sigma} &= \partial_\rho^*(\widetilde{-1})^{\tau_\sigma(\tau_\rho+\tau_\sigma)}\partial_\sigma^*(\widetilde{-1})^{\tau_\rho(\tau_\rho+\tau_\sigma)}(-1)^{\partial_\rho\smile\partial_\sigma}(\partial_\rho^*\widetilde{\zeta}_\sigma)(\partial_\sigma^*\widetilde{\zeta}_\rho)\mathrm{f}_\rho\mathrm{f}_\sigma.
\end{aligned}$$

The first four are, in the context of orthogonal representations $\rho : G \to \mathbf{O}(V, f)$ and $\sigma : G \to \mathbf{O}(W, g)$, simply the well-known relationships

$$\begin{aligned}\dim_{/2}(V \perp W) &= \dim_{/2} V + \dim_{/2} W \\ \operatorname{disc}(f \perp g) &= (-1)^{nm}(\operatorname{disc} f)(\operatorname{disc} g) \\ \mathrm{c}(f \perp g) &= \mathrm{c}(f)\mathrm{c}(g)\left((-1)^{n(n+m)}\operatorname{disc} f,\ (-1)^{m(n+m)}\operatorname{disc} g\right)_K \\ \det(\rho(s) \perp \sigma(s)) &= (\det \rho(s))(\det \sigma(s))\end{aligned}$$

where $\dim V = n$ and $\dim W = m$.

Let $\rho : G \to \mathbf{O}(V, h)$ be the hyperbolic representation based on $\rho_0 : G \to \mathbf{GL}(V_0)$, where $\dim V_0 = n$, say. Thus $V = V_0 \oplus V_0^*$, $h(V_0, V_0) = 0 = h(V_0^*, V_0^*)$ and $h(v, v^*) = \langle v, v^* \rangle$ for all $v \in V_0$ and $v^* \in V_0^*$. We will identify the subalgebra of the Clifford algebra $\mathrm{C}(V, h)$, generated by the vectors of V_0, with $\mathrm{C}(V_0, 0)$ – it is in fact the exterior algebra of V_0. Let $v_1, \ldots, v_n$ be a basis of V_0 and $w_1, \ldots, w_n$ a basis of V_0^*. Let $x = w_1 w_2 \ldots w_n$. We leave it as an exercise to show that the left ideal $I = \mathrm{C}(V, h)x$ is $= \mathrm{C}(V_0, 0)x$. Since the products $v_{i_1} v_{i_2} \ldots v_{i_k} w_{j_1} w_{j_2} \ldots w_{j_l}$, where $0 \le k \le n$, $0 \le l \le n$, $i_1 < i_2 < \ldots < i_k$ and $j_1 < j_2 < \ldots < j_l$, form a basis of $\mathrm{C}(V, h)$, $\dim_K I = 2^n$, the dimension of the central simple algebra $\mathrm{C}(V, h)$ is 2^{2n}, I is a minimal left ideal and $\mathrm{C}(V, h) \cong \mathrm{M}(2^n, K)$ is split. In fact even more is true:

Lemma 6.4.1 *If ρ is a hyperbolic representation, the Clifford algebra $\mathrm{C}(\rho)$ is a trivial ECSGA.*

Let $I_i = \mathrm{C}_i(V_0, 0)x$ for $i = 0, 1$. Then $I = I_0 \oplus I_1$ and it is clear that $\mathrm{C}_j(V, h)I_i \subset I_{i+j}$ (indices mod 2). Thus the canonical isomorphism $\mathrm{C}(V, h) \to \mathrm{End}_K I$, $y \to y_I$, is in fact a *graded* isomorphism

$$\mathrm{C}(V, h) \to \mathrm{End}(I_0, I_1). \tag{6.28}$$

Since V_0 and V_0^* are stable under ρ (that is to say, under all $\rho(s)$), $\mathrm{C}_i(V_0, 0)$ and Kx are stable under ρ_C. It follows that I_0 and I_1 are stable under ρ_C, making $\mathrm{End}(I_0, I_1)$ into a trivial ECSGA.

Therefore the lemma will follow if we show that the graded homomorphism (6.28) is equivariant. Let $y \in \mathrm{C}(V, h)$. The result of $s \in G$ acting on y_I is $\rho_C(s) y_I \rho_C(s)^{-1}$. Now I_i is spanned by suitable products $u_1 \ldots u_k$ where the $u_j \in V_0 \cup V_0^*$, and

$$\begin{aligned}(\rho_C(s) y_I \rho_C(s)^{-1})(u_1 \cdots u_k) &= \rho_C(s) y_I (\rho(s)^{-1} u_1 \cdots \rho(s)^{-1} u_k) \\ &= \rho_C(s)(y \rho(s)^{-1} u_1 \cdots \rho(s)^{-1} u_k) \\ &= \rho_C(s)(y) \rho(s) \rho(s)^{-1} u_1 \cdots \rho(s) \rho(s)^{-1} u_k \\ &= \rho_C(s)(y) u_1 \cdots u_k\end{aligned}$$

which implies that $(\rho_C(s)(y))_I = \rho_C(s) y_I \rho_C(s)^{-1}$, so (6.28) is an equivariant isomorphism. □

This lemma implies that the (classes of) hyperbolic orthogonal representations are in the kernel of the map in (6.27), and so we get a homomorphism $\mathrm{W}_1(K, G) \to$

BW(K,G), which we also label BW, making

$$\begin{array}{ccccccc} \mathrm{K}_0(KG) & \xrightarrow{\mathrm{H}} & \hat{\mathrm{W}}_1(K,G) & \longrightarrow & \mathrm{W}_1(K,G) & \longrightarrow & 0 \\ & & & \overset{\mathrm{BW}}{\searrow} & \downarrow{\scriptstyle \mathrm{BW}} & & \\ & & & & \mathrm{BW}(K,G) & & \end{array}$$

commutative.

6.5 A formula for Fröhlich's cohomology class

Let $\rho : G \to \mathbf{O}(V,b)$ be an orthogonal representation and $\Gamma \subset \mathrm{C}(V,b)$ the Clifford group. The vector representation

$$\pi : \Gamma \to \mathbf{O}(V,b), \quad \pi(\gamma)(v) = \nu(\gamma)v\gamma^{-1}, \tag{6.29}$$

is an epimorphism with kernel $\dot{K}$ (cf. p. 260). Choose a lifting

$$\widetilde{\rho} : G \to \Gamma$$

of ρ, $\rho = \pi\widetilde{\rho}$. Then (p. 260)

$$\begin{aligned} \widetilde{\rho}(s) &\in \mathrm{C}_0(V,b) \quad \text{if} \quad \rho(s) \in \mathbf{SO}(V,b), \\ \widetilde{\rho}(s) &\in \mathrm{C}_1(V,b) \quad \text{if} \quad \rho(s) \in \mathbf{O}(V,b) \setminus \mathbf{SO}(V,b), \end{aligned}$$

in other words,

$$\widetilde{\rho}(s) \in \mathrm{C}_{\partial(s)} \tag{6.30}$$

where $\partial = \partial_\rho$, and

$$\nu(\widetilde{\rho}(s)) = (-1)^{\partial(s)}\widetilde{\rho}(s) = (\det \rho(s))\widetilde{\rho}(s). \tag{6.31}$$

Define

$$\hat{\rho}(s) = z_{\mathrm{C}}^{\partial(s)}\widetilde{\rho}(s).$$

Suppose first that $\dim V$ is even. If $v \in V \subset \mathrm{C}(V,b)$,

$$\begin{aligned} \hat{\rho}(s)v\hat{\rho}(s)^{-1} &= z_{\mathrm{C}}^{\partial(s)}\widetilde{\rho}(s)v\widetilde{\rho}(s)^{-1}z_{\mathrm{C}}^{-\partial(s)} \\ &= z_{\mathrm{C}}^{\partial(s)}(-1)^{\partial(s)}\rho(s)(v)z_{\mathrm{C}}^{-\partial(s)} \quad \text{by (6.29) and (6.31)} \\ &= \rho(s)(v) \quad \text{by (6.5).} \end{aligned}$$

Thus for any $a \in \mathrm{C}(V,b)$

$$\hat{\rho}(s)a\hat{\rho}(s)^{-1} = \rho_{\mathrm{C}}(s)(a). \tag{6.32}$$

Now suppose that $\dim V$ is odd. Then the same calculation yields

$$\hat{\rho}(s)v\hat{\rho}(s)^{-1} = (-1)^{\partial(s)}\rho(s)(v)$$

since z_{C} is in the center of $\mathrm{C}(V,b)$, so (6.32) holds in this case for any $a \in \mathrm{C}_0(V,b)$. We note also that $\hat{\rho}(s) \in \mathrm{C}_0(V,b)$ in this case since $z_{\mathrm{C}} \in \mathrm{C}_1(V,b)$, and $\partial(s) = 1$ if and only if $\widetilde{\rho}(s) \in \mathrm{C}_1(V,b)$ by (6.30).

Thus (cf. pp. 265*ff*) we can use $\hat{\rho}$ to calculate a cocycle belonging to the ECSGA $\mathrm{C}(V,b)$:

$$\begin{aligned} f_{\mathrm{C}}(s,t) &= \hat{\rho}(s)\hat{\rho}(t)\hat{\rho}(st)^{-1} \\ &= z_{\mathrm{C}}^{\partial(s)}\widetilde{\rho}(s)z_{\mathrm{C}}^{\partial(t)}\widetilde{\rho}(t)\widetilde{\rho}(st)^{-1}z_{\mathrm{C}}^{-\partial(st)} \\ &= z_{\mathrm{C}}^{\partial(s)+\partial(t)-\partial(st)}(-1)^{(1+\tau_{\mathrm{C}})\partial(s)\partial(t)}\widetilde{f}(s,t) \\ &= ((-1)^{(1+\tau_{\mathrm{C}})}\operatorname{disc} b)^{\partial(s)\partial(t)}\widetilde{f}(s,t) \end{aligned} \tag{6.33}$$

where

$$\widetilde{f}(s,t) := \widetilde{\rho}(s)\widetilde{\rho}(t)\widetilde{\rho}(st)^{-1} \in \mathrm{Z}^2(G, \dot{K}). \tag{6.34}$$

As in the proof of Lem. 6.3.3, one must treat the values of ∂ in the exponent $\partial(s)+\partial(t)-\partial(st)$ of z_{C} as integers rather than elements of $\mathbb{Z}/2$; thus the exponent is 0 unless $\partial(s) = \partial(t) = 1$, in which case it is 2 and $z_{\mathrm{C}}^2 = \zeta_{\mathrm{C}} = \operatorname{disc} b$.

To prove that $f_{\mathrm{C}}(s,t)$ is given by (6.33), we note that $\widetilde{\rho}(s)z_{\mathrm{C}}^{\partial(t)} = z_{\mathrm{C}}^{\partial(t)}\widetilde{\rho}(s)$ unless $\partial(t) = 1$, C is even, i.e. $\tau_{\mathrm{C}} = 0$, and $\widetilde{\rho}(s) \in \mathrm{C}_1$ (cf. (6.5)), i.e. $\partial(s) = 1$, and when this is the case $\widetilde{\rho}(s)z_{\mathrm{C}}^{\partial(t)} = -z_{\mathrm{C}}^{\partial(t)}\widetilde{\rho}(s)$.

Theorem 6.5.1 *The cohomology class* $\mathrm{f_C} \in \mathrm{H}^2(G, \dot{K})$ *of* $\mathrm{C}(V,b)$ *as an ECSGA is related to the cohomology class* $\widetilde{\mathrm{f}}$ *of a lifting* $\widetilde{\rho} : G \to \Gamma(V,b)$ *of* $\rho : G \to \mathbf{O}(V,b)$ *by the formula*

$$\mathrm{f_C} = [((-1)^{1+\tau_{\mathrm{C}}}\operatorname{disc} b)^{\partial_\rho(s)\partial_\rho(t)}]\,\widetilde{\mathrm{f}}. \qquad \square$$

The cohomology class $\widetilde{f}$ is called the Clifford invariant of ρ in [15].

We note in passing that $\widetilde{f}$ can be calculated as follows: For each s, write $\rho(s)$ as a product of symmetries $\sigma_{v_1}\sigma_{v_2}\cdots\sigma_{v_k}$ – this can be done effectively by the methods in [73]. Then $\widetilde{\rho}(s) = v_1v_2\cdots v_k$ is a lifting of $\rho(s)$ to Γ and the definition (6.34) can be applied.

6.6 Notes on Chapter 6

1. It was C.T.C. Wall [70] who first recognized the importance of considering the Clifford algebra as a graded algebra, and completely determined the structure and nature of CSGAs, as well as (what is now called) the Brauer-Wall group.

On the other hand, A. Fröhlich [26] defined the equivariant Brauer group $\mathrm{B}(K,G)$ for a field K (of characteristic $\neq 2$) and a group G, and related it to the Brauer group $\mathrm{B}(K)$ through the split exact sequence

$$0 \to \mathrm{H}^2(G,\dot{K}) \to \mathrm{B}(K,G) \to \mathrm{B}(K) \to 0.$$

Our treatment here combines these two points of view and follows my paper [62].

2. In the case when G is both Abelian and finite, the results in this chapter are related to those obtained in the context of dimodule algebras and the Brauer-Long group $\mathrm{BD}(R,G)$ over a commutative ring R ([46]). To apply these results to the case considered here, one considers the subgroup of $\mathrm{BD}(K, G\times\mathbb{Z}/2)$ of classes represented by algebras with trivial G-grading and $\mathbb{Z}/2$-action induced by the $\mathbb{Z}/2$-grading, namely $\sigma a = (-1)^{\partial a}a$, where σ is the nontrivial element of $\mathbb{Z}/2$. Many of the explicit results on $\mathrm{BD}(R,G)$ (e.g. [46], [45], [8]) assume (in addition to conditions on R) that G is cyclic, and so do not generally apply to our situation. On the other hand S. Caenepeel and M. Beattie used a cohomological approach (with mild assumptions on R, including the invertibility of $|G|$ in R) to give a description of $\mathrm{BD}(R,G)$ (Ths. 2.7 and 3.4, [13]), up to the cocycle involved in the central extension in Th. 3.4. This provides a description of $\mathrm{BW}(K,G)$ for G finite and Abelian in the case when $\operatorname{char} K \not| |G|$, up to the cocycle implicit in the aforementioned exact sequence. One result that coincides (when R is a field) with a special case of those here is the explicit computation of $\mathrm{BD}(R,\mathbb{Z}/2)$ (so $G=\mathbb{Z}/2$) by F. DeMeyer and T. Ford [20].

Bibliography

1. A. Adem and R.J. Milgram, *Cohomology of Finite Groups*, Grund. d. math. Wiss., vol. 309, Springer-Verlag, New York, Heidelberg, Berlin, 1994.
2. E. Artin, *Geometric Algebra*, Wiley Classics Library, John Wiley and Sons Inc., New York, 1988.
3. Hans-Jochen Bartels, *Invarianten hermitescher Formen über Schiefkörpern*, Math. Ann. **215** (1975), 269–288.
4. E. Bayer Fluckiger, *Self-dual normal bases*, Nederl. Akad. Wetensch. Indag. Math. **51** (1989), no. 4, 379–383.
5. E. Bayer Fluckiger and H.W. Lenstra, *Forms in odd degree extensions and self-dual normal bases*, Amer. J. Math. **112** (1990), no. 3, 359–373.
6. E. Bayer Fluckiger and R. Parimala, *Galois cohomology of the classical groups over fields of cohomological dimension* ≤ 2, Invent. Math. **122** (1995), 359–373.
7. E. Bayer Fluckiger and J.-P. Serre, *Torsions quadratiques et bases normales autoduales*, Amer. J. Math. **116** (1994), 1–64.
8. M. Beattie and S. Caenepeel, *The Brauer-Long group of* $\mathbb{Z}/p^t\mathbb{Z}$*-dimodule algebras*, J. Pure Appl. Algebra **60** (1989), 219–236.
9. N. Bourbaki, *Algèbre (Modules et Anneaux Semi-simples)*, Actualités Scientifiques et Industrielles, no. 1261, Hermann, 1958.
10. N. Bourbaki, *Éléments de mathématique. 23. Première partie: Les structures fondamentales de l'analyse. Livre II: Algèbre. Chapitre 8: Modules et anneaux semi-simples*, Actualités Sci. Ind. no. 1261, Hermann, Paris, 1958.
11. ______, *Elements of mathematics. Algebra, Part I: Chapters 1-3*, Hermann, Paris, 1974, Translated from the French. MR MR0354207 (50 #6689)
12. Kenneth S. Brown, *Cohomology of Groups*, Graduate Texts in Mathematics, vol. 87, Springer-Verlag, New York, Heidelberg, Berlin, 1982.
13. S. Caenepeel and M. Beattie, *A cohomological approach to the Brauer-Long group and the groups of Galois extensions and strongly graded rings*, Trans. Amer. Math. Soc. **324** (1991), 747–775.
14. J.W.S. Cassels and A. Fröhlich, *Algebraic Number Theory*, Conference Proceedings, Academic Press Inc., London, 1967.
15. Ph. Cassou-Noguès and M.J. Taylor, *Invariant de Clifford équivariant d'un caractère orthogonal*, Journal of Algebra **99** (1986), 167–180.
16. Claude Cibils, *Groupe de Witt d'une algèbre avec involution*, L'Enseignement Mathématique **29** (1983), 27–34.
17. P. Conner and R. Perlis, *A Survey of Trace Forms of Algebraic Number Fields*, World Scientific, Singapore, 1984.
18. Charles W. Curtis and Irving Reiner, *Representation Theory of Finite Groups and Associative Algebras*, Pure and Applied Mathematics, vol. XI, Interscience Publishers, John Wiley and Sons, New York, 1962, reprinted by AMS Chelsea Publishing.
19. ______, *Methods of Representation Theory, with Applications to Finite Groups and Orders*, vol. 1, John Wiley and Sons, New York, 1981.
20. F. Demeyer and D. Ford, *Computing the Brauer-Long group of* $\mathbb{Z}/2$*-dimodule algebras*, J. Pure Appl. Algebra **54** (1988), 197–208.
21. Francois Digne and Jean Michel, *Representations of Finite Groups of Lie Type*, LMS Student Texts, no. 21, Cambridge University Press, Cambridge, 1991.

22. Martin Epkenhans, *Spurformen über Lokalen Körpern*, Schriftenreihe der Mathematischen Institut der Universität Münster, 1987.
23. ———, *Trace forms of normal extensions over local fields*, Linear and Multilinear Algebra **24** (1989), 103–116.
24. Walter Feit, *Characters of Finite Groups*, Mathematics Lecture Notes, W.A.Benjamin Inc., New York, 1967.
25. G. Frobenius, *Über die cogredienten Transformationen der bilinearen Formen*, S.-B. Preuss. Akad. Wiss. (Berlin) (1896), 7–16.
26. A. Fröhlich, *Orthogonal and symplectic representations of groups*, Proceedings of the L.M.S. **24** (1972), 470–506.
27. A. Fröhlich and A.M. McEvett, *Forms over rings with involution*, Journal of Algebra **12** (1969), 79–104.
28. ———, *The representations of groups by automorphisms of forms*, Journal of Algebra **12** (1969), 114–133.
29. V.P. Gallagher, *Local trace forms*, Linear and Multilinear Algebra **7** (1979), no. 2, 167–174.
30. N. Jacobson, *A note on Hermitian forms*, Bull. A.M.S. **46** (1940), 264–268.
31. Nathan Jacobson, *Basic algebra. II*, second ed., W. H. Freeman and Company, New York, 1989.
32. Gordon James and Adalbert Kerber, *The Representation Theory of the Symmetric Group*, Encyclopedia of Mathematics and Its Applications, vol. 16, Addison-Wesley, Reading, MA, 1981.
33. Gerald J. Janusz, *The Schur subgroup of an algebraic number field*, Annals of Mathematics (2) **103** (1976), no. 2, 253–281.
34. Bruno Kahn, *Représentations orthogonales et symplectiques sur un corps de caractéristique différente de 2*, Communications in Algebra **31** (2003), no. 1, 133–196.
35. Max-Albert Knus, *Quadratic and Hermitian forms over Rings*, Grundlehren der mathematischen Wissenschaften, vol. 294, Springer-Verlag, Berlin-Heidelberg, 1991.
36. Max-Albert Knus, Alexander Merkurjev, Markus Rost, and Jean-Pierre Tignol, *The Book of Involutions*, Colloquium Publications, vol. 44, American Mathematical Society, Providence, R.I., 1998.
37. H. Koch, *Number Theory II*, Encyclopaedia of Mathematical Sciences, vol. 62, Springer-Verlag, 1992.
38. T.-Y. Lam, *A First Course in Noncommutative Rings*, Graduate Texts in Mathematics, vol. 131, Springer-Verlag, Berlin-Heidelberg, 1991.
39. ———, *Introduction to Quadratic Forms over Fields*, Graduate Studies in Mathematics, vol. 67, American Mathematical Society, Providence, R.I., 2004.
40. S. Mac Lane, *Categories for the Working Mathematician*, Graduate Texts in Mathematics, vol. 5, Springer-Verlag, New York, Heidelberg, Berlin, 1971.
41. Serge Lang, *Algebra*, third ed., Addison-Wesley Publishing Co., Reading, MA, 1993.
42. D.W. Lewis, *Exact sequences of Witt groups of equivariant forms*, L'Enseignement Mathematique **29** (1983), 45–51.
43. D.W. Lewis, *Involutions and anti-automorphisms of algebras*, Bull. London Math. Soc. **38** (2006), 529–545.
44. D.W. Lewis and T. Unger, *Hermitian morita theory: a matrix approach*, Irish Math. Soc. Bull. (2008), no. 62, 37–41.
45. F. Long, *The Brauer group of dimodule algebras*, J.Algebra **30** (1974), 559–601.
46. ———, *A generalization of the Brauer group of graded algebras*, Proc. London Math. Soc. (2) **29** (1974), 237–256.
47. M.G. Mahmoudi, *A remark on transfers and hyperbolicity*, Comm. in Algebra **32** (2004), no. 12, 4733–4739.
48. A. I. Malcev, *On semisimple subgroups of Lie groups*, Translations of the A.M.S. (1950), no. 33, 172–213.
49. A.M. McEvett, *Forms over semisimple algebras with involution*, Journal of Algebra **12** (1969), 105–113.
50. John Milnor, *On isometries of inner product spaces*, Invent. Math. **8** (1969), 83–97.
51. Gabriele Nebe, *Invariants of orthogonal G-modules from the character table*, Experiment. Math. **9** (2000), 623–629.
52. ———, *Orthogonal Frobenius reciprocity*, Journal of Algebra **225** (2000), 250–260.

53. Jürgen Neukirch, *Algebraic Number Theory*, Springer-Verlag, Berlin, Heidelberg, New York, 1999.
54. O.T. O'Meara, *Introduction to Quadratic Forms*, Grund. d. math. Wiss., vol. 117, Springer-Verlag, New York, Heidelberg, Berlin, 1963.
55. I. Reiner, *Maximal Orders*, L.M.S. Monographs, Academic Press, Inc., London, New York, San Francisco, 1975.
56. Carl Riehm, *The corestriction of algebraic structures*, Inventiones Mathematicae **11** (1970), 73–98.
57. ———, *The equivalence of bilinear forms*, Journal of Algebra **31** (1974), 45–66.
58. ———, *The Schur subgroup of the Brauer group of a local field*, Enseign. Math. **34** (1988), no. 2, 1–11.
59. ———, *The quadratic Schur subgroup over local and global fields*, Math. Ann. **283** (1989), 479–489.
60. ———, *Orthogonal, symplectic, and unitary representations of finite groups*, Trans. of the American Mathematical Society **353** (2001), 4687–4727.
61. Carl Riehm and Maurice Schrader-Frechette, *The equivalence of sesquilinear forms*, Journal of Algebra **42** (1976), 495–530.
62. Carl R. Riehm, *Equivariant central simple graded algebras and the equivariant Brauer-Wall group*, J. Algebra **287** (2005), no. 2, 501–520.
63. J urgen Ritter, *On orthogonal and orthonormal characters*, J. Algebra **76** (1982), no. 2, 519–531.
64. W. Scharlau, *Induction theorems and the structure of the Witt group*, Invent. Math. **11** (1970), 37–44.
65. ———, *Quadratic and Hermitian forms*, Grund. d. math. Wiss., vol. 270, Springer-Verlag, New York, Heidelberg, Berlin, 1985.
66. J.-P. Serre, *Corps Locaux*, Publ. de l'Inst. de Math. de l'Univ. de Nancago, vol. VIII, Hermann, Paris, 1962.
67. ———, *Cohomologie Galoisienne*, Lecture Notes in Mathematics, no. 5, Springer-Verlag, Berlin, Heidelberg, New York, 1973.
68. ———, *Linear Representations of Finite Groups*, Graduate Texts in Mathematics, vol. 42, Springer-Verlag, New York, Heidelberg, Berlin, 1977.
69. ———, *Local Fields*, Springer-Verlag, New York, 1979, Translated from the French by Marvin Jay Greenberg.
70. C.T.C. Wall, *Graded Brauer groups*, Journal für die Reine und Angewandte Mathematik **213** (1964), 187–199.
71. ———, *Surgery on Compact Manifolds*, London Mathematical Society Monographs, vol. 1, London Mathematical Society, London-New York, 1970.
72. ———, *Classification of Hermitian Forms. VI. Group rings*, Ann. of Math. (2) **103** (1976), no. 1, 1–80.
73. G.E. Wall, *On the conjugacy classes in the unitary, symplectic and orthogonal groups*, J. Austral. Math. Soc. **III** (1963), 1–62.
74. T. Yamada, *The Schur Subgroup of the Brauer Group*, Lecture Notes in Mathematics, vol. 307, Springer-Verlag, Berlin, Heidelberg, New York, 1974.

Glossary

$[A, ^-]$	involutory Brauer class of $(A, ^-)$ in $\mathrm{Br}(K, ^-)$, 185
$[A]$	Brauer class of A in $\mathrm{Br}(K)$, 17
$(A, ^-)_+, A_+$	symmetric elements of algebra A with respect to involution $^-$, 23
$A\widehat{\otimes}_K B$	graded tensor product of graded algebras, 254
A^L, A^C	$L \otimes_K A, C \otimes_K A$, 11
A^o, D^o	opposite algebra of A or D, 14
$(\alpha, \beta)_{K_\mathfrak{p}}$	component in $\mathrm{Br}(K_\mathfrak{p})$ of Brauer class of $(\alpha, \beta)_K$, 18
$(\alpha, \beta)_K$	quaternion algebra over K or its Brauer class, 16
$\mathrm{BW}(\sigma)$	equivariant Brauer-Wall class of $\sigma : G \to \mathbf{O}(V, b)$, 262
$\mathrm{BW}(K, G)$	equivariant Brauer-Wall group, 262
$\mathrm{Br}(K)$	Brauer group, 17
$\mathrm{Br}(K, ^-)$	involutory Brauer group, 185
$\mathrm{Br}_2(K)$	Brauer classes of order 1 or 2, 53
$\mathrm{b}(h, g)$	Bartels invariant of quaternionic skew Hermitian forms, 63
b_A	Brauer class of a CSGA, 257
$\mathbb{C}$	field of complex numbers, 18
CSA	central simple algebra, 255
CSGA	central simple graded algebra, 255
$\mathrm{c}(b)$	Witt invariant, 54
χ, χ_ρ	character of representation ρ, 71
δ_{ij}	Kronecker delta, 222
$\det L/K$	determinant of field extension L/K, 57
$\det_K(^-)$	determinant of orthogonal or unitary involution, 31, 32
$\det f$	determinant of sesquilinear form f, 45
$\dim_{/2}$	dimension mod 2 in the Witt group, 51
$\mathrm{disc}\, L/K$	discriminant of field extension L/K, 57
$\mathrm{disc}\, h$	discriminant of reflexive form h, 45
$\mathrm{disc}_K(^-)$	discriminant of orthogonal involution, 31
∂	degree function on equivariant Brauer-Wall group, 263
$\partial_\rho = \partial_A$	degree function on G, 263
$\mathrm{End}(V_0, V_1)$	graded algebra of graded endomorphisms, 254

ECSGA	equivariant central simple graded algebra, 261
$=_{\mathrm{N}}$	equal mod norms, 45
$=_2$	equal mod squares, 45
$\sim_W$	Witt equivalence of symmetric forms, 227
$\bar{f}$	reflection of f, 92
f_A	class in $\mathrm{H}^2(G, \dot{K})$ associated with ECSGA A, 265
$f \cong g$	equivalence of sesquilinear forms, 48
f^L	extension of scalars of sesquilinear form f, 48, 110
f^o	opposite form, 105
f_l	left adjoint of sesquilinear form f, 44, 93
f_r	right adjoint of sesquilinear form f, 44, 93
$\mathbf{GL}(n, R)$	general linear group of $n \times n$ invertible matrices over R, 1
$\Gamma(b)$	Clifford group of symmetric space (V, b), 259
$\mathbf{GL}(V)$	general linear group of vector space V, 68
$\mathcal{G}$	Galois group of L/K, 221
g_K	unique quaternary anisotropic form over local field K, 227
$\mathrm{H}(U), h_{\mathrm{H}(U)}$	hyperbolic space, form, based on U, 97
$\mathrm{H}(\rho)$	hyperbolic representation based on ρ, 89
$\mathrm{H}_\varepsilon(K, A)$	Grothendieck group of strongly hyperbolic $(A, ^-)$-invariant forms, 153
$\mathrm{H}(A)$	the hyperbolic involution algebra based on A, 25
$\mathbb{H}$	quaternion algebra $(-1,-1)_{\mathbb{R}}$, 17
$\mathbf{H}$	quaternion algebra $(-1,-1)_{\mathbf{R}}$, 17
${}^B\mathcal{H}_{\varepsilon,A}$	B-invariant forms over A on (B, A)-bimodules, 105
${}_A\mathcal{H}_\varepsilon^B, {}_{(A,^-)}\mathcal{H}_\varepsilon^{(B,^-)}$	$(B, ^-)$-invariant forms, 105
${}_{(A,^-)}\mathcal{H}_\varepsilon, {}_A\mathcal{H}_\varepsilon$	ε-Hermitian forms over $(A, ^-)$, 85
${}_{(A,^-)}\mathcal{H}_\varepsilon^*, {}_A\mathcal{H}_\varepsilon^*$	nonsingular ε-Hermitian forms, 93
$\hat{h}$	$(KG, ^-)$-form associated to the invariant form h, 86
$h \simeq g$	similarity of forms, 188
$h \otimes g$	tensor product of sesquilinear forms, 193
$h_1 \perp h_2$	orthogonal sum of forms, 94
$\mathcal{I}(K, ^-)$	isomorphism classes of involution algebras, 194
$\mathcal{I}(\mathbf{b})$	isomorphism classes of involution algebras, 189
$\mathrm{Ind}_H^G \rho$	standard induced representation, 162
$\mathbf{I}(V, h)$	isometry group of (V, h), 49
$\leftrightarrow$	interchange involution, 26
inn_c	inner automorphism $a \to cac^{-1}$ of ring or group, 10
KG	group algebra of G over K, 15
K_c	subcyclotomic closure of $\mathbb{Q}$ in K, 186
$\mathrm{K}(\mathbf{M})$	Grothendieck group of Abelian monoid $\mathbf{M}$, 64
$\mathrm{K}_0(KG)$	ring K_0 of a group algebra, 70
$\mathrm{K}_0(R)$	K_0 of ring R, 67

$L \otimes_K (A, ^-)$	scalar extension of involution algebra, 24
L	linearization, 155
$\text{len}_R M$	length of R-module M, 1
M^*	dual R-module M, 3
M_S	isotypic component of type S of module M, 6
$\mathrm{M}(n, (R, ^-))$	$n \times n$ matrix ring over $(R, ^-)$, vii
$\mathrm{M}(n, R)$	ring of $n \times n$-matrices over R, 1
$\mathrm{M}(n_0, n_1, K)$	graded matrix algebra, 254
$_RM, M_R$	left R-module, right R-module, 1
$_RM_S$	(R, S)-bimodule, 1
$_D\mathbf{M}_m$	equivalence classes of reflexive forms of rank m, 220
N^o	annihilator in M^* of submodule N of M, 3
$\mathrm{N}\dot{K}$	"norms" $\delta\bar{\delta}$ in K/K_0, 45
$\mathrm{N}_{L/K}$	norm in the finite field extension L/K, 13
n_{K/K_0}	modified norm when $(K : K_0) \leq 2$, 124
$\mathrm{nrd}_{A/K}$	reduced norm in the simple or separable algebra A, 13
ν	graded automorphism of $\mathrm{C}(V, b)$, 259
$\mathbf{O}(V, f)$	orthogonal group, 49
Pf_n	the Pfaffian, 45
$\mathbb{Q}$	field of rational numbers, 18
$\mathbb{Q}_p$	rational p-adic numbers, 17
$(R, ^-)$	ring with involution, 4
$R^\times$	group of units of ring R, vii
R^o	opposite ring, 14
$R^{n \times m}$	$n \times m$ matrices over R, 1
$\mathbf{R}$	real closed field, 17
$\mathrm{R}_H^G V$	representation of G induced from KH-module V, 161
$\bar{\rho}^*$	twisted contragredient representation, 74
ρ^L	scalar extension of representation ρ, 70, 89
$\rho_1 \otimes \rho_2$	tensor product of representations, 70
$\mathrm{S}_\varepsilon(K, G)$	signature group, 205
$\mathrm{Sim}_\varepsilon(A, ^-)$	similarity classes of ε-Hermitian forms over $(A, ^-)$, 189
$\mathbf{SO}(V)$	special orthogonal group, 49
$\mathbf{SU}(V)$	special unitary group, 49
$\mathbf{Sp}(V, f)$	symplectic group, 49
$\mathrm{Sch}(K)$	Schur subgroup of the Brauer group, 184
$\mathrm{Sch}(K, ^-)$	involutory Schur subgroup, 185
$_{(A,^-)}\mathcal{S}, {}_A\mathcal{S}$	sesquilinear forms over $(A, ^-)$, 91
$_{(A,^-)}\mathcal{S}^*, {}_A\mathcal{S}^*$	nonsingular sesquilinear forms, 93
$\mathrm{s}(^-)$	Hasse invariant of orthogonal involution, 32

$\mathrm{s}_{\mathfrak{p}}(^-)$	Hasse invariant of involution at prime $\mathfrak{p}$, 32
$\mathrm{sgn}(f)$	signature of symmetric or Hermitian form f, 37, 41
σ_ε	signature map, 206
$s(K)$	level of field K, 36
$\mathrm{sgn}_i(b)$	isotypic signature, 179
$\mathrm{s}(b)$	Hasse invariant of symmetric form b, 53
$\mathrm{Tr}_{KG/K}$	canonical trace of KG, 15
$\mathrm{Tr}_{L/K}$	trace in the finite field extension L/K, 13
tr	matrix trace, 13
τ	type of ECSGA or of Brauer-Wall class, 264
τ_A	type of the CSGA A, 255
$\times_{\mathrm{z}}$	central extension with respect to 2-cocycle z, 51
$\times_{1\le i\le l} h_i$	vector sum of forms, 95
φ^t	transpose of φ, 4, 92
$\mathrm{trd}_{A/K}$	reduced trace in the simple or separable algebra A, 13
$\mathrm{tr}_{A/K} a$	algebra trace in K-algebra A, 14
$U \perp W$	orthogonal direct sum of U and W, 46
$U^\perp$	orthogonal complement of U, 45, 94
$\mathbf{U}(V,f)$	unitary group, 49, 91
V^L	extension of scalars of vector space V, 48
$\bar{V}$	the A-module V twisted by the involution $^-$ on A, 4
$\hat{\mathrm{W}}_\varepsilon(K,A)$	Grothendieck group of $(A,^-)$-invariant ε-Hermitian forms, 153
$\hat{\mathrm{W}}_\varepsilon(K,G)$	Witt-Grothendieck group of isometry representations, 152
$\hat{\mathrm{W}}_\varepsilon(A,^-), \hat{\mathrm{W}}_\varepsilon(A)$	Witt-Grothendieck group of ε-Hermitian forms, 151
$\hat{\mathrm{W}}_\varepsilon(K,^-)$	Witt-Grothendieck group, 50
$\mathrm{W}_\varepsilon(D,^-)$	Witt group over division algebra D, 50
$\mathrm{W}_\varepsilon(K,A)$	Witt group of $(A,^-)$-invariant ε-Hermitian forms, 153
$\mathrm{W}_\varepsilon(K,G)$	Witt group of isometry representations, 152
$\mathrm{W}_\varepsilon(A,^-), \mathrm{W}_\varepsilon(A)$	Witt group of ε-Hermitian forms, 151
$\mathrm{W}(K,^-)$	graded Witt ring, 50
$\mathrm{w}(^-)$	Hasse-Witt invariant of orthogonal involution, 32
$\mathrm{w}(b)$	Hasse-Witt invariant, 54
X_d	orderings of K_0 at which K/K_0 is definite, 205
$\mathfrak{X}_d$	real primes of K_0 at which K/K_0 is definite, 207
$\mathrm{Z}_R(X)$	centralizer in the ring R of the subset X of R, 7
$\widehat{\mathrm{Z}}_A$	graded center of graded algebra, 255
$\mathrm{Z}(R)$	center of the ring R, 7
ζ_A	quadratic invariant of CSGA A, 256
z_A	characteristic element of a CSGA, 256

Index

Titles in This Series

28 **Carl R. Riehm,** Introduction to orthogonal, symplectic and unitary representations of finite groups, 2011

27 **Levent Tunçel,** Polyhedral and semidefinite programming methods in combinatorial optimization, 2010

26 **Clifton Cunningham and Monica Nevins, Editors,** Ottawa lectures on admissible representations of reductive p-adic groups, 2009

25 **Eric T. Sawyer,** Function theory: Interpolation and corona problems, 2009

24 **Kenji Ueno,** Conformal field theory with gauge symmetry, 2008

23 **Marcelo Aguiar and Swapneel Mahajan,** Coxeter groups and Hopf algebras, 2006

22 **Christian Meyer,** Modular Calabi-Yau threefolds, 2005

21 **Arne Ledet,** Brauer type embedding problems, 2005

20 **James W. Cogdell, Henry H. Kim, and M. Ram Murty,** Lectures on automorphic L-functions, 2004

19 **Jeremy P. Spinrad,** Efficient graph representations, 2003

18 **Olavi Nevanlinna,** Meromorphic functions and linear algebra, 2003

17 **Vitaly I. Voloshin,** Coloring mixed hypergraphs: theory, algorithms and applications, 2002

16 **Neal Madras,** Lectures on Monte Carlo Methods, 2002

15 **Bradd Hart and Matthew Valeriote, Editors,** Lectures on algebraic model theory, 2002

14 **Frank den Hollander,** Large deviations, 2000

13 **B. V. Rajarama Bhat, George A. Elliott, and Peter A. Fillmore, Editors,** Lectures in operator theory, 2000

12 **Salma Kuhlmann,** Ordered exponential fields, 2000

11 **Tibor Krisztin, Hans-Otto Walther, and Jianhong Wu,** Shape, smoothness and invariant stratification of an attracting set for delayed monotone positive feedback, 1999

10 **Jiří Patera, Editor,** Quasicrystals and discrete geometry, 1998

9 **Paul Selick,** Introduction to homotopy theory, 1997

8 **Terry A. Loring,** Lifting solutions to perturbing problems in C^*-algebras, 1997

7 **S. O. Kochman,** Bordism, stable homotopy and Adams spectral sequences, 1996

6 **Kenneth R. Davidson,** C*-Algebras by example, 1996

5 **A. Weiss,** Multiplicative Galois module structure, 1996

4 **Gérard Besson, Joachim Lohkamp, Pierre Pansu, and Peter Petersen Miroslav Lovric, Maung Min-Oo, and McKenzie Y.-K. Wang, Editors,** Riemannian geometry, 1996

3 **Albrecht Böttcher, Aad Dijksma and Heinz Langer, Michael A. Dritschel and James Rovnyak, and M. A. Kaashoek Peter Lancaster, Editor,** Lectures on operator theory and its applications, 1996

2 **Victor P. Snaith,** Galois module structure, 1994

1 **Stephen Wiggins,** Global dynamics, phase space transport, orbits homoclinic to resonances, and applications, 1993